THE AMER

Beginning in the 1870s, migrant groups from Russia's steppes settled in the similar environment of the Great Plains. Many were Mennonites. They brought plants, in particular grain and fodder crops, trees, and shrubs, as well as weeds. Following their example, and drawing on the expertise of émigré Russian–Jewish scientists, the U.S. Department of Agriculture introduced more plants, agricultural sciences, especially soil science, and methods of planting trees to shelter the land from the wind. By the 1930s, many of the grain varieties in the Great Plains had been imported from the steppes. The fertile soil was classified using the Russian term "Chernozem." The U.S. Forest Service was planting shelterbelts using techniques pioneered in the steppes. And, tumbling across the plains was an invasive weed from the steppes: tumbleweed. Based on archival research in the United States, Russia, Ukraine, and Kazakhstan, this book explores the unexpected Russian roots of Great Plains agriculture.

DAVID MOON is Visiting Professor at Nazarbayev University in Nur-Sultan, Kazakhstan, and Anniversary Professor in History at the University of York, UK. He is the author of *The Plough that Broke the Steppes: Agriculture and Environment on Russia's Grasslands, 1700–1914* (Oxford: Oxford University Press, 2013).

Studies in Environment and History

Editors

J. R. McNeill, *Georgetown University*
Ling Zhang, *Boston College*

Editors Emeriti

Alfred W. Crosby, *University of Texas at Austin*
Edmund P. Russell, *Carnegie Mellon University*
Donald Worster, *University of Kansas*

Other Books in the Series

James L. A. Webb, Jr. *The Guts of the Matter: A Global Environmental History of Human Waste and Infectious Intestinal Disease*
Maya K. Peterson *Pipe Dreams: Water and Empire in Central Asia's Aral Sea Basin*
Thomas M. Wickman *Snowshoe Country: An Environmental and Cultural History of Winter in the Early American Northeast*
Debjani Bhattacharyya *Empire and Ecology in the Bengal Delta: The Making of Calcutta*
Chris Courtney *The Nature of Disaster in China: The 1931 Yangzi River Flood*
Dagomar Degroot *The Frigid Golden Age: Climate Change, the Little Ice Age, and the Dutch Republic, 1560–1720*
Edmund Russell *Greyhound Nation: A Coevolutionary History of England, 1200–1900*
Timothy J. LeCain *The Matter of History: How Things Create the Past*
Ling Zhang *The River, the Plain, and the State: An Environmental Drama in Northern Song China, 1048–1128*
Abraham H. Gibson *Feral Animals in the American South: An Evolutionary History*
Andy Bruno *The Nature of Soviet Power: An Arctic Environmental History*
David A. Bello *Across Forest, Steppe, and Mountain: Environment, Identity, and Empire in Qing China's Borderlands*
Erik Loomis *Empire of Timber: Labor Unions and the Pacific Northwest Forests*
Peter Thorsheim *Waste into Weapons: Recycling in Britain during the Second World War*
Kieko Matteson *Forests in Revolutionary France: Conservation, Community, and Conflict, 1669–1848*
Micah S. Muscolino *The Ecology of War in China: Henan Province, the Yellow River, and Beyond, 1938–1950*
George Colpitts *Pemmican Empire: Food, Trade, and the Last Bison Hunts in the North American Plains, 1780–1882*
John L. Brooke *Climate Change and the Course of Global History: A Rough Journey*
Paul Josephson et al. *An Environmental History of Russia*
Emmanuel Kreike *Environmental Infrastructure in African History: Examining the Myth of Natural Resource Management*
Gregory T. Cushman *Guano and the Opening of the Pacific World: A Global Ecological History*
Sam White *The Climate of Rebellion in the Early Modern Ottoman Empire*
Edmund Russell *Evolutionary History: Uniting History and Biology to Understand Life on Earth*

Alan Mikhail *Nature and Empire in Ottoman Egypt: An Environmental History*
Richard W. Judd *The Untilled Garden: Natural History and the Spirit of Conservation in America, 1740–1840*
James L. A. Webb, Jr. *Humanity's Burden: A Global History of Malaria*
Myrna I. Santiago *The Ecology of Oil: Environment, Labor, and the Mexican Revolution, 1900–1938*
Frank Uekoetter *The Green and the Brown: A History of Conservation in Nazi Germany*
Matthew D. Evenden *Fish versus Power: An Environmental History of the Fraser River*
Alfred W. Crosby *Ecological Imperialism: The Biological Expansion of Europe, 900–1900, Second Edition*
Nancy J. Jacobs *Environment, Power, and Injustice: A South African History*
Edmund Russell *War and Nature: Fighting Humans and Insects with Chemicals from World War I to Silent Spring*
Adam Rome *The Bulldozer in the Countryside: Suburban Sprawl and the Rise of American Environmentalism*
Judith Shapiro *Mao's War against Nature: Politics and the Environment in Revolutionary China*
Andrew Isenberg *The Destruction of the Bison: An Environmental History*
Thomas Dunlap *Nature and the English Diaspora*
Robert B. Marks *Tigers, Rice, Silk, and Silt: Environment and Economy in Late Imperial South China*
Mark Elvin and Tsui'jung Liu *Sediments of Time: Environment and Society in Chinese History*
Richard H. Grove *Green Imperialism: Colonial Expansion, Tropical Island Edens and the Origins of Environmentalism, 1600–1860*
Thorkild Kjærgaard *The Danish Revolution, 1500–1800: An Ecohistorical Interpretation*
Donald Worster *Nature's Economy: A History of Ecological Ideas, Second Edition*
Elinor G. K. Melville *A Plague of Sheep: Environmental Consequences of the Conquest of Mexico*
J. R. McNeill *The Mountains of the Mediterranean World: An Environmental History*
Theodore Steinberg *Nature Incorporated: Industrialization and the Waters of New England*
Timothy Silver *A New Face on the Countryside: Indians, Colonists, and Slaves in South Atlantic Forests, 1500–1800*
Michael Williams *Americans and their Forests: A Historical Geography*
Donald Worster *The Ends of the Earth: Perspectives on Modern Environmental History*
Robert Harms *Games against Nature: An Eco-Cultural History of the Nunu of Equatorial Africa*
Warren Dean *Brazil and the Struggle for Rubber: A Study in Environmental History*
Samuel P. Hays *Beauty, Health, and Permanence: Environmental Politics in the United States, 1955–1985*
Arthur F. McEvoy *The Fisherman's Problem: Ecology and Law in the California Fisheries, 1850–1980*
Kenneth F. Kiple *The Caribbean Slave: A Biological History*

THE AMERICAN STEPPES

The Unexpected Russian Roots of Great Plains Agriculture, 1870s–1930s

DAVID MOON

University of York, UK

Nazarbayev University, Kazakhstan

CAMBRIDGE
UNIVERSITY PRESS

University Printing House, Cambridge CB2 8BS, United Kingdom

One Liberty Plaza, 20th Floor, New York, NY 10006, USA

477 Williamstown Road, Port Melbourne, VIC 3207, Australia

314-321, 3rd Floor, Plot 3, Splendor Forum, Jasola District Centre, New Delhi - 110025, India

103 Penang Road, #05-06/07, Visioncrest Commercial, Singapore 238467

Cambridge University Press is part of the University of Cambridge.

It furthers the University's mission by disseminating knowledge in the pursuit of education, learning and research at the highest international levels of excellence.

www.cambridge.org
Information on this title: www.cambridge.org/9781107503205
DOI: 10.1017/9781316217320

First published 2020
First paperback edition 2022

A catalogue record for this publication is available from the British Library

ISBN 978-1-107-10360-3 Hardback
ISBN 978-1-107-50320-5 Paperback

Dedicated by Permission to the Leverhulme Trust for its generous support for original and distinctive research.

A traveler on the plains of Kansas, if suddenly transported while asleep to southern Russia . . . would discover very little difference in his surroundings except as to the people and the character of the farm improvements and live stock. . . Even these last would be of the same kind if he were transported from certain localities in Kansas where Russian immigrants [Mennonites] now live.

Mark Alfred Carleton, "Hard Wheats Winning Their Way," *Yearbook of the United States Department of Agriculture 1914* (Washington, DC: USDA, 1915), pp. 398–9

Contents

Figures, Maps, and Tables

Figure

Maps

Table

Preface

In the early spring of 2007, I was driving west from Lincoln, Nebraska, in the Great Plains, to visit some shelterbelts of trees planted in the "dirty thirties" to protect the land from the drying and erosive force of the wind. As I approached the 98th meridian, the landscape opened up, the plains got flatter, there were fewer trees (outside shelterbelts and river valleys), fewer and smaller human settlements, larger fields. And the sky got bigger and overwhelmed me. Interstate-80 stretched out in an almost straight line that extended towards a pinprick on a distant horizon where I could scarcely distinguish between the vast sky and the flat land beneath it. It took me back to how I'd felt four years earlier when I had explored the steppes of southeastern Russia. At first, I had also felt dwarfed by the boundless flat landscapes with few trees dominated by enormous skies. Over the years and several trips to both grasslands, I have gradually got accustomed to large, flat landscapes, and to appreciate nuances and diversity within them as well as similarities between them.

I went to the Great Plains in 2007 in part to see if the landscape resembled the steppes. It certainly did, except where human artifacts, such as I-80, 18-wheelers and pick-up trucks, strip malls and fast-food restaurants, distinctive fencing, buildings, and hi-tech agricultural machinery gave away the location. I was also exploring connections that linked the Great Plains of North America with the steppes of Eurasia. These were transfers and influences that did not, as many inhabitants of the western world might expect, travel from a highly advanced United States to Russian and Soviet states that were trying to learn from scientific and economic achievements across the Atlantic (of which there were examples). But, the transfers and influences I was interested in went the other way. Perhaps counterintuitively, I was looking for what Americans in the Great Plains had learned from Russians, Ukrainians, Kazakhstanis, and Germanic peoples in the steppes.

One of these influences was the forestry science and techniques behind the shelterbelts. While there was American experience of planting windbreaks in the Great Plains, it was Russian and Soviet studies that played a decisive role in the decision to launch the American Shelterbelt Project in 1934. Another influence was the science of conceptualizing, studying, and classifying soils. Beneath my feet in Nebraska was very fertile, deep, black soil that had formed over millennia in a flat grassland and a semi-arid climate. This was the innovative theory of soil formation devised by pioneering Russian soil scientists led by Vasilii Dokuchaev during field work in the steppes in the 1870s. It was later adapted for American soils by the U.S. Soil Survey under Curtis Marbut. I came across another influence from the steppes when I visited neighboring Kansas. In Marion County, I met members of a community of Mennonites, whose forebears had moved there from the south of today's Ukraine in the 1870s. My hosts proudly related how the migrants had brought with them seeds of a hard, red, winter wheat that thrived in the central plains.

I have given talks to audiences, including historians, geographers, scientists, and the general public, in the United States, Canada, Russia, and Kazakhstan. I have shown them photographs of the two grasslands and asked them to identify where they were taken.[1] The results have been interesting. Most people in all these countries have struggled to distinguish between images of the two regions or have guessed wrongly. For example, when I showed a photograph I had taken in the Flint Hills of central Kansas to an audience about 50 miles away, some were sure it was the Molotschna River valley in Ukraine, which some had visited. My audience in Fargo, North Dakota, convinced themselves that a photograph of a large expanse of grassland in their home state was the Russian steppes. However, when I showed the same picture to geographers at Moscow State University, one explained how it could not be the Russian steppes, since there was no such large area of unplowed grassland left. A botany student at Karaganda State University in Kazakhstan identified the location of one of my photographs from the species of steppe grasses. These informed, correct answers were exceptional (a few Americans also correctly identified my photographs). While researching this book, I came across many examples of people familiar with either the American or Eurasian grasslands commenting on the parallels when visiting the other or expressing uncertainty over where they were. More important for the argument of

[1] Readers wishing to replicate my experiment can search online for images of the "steppes" and the "Great Plains."

this book is the use made of the parallels by people on both continents who realized they could share and benefit from experience and expertise appropriate to such similar environments.

During several visits to the Great Plains, I became aware of other connections with the steppes. Descendants of migrants from the steppes, mostly Mennonites and other Germanic peoples, have maintained a keen interest in their history, running museums, libraries, and a historical society. As well as wheat and other crops the migrants also brought their cuisine. On my visit to the shelterbelts in Nebraska, my guide suggested we have a local delicacy for lunch: runzas. I laughed when I saw them as I was familiar with these bread rolls with savory fillings from Russia and Ukraine as *pirogi*. It transpired that the Germanic inhabitants of the steppes had adopted them from their Slav neighbors and then taken them across the Atlantic. Later on my drive across the plains in the spring of 2007, I came across another influence from "Russia": the Minuteman Missile National Historic Site, at Philip, South Dakota. The site commemorates the arsenal of intercontinental ballistic missiles armed with nuclear warheads that were deployed, "hidden in plain sight," beneath the Great Plains as part of the United States' nuclear deterrent against the perceived threat from the Soviet Union.[2]

[2] "Hidden in Plain Sight," available online at www.nps.gov/mimi/index.htm, accessed July 23, 2018.

Acknowledgements

A historian from the mild, humid, and green North East of England who trained a specialist on Russia could not have attempted to write a transnational environmental history of the semi-arid Great Plains of the United States without a lot of help. The length of these acknowledgements is testament to the generosity of many organizations and people, without whom there would be no book.

I would like to thank several institutions for fellowships and grants. The Leverhulme Trust has been exceptionally generous (hence the dedication). The Trust awarded me a Major Research Fellowship for 2015–17, during which I completed the research and wrote the first draft. A grant for a Leverhulme International Network on Russian environmental history from 2013 to 2016 funded some of the research in Russia. Back in 2003, I held a Leverhulme Study Abroad Fellowship in Rostov-on-Don, in the steppe region of Russia, where I carried out research and field work for my previous book on the steppes. I held a fellowship at the Institute for Advanced Study in Princeton in 2008–9, when I conducted some of the research and tried out some of the ideas for this book. The pilot project was supported by a short-term grant from the Kennan Institute in Washington, DC, in the spring of 2007. I am grateful also to the British Academy for three Small Research Grants, to Santander Universities for an International Connections Gold Award, and to the Rachel Carson Center in Munich for a short-term fellowship in July 2017. I would also like to thank Nazarbayev University in Nur-Sultan (Astana), Kazakhstan, for appointing me to a visiting professorship in 2018–20, which enabled me to visit Kazakhstani archives and complete this book in the heart of the Eurasian steppe. Thanks are due also, for a two-year appointment at the start of my career, to the History Department of the University of Texas at Austin, where I encountered the southern end of the Great Plains and started to think about a Russian-American research project.

Just as important has been the assistance of archivists and librarians. Too many have provided invaluable help for me to thank them individually.

I would like to express my gratitude to the staffs of the following archives, listed in alphabetical order. In the United States: The Bancroft Library, University of California, Berkeley; The Duane G. Meyer Library Special Collections and Archives, Missouri State University, Springfield; Georgetown University Library Special Collections Division, Washington, DC; The Kansas Historical Society State Archives, Topeka; The Library of Congress Manuscript Division, Washington, DC; The Mennonite Library and Archives, Bethel College, North Newton, Kansas; The Minnesota Historical Society, St. Paul; The National Agricultural Library Special Collections, Beltsville, Maryland; The National Archives at College Park, Maryland; The National Archives at Kansas City, Missouri; The Nebraska State Historical Society State Archives and Manuscript Division, Lincoln; North Dakota State University Archives, Fargo; The State Historical Society of Missouri Manuscript Collection, Columbia; South Dakota State Agricultural Heritage Museum, Brookings; South Dakota State University Archives and Special Collections, Hilton M. Briggs Library, Brookings; The University of Minnesota Archives, Minneapolis; Wichita State University Libraries Special Collections and University Archives, Wichita, Kansas. In Kazakhstan: the Central State Archive of the Republic of Kazakhstan and the Central State Archive of Scientific and Technical Documentation of the Republic of Kazakhstan, both in Almaty. In the Russian Federation: the Archive of the Russian Academy of Sciences in Moscow and the branch of the archive in St. Petersburg; the State Archives of Rostov and Samara regions; and the Russian State Historical Archive and the Central State Archive of Scientific and Technical Documentation, both in St. Petersburg. And in Ukraine: the State Archive of Odessa Region.

I am grateful to the librarians at, in the United States: The American Historical Society of Germans from Russia, Lincoln, Nebraska; The Ellis Library of the University of Missouri in Columbia; The Hilton M. Briggs Library of South Dakota State University; The Library of Congress; The Love and C. Y. Thompson libraries of the University of Nebraska-Lincoln (especially the librarian in the latter who located a misshelved copy of a rare publication on the history of the shelterbelt project by E. L. Perry, which proved extremely useful); North Dakota State University Library, including the Germans from Russia Heritage Collection, in Fargo; the University of Minnesota Library, Minneapolis; in Canada: the University Library of the University of Saskatchewan, Saskatoon; in the Russian Federation: the Library of the Academy of Sciences and the Russian National Library, both in St. Petersburg; the Russian State Library (formerly the Lenin Library), Moscow; and in Finland: The Slavonic Library of the National Library of Finland in Helsinki.

I visited many museums, state historical societies, state parks, preserves, national grasslands, tree nurseries, and research stations, too numerous to list in full, all of which deepened my understanding of the region, its environment, and history. I am particularly grateful to: Annette LeZotte of the Kauffman Museum, North Newton, Kansas; Gwen McCausland of the South Dakota Agricultural Heritage Museum, Brookings; Karen Penner of the Bernhard Warkentin House Museum in Newton, Kansas; Darlene Schroeder of the Mennonite Heritage Museum, Goessel, Kansas; and to the staff and volunteers at the Cimarron Heritage Center, Boise City, Oklahoma; and the Panhandle–Plains Historical Museum, Canyon, Texas. Thanks are due also to my guides at the Konza Prairie Biological Station, near Manhattan, Kansas; the Tallgrass Prairie National Preserve near Strong City, Kansas; and the Charles E. Bessey Nursery in the Nebraska Sandhills. North of the 49th parallel, I am grateful to my guides at the Mennonite Heritage Village in Steinbach, Manitoba, the Seager Wheeler Experimental Farm National Historic Site, Rosthern, Saskatchewan; the Saskatoon Forestry Farm (previously the Dominion, later Sutherland, Forest Nursery Station); and the Genebank of the Federal Plant Gene Resources of Canada on the campus of the University of Saskatchewan. I am grateful also to many farmers with whom I had conversations in Kansas, Nebraska, and Manitoba for sharing their knowledge and experience with a curious visitor from England.

Since this book builds on my previous monograph on the steppes of Russia and Ukraine, I would like to repeat my thanks to the people and institutions I acknowledged there, many of whom have continued to offer assistance. I would like to acknowledge the visa support provided by the European University and Higher School of Economics in St. Petersburg, and the advice of scholars in that city, in particular Alexandra Bekasova, Anastasia Fedotova, Julia Lajus, Marina Loskutova, and Alexander Semyonov, and in Moscow, Nikolai Dronin and his colleagues in the Geography Faculty at Moscow State University.

My thanks to all who invited, and in some cases funded, me to deliver papers, talks, and lectures and who came to hear me speak, ask questions, make comments, improve my knowledge, refine my argument (and join in trying to guess the location of my photographs). In the Great Plains and prairies, from south to north, at: The University of Texas History of Science Colloquium, Austin; The Kauffman Museum, Bethel College, North Newton, Kansas; The Center for Russian, East European and Eurasian Studies, University of Kansas, Lawrence; The Center for Great Plains Studies, University of Nebraska-Lincoln; Southwest Minnesota

State University, Marshall; North Dakota State University, Fargo; The University of Winnipeg, Manitoba, Canada; and the Doukhobor Prayer Home in Blaine Lake, Saskatchewan, Canada, during a symposium organized by the University of Saskatchewan. Elsewhere in North America, I spoke to the Institute for Advance Study, Princeton; the Institute of Slavic, East European, and Eurasian Studies at the University of California, Berkeley; and Pomona College in Claremont, California. In Russia, I gave talks at the School of History of the Higher School of Economics, St. Petersburg, and the Faculty of Geography, Moscow State University. In Kazakhstan, I presented my work at: Nazarbayev University and Eurasian National University, both in Nur-Sultan (Astana); the Central Asian Studies Center, KIMEP University, The Institute of History and Ethnology of the Academy of Sciences of the Republic of Kazakhstan, and the Archive of the President of the Republic of Kazakhstan, all in Almaty; Karaganda State University; and Kazakh Humanitarian, Law, Innovation University in Semey. Elsewhere, I spoke at the École des hautes études en sciences sociales, Paris, France; and at the universities of Durham, Essex, Glasgow, Hull, Leeds, Northumbria, Oxford, York, and University College London in the UK.

I have learned a great deal at conferences. I presented papers at the following: The European Association for Environmental History (UK Branch), St. Antony's College, Oxford, 2009; the 79th Anglo-American Conference, "Environments," at the Institute for Historical Research, London, 2010; the British Agricultural History Society, Winter Conference, London, 2012; the conferences of the European Society for Environmental History in Munich in 2013 and Versailles in 2015; the Annual Conference of the American Society for Environmental History, San Francisco, 2014; and the Annual Convention of the Association for Slavic, East European, and Eurasian Studies, in Philadelphia, in 2015; the British Association for Slavonic and East European Studies, Cambridge, 2016; a Global History Conference on "Mennonites, Land and the Environment," at the University of Winnipeg, Canada in 2016; and the Canadian History and Environment Summer Symposium at the University of Saskatchewan, Saskatoon, Canada, in 2018. Regrettably limitations of funding and time prevented me from attending further specialist conferences in North America from which I would have greatly benefitted.

My guide when I visited the shelterbelts in Nebraska was Bruce Wight, then of the National Agroforestry Center in Lincoln, Nebraska. I am grateful to Steven Anderson of the Forest History Society for putting us in touch. Many American and Canadian scholars have been generous with advice and assistance when I have been in North America, including: Tony Amato of Southwest Minnesota State University; Charles Braithwaite and his colleagues at the Center for Great Plains Studies of the University of Nebraska-Lincoln; Douglas Helms, formerly historian at the United States Department of Agriculture; Tom Isern and Suzzanne Kelley in Fargo, North Dakota, who introduced me to many people, including Michael Miller; Kevin D. Kephart, Vice President, South Dakota State University, who shared his knowledge of Niels Hansen; Bonnie Lynn Sherow and James Sherow, who extended hospitality and advice in Manhattan, Kansas, when I was embarking on this project; Don and Beverly Worster, who hosted me near Lawrence, Kansas. When I ventured north to Canada, Royden Loewen and his colleagues at the University of Winnipeg and participants at their conference were very helpful to an "accidental Mennonite historian," and Ashleigh Androsoff, Jim Clifford, Geoff Cunfer, Laura Larsen, Cheryl Troupe, and Andrew Watson were generous hosts during their field trip and symposium at the University Saskatchewan. My guides in Korgalzhyn in the near identical environment of northern Kazakhstan were Timur Iskakov and Ruslan Urazaliyev. Nikolay Tsyrempilov, Aidyn Zhuniskhanov, and Ulan Zhangaliev invited me to join their expedition to see the steppes and semi-desert of Eastern Kazakhstan Region. I have benefitted from advice on Mennonite historiography from James Urry, on soil science from Edward Landa, on American history from David Huyssen, and on many topics, as ever, from Denis Shaw and Jonathan Oldfield.

To include everyone who helped me would produce a very long list. I apologize to anyone I've omitted who feels they deserve thanks, and a sincere thank you to everyone I met and who made my times in the plains (as well as the east and west coasts, which I didn't always just fly over to get to the plains) as memorable as my visits to Russia, Ukraine, and Kazakhstan. I am conscious that if I thank too many people I will incur – as I did after I included a long list in my last book – reprimands from review editors of journals looking for reviewers who have not been implicated by the author.

Unusually for an environmental historian, I feel I should acknowledge several car rental companies, automotive manufacturers, and oil companies for enabling me to explore the Great Plains and prairies. I am not sure I was supposed to take the rental cars off road in state parks, preserves, national grasslands, and farms, but I did my best to clean up them afterwards. A word also for the Russian country and western group Kukuruza (Corn), whose music I discovered when researching this book and seemed an appropriate accompaniment to some of my writing. Their bluegrass version of the Stenka Razin song is particularly memorable.

I am especially grateful to the editors of the Cambridge Studies in Environment and History series and Cambridge University Press for trusting a British historian of Russia to write a book about North America. Thanks are due also to anonymous readers of my book proposal and manuscript, and the production team at the Press. While I have benefitted more than I can say from the support, help, and advice of the many people and institutions I have thanked here, responsibility for its contents, especially any errors of fact or interpretation, lies solely with the author.

Notes on the Text

Writing transnational history entails addressing issues of naming, language, translation, transliteration from other alphabets, formats for archival citations, and differing conventions for some of the above among the countries under investigation. Conscious that readers will include specialists on both North America and on the territory of the former Soviet Union and Russian Empire, but with less expertise on the other, I have endeavored to explain in a series of notes how I have dealt with these issues to make the text as accurate, consistent, and clear as possible.

Place Names

For clarity I have used names for places in the Russian Empire and Soviet Union (USSR), and the independent states that emerged from the latter's collapse in 1991, in the anglicized forms most familiar to readers in the Anglophone world. Thus, St. Petersburg is used in preference to Sankt-Peterburg, Kazakhstan instead of Qazaqstan, etc. Even this is problematic, since the spellings of place names in Ukraine are generally more familiar to international audiences from their Russian rather than Ukrainian spellings. Thus, and with no desire to cause offense, I have preferred Odessa to Odesa and Dnepr instead of Dnipro. I have sometimes used the term Turkestan, which was used in the nineteenth century and into the 1930s, for the southern part of present-day Kazakhstan and the rest of Russian/Soviet Central Asia. On the other hand, when writing about Mennonites, I have used their German names for their settlements in today's Ukraine. Thus, I use Molotschna, rather than the Russian and Ukrainian Molochna, for an important group of settlements that surrounded the present-day Ukrainian town of Molochans'k (Ukrainian spelling).

In the interests of accuracy, I have generally used the names and the administrative statuses of places at the time I am writing about them, but have sometimes given their present-day names for clarity. At the start of our period the steppes lay in the south of the Russian Empire, which was ruled by the tsars from St. Petersburg. The Russian Empire and tsarist rule collapsed in March 1917. Following the Bolshevik revolution in October 1917 the Soviet state was founded. The Soviet Union (USSR), which was formally established in December 1922, covered much of the territory, including most of the steppe region, of the empire it succeeded. Thus, to March 1917 I refer to the Russian Empire; from October 1917 to December 1922, Soviet Russia; and from December 1922 until its collapse in December 1991, the Soviet Union. Aware of their serious limitations, and again with no intention of causing offense, for reasons of style I have sometimes resorted to the terms "Russia" and "Russian" as unsatisfactory short hands for the larger states centered on Russia and their inhabitants when referring to periods across the 1917 divide. The break up of the Soviet Union in 1991 left the steppe region divided between, from west to east: Moldova, Ukraine, the Russian Federation, and Kazakhstan.

A significant part of the steppe region is located in the territory of today's independent Ukraine. The term Ukraine did not have official status before 1917. From 1917 to 1922, Ukraine was at least nominally independent and, between 1922 and 1991, it was a constituent republic of the Soviet Union. A further complexity is added by the term "New Russia" (Novorossiia). In the period covered by this book, it referred to the region immediately north of the Black Sea and Sea of Azov, including the Crimean peninsula, in the south of today's Ukraine. The term New Russia was first used following the annexation of this region by the Russian Empire from the Ottoman Empire in the late eighteenth century. This part of the steppe region features prominently in this book as it was the home of the Mennonite communities, many of whose members migrated to the Great Plains from the 1870s.

It is not just countries that have changed their names. For example, the tsars' capital of St. Petersburg became Petrograd between 1914 and 1924, Leningrad from 1924 to 1991, when it reverted to St. Petersburg. (It was replaced as capital by Moscow in 1918.) I use the name of the city for the time I am writing about. Another city that has changed its name is present-day Gdańsk on the Baltic coast of modern Poland. In the period covered by this book, it was generally known by its German name of Danzig. It appears in this book, because the Mennonites (whose role in this book has created several issues for place names), who lived in New Russia had previously lived near Danzig before moving to the steppes in the late eighteenth and early nineteenth centuries. The city in which I completed the manuscript of this book underwent a name change while I was there:

in March 2019, Astana, the capital of Kazakhstan, was renamed Nur-Sultan (which is the city's fifth name since the foundation of Akmolinsk in 1832, it was called Tselinograd from 1961 and Aqmola/Akmola from 1992 to 1998).

Matters are less complicated for the Great Plains. I use this term for the semi-arid grassland region, where the natural vegetation is largely short and mixed grasses, extending from northwards from central Texas to North Dakota and west to the foothills of the Rocky Mountains. I generally use the term "prairies" for the grasslands immediately to the east and also, following standard usage, for the northern extension of the Great Plains in Canada. When quoting from or referring directly to contemporary sources, however, I have used the terms the authors used, but if necessary have explained what part of the region was intended. I have used the term "North America" for both the United States and Canada, but "American" as a shorthand for inhabitants of the United States.

The only real complexity when writing about place names in the Great Plains is the creation of territories that were later admitted as states to the Union. I have used their names and administrative statuses, and their geographical extent, at the time I am writing about them, with brief explanatory notes where necessary. For example, the Dakota Territory formed in 1861 comprised for a time lands that became the states of North and South Dakota, as well parts of Wyoming and Montana, all of which were granted statehood in 1889–90. I have done the same for counties that underwent changes of name or territory.

Names of Plants

An important part of this book is trying to identify species of plants in both the Great Plains and the steppes and those that made the journey from one to the other. The systems for classifying varieties of cultivated crops used by agricultural scientists and institutions such as the U.S. Department of Agriculture and its counterparts in Russia were evolving over the period covered by this book and differed between the two countries. There were also a variety of folk names for plants in both countries. In an effort to cut through possible confusions, and to explain how I reached my decisions on identification, where appropriate I have given the common names for plants in English and Russian (and on occasions also German) as well as the scientific binominal nomenclatures. Where these names, including scientific names, have changed, or do not necessarily correspond with each other, I have endeavored to give the alternatives and explain these issues and changes. I have followed the American usage of "corn" for the grain known as "kukuruza" in Russian and maize in other varieties of English.

Transliteration from Cyrillic

In references to works in Russian and Ukrainian and in all places where I have transliterated words from the Cyrillic alphabet I have used the simplified Library of Congress of system. I have retained what look like quotation marks to transliterate two Cyrillic letters (the hard and soft signs). However, for much of the period covered by this book, other transliteration systems were in use, including some devised by individual writers. Different systems are and have been used for transliteration from Slavic languages into other European languages, in particular German, which will appear periodically in this book. For sake of consistency, when referring to people and terms, I have generally used the spellings used by the people and in the publications I am referring to. In an attempt to limit possible confusion, where appropriate, I have given also the transliterations in the Library of Congress system. For example, the Russian émigré American soil scientist spelled his name, when writing in English, as Constantine Nikiforoff. But, when citing his pre-emigration Russian publications and referring to him in the country of his birth, I have transliterated his name from Cyrillic in its Russian version as Konstantin Nikiforov. There are other such examples. Another potential source of confusion is the Russian terms for types of soils that were rendered into German by Russian scientist Konstantin Glinka and then, from the German transliterations, made their way into American publications of the 1910s to the 1930s. Thus, Chernozem (black earth) sometimes appears in its German transliteration from Cyrillic as "Tschernosem." Explanatory notes are provided where necessary.

Archival References

References to sources in Kazakhstani, Russian, and Ukrainian archives are given in the standard format for archives in the post-Soviet world: Name of Archive, fond (collection), opis' (inventory), delo (file), list (folio) in abbreviated form (see List of Abbreviations). Contrary to the standard system for archival citations in these countries, but in line with American systems, where appropriate I have included information on the document, e.g., the authors and recipients of letters and the dates. For example, a letter from Soviet soil scientist Leonid Prasolov to American soil scientist Curtis Marbut in the Archive of the Russian Academy of Sciences is cited as: ARAN, f[ond]. 687, op[is'] 4, d[elo]77, l[ist].3, L. Prasolov to C. Marbut, April 1, 1922.

For consistency and for ease of locating the documents should anyone wish to follow up the references in this book, I have adapted the standard post-Soviet system for references to documents in American archives. For example, a letter from Mark Carleton to A. F. Woods at the USDA in Washington, DC, in the National Archives in College Park, Maryland, is given as: NARA CP, RG [Record Group] 54, Finding Aid A1, Entry 58, Division of Vegetable Pathology and Physiology: Correspondence of M. A. Carleton, 1891–1900, Folder, M. A. Carleton – 1900, Carleton to A. F. Woods, June 3, 1900.

A full list of archival collections cited is provided at the back of this book.

Dramatis Personae

Several key people recur in several chapters. I have introduced them in detail in the chapter in which they play their starring roles. Since some are referred to in supporting roles earlier in the book, I have provided brief biographies here, rather than distract attention from the wider argument when they first appear (and in preference to hiding this information in footnotes). The information here is based on the sources cited at the points in the book where these people are introduced properly.

Mark Alfred Carleton (1866–1925) American crop scientist and plant explorer. Educated at Kansas State Agricultural College, Manhattan. Employed by the United States Department of Agriculture (USDA), 1894–1918. Worked with Mennonites immigrants, including Bernhard Warkentin, in Kansas to test their wheat varieties. Visited the Russian Empire twice, in 1898–9 and 1900, to collect plants, in particular types of wheat, that could be useful in the Great Plains. Vigorously promoted the crop varieties he introduced to the United States.

Vasilii Dokuchaev (1846–1903) Russian soil scientist. Devised new theory of soil formation during field work in steppe region of the Russian Empire in 1870s and 1880s. This was the starting point for a new science of understanding soils that is widely considered the basis for modern soil science. Led special scientific expedition to steppe region in 1892–7 to research ways of making crop cultivation more reliable in wake of drought, crop failure, and famine of 1891–2.

Konstantin Glinka (1867–1927) Russian soil scientist. Former student and later colleague of Dokuchaev. Led soil science expeditions in the steppe region. In 1914 he published a book on Russian soil science in German that brought the innovative Russian work to wider international attention, including in the United States. Leader of Soviet delegation at the 1st International Congress of Soil Science in Washington, DC, in 1927.

Niels Hansen (1866–1950) Danish-born American botanist and plant explorer for the USDA and state of South Dakota. Worked for South

Dakota State Agricultural College, Brookings, 1895 to his retirement. Visited Eurasia, including the steppes, six times between 1897 and 1934 to collect a wide variety of plants for use in the northern plains.

Eugene W. Hilgard (1833–1916) German-born American soil scientist. Educated in Europe. Worked for the state of Mississippi and the Universities of Mississippi (1855–73), Michigan (1873–5) and California, Berkeley (1875–1904). Worked out the relationship between soils and climate and his work to some extent resembled that of Dokuchaev. Corresponded with Russian soil scientists, including Nikolai Tulaikov, but not Dokuchaev.

Curtis F. Marbut (1863–1935) American soil scientist. Director of the Soil Survey Division of the USDA, 1913–35. Marbut, a geologist by training, was converted to Russian soil science after reading Glinka's book in German. He painstakingly and despite opposition from his chief, Milton Whitney, drew on the Russian innovative soil science to revise the approach of the U.S. Soil Survey. Participated in the first two international Congresses of Soil Science in Washington, DC, in 1927 and in Moscow and Leningrad in the Soviet Union in 1930. Died in Harbin, China, in 1935 en route to China.

Gifford Pinchot (1865–1946) American forester, conservationist, and political figure. Chief of the USDA's Division of Forestry, 1898–1905, and its successor the U.S. Forest Service, 1905–10. Mentor to Raphael Zon. Served for two terms as Governor of Pennsylvania (1923–7, 1931–5).

Nikolai Tulaikov (1875–1938) Russian/Soviet agronomist and soil scientist. Born in peasant family in Simbirsk province on the edge of the steppes. Spent 1908–9 in the United States, based at the University of California, Berkeley, where he met Hilgard. Directed steppe agricultural experiment stations. Participated in the first two international Congresses of Soil Science in the United States in 1927 and the Soviet Union in 1930. Leading Russian/Soviet specialist on American agriculture. Arrested during the terror in 1937 and executed in 1938.

Nikolai Vavilov (1887–1943) Russian/Soviet botanist, agronomist, and geneticist. Director of the Institute of Plant Industry and its successors in Petrograd/Leningrad, 1921–40. Led expeditions all over the world to collect plants for his seed bank to assist in improving crop varieties in the interests of food security. Had extensive international scientific contacts and visited the United States several times. Arrested in 1940 following a dispute with pseudo-scientist Trofim Lysenko. Tragically and pointlessly died of starvation in prison in Saratov in 1943. (Vavilov is far more important than his supporting role in this book might suggest.)

Georgii Vysotskii (1865–1940) Russian/Soviet forestry scientist. Specialist in steppe forestry and shelterbelts, about which he was more skeptical than most of his colleagues. Participated in Dokuchaev's special expedition to steppes in 1892–7. Contributed to debate over American Shelterbelt Project in 1935.

Bernhard Warkentin (1847–1908) American miller and seed importer. Born in Mennonite community in southern steppes of today's Ukraine. In 1872 he moved to the United States, where he settled in Kansas, and established his businesses. Promoted Mennonite immigration to the United States and the cultivation of Turkey Red (hard red winter wheat) from the southern steppes in the Great Plains. Worked with Mark Carleton.

Raphael Zon (1874–1956) Russian-born American forestry scientist. Born into Jewish family in Simbirsk on the edge of the steppes. He left Russia in 1896 to escape punishment for radical activities. Moved to the United States in 1898, where he trained as a forestry scientist. Mentee of Gifford Pinchot. Worked for U.S. Division of Forestry and Forest Service in Washington, DC, 1901–23 and was director of the Lakes States Forest Experiment Station in St. Paul, Minnesota, from 1923 until his retirement in 1944. Played key role in launching Shelterbelt Project in Great Plains in 1934. Zon, together with other Russian–Jewish immigrants, served as a conduit for Russian and Soviet sciences in the United States.

Abbreviations

AAAG	*Annals of the Association of American Geographers*
AH	*Agricultural History*
ARAN	Arkhiv Rossiiskoi Akademii nauk, Moscow
BAASSW	*Bulletin of the American Association of Soil Survey Workers*
BASSA	*Bulletin of the American Soil Survey Association*
d.	delo (file) (Russian and Kazakhstani archives)
EH	*Environmental History*
E and H	*Environment and History*
ES	*Entsiklopedicheskii slovar'*, 82 vols. (Spb: Brokgauz and Efron, 1890–1907)
ESR	*Experiment Station Record*
f.	fond (collection) (Russian, Ukrainian, and Kazakhstani archives)
f.	folio (American archives)
GE	*Global Environment*
GPQ	*Great Plains Quarterly*
GR	*Geographical Review*
JAH	*Journal of American History*
JHG	*Journal of Historical Geography*
JMH	*Journal of Modern History*
JoF	*Journal of Forestry*
KHQ	*Kansas Historical Quarterly*
KSHS SA	Kansas Historical Society, State Archives, Topeka, Kansas
l.	list (folio) (Russian, Ukrainian, and Kazakhstani archives)
LoC MD	Library of Congress, Manuscript Division, Washington, DC

MLA	Mennonite Library and Archives, Bethel College, North Newton, Kansas
MHS	Minnesota Historical Society, St. Paul, MN
MSU DGML SCA	Missouri State University, Duane G. Meyer Library, Special Collections and Archives
NAL	National Agricultural Library, USDA, Beltsville, Maryland
NARA CP	National Archives and Records Administration, College Park, Maryland
NARA KC	National Archives and Records Administration, Kansas City, Missouri
NDSUA	North Dakota State University Archives, Fargo, ND
NSHS SAMD	Nebraska State Historical Society, State Archives and Manuscript Division, Lincoln, NE
NYT	*New York Times, The*
op.	opis' (inventory) (Russian, Ukrainian, and Kazakhstani archives)
PHR	*Pacific Historical Review*
RG	Record Group (American archives)
RGIA	Rossiiskii Gosudarstvennyi Istoricheskii arkhiv, St. Petersburg
RR	*Russian Review*
SDSU ASC	South Dakota State University Archives and Special Collections, Hilton M. Briggs Library, Brookings, South Dakota
SEER	*Slavonic and East European Review*
SHSM MC	The State Historical Society of Missouri Manuscript Collection, Columbia, Missouri
SKhiL	*Sel'skoe khoziaistvo i lesovodstvo*
Spb	St. Petersburg
SS	*Soil Science*
TsGA RK	Tsentral'nyi gosudarstvennyi arkhiv Respubliki Kazakhstan, Almaty
TsGANTD RK	Tsentral'nyi gosudarstvennyi arkhiv nauchno-tekhnicheskoi dokumentatsii Respubliki Kazakhstan, Almaty
TsGANTD Spb	Tsentral'nyi gosudarstvennyi arkhiv nauchno-tekhnicheskoi dokumentatsii, Sankt-Peterburg

UMA	University of Minnesota Archives, Minneapolis, Minnesota
USDA	United States Department of Agriculture
WHQ	*Western Historical Quarterly*
WPA	Works Progress Administration
WSUL SCUA	Wichita State University Libraries, Special Collections and University Archives, Wichita, Kansas
ZhMGI	*Zhurnal Ministerstva gosudarstvennykh imushchestv*

Map 1 Map of grasslands around the world.
Adapted from "Extent of temperate grasslands, savannas and shrublands," David Moon, "The Saskatchewan Steppe in a Comparative and Transnational Perspective," Network in Canadian History & Environment\Nouvelle initiative Canadienne en histoire de l'environnement, http://niche-canada.org/2018/06/21/the-saskatchewan-steppe-in-a-comparative-and-transnational-perspective, accessed January 8, 2020

Map 2 Map of the Great Plains.
Reproduced with Permission from the Center for Great Plains Studies, Lincoln, Nebraska.
www.unl.edu/plains/about/map.shtml, accessed January 8, 2020

Map 3 Map of the steppes and other environmental regions of Eurasia

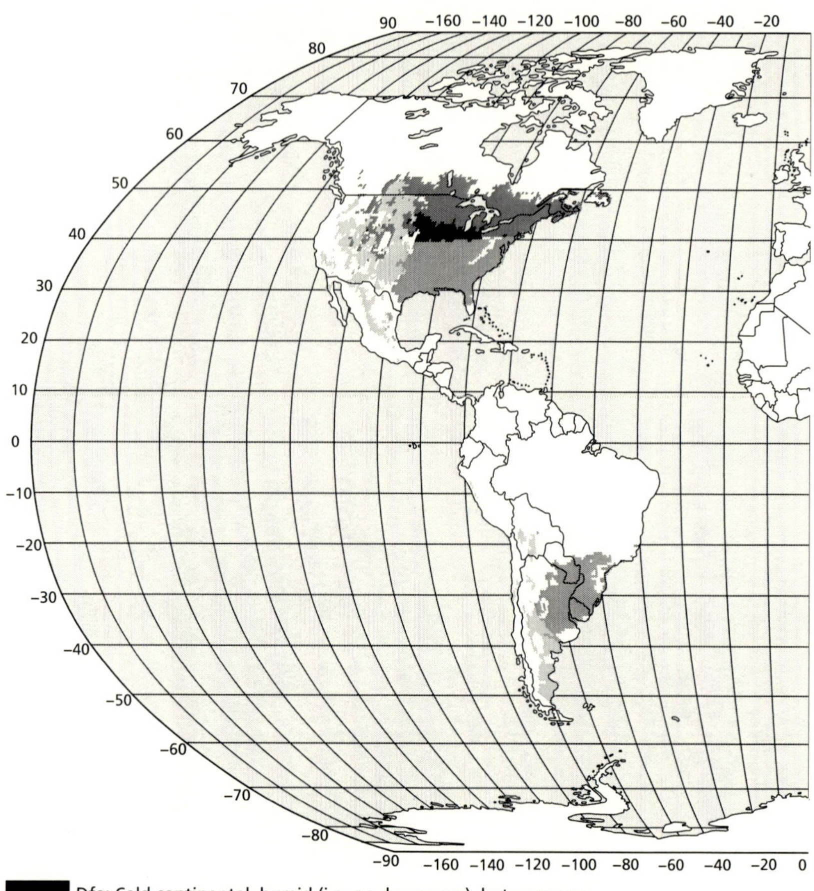

Map 4 World map of Köppen–Geiger climate classification for 1901–25.
Revised and reproduced with permission from: Franz Rubel and Markus Kottek, "Observed and projected climate shifts 1901–2100 depicted by world maps of the Köppen-Geiger climate classification," Meteorologische Zeitschrift 19, no. 2 (2010): 136, Figure 1, www.schweizerbart.de/papers/metz/detail/19/74932/Observed_and_projected_climate_shifts_1901_2100_depicted_by_world_maps_of_the_Koppen_Geiger_climate_classification, accessed January 8, 2020

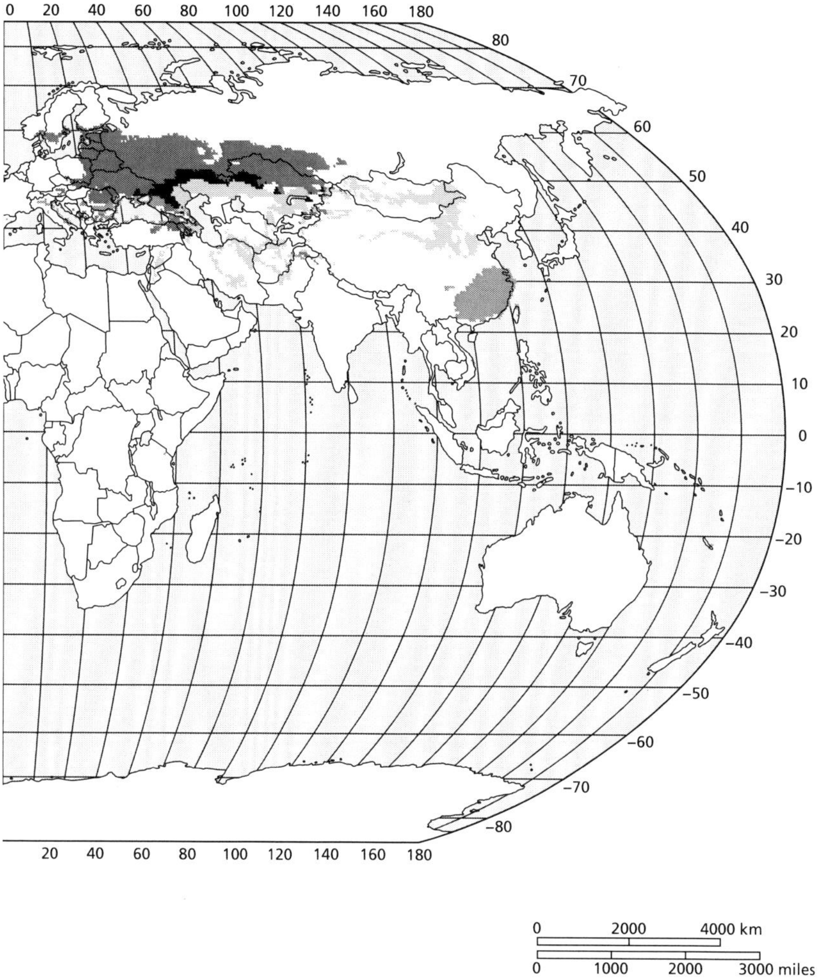

0 20 40 60 80 100 120 140 160 180
80
70
60
50
40
30
20
10
0
−10
−20
−30
−40
−50
−60
−70
−80
20 40 60 80 100 120 140 160 180
0 2000 4000 km
0 1000 2000 3000 miles

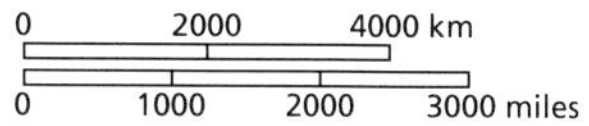

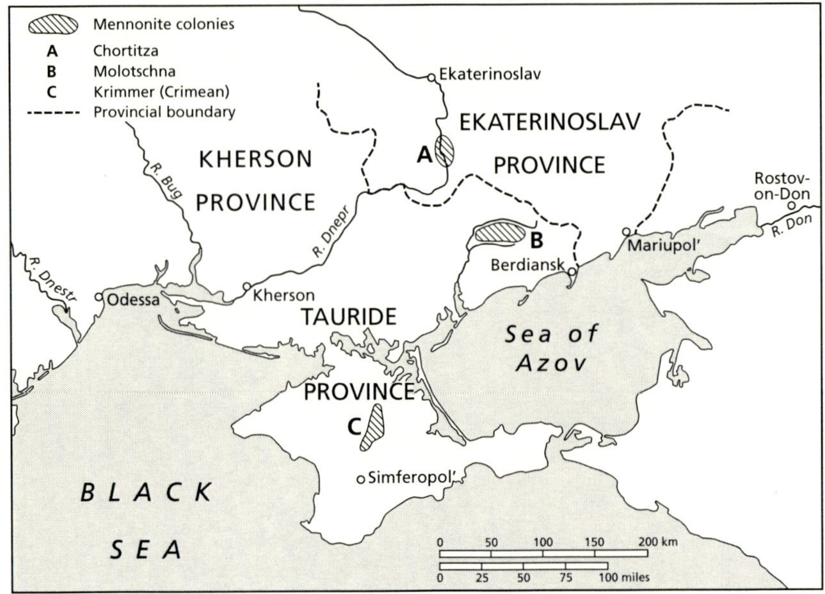

Map 5 Map of the Mennonite colonies in "New Russia" (S. Ukraine) in 1875.
Based on original map produced by Walter Unger (1927–2017): https://sites.ualberta.ca/~german/AlbertaHistory/Odessa.htm, accessed January 8, 2020

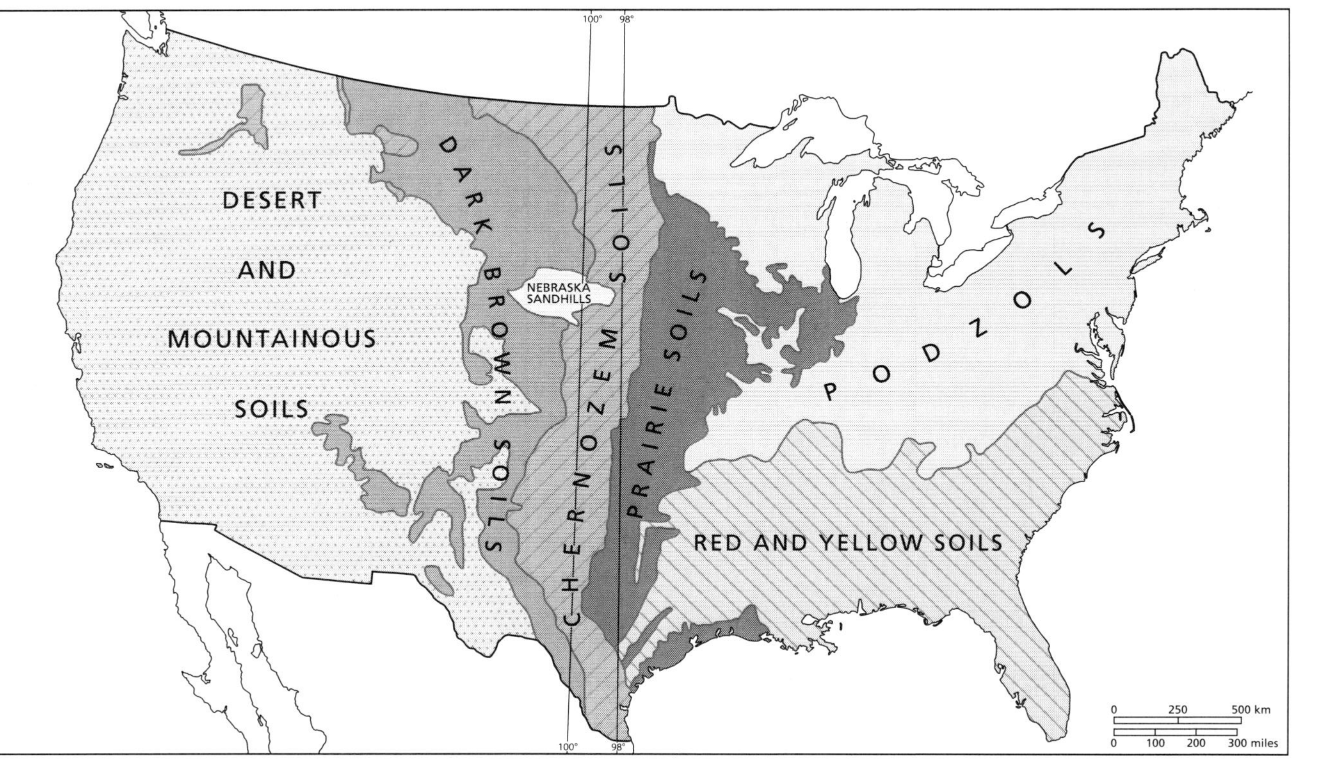

Map 6 Map of the distribution of the great soil groups, Curtis Marbut, 1935 (simplified).
Based on C. F. Marbut, "Soils of the United States," in Atlas of American agriculture: Physical Basis including Land Relief, Climate, Soils, and Natural Vegetation of the United States, ed. O.E. Baker (Washington: USDA, 1936), Part 3, Plate 2, following p. 14: https://archive.org/details/AtlasOfAmericanAgriculture/page/n127, accessed January 8, 2020

Introduction

This book seeks to explain how and why aspects of agriculture in the Great Plains of the United States have, perhaps unexpected, roots in the steppes that lie part of the way around the globe in present-day Ukraine, the Russian Federation, and Kazakhstan.[1] Between the 1870s and 1930s, there were a series of transfers from the Eurasian steppes to the American plains. They included plants, in particular varieties of grain, fodder crops, trees and shrubs, as well as weeds; of agricultural sciences, especially the science of understanding the soils of such grasslands; and of agricultural techniques, for example planting shelterbelts of trees to protect the land from the wind.[2] They replaced or supplemented plants, sciences, and techniques already in the American plains, creating what I have called, with intentional irony, the "American steppes." These transfers from the steppes came after Euro-American agricultural settlers backed by the federal government and U.S. Army had dispossessed the Native Americans and largely ended the ways they had devised to live in the plains environment.

The transfers from the steppes to the Great Plains were facilitated by movements of people. Between the early nineteenth and early twentieth centuries there was mass migration from Europe to the United States. From the 1870s, the migrants included thousands of farmers from the steppes who moved to the similar environment of the plains. They took with them some of their plants and their experience of farming in semi-arid grasslands.

[1] See "Notes on the Text: Place Names." For an environmental history of steppe agriculture, see David Moon, *The Plough that Broke the Steppes: Agriculture and Environment on Russia's Grasslands, 1700–1914* (Oxford: Oxford University Press, 2013).

[2] The term "agriculture" is used in this book to include crops, agricultural sciences and techniques. See Frieda Knobloch, *The Culture of Wilderness: Agriculture as Colonization in the American West* (Chapel Hill: University of North Carolina Press, 1996), pp. 1–5. For similar dictionary definitions, see "Agriculture," in *Oxford Dictionary of English*, ed. Angus Stevenson (Oxford: Oxford University Press, 2010), available online at www.oxfordreference.com/view/10.1093/acref/9780199571123.001.0001/m_en_gb0014280, accessed June 6, 2018; Merriam-Webster Dictionary, available online at www.merriam-webster.com/dictionary/agriculture, accessed June 6, 2018.

From around the same time, American and Russian agricultural scientists began to visit each other's countries, including the grasslands, and learn from each other's experience. Contacts between agricultural scientists allowed them to share expertise and seeds of crop varieties. These contacts largely halted, for logistical reasons, between the Russian Revolution of 1917 and the end of the Russian Civil War in 1921. Contacts were quickly re-established and continued throughout the inter-war period, despite the absence of diplomatic relations between the United States and the new Soviet Union until 1933. Among the American scientific community were Jewish émigrés who had fled the Russian Empire in the decades before the revolution. Their origins and language skills assisted contacts between the two countries' scientists and the assimilation of Russian sciences in the United States.

Thus, this book opens in the 1870s, with the arrival in the Great Plains of the migrants from the steppes and the start of contacts between agricultural scientists interested in our two regions. It ends in the late 1930s. The contacts and transfers were sharply reduced when Stalin's terror consumed some of the Soviet scientists with international connections and created an atmosphere of suspicion of foreigners. The onset of the Cold War in the late 1940s, following the wartime alliance, prevented a resumption of contacts and exchanges on the same scale as earlier. The end of our story, in the late 1930s, coincides with the last part of the Dust Bowl, during which plains farmers, agricultural scientists, and state and federal governments grappled with the ecological crisis.

It is more complicated to explain why these transfers took place, and why some proved so important in the Great Plains. The two regions – in the centers of the North American and Eurasian continents – are very far apart and are not connected by convenient communication routes. Several millennia earlier, the Native Americans' ancestors had made their way from Eurasia, via the Beringia land bridge, to North America.[3] But, at the end of the last Ice Age, Beringia was flooded by the north Pacific, separating North America from Eurasia. Until the nineteenth century, moreover, many Americans and Russians were unfamiliar with their own countries' grasslands, since at that time they lay near or beyond their expanding borders. For rather longer, most Americans and Russians knew even less about such regions on other continents. Both familiarity with their own grasslands and, for some, with similar regions overseas,

[3] See John F. Hoffecker, Scott A. Elias, and Dennis H. O'Rourke, "Out of Beringia?" *Science* 343, no. 6174 (28 February 2014), 979–980.

developed over the nineteenth century, paving the way for exchanges between them.

A further complication was the two countries' contrasting political and economic systems. Throughout our period, the United States was a federal republic with a constitution that shared powers between an elected president, elected legislature, an independent judiciary, and its constituent states. The economy was capitalist, albeit one in serious crisis in the 1930s. The Russian Empire was an autocracy where, in theory, power was held by the tsars and only slightly curtailed by elected dumas (parliaments) between 1906 and 1917. A capitalist economy was developing in the decades prior to 1917. After 1917, the Bolshevik (from 1918, Communist) Party sought rapidly to transform a largely agricultural country into an advanced, industrial, socialist state. Regardless of their contrasting systems, both the United States and the Russian/Soviet states were vast, continental countries with a diversity of environments, including semi-arid grasslands that offered opportunities for large-scale grain cultivation, if the environmental challenges could be overcome.

In explaining the transfers from the steppes to the Great Plains it is very important that they share similar environments: both are flat, semi-arid grasslands, prone to droughts, with continental climates, high winds, fertile soils, but few trees. But these similarities are not sufficient to explain why the transfers took place. Over the millennia that humans have inhabited the Great Plains and steppes, their environments have supported many different ways of life and economic activities. These have included: hunting, herding, and ranching wild or domesticated animals; extracting mineral resources, for example, shale oil in North Dakota and coal in the Donbas region in the east of present-day Ukraine; industries, including aeronautics engineering in Wichita, Kansas, and making rockets for the space program in Samara, Russia; as well as cultivating crops, from the corn, beans, and squash (the "three sisters") of Native American farming in the river valleys, to the wheat, soybeans, and canola, as well as corn, of modern farmers. Since there have been so many possibilities, there is no suggestion here that the two regions' similar environments in any way determined their inhabitants' choices about how to make their livings or the transfers of plants, sciences, and techniques between them.

What matters is that, regardless of the differing political and economic systems, from the mid eighteenth century in the Russian Empire and the mid nineteenth century in the United States, with government support, large numbers of migrants settled in their countries' grasslands, displaced the indigenous peoples, and supplanted their ways of life by plowing up large expanses of land to cultivate crops. In both regions, agricultural settlers encountered environmental conditions that for most, but not all, differed sharply from their previous homes. Most were accustomed to less monotonous landscapes, more and more reliable rainfall, more temperate climates, weaker winds, less fertile soils, and more trees. The exceptions – who are important for our story – were the farmers who moved to the Great Plains from the steppes in the 1870s and afterwards. Many were Germanic and included Mennonites whose ancestors originated in the Low Countries. Germanic settlers moved to the steppes in the late eighteenth and early nineteenth centuries on the invitation of, and in return for land and privileges from, the Russian government. Thus, when they moved again from the 1870s to the grasslands of North America, they encountered a familiar environment on another continent. Most of the settlers in the grasslands of both countries, however, came from quite different environments and were trying to farm in unfamiliar conditions. They learned by trial and error, risking failure, but, with the exception of some crops, rarely drew on the experience of the indigenous populations.

In both countries, agricultural settlers received advice from their governments and institutions established to produce scientific expertise to support the agricultural development of the grasslands. The main institutions were the United States Department of Agriculture (USDA), agricultural departments at land-grant and state universities, agricultural and forestry experiment stations, and their counterparts in the Russian Empire and Soviet Union. Since many agricultural scientists in both countries were also unfamiliar with the environments of such regions, they carried out field work and studies to assist them in understanding the grasslands and how to farm in them. Jeremy Vetter has emphasized the importance of field work by American scientists in the distinctive environment of the American West, including the Great Plains, in the production of new scientific knowledge about the region between the 1860s and 1910s. A similar story played out in the steppes, where the Russian government and scientific organizations supported naturalists and scientists in conducting field work to assist them in understanding the environment and how it could be utilized for agriculture. A key difference was that in the steppes, the production of scientific expertise to support farming in semi-arid grasslands began around a century earlier than in the Great Plains as agricultural settlement began much earlier.[4]

[4] Jeremy Vetter, *Field Life: Science in the American West during the Railroad Era* (Pittsburgh: University of Pittsburgh Press, 2016); David Moon, "The Russian Academy of Sciences Expeditions to the

Some Americans became aware of this Russian prior experience, recognized it could be useful in the Great Plains, and began to learn from steppe agriculture. This took time since there were barriers to influences from the Russian Empire and Soviet Union in the United States. A significant barrier was a widespread perception among Americans that their progressive society, which saw itself at the forefronts of scientific, technological, economic, political, and other developments, could have little to learn from a country on the fringes of Europe about which many knew little, besides stereotyped images of "backwardness," poverty, and oppression of its population. At the same time as the transfers from the steppes to the Great Plains discussed in this book, there were also large-scale transfers of technology the other way. From the late nineteenth century, there were major American influences in Russian and Soviet agriculture, especially machinery, and in industry. The first Soviet Five-Year Plan (1928–32) for economic development was based heavily on American technology. American designs and equipment were used for prestige projects, including the Stalingrad Tractor Factory, the dam and hydroelectric power station on the Dnepr River in southern Ukraine (DneproGES), and the iron and steel works at Magnitogorsk in the southern Urals, all of which were in the steppe region.[5] These transfers from west to east fit better widely held perceptions of American "superiority" and Russian "backwardness." These perceptions have continued to color some American and western views of Russia: hence the "unexpected" nature of the transfers from the Russian Empire and Soviet Union conveyed in the title of this book.

The plants, agricultural sciences and techniques that were transferred from the steppes proved effective in the American plains precisely because they came from a region with similar environmental conditions and where farmers from similar, European, backgrounds were trying to do similar things: engage in European-style farming. For the same reason, many of the farmers who moved from the steppes to the Great Plains were

Steppes in the Late Eighteenth Century," *SEER* 88 (2010), 204–36; Moon, *Plough*, pp. 46–88; David Moon, "Scientific Innovation in the Russian Empire: The Case of Genetic Soil Science," in *Science and Empire in Eastern Europe: Imperial Russia and the Habsburg Monarchy in the 19th Century*, ed. Jan Arend (Munich: Vandenhoeck and Ruprecht, 2020), pp. 305–20. On field work, see also Henrika Kuklick and Robert E. Kohler, "Introduction," *Osiris* 11 (1996), 1–14.

[5] See, for example, Dana G. Dalrymple, "The American Tractor Comes to Soviet Agriculture: The Transfer of a Technology," *Technology and Culture* 5 (1964), 191–214; Dalrymple, "American Technology and Soviet Agricultural Development, 1924–1933," *AH* 40 (1966), 187–206; Kendall E. Bailes, "The American Connection: Ideology and the Transfer of American Technology to the Soviet Union, 1917–1941," *Comparative Studies in Society and History* 23 (1981), 421–48. Technology transfers from the United States to Russia and the Soviet Union will not be considered in detail as they have been the subject of much research.

successful. Thus, the similarities in the environments were necessary for the transfers from the steppes to be viable in the Great Plains, but are not sufficient to explain why aspects of Great Plains agriculture have Russian roots. This requires a human dimension involving the choices made by people in both countries to settle in the unfamiliar environments of their grasslands with the intention of farming, and by their governments to support agricultural settlement and development in these regions.[6]

The transfers from the steppes were not simply contributions to the development of agriculture in the Great Plains that could have come from elsewhere and made little difference. It was significant that the transfers came from another settler society in a region with a similar environment. Some of the transfers, in particular the varieties of wheat and other crops, soil science, and the weeds, fundamentally transformed Great Plains agriculture.[7] These transfers were part of wider exchanges of techniques, knowledge, and plants between settler societies and the overseas colonies of European states in this period. In most, knowledge produced by scientists and others with "specialist" training from other, similar, societies was privileged over the local knowledge of indigenous peoples.[8] In both the steppes and the Great Plains the colonizing powers – the Russian Empire and United States – constructed similar, negative stereotypes of the indigenous populations as "backward, uncivilized, wandering, [and] primitive," in part to justify taking their land. They deemed the nomadic ways of life of the pastoralists of the steppe and hunters of the plains inferior to "civilized," settled arable farming, and certainly not models to learn from.[9]

Comparing the Great Plains and the Steppes

The Great Plains and the steppes are semi-arid grasslands: an ecosystem found in different parts of the world.[10] (See Map 1: Map of grasslands around the world.) The Great Plains extend from Texas in the south up

[6] For a similar point, see Edward Dallam Melillo, *Strangers on Familiar Soil: Rediscovering the Chile–California Connection* (New Haven: Yale University Press, 2015), pp. 6–8.

[7] See Melillo, *Strangers on Familiar Soil*, p. 11.

[8] Richard Grove found evidence that European naturalists were more open to indigenous knowledge in colonial contexts before the mid nineteenth century. Grove, *Green Imperialism: Colonial Expansion, Tropical Island Edens and the Origins of Environmentalism, 1600–1860* (Cambridge, England: Cambridge University Press, 1995).

[9] Steven Sabol, *"The Touch of Civilization": Comparing American and Russian Internal Colonization* (Boulder, CO: University Press of Colorado, 2017), pp. 14, 33–59.

[10] See Andrew C. Isenberg, "Seas of Grass: Grasslands in World Environmental History," in *The Oxford Companion of Environmental History*, ed. Isenberg (Oxford: Oxford University Press, 2014), pp. 133–53. The grassland that most closely resembles the Great Plains and steppes is the pampas of

the middle the United States, through Oklahoma, Kansas, Nebraska, the Dakotas, and into the Canadian prairie provinces. The plains encompass much of the upper and middle basin of the Missouri–Mississippi river system, and a smaller area in the basin of the Red River of the North. In the west, the plains run up against the foothills of the Rocky Mountains, and include eastern Montana, Wyoming, Colorado, and New Mexico. In the east, they merge into the prairies roughly along the eastern borders of North and South Dakota and Nebraska, in eastern Kansas, central Oklahoma, and central Texas. The plains extend into southwestern Minnesota, but the prairies in most of Minnesota as well as Iowa, Missouri, Illinois, Indiana, and Ohio, are largely beyond the scope of this book as the conditions differed from those in the steppes. Much of the attention will be on the northern and central plains (North and South Dakota, Nebraska, and Kansas). The United Stated acquired much of the Great Plains from France by the Louisiana Purchase of 1803. The smaller area in the Red River basin was transferred from British North America by the Treaty of 1818 that established the 49th parallel as the international border west of the Great Lakes. At the start of our story in the 1870s, therefore, the Great Plains region was a relatively recent acquisition to the United States.[11] (See Map 2: Map of the Great Plains.)

The conquest and annexation of much of the Eurasian steppe by the Russian state began a lot earlier, in the late fifteenth century, and took longer, into the early nineteenth century.[12] The steppe region of the Russian Empire and Soviet Union (and its successor states) is orientated – in contrast to the Great Plains – from west to east. It encompasses the lower parts of the basins of the rivers Dnestr, Bug, Dnepr, Don, Volga, and Ural extending to the northern shores of the Black, Azov, and Caspian seas and the Caucasus Mountains, and part of the basin of the Irtysh (Ertis) river system in southern Siberia and northern Kazakhstan. The steppe continues beyond the Altai Mountains into northern Mongolia and

South America. See Adrián Gustavo Zarrili, "Capitalism, Ecology, and Agrarian Expansion in the Pampean Region, 1890–1950," *EH* 6 (2001), 561–83. See also C. E. Solberg, *The Prairies and the Pampas: Agrarian Policy in Canada and Argentina, 1880–1930* (Stanford CA: Stanford University Press, 1987); Jeremy Adelman, *Frontier Development: Land, Labour, and Capital on the Wheatlands of Argentina and Canada, 1890–1914* (Oxford: Oxford University Press, 2004).

[11] See Peter J. Kastor, *The Nation's Crucible: The Louisiana Purchase and the Creation of America* (New Haven: Yale University Press, 2004); S. Anderson, "The North-American Boundary from the Lake of the Woods to the Rocky Mountains," *Journal of the Royal Geographical Society of London* 46 (1876), 228–62.

[12] See Michael Khodarkovsky, *Russia's Steppe Frontier: The Making of a Colonial Empire, 1500–1800* (Bloomington: Indiana University Press, 2002).

Manchuria.[13] The significance of the different orientations of the two grasslands – the Great Plains run from north to south, while the steppes extend from east to west – will be considered later. (See Map 3: Map of the Steppes and other environmental regions of Eurasia.)

The boundaries of the two regions have been defined in various ways and with varying degrees of precision. In this book they are defined in two ways. The first is the definitions used in the sources the book draws on, which therefore differ, and will be explained when necessary. The second is the regions' environmental conditions, in particular topography, average annual precipitation, vegetation, and soil. Both regions are largely flat, but also contain undulating countryside, and thus end when they come up against mountains or seas. The two regions receive less precipitation than the lands to the east (in North America) and the northwest (in Eurasia), where many of the agricultural setters came from. The eastern boundary of the Great Plains has been defined as the 98th or 100th meridians. In the northern plains these coincide roughly with the isohyet – a line connecting points with the same yearly average rainfall – of 20 inches (c. 500 mm). Other specialists have considered the eastern boundary to be the 28 inch (700 mm) isohyet, which approximates to the 98th or 100th meridian in the southern plains, where the climate is warmer and evaporation higher.[14] In Eurasia, the northwestern boundary of the treeless steppe is generally considered to be around the 16 inch (400 mm) isohyet, although parts in the west of Ukraine and the North Caucasus receive more rainfall.[15]

The climates of the two regions can be compared with reference to the Köppen–Geiger climate classification, which designates different climates by letters that refer to the main climate group and seasonal distributions of precipitation and heat.[16] According to data for 1901–25, the climate of the

[13] See A. A. Chibilev, *Stepi Severnoi Evrazii: ekologo-geograficheskii ocherk i bibliografiia* (Ekaterinburg: UrO RAN, 1998); Moon, *Plough*, pp. 6–10.

[14] See Walter Prescott Webb, *The Great Plains*, new edition [1st published 1931] (Lincoln: University of Nebraska Press, 1981), pp. 3–44; Geoff Cunfer, *On the Great Plains: Agriculture and Environment* (College Station: Texas A&M University Press, 2005), p. 242. See also Frederick C. Luebke, "Regionalism and the Great Plains: Problems of Concept and Method," *WHQ* 15 (1984), 19–38. For a map showing "as least fifty versions of the Great Plains regional boundary," see Douglas Hurt, *The Big Empty: The Great Plains in the Twentieth Century* (Tuscon: University of Arizona Press, 2011), p. xvii.

[15] See Moon, *Plough*, pp. 7, 70–2, and works cited therein.

[16] See Franz Rubel and Markus Kottek, "'The Thermal Zones of the Earth' by Wladimir Köppen (1884)," *Meteorologische Zeitschrift* 20, no. 3 (2011), 361–5. Köppen was born in St. Petersburg, Russia, attended Simferopol' gymnasium in the Crimea and St. Petersburg University, before continuing his education at Heidelberg and Leipzig universities. He worked for a short time in Russia, but from 1875 lived in Germany and Austria. "Wladimir Köppen, 1846–1940" [obituary], *Bulletin of the American Meteorological Society*, 21–2 (1940), 81.

two regions has been classified as follows. The eastern and central parts of the Canadian prairies, North Dakota, and northern South Dakota were designated Dfb. This signified that the climate was: cold/continental, humid (i.e. no dry season), and with warm summers. The climate of the eastern and central parts of southern South Dakota and Nebraska was assigned the letters Dfa, which stand for cold/continental, humid, and hot summers. In eastern and central Kansas, Oklahoma, and Texas the climate was designated Cfa: warm temperate, humid, hot summers. Much of the western plains from Alberta to West Texas, however, were classified BSk: cold, semi-arid, steppe. The same four classifications designated the climate of the Eurasian steppe for these years. The climate of the northern steppes, both west of the Urals and in southern Siberia and northeastern Kazakhstan, was classified as Dfb (similar to the northern plains and Canadian prairies). Further south and west, in southern Siberia, northwestern Kazakhstan, and across the Ural and Volga rivers into the north-central Caucasus, the climate was Dfa (the same as southern South Dakota and Nebraska). Along the coasts of the Black and Azov Seas in southern Ukraine, including the Crimean peninsula, and the northwestern Caucasus, the climate was defined as Cfa (the same as eastern and central Kansas). Further south and east, to the north of Caspian Sea and in central Kazakhstan, the climate was cold, semi-arid, steppe (BSk) – the same as the western plains.[17] (See Map 4: World map of Köppen–Geiger climate classification for 1901–25.)

These broad classifications conceal significant differences between the two regions' climates. The Great Plains as a whole is slightly warmer than the steppes in their entirety. However, the temperatures in the northern plains and Canadian prairies resemble southern Siberia and northern Kazakhstan. The average, annual precipitation in the Great Plains is slightly higher than in the steppes. But, due to the higher temperatures,

[17] Franz Rubel and Markus Kottek, "Observed and Projected Climate Shifts 1901–2100 Depicted by World Maps of the Köppen–Geiger Climate Classification," *Meteorologische Zeitschrift* 19, no. 2 (2010), 136; M. C. Peel, B. L. Finlayson, and T. A. McMahon, "Updated World Map of the Köppen–Geiger Climate Classification," *Hydrological Earth System Science* 11 (2007), 1633–44. In both regions, the climate is getting harsher, and is projected to get more extreme, as a result of climate change. Rubel and Kottek, "Observed and Projected Climate Shifts 1901–2100," 137; "A North American Climate Boundary Has Shifted 140 Miles East Due to Global Warming," *Yale Environment 360 Digest*, (April 11, 2018), available online at https://e360.yale.edu/digest/a-north-american-climate-boundary-has-shifted-140-miles-east-due-to-global-warming, accessed May 14, 2018. Plains ecosystems "have shifted hundreds of miles northward in the past 50 years, driven by climate change, wildfire suppression, energy development, land use changes, and urbanization." "Great Plains' Ecosystems Have Shifted 365 Miles Northward Since 1970," *Yale Environment 360 Digest* (July 2, 2019), available online at https://e360.yale.edu/digest/great-plains-ecosystems-have-shifted-365-miles-northward-since-1970/, accessed July 3, 2019.

more moisture evaporates than in the steppes. The thermal and moisture conditions combined of Pierre, in southern South Dakota, resemble those in Odessa on the north coast of the Black Sea and Rostov-on-Don near the Sea of Azov. These are the parts of the steppes with the most favorable conditions for agriculture, while the same cannot be said of southern South Dakota in the Great Plains.[18] The generally harsher climate of the steppes, with greater extremes to the east in southern Siberia and the north of present-day Kazakhstan, will play an important part in our story.

The climates of the Great Plains and the steppes resemble each other, however, in that the rainfall fluctuates from year to year and there are recurring droughts.[19] Long into the nineteenth century, many Americans considered the Great Plains unsuitable for agriculture and called it the "Great American Desert."[20] A further similarity is that in both regions dust storms are whipped up by the high winds that blow across the flat landscapes. Droughts and dust storms created terrifying experiences that are well known from accounts of the Dust Bowl in the United States, but have also been a recurring experience in the steppes.[21]

The climate, especially the available moisture, influences the regions' vegetation. In both, the predominant natural vegetation is grasses. This contrasts with trees in the more humid regions east of the Great Plains and north of the steppes, and desert vegetation in the more arid regions west of the Great Plains and southeast of the steppes. In the drier western plains, the natural vegetation is short grasses, while in the more humid east, it is tall grasses. In between is an ecotone (transition area) of mixed grasses. There are similar variations in the steppes, with shorter grasses in the more arid south and southeast. Topography, climate, and vegetation are three of

[18] See N. C. Field, "Environmental Quality and Land Productivity: A Comparison of the Agricultural Land Base of the USSR and North America," *Canadian Geographer* 12 (1968), 1–14. See also David Moon, "The Saskatchewan Steppe in a Comparative and Transnational Perspective," *Network in Canadian History & Environment/Nouvelle initiative Canadienne en histoire de l'environnement*, (June 21, 2018), available online at http://niche-canada.org/2018/06/21/the-saskatchewan-steppe-in-a-comparative-and-transnational-perspective/, accessed June 21, 2018.

[19] See Kevin Z. Sweeney, *Prelude to the Dust Bowl: Drought in the Nineteenth-Century Southern Plains* (Norman: University of Oklahoma Press, 2016); Donald Worster, *Dust Bowl: The Southern Plains in the 1930s* (New York: Oxford University Press, 2004), pp. 10–12, 75; Moon, *Plough*, pp. 65–8; Nikolai M. Dronin and Edward G. Bellinger, *Climate Dependence and Food Problems in Russia, 1900–1990* (Budapest: CEU Press, 2005), pp. 44–53, 80–9, 123–35, 200–7, 245–52, 293–305.

[20] B. H. Baltensperger, "Plains Boomers and the Creation of the Great American Desert myth," *JHG* 18 (1992), 59–73.

[21] James C. Malin, "Dust Storms, 1850–1900," *KHQ* 14 (1946), 129–44, 265–96, 391–413; Moon, *Plough*, pp. 144–8; Marc Elie, "The Soviet Dust Bowl and the Canadian Erosion Experience in the New Lands of Kazakhstan, 1950s–1960s," *GE* 8 (2015), 274–8.

the factors that contribute to the formation of particular types of soils. Thus, the two regions share similar soils, for the most part, deep, dark, and fertile, that are rich in humus and mineral nutrients. In the more arid parts of both regions, in the northern and western Great Plains and the eastern and southern steppes, are shallower, chestnut-colored, soils. Soils, and how scientists understood and classified them, are among the main topics of this book.[22]

The environment of the Great Plains and the steppes provided opportunities and challenges for agricultural settlers from outside the regions. The soils are among the most fertile in the world and are ideal for growing grain. As Russian soil scientists worked out in the late nineteenth century, however, the fertile soil had formed in the very conditions of relatively low and unreliable rainfall that proved so difficult for farmers growing crops. A further challenge was the winds that dried out the land and periodically blew away the top soil.

American and Russian Attitudes to Their and Other Grassland Regions

Into the nineteenth century, their semi-arid grasslands were alien to most Americans and Russians: quite different from the humid, forested landscapes of their homelands to the east and northwest, respectively.[23] In 1832, Washington Irving went on a hunting expedition in "the prairies" (mostly in present-day Oklahoma). The native New Yorker had mixed feelings about the landscape: "to our infinite delight," he wrote, "[we] beheld 'the Great Prairie' stretching to the right and left before us ... The landscape was vast and beautiful."[24] Earlier, he expressed a preference for familiar wooded landscapes.[25] A further reason for his preference was fear of Indian attack in the exposed prairies that lacked woodlands for refuge.[26]

[22] For analyses of the Great Plains climate, vegetation, soils, and agriculture in the early 1920s, see: Joseph B. Kincer, "The Climate of the Great Plains as a Factor in Their Utilization," *AAAG* 13, 2 (1923), 67–80; H. L. Shantz, "The Natural Vegetation of the Great Plains Region," *AAAG* 13, 2 (1923), 81–107; C. F. Marbut, "Soils of the Great Plains," *AAAG* 13, no. 2 (1923), 41–66; O. E. Baker, "The Agriculture of the Great Plains Region," *AAAG* 13, 3 (1923), 109–67. See also James E. Sherow, *The Grasslands of the United States: An Environmental History* (Santa Barbara, CA: ABC-CLIO, 2007). On the steppe environment, see Chibilev, *Stepi Severnoi Evrazii.*

[23] Webb paid great attention to this in *The Great Plains.* On the value of travel writing as a genre for understanding perceptions of the American West by Americans and others, see David M. Wrobel, *Global West, American Frontier: Travel, Empire, and Exceptionalism, from Manifest Destiny to the Great Depression* (Albuquerque, NM: University of New Mexico Press, 2013).

[24] Washington Irving, *A Tour on the Prairies* (Paris: A. and W. Galignani, 1835), 158.

[25] Irving, *A Tour on the Prairies*, pp. 96, 101.

[26] Irving, *A Tour on the Prairies*, pp. 62, 83.

In a similar manner, in the early twentieth century, the Russian historian Vasilii Kliuchevskii noted that before the mid eighteenth century, most Russians had lived in the forest region, and until that time the steppe had featured in Russian history mainly in malevolent events such as raids by nomadic Tatars or rebel cossacks.[27]

Many visitors to both grasslands compared the landscapes to the sea or ocean: the only other panorama of endless flatness many had previously encountered. Irving noted: "An immense extent of grassy, ... 'rolling' country, with here and there a clump of trees dimly seen in the distance like a ship at sea." Recalling a journey from the Missouri line to California, Mark Twain thought the "grassy carpet greener and smoother than any sea."[28] In 1835, Nikolai Gogol' imagined the unplowed steppe of his native Ukraine in the sixteenth century: "No plough had ever passed over the immeasurable waves of wild growth ... The whole surface resembled a golden-green ocean, upon which were sprinkled millions of different flowers."[29] Some Russian travelers transferred the metaphor to America's grasslands. Aleksandr Lakier, a native of the southern steppes who visited the United States in 1857–8, described Illinois: "Everywhere dense grasses and flowers, and in places where the land had been plowed, corn [*kukuruza*] and wheat grow. It is an entire sea of plants ..."[30] Two decades later, Mikhail Vladimirov visited Nebraska and saw the "boundless expanse of a sea of grass."[31]

In both countries, over the nineteenth century, the once alien grasslands previously inhabited by "hostile" peoples were incorporated into "national" landscapes after the indigenous populations had been displaced by incoming farmers who plowed up their land. The steppes became a symbol of "Russian" culture and the "Great American Desert" the setting for the story of the "winning of the West."[32] By the late nineteenth

[27] V. O. Kliuchevskii, *Sochineniia v deviati tomakh*, 9 vols. (Moscow: Mysl', 1987–1990), vol. 1, pp. 83–5.

[28] Irving, *A Tour on the Prairies*, pp. 94; Mark Twain, *The Innocents Abroad, or The New Pilgrims' Progress* (Ware, UK: Wordsworths Classics, 2012 [1st published 1869]), p. 67. See also Kenneth D. Rose, *Unspeakable Awfulness: America through the Eyes of European Travelers, 1865–1900* (New York: Routledge, 2014), p. 222.

[29] Nikolai Vasilievich Gogol, *Taras Bulba and Other Tales* (London: J. M. Dent, 1952), p. 19.

[30] Aleksandr Lakier, *Puteshestvie po Severo-Amerikanskim Shtatam, Kanade i ostrovu Kube*, 2 vols. (Spb: K. Vul'f, 1859), vol.1, p. 130.

[31] M. M. Vladimirov, *Russkii sredi Amerikantsev, 1872–1876* (Spb: Obshchestvennaia pol'za, 1877), p. 105.

[32] See Christopher Ely, *This Meagre Nature: Landscape and National Identity in Imperial Russia* (DeKalb, IL: Northern Illinois University Press, 2002); A. A. Chibilev, ed., *Stepnye shedevry* (Orenburg: Orenburgskoe knizhnoe izd-vo, 2009); William W. Stowe, "'Property in the

century, therefore, for some Russian and American travelers to each other's grasslands, encountering similar landscapes in the other country created a feeling of familiarity. For most other Europeans, however, including some of the settlers in the Great Plains, the American West was quite different from landscapes they had previously experienced.[33]

In both countries, moreover, there were similar reactions to the periodic droughts, winds, and dust storms: anxiety as fertile soil blew away in clouds that obscured the Sun and concerns that people had caused such catastrophes by plowing up the grasslands. One writer asked: "Shall our lands go the way of China?" He wrote of the "pitiable struggle to retain and cultivate the scant remains of once ample soils" and of the "large tracts of almost bare soil on which stand ruins implying former flourishing populations." The writer compared the ravaged grasslands with the ruins of "splendid Roman cities" in North Africa that had been destroyed by soil erosion. In a similar vein, another writer was concerned that agricultural land was "threatened with being transformed into an uninhabited desert" as a result of plowing and the removal of native woodland, and raised the specter of the fate of Central Asia: "a desert with a few oases – [which] was once a flourishing, well populated country with one of the main centers of Islamic education." The author of the first passage was Morris L. Cooke, of the U.S. Rural Electrification Administration, in August 1935, during the Dust Bowl.[34] The second was produced by the Russian philosopher Vladimir Solov'ev in October 1891, following a terrible drought in the steppe region.[35] These two passages illustrate the similarities both between the two grassland regions and the responses to the vagaries of the environment by people, an American and a Russian, who were more familiar and comfortable with environments that did not experience such phenomena. The similarities in responses by Americans and Russians to their grasslands provide a context for the transfers between them.

Horizon': Landscape and American Travel Writing," in *The Cambridge Companion to American Travel Writing*, ed. Alfred Bendixen and Judith Hamera (New York: Cambridge University Press, 2009), pp. 32–3; Ian Frazier, *Great Plains* (New York: Picador, 1989).

[33] See Ray Allen Billington, *Land of Savagery, Land of Promise. The European Image of the American Frontier in the Nineteenth Century* (New York: W. W. Norton & Co, 1981).

[34] Library of Congress, Washington, DC: Manuscript Division, Gifford Pinchot Papers (hereafter Pinchot Papers), Box 588, File: "Soil Erosion," Pinchot to Morris L. Cooke, September 5, 1935 and copy of article from the [London] *Times*, August 9, 1935.

[35] Vladimir Solov'ev, "Narodnaia beda i obshchestvennaia pomosh'," *Vestnik Evropy* 10 (October 1891), 781.

Transfers between the Steppes and the Great Plains

Between the 1870s and 1930s, plants, agricultural sciences and techniques were transferred from the steppes to the Great Plains. The transfers assisted plains farmers and the people who advised them put agriculture on a more sustainable basis and helped in dealing with the Dust Bowl at the end of our period. Analysis of these transfers will be the main subject of this book. Much attention will be on the institutions, governmental and scientific, that facilitated and promoted the transfers. To some extent, therefore, it is an institutional history. I have traced the background to each of the transfers in the steppes and how, in spite of various barriers, they made their part of the way round the globe to the Great Plains. The reception of the transfers among American agricultural scientists and other specialists is a large part of our story. How the plants, sciences, and techniques from the steppes were received among the settler population in the plains, however, is mostly beyond the scope of this book. An exception is the first transfer, varieties of wheat, where a leading role was probably played by migrants who brought them to the Great Plains from the steppes in the 1870s. They cultivated these varieties successfully in their new homes, where they attracted interest among their neighbors.

Wheat varieties were among the plants that traveled from one grassland to the other. The most famous is the hard, red winter wheat, "Turkey Red," that thrived in dry years when other types of wheat did not. It seems first to have been brought over from the steppes by Mennonite immigrants in the 1870s. Some of them, in particular Bernhard Warkentin, went on to import larger quantities. The Mennonites' wheat came to the attention of Mark Carleton, a USDA crop scientist, during the dry 1890s. He visited the steppes twice as a "plant explorer" at the end of the 1890s and brought back more crop varieties. They included very hard spring wheats, known as durum wheats, that proved well suited to the northern plains. Another import from the steppes was used by Canadian crop scientists to create the hybrid Marquis, a hard, red spring wheat that grew well in the Canadian prairies and northern plains. American crop scientists used contacts with their Russian and Soviet counterparts to obtain more types of wheat, and used them to create further hybrids. Varieties imported from steppes, and hybrids derived from them, became the mainstays of wheat culture in the plains.

In the 1920s and 1930s, there was a "Russian revolution" in the bureau of the USDA charged with surveying the soils of the United States to underpin agronomical advice to farmers. Despite significant opposition, the head of the Soil Survey, Curtis Marbut, oversaw a fundamental change in the science and methodology for understanding, surveying, and classifying soils. It was based on an innovation devised by Russian scientists led by Vasilii Dokuchaev during field work analyzing the fertile chernozem, or black earth, in the steppe region half a century earlier. The Russians analyzed cross sections of soils, observing how they had formed from the parent rock at the bottom, under the influence of the vegetation, climate, and topography, over time. The Russian theory of soil formation allowed the classification of soils into overarching types that were global in application. Marbut learned about the new soil science from a book written in German by Konstantin Glinka, one of Dokuchaev's students. After testing the Russian science in the Great Plains, Marbut concluded that the soils of the Great Plains were fundamentally the same as those of the steppes, and that the Russian science could be adapted to the soils of the United States. Marbut was not the first American scientist to realize the importance of the Russian work, but it was he who overcame the opposition to its application by the U.S. Soil Survey. Relations between American and Russian soil scientists were placed on firmer ground at the first two International Congresses of Soil Science held in the United States in 1927 and the Soviet Union in 1930.

In 1934, based largely on claims by Russian and Soviet forestry scientists, President Franklin D. Roosevelt approved a project to plant shelterbelts of trees across the Great Plains to protect the land from wind erosion. The technique of cultivating trees and shrubs to break the wind was not new to the Great Plains. But, since the agricultural settlement of the steppes had started first and steppe farmers had been trying for longer to deal with the winds, there was more experience of shelterbelts in the steppes. Both the tsarist and Soviet governments had invested money in the technique. When facing a dual crisis of the Dust Bowl and the prospect of cuts to their funding in the 1930s, the U.S. Forest Service pressed for a federal government-sponsored project to plant shelterbelts in the drought-stricken region. The Shelterbelt Project is important in this book for another reason. One of the chief advocates was the American forestry scientist Raphael Zon. He was a Russian–Jewish immigrant who had moved from a city on the edge of the steppe region to the United States in the 1890s. He acted as a conduit for Russian forestry science in his adopted homeland.

The transfers from the steppes to the Great Plains analyzed in this book are examples. The selection was based on the significance of the transfers, the availability of sources, to show the roles of different actors (migrant farmers, émigré and American-born scientists, governmental, and scientific institutions), and to present various types of transfers: plants, agricultural sciences and techniques. Many more plants than are discussed in depth in this book were introduced to the Great Plains from the steppes by migrants, agricultural explorers working for the USDA and plains states, and through exchanges between crop scientists, for example: further varieties of wheat and other grains; hardy fodder crops, such as alfalfa and clover; drought-resistant grasses, in particular crested wheat grass; fruit trees and ornamental plants.[36]

Two areas of science and agricultural techniques that are important in both regions are not discussed in this book, because I could not find sufficient evidence that they were transferred from the steppes to the Great Plains. In both cases, American and Russian specialists seem to have worked on similar problems largely independently. These are climate science, in particular cycles of drought, and techniques for growing crops without artificial irrigation in semi-arid environments, known as dry- or dryland farming.

Both regions experienced recurring droughts, which prompted much anxiety, speculation, and scientific research. In the steppe region from the early nineteenth century, landowners, government officials, and scientists debated the nature and causes of the regular droughts. Until the late nineteenth century, many believed (with little hard evidence) that the steppe climate was becoming progressively drier and droughts more frequent. They attributed this apparent change to the felling of much of the region's small areas of woodland. Towards the end of the nineteenth century, however, a consensus emerged that droughts were a cyclical phenomenon caused by outside factors such as solar activity. This was based on observations of the steppe climate over the previous decades and drew on the latest international research.[37] Climate scientists in the Russian Empire were influenced by an essay on changes in the climate since 1700 published in Vienna in 1890 by the German scientist Eduard Brückner. Drawing on data from around the world, including the Russian Empire, he calculated that wet and cool periods alternated with dry and warm periods in cycles of 35 years.[38]

[36] Mark Alfred Carleton, "Russian Cereals adapted for cultivation in the United States," *USDA Division of Botany Bulletin*, no. 23 (1900); Niels E. Hansen, "Fifty Years Work as Agricultural Explorer and Plant Breeder," *Transactions of the Iowa State Horticultural Society* 79 (1944), 28–49.

[37] See David Moon, "The Debate over Climate Change in the Steppe Region in Nineteenth-Century Russia," *RR* 69 (2010), 251–75.

[38] *Eduard Brückner: The Sources and Consequences of Climate Change and Climate Variability in Historical Times*, eds. Nico Stehr and Hans von Storch (Dordrecht: Kluwer Academic Publishers, 2000), pp. 47–191, 219–42, 255–68.

American farmers, government figures, and scientists took a keen interest in the droughts that recurred in the Great Plains, that seemed to culminate in the Dust Bowl of the 1930s. They traced fluctuations in the plains climate since the start of Euro-American settlement in the 1860s. Years with good rains in the mid 1870s and first half of the 1880s had alternated with dry years in the early 1870s, late 1880s, and mid 1890s. The drought-stricken 1930s had been preceded by wetter years in the 1920s. American scientists speculated on the causes of the cyclical changes, but none that I could find referred to Russian work on the same phenomenon in the steppes. My sample included authors familiar with Russian scientific work in other fields, for example, the ecologist Frederic Clements, forester Carlos Bates, who was a colleague of the Russian-born Zon, and C. Warren Thornwaite, who worked for the USDA, which had extensive Russian contacts.[39] Why did American scientists not draw on Russian work on climate when they did so in other fields? Aleksandr Voeikov, a Russian climate scientist had visited the United States in the 1870s and corresponded with American scientists for several decades, but he had not focussed on recurring droughts in the steppe region, and was skeptical about Brückner's cycles.[40] A further reason is that, while Russian and Soviet climate scientists continued their research on droughts, it was not until the second half of the twentieth century that their work came to significant international attention.[41] In the 1930s, the most well-known international work on climate cycles was still that of Brückner, which was

[39] Frederic Clements, "Climatic Cycles and Human Populations in the Great Plains," *The Scientific Monthly* 47, 3 (1938), 193–210; Carlos G. Bates, "Climatic Characteristics of the Plains Region," in *Possibilities of Shelterbelt Planting in the Plains Region* (Washington, DC: U.S. Forest Service, Lake States Forest Experiment Station, 1935), pp. 83–110; C. Warren Thornwaite, "Climate and Settlement in the Great Plains," in *Climate and Man: The 1941 Yearbook of Agriculture*, ed. Gove Hambidge (Washington, DC: USDA, 1941), pp. 177–87. On Clements' familiarity with Russian work on steppe vegetation, see Moon, *Plough*, pp. 112–13. On Bates' knowledge of Russian work on steppe forestry, see pp. 304, 307.

[40] See G. P. Kuropiatnik, "Russians in the United States: Social, Cultural, and Scientific Contacts in the 1870s," in *Russian–American Dialogue on Cultural Relations 1776–1914*, eds. Norman E. Saul and Richard D. McKinzie (Columbia, MO: University of Missouri Press, 1997), pp. 148–50; Moon, "Debate over Climate Change," 252, 256–7, 267–9, 272. On Voeikov's correspondence with Eugene W. Hilgard, see pp. 72 n.64, 74, 116, 221.

[41] See Marc Elie, "Desiccated Steppes: Droughts and Climate Change in the USSR, 1960s–1980s," in *Eurasian Environments: Nature and Ecology in Imperial Russian and Soviet History*, ed. Nicholas Breyfogle (Pittsburgh, PA: University of Pittsburgh Press, 2018), pp. 75–93; J. D. Oldfield, "Imagining Climates Past, Present and Future: Soviet Contributions to the Science of Anthropogenic Climate Change, 1953–1991," *JHG* 60 (2018), 41–51; Oldfield, "Mikhail Budyko's (1920–2001) Contributions to Global Climate Science: From Heat Balances to Climate Change and Global Ecology," *WIREs Climate Change* 7 (2016), 682–92.

cited by Clements, Bates, and other Americans.[42] Brückner had experience in the Russian Empire. As a child he had lived in Odessa and Dorpat (today's Tartu in Estonia), and was educated at universities in Russia as well as Germany. But he spent his career in Switzerland, Germany, and Austria, and so the reception of his work in the United States cannot be considered another case of "Russian" influences.[43]

Investigation of "dry-farming techniques" in both the steppes and the Great Plains suggests that, while they were similar, they may have been devised independently. Dry-farmers cultivated the soil in ways to retain scarce water. They plowed deeply down to around 7–8 inches (18–20 cm) in the fall to ensure rain and snow-melt water was absorbed into the soil. They cultivated the soil between rows of crops with harrows after rainfall to reduce evaporation. And they kept fallow fields clean of vegetation that would use up moisture. In the Great Plains, one of the main promoters of "dry farming," including all these techniques, was Hardy Webster Campbell. He also invented and marketed a new implement, a "subsurface packer," to compact the subsoil to retain moisture.[44] In 1905, the USDA established an Office of Dryland Agriculture headed by South Dakota agronomist Ellery Channing Chilcott. The Office, which built on work by USDA crop scientist Carleton, set up field stations throughout the plains.[45]

In the early twentieth century, Russian specialists familiar with agriculture in the American plains, for example, Nikolai Tulaikov, noted similarities in the techniques used by farmers in the two regions. They singled out deep plowing, regular cultivation of the soil, and keeping the fallow field clear of vegetation.[46] Steppe farmers, for example, Mennonites, had

[42] Clements, "Climatic Cycles," 193; Bates, "Climatic Characteristics," 95; Alfred J. Henry, "The Bruckner Cycle of Climatic Oscillations in the United States," *AAAG* 17, no. 2 (1927), 60–71.

[43] See Ernst, Milkutat, "Brückner, Eduard," in *Neue Deutsche Biographie* 2 (1955), 656, available online at www.deutsche-biographie.de/pnd116741910.html#ndbcontent, accessed April 22, 2019

[44] Mary W. M. Hargreaves, *Dry Farming in the Northern Great Plains, 1900–1925* (Cambridge, MA: Harvard University Press, 1957); Hargreaves, "The Dry-Farming Movement in Retrospect," *AH* 51 (1977), 149–65; Hargreaves, "Hardy Webster Campbell (1850–1937)," *AH* 32 (1958), 62–5.

[45] Jeremy Vetter, "Regionalizing Knowledge: The Ecological Approach of the USDA Office of Dryland Agriculture on the Great Plains," in *New Perspectives on the History of Life Sciences and Agriculture*, eds. D. Phillips and S. Kingsland (Cham, Switzerland: Springer Verlag, 2015), pp. 277–96.

[46] N. Tulaikov, "'Sukhoe' zemledelie (sistema Kembellia)," *Polnaia entsiklopediia russkogo sel'skogo khoziaistva i soprikasaiushchikhsia s nimi nauk*, vol. 12 (Spb: Devrien, 1912), 1262–7; V. Benzin, "Sukhoe zemledelie," *SKhiL* 241 (1913), 492–504; (1913), 691–708; 242 (1913), 3–19; (1913), 137–61. Some "subsurface packers" were imported to the steppe region. In 1911 agronomists in Akmolinsk region in today's northern Kazakhstan possessed two "podpoverkhnotnye uplotniteli Kempbelia" and tested Campbell's techniques. TsGA RK, f.317, op.1, d.49, ll.21–23ob.

used such techniques since the 1830s.[47] Did American dry-farmers acquire their techniques from steppe farmers? Many Mennonites migrated from the steppes to the Great Plains from the 1870s and brought their techniques as well as their wheat. Carleton attributed part of their success in the Great Plains in the dry 1890s to their farming methods.[48] Other Germanic migrants from the steppes brought similar techniques with them to the Great Plains.[49] Mary Hargreaves noted that in the 1890s, Americans recognized that dry farming was not new, that similar methods were used elsewhere, including Russia, and that Mennonite immigrants had brought their crops with them. But she did not explore the link in greater detail.[50] On the other hand, in 1909, Iosef Rozen, who was the agent in the United States for a provincial council (zemstvo) in the southern steppes, attended the Dry Farming Congress in Cheyenne, Wyoming. He noted that Americans were surprised to hear that experiment farms in the Russian Empire had been testing dry farming methods for two decades.[51] In the same year, the Russian agronomist A. N. Chelintsev suggested that American and Russian farmers had devised similar methods by themselves.[52] These claims seem plausible, since in both countries, farmers familiar with more humid climates were facing the same problem: growing crops in regions with low and unreliable rainfall. In both, moreover, agronomists and some farmers were familiar with the same European agricultural literature, including the writings of the Englishman Jethro Tull, which emphasized cultivation techniques.[53]

[47] David Moon, "Cultivating the Steppe: The Origins of Mennonite Farming Practices in the Russian Empire," *Journal of Mennonite Studies* 35 (2017), 251–78.

[48] Mark Alfred Carleton, "Successful Wheat Growing in Semiarid Districts," *Yearbook of the United States Department of Agriculture 1900* (Washington, DC: USDA, 1900), pp. 538–42; Carleton, *The Small Grains* (New York: Macmillan, 1916), p. 351.

[49] J. S. Otto, "From the Russian Steppes to the North Dakota Prairies: The Agricultural Practices of a Russian-German Family," *Journal of the American Historical Society of Germans from Russia* 7, no. 1 (1984), 37–44.

[50] Hargreaves, *Dry Farming in the Northern Great Plains, 1900–1925*, pp. 3, 84, 411.

[51] I. B. Rozen, "Otchet starshego agenta Ekaterinoslavskogo zemstva v Soed. Shtatakh s 1-go oktiabri 1908 g. po 1 marta 1909 g.," *Izvestiia zemskoi sel'sko-khoziaistvennoi agentury v Soedinennykh Shtatakh*, no. 7 (Ekaterinoslav: tip. Gubernskogo zemstva, 1909), p. 99. On the Ekaterinoslav zemstvo agency in the United States, see p. 109.

[52] A. N. Chelintsev, "Obzor russkoi literatury po sel'skomu khoziaistvu," *SKhiL* 229 (1909), 622.

[53] See Webb, *Great Plains*, p. 367, Hargreaves, *Dry Farming in the Northern Great Plains, 1900–1925*, p. 153; Hargreaves, "Hardy Webster Campbell," 63; Ol'ga Iur'evna Elina, *Ot tsarskikh sadov do sovetskikh polei: istoriia sel'sko-khoziaistvennykh opytnykh uchrezhdenii XVIII-20-e gody XX v.* (Moscow: Rossiiskaia Akademiia Nauk, 2008) vol. 1, pp. 162–7. See also D. Mun [Moon], "Agronomiia stepi: razvitie 'sukhogo zemledeliia' v Rossiiskoi imperii," *Vestnik Sankt-Peterburgskogo Universiteta, Seriia 2, Istoriia* 63, no. 2 (2018), 378–96.

This book will focus, therefore, on plants, sciences, and techniques for which there is clear evidence that they were introduced to the Great Plains from the steppes. Before these transfers could take place, however, there were barriers to overcome and bridges to be built. The barriers included: competition between the United States and Russia in the world grain market that could have hindered cooperation; American perceptions of Russian "backwardness" that obscured Russian advances; American ignorance of Russian agricultural sciences before the turn of the twentieth century; the language barrier; American suspicion of Soviet communism after the revolution of 1917; and in some quarters a more general resistance to change. Over time and once specialists in both countries recognized the similarities between their grasslands and the challenges facing farmers, transfers did take place. Important roles were played by migrants from the Russian Empire to the United States, in particular Germanic farmers and Jewish people such as Zon, and growing contacts between scientists.

Sources and Methodology

The main sources for this book include materials generated by the institutions that promoted the agricultural settlement of the Great Plains and provided settlers with "expert" advice, and their counterparts in the Russian Empire and Soviet Union. Thus, important sources are the records and publications of the USDA and its relevant bureaus, in particular the Bureau of Plant Industry, the Bureau of Soils, and the Forest Service (all were reorganized and changed their names over the period), and the Russian Ministry of State Domains and Agriculture (which was also reorganized and renamed) that was succeeded by the People's Commissariat of Agriculture (NarKomZem) in the Soviet Union. Also important were sources produced by agricultural colleges, universities, research institutes, agricultural, and forest experiment stations, and by scientific societies, national and international. Scientific journals have been rich sources, but just as important were professional publications of scientific societies where their members reported on their activities. For example, the *Bulletin of the American Association of Soil Survey Workers* (renamed the *Bulletin of the American Soil Survey Association*), and the Soviet equivalent, *Biulleteni Pochvoveda* (Soil Scientist's Bulletin) were just as useful as the scientific journals *Soil Science* and *Pochvovedenie* (Soil Science).

The papers of some of the main human protagonists were very useful, including: Bernhard Warkentin, a Mennonite migrant from the steppes, who established himself as a miller, seed importer, and entrepreneur in

Kansas; Curtis Marbut, who headed the U.S. Soil Survey; and Raphael Zon, the maverick head of a Forest Experiment Station and mastermind, or so he claimed, behind the Shelterbelt Project. Regrettably no such collection exists for Mark Carleton, the crop scientist and plant explorer who worked with Warkentin and visited the steppes in the late 1890s. These personal archives and the official collections of the institutions they worked for contain wealths of correspondence, personal, official, and political, between the key actors. Because this book covers a period when people devoted much time and energy to expressing their thoughts, ideas, and exasperations in letters, I felt I almost got to know some of the main characters. Much of the correspondence was between Americans or between Russians. Of great interest were letters between American and Russian scientists. I managed after archival visits to both countries to piece together exchanges of letters. Among the correspondence I fitted together were letters in filed under "Russia" in the foreign correspondence of the Bureau of Plant Industry of the USDA in the National Archives in College Park, Maryland,[54] those of the Agricultural Scientific Committee of the tsarist Ministry of Agriculture in the Russian State Historical Archive in St. Petersburg,[55] and Nikolai Vavilov's Institute of Plant Industry in Petrograd/Leningrad in the Archive of Scientific and Technical Documentation in St. Petersburg.[56] The other side of the some of the correspondence in Marbut's papers in the Missouri State Historical Society[57] is in the Archive of the Academy of Sciences in Moscow.[58]

From the perspective of methodology, I started looking for "the Russian roots of Great Plains agriculture" while researching the parallel environmental history of the steppes. Thus, I read my American sources through the lens of my knowledge of Russian work in the same areas. I was consciously looking in American sources for influences, acknowledged or not, from prior Russian experience in the steppes. The acknowledged influences were straightforward to identify. In other cases, I found in American sources references to crop varieties, sciences, and techniques that closely resembled those I was familiar with in the steppe region, but in which their Russian origins were not acknowledged. In such cases, I delved further into other sources to try to locate information on their origins. I looked for references to Russian work in American publications cited in

[54] NARA CP, RG 54, Finding Aid PI-66, Entry 30, Box 8, Files Russia A–K and L–Z.
[55] RGIA, f.382, op.9, 1911–17, d.290. [56] TsGANTD Spb, f.318.
[57] SHSM MC, Marbut, Curtis Fletcher (1863–1935), Papers, 1852–1963 (C3720). [Hereafter Marbut Papers (C3720).]
[58] ARAN, ff.487, 582, 687.

my original sources. I looked for English translations of Russian work and citations of them by American scientists. I identified who had attended particular scientific meetings where American and Russian specialists were present, and consulted accounts of such meetings to discover who met whom and exchanged ideas and publications. Thus, I traced some of the routes of transmission for the transfers. While the focus of much of this book is on people and institutions, I have also paid attention to the plants, sciences, and techniques that made their way from one region to the other.

Setting this book apart from some older works on the Great Plains, and drawing on environmental history, I have considered the two grasslands as among the leading characters and not passive backdrops for human history. The environmental conditions are active, if unconscious, actors in this story: the vegetation, climate, and soils of the Great Plains and the steppes have their own histories that unfolded both independently from and in processes of interaction with human history. In seeking to use crops, sciences, and techniques from the steppes in the Great Plains, the human actors encountered and worked to understand a dynamic non-human world. The crop varieties from the steppes proved effective in the Great Plains if they could withstand the periodic fluctuations in the climate and thrive in the soil. The soils of the American plains had formed, as Russian scientists had worked out in the steppes, over time in processes of interaction between the component parts of the environment. American soil scientists were applying the Russian understanding to their soils at a time when the Great Plains were hit by serious wind erosion during the Dust Bowl, which American scientists were struggling to comprehend. In addressing the problem, they drew on Russian experience of planting shelterbelts of trees. However, they needed to select species of trees that could grow in the region's soils at a time when they were only starting to understand these soils. American agricultural scientists were also trying to understand why invasive species that reached the Great Plains from the steppes seemed to thrive, in spite of their efforts to stop them, in the very conditions that proved harmful for the plants, crops, and trees, they wanted to promote.

Leading environmental historians of the plains James Sherow and Dan Flores have stressed the importance of gaining a sense of place by personal experience to understand how humans are a part of nature and as essential to studying the region's environmental history.[59] In my previous research

[59] James E. Sherow, "Introduction," in *A Sense of the American West: An Anthology of Environmental History*, ed. Sherow (Albuquerque: University of New Mexico Press, 1998), pp. 3–6, 18–21; Dan Flores, "A Spirit of Place and the Value of Nature in the American West," in *Sense of the American West*, ed. Sherow, pp. 31–8.

on the steppes of Russia and Ukraine, I explored the region to develop an understanding of the environment.[60] While I have not replicated even in part Flores' and Sherow's experiences in the Great Plains, from my home in England, I explored the Great Plains and prairies, from Texas to Saskatchewan, but perhaps especially Kansas. I discovered how they resemble the steppes. I got a sense of how the history of the Great Plains' agricultural settlement, the heritages of the different peoples who have lived there, including Native Americans as well as settlers, and human activities in the region are presented in museums. I learned how scientists study the region's environment, how conservationists seek to preserve it, and how, in reintroducing bison and controlled burning, scientists are trying to replicate processes that shaped the plains environment that was encountered by the first Euro-American visitors. From conversations with farmers in the plains and prairies, I have learned how farmers cultivate crops in the region today. My experiences exploring the Great Plains and the steppes will inform, mostly unobtrusively, much of what follows.

Historiography

This book aims to contribute to historical literatures in several areas. Most directly, it seeks to modify existing environmental and agricultural histories of the Great Plains by directing more attention to the "Russian" roots of aspects of the region's agriculture. Most studies of agriculture and environment in the Great Plains have presented a predominantly American story, albeit one aware of the overseas origins of some of the settlers and crops.[61] In contributing to explanations of how Euro-American settlers in the plains adapted to what was for most an unfamiliar environment, this book, to some extent, follows the classic work of Walter Prescott Webb published in 1931. Webb described the Great Plains as a semi-arid grassland with few trees inhabited by hostile native peoples. He explained how white American settlers overcame these obstacles with a series of innovations. They used six-shooter revolvers to fight on horseback against the formidable Comanches and other Plains Indians. Railroads

[60] Moon, *Plough*, pp. ix–x, 301–3; Moon, "Steppe by Steppe: Exploring Environmental Change in Southern Ukraine," *GE* 9 (2016), 414–39. On a project based on exploring environments in Britain, see Peter Coates, David Moon, and Paul Warde, eds., *Local Places, Global Processes: Histories of Environmental Change in Britain and Beyond* (Oxford: Windgather Press, 2016).

[61] For examples of the evolving historiography of the environmental and agricultural history of the Great Plains, see Webb, *Great Plains*; Worster, *Dust Bowl*; a symposium on "Agriculture in the Great Plains, 1876–1936," *AH* 51, no. 1 (1977); William Cronon, "A Place for Stories: Nature, History, and Narrative," *JAH* 78 (1992), 1347–76; Cunfer, *On the Great Plains*.

brought large numbers of settlers to the plains. To keep cattle off their land so they could grow crops without them being trampled or eaten, they fenced it with barbed wire. This was invented in the 1870s to compensate for the shortage of timber to make fences. Settlers partly overcame the shortage of water with windmills to pump water from underground. These innovations were followed by irrigation, where it was practicable, dry-farming methods to conserve moisture in the soil elsewhere, and drought-resistant crops.[62] Webb mentioned wheat varieties from Russia, but I have paid far more attention to what the settlers and their advisors learned from earlier developments in the similar environment of the steppes.

My book may be closer to the work of James Malin, a Kansas historian who pioneered ecological approaches to history in the 1930s–60s, in that it seeks to understand that environment.[63] Malin's fierce pride in the plains and their inhabitants led him to dispute the Russian origins of the hard, red winter wheat and soil science discussed in this book. For this reason, he appears as a protagonist. This is a shame, since the originality of his work and his conviction in presenting it were among the inspirations for my own research. Malin's ecological approach showed the importance drawing on scientific as well as historical studies. Although my book is not specifically a work on the history of science, it considers the construction of scientific knowledge through field work in the grasslands of both Eurasia and North America,[64] and the transfer of such knowledge between them.[65]

[62] Webb, *Great Plains*, pp. 9–44, 167–79, 295–318, 333–74.

[63] See James C. Malin, "The Adaptation of the Agricultural System to Sub-Humid Environment: Illustrated by the Activities of the Wayne Township Farmers' Club of Edwards County, Kansas, 1886–1893," *AH* 10, no. 3 (1936), 118–141; Malin, *Winter Wheat in the Golden Belt of Kansas: A Study in Adaption to Subhumid Geographical Environment* (Lawrence: University of Kansas Press, 1944); Malin, "Dust Storms"; Malin, *The Grasslands of North America: Prolegoma to Its History* (Lawrence, KS: privately published, 1947). On Malin, see Robert W. Johanssen, "James C. Malin: An Appreciation," *KHQ* 38 (1972), 457–66; Bonnie Lynn-Sherow, "Beyond Winter Wheat: The USDA Extension Service and Kansas Wheat Production in the Twentieth Century," *Kansas History: A Journal of the Central Plains* 23, nos. 1–2 (2000), 100–111; Robert P. Swierenga, "The Malin Thesis of Grassland Acculturation and the New Rural History," in *Canadian Papers in Rural History*, ed. Donald H. Akenson (Gananoque, ON: Langdale Press, 1986), pp. 11–22.

[64] See Vetter, *Field Life*; Jonathan Oldfield and Denis J. B. Shaw, *The Development of Russian Environmental Thought: Scientific and Geographical Perspectives on the Natural Environment* (London: Routledge, 2016).

[65] On soil science, considered in detail later in this book, see Jan Arend, *Russlands Bodenkunde in der Welt: Eine ost-westlicheTransfergeschichte,1880–1945* (Göttingen: Vandenhoeck und Ruprecht, 2017); Arend, "Russian Science in Translation: How *pochvovedenie* was brought to the West, c. 1875–1945," *Kritika* 18 (2017), 683–708; on other agricultural sciences, see Mark R. Finlay, "Transnational Exchanges of Agricultural Scientific Thought from the Morrill Act through the Hatch Act," in *Science as Service: Establishing and Reformulating Land-Grant Universities, 1865–1930*, ed. Alan I. Marcus (Tuscaloosa, AL: University of Alabama Press, 2015), pp. 33–60.

This book has been influenced by the "New Western History" that has transformed our understanding of the American West, including the Great Plains, and has largely supplanted older, progressive, accounts, such as Webb's, of people, mostly white men, conquering nature and native peoples with the aid of science and technology.[66] Following new western historians, the present work studies the environment in a process of interaction with human societies in which it was also changed. Among other issues in the new history of the West are race, ethnicity, and gender. Explicit attention is paid here to the ethnic diversity of the immigrants to the Great Plains, in particular those of Germanic backgrounds from the steppes, and in the United States in general by highlighting the importance of Russian–Jewish immigrants as mediators for Russian science. The experience accumulated by generations of Native Americans who devised various ways of living in the Great Plains is the elephant in the room – or perhaps the bison in the plains – as it was largely ignored by the incoming Euro-American settlers, the scientists who advised them, the state and federal governments who supported them, and is thus largely absent from the sources this book is based on. Silence in these sources does not signify a lack of significance as it was the disregard for local knowledge that created the space for Americans to learn from the Russian experience from the steppes.

I am very conscious that this book also has little to say about gender, or rather has a lot say indirectly by telling a story of a time when most public and senior roles were filled by men. Women were "hidden figures," but no less important for being largely undocumented. On farms in both the Great Plains and the steppes, the heads of households who ran them were mainly, but not solely, men. A lot of the labor in farming communities in both regions was done, with little outward recognition at the time, by women. The memoir by Rachel Calof, a Jewish immigrant from the Russian Empire who homesteaded in North Dakota, gives us insights into the enormous contribution of women to the development of an

[66] For a few examples of this large literature, see the special issue of *Western Historical Quarterly*, "The WHA at Fifty: Essays on the State of Western History Scholarship. A Commemoration," *WHQ* 42 (2011); Renee M. Laegreid and Sandra K. Mathews, eds., *Women on the North American Plains* (Lubbock, TX: Texas Tech University Press, 2011); William Cronon, George Miles, and Jay Gitlin, eds., *Under an Open Sky: Rethinking America's Western Past* (New York: Norton, 1992); Richard White, *"It's Your Misfortune and None of My Own": A History of the American West* (Norman: University of Oklahoma Press, 1991); Donald Worster, *Under Western Skies: Nature and History in the American West* (New York: Oxford University Press, 1992).

agricultural way of life in the plains.[67] Women's work continues to play a major role in farming communities in the Great Plains.[68]

In the scientific and governmental institutions that feature prominently in this book, much of the crucial administrative and support work behind the scenes was carried out by women, in many cases, young, unmarried women, who could never aspire to careers on the same terms as their male co-workers. Most go unnamed. There is one significant exception: Grace Cramer (who, fittingly, was from Kansas). She was an administrator who worked with David Fairchild in the Bureau of Plant Industry in the USDA in Washington, DC. Fairchild singled her out for praise in his memoir (but he did not include her in the index): "Grace entered into the spirit of the work with great enthusiasm and for many years carried the burden of correspondence and clerical management . . . Her intense and quick adaptability soon made her a most important factor in the organization."[69] Beyond her professional duties, she befriended the plant explorer Frank Meyer and in her official correspondence tried to keep his spirits up during his long, and lonely, expeditions in Eurasia.[70] There will have been many more Grace Cramers and, in Russia, Daria Kuptsovas, working in the back offices of the institutions that played such large roles in the transfers between the steppes and the Great Plains.

Among the soil scientists attending the international congresses in 1927 and 1930 there were a few women. At the second Congress in the Soviet Union in 1930, one of the British delegation of 10 was female: Rachel Mary Flemming. There was one woman in the American scientific delegation: Rebekah A. Lockwood. Among the larger female American contingent, however, she was outnumbered by the wives of her male colleagues. The Soviet delegation of around 400 included a little over 30 female scientists, who were more numerous than the wives listed. None of the female soil scientists was allowed a prominent role in the

[67] Rachel Calof, *Rachel Calof's Story: Jewish Homesteader on the Northern Plains*, ed. J. Sanford Rikoon (Bloomington: Indiana University Press, 1995). There has been much work on women homesteaders, for example, H. Elaine Lindgren, *Land in Her Own Name: Women as Homesteaders in North Dakota* (Norman, OK: University of Oklahoma Press, 1996). On women in Russian agriculture, see Beatrice Farnsworth and Lynne Viola, eds., *Russian Peasant Women* (Oxford: Oxford University Press, 1992) and David Moon, *The Russian Peasantry 1600–1930: The World the Peasants Made* (London and New York: Addison Wesley Longman, 1999), passim.

[68] See Sarah Smarsh, *Heartland: A Memoir of Working Hard and Being Broke in the Richest Country on Earth* (New York: Scribner, 2018), pp. 209–10 and passim.

[69] David Fairchild, *The World Was My Garden: Travels of Plant Explorer* (New York: Scribner's Sons, 1938), pp. 107–8.

[70] Isabel Shipley Cunningham, *Frank N. Meyer: Plant Hunter in Asia* (Ames, IA: Iowa State University Press, 1984), pp. 297–8.

Congresses. The women, scientists and spouses, were treated by their male counterparts with the gallantry customary of the period.[71] Another woman who made an important contribution to this book was Louise Marbut Moomaw (1892–1956). She was the eldest daughter of Curtis Marbut, head of the U.S. Soil Survey from 1913. She became her father's confidant following the death of his wife, and her mother, in 1909. After Curtis moved from the family home in Columbia, Missouri, to Washington, DC, he and Louise wrote to each other regularly. He confided in her about his professional position, tensions with his colleagues, and his evolving ideas about the science of soils under the influence of the work of Russian scholars. She used their correspondence as the basis for two memoirs of her father: a longer and franker, unpublished version and a shorter published account. Their letters and Louise's memoirs were major sources for the chapters on soil science.[72]

This book makes direct contributions to the history of the United States in comparative and, especially, transnational contexts. Comparative history has a longer pedigree and there have been several such studies of the United States and Russia.[73] While comparisons of the Great Plains and the steppes reveal similarities as well as contrasts,[74] this book's primary purpose is not to compare the two regions, except when the authors of the sources did so or when it is necessary to contextualize the arguments. Rather, the purpose is to explain the transfers and influences between them, especially from the steppes to the Great Plains, following

[71] *Trudy II Mezhdunarodnogo Kongressa Pochvovedov, Leningrad–Moscow, July 20–31, 1930, Protokoly, Ekskursii*, eds. D. G. Vilenskii, V. A. Kovda, and A. A. Iarilov (Moscow: Gos. izdatel'stvo Sel'sko-Khoziaistvennoi literatury, 1935), pp. 153–7, available online at www.iuss.org/index.php?article_id=27, accessed November 7, 2019.

[72] Marbut Papers (C3720), folders 5–73 (correspondence); folders 145–52, Louise Moomaw Manuscript, c. 1940; Louise Marbut Moomaw, "Curtis Fletcher Marbut," in *Life and Work of C.F. Marbut, Soil Scientist: A Memorial Volume*, ed. H. H. Krusekopf (Columbia, MO: Soil Science Society of America, 1942), pp. 11–27.

[73] See, for example, Mark Bassin, "Turner, Solov'ev, and the 'Frontier Hypothesis': The Nationalist Signification of Open Spaces," *JMH* 65 (1993), 473–511; Dorothy Zeisler-Vralsted, *Rivers, Memory, and Nation-Building: A History of the Volga and Mississippi rivers* (New York: Berghahn Books, 2015); Colin White, *Russia and America: The Roots of Economic Divergence* (London: Crook Helm, 1987); Peter Kolchin, *Unfree Labor: American Slavery and Russian Serfdom* (Cambridge, MA: Harvard University Press, 1987).

[74] For example, see Alexander Chibilev and Sergei Levykin, "Virgin Lands divided by an Ocean: The Fate of Grasslands in the Northern Hemisphere," trans. David Moon, *Nova Acta Leopoldina, Neue Folge*, 114, 390 (2013), 91–103; David Moon, "The Grasslands of North America and Russia," in *A Companion to Global Environmental History*, eds. J. R. McNeill and Erin Stewart Mauldin (Chichester, UK: Wiley-Blackwell, 2012), pp. 247–62; Sabol, *"The Touch of Civilization."* Sabol compared the impact on the Sioux and the Kazakhs of the American and Russian colonization of the Great Plains and steppes.

the start of Euro-American agricultural settlement. One of the pioneers of transnational American history, Ian R. Tyrrell, defined this field as "the movement of peoples, ideas, technologies, and institutions across national boundaries." He stressed the "profound connectedness" of American history to world history in the nineteenth century.[75] An important implication of comparative and transnational histories is to confront notions of American exceptionalism, such as those advanced in the 1890s by Frederick Jackson Turner. David Wrobel, for example, challenged such ideas in his transnational analysis of travel writing by visitors to the American West and other "frontiers" around the globe.[76] In examining connections and transfers between the Great Plains and the steppes, which came about partly because of the similarities between them, this book also implies questions about the exceptional nature of the history of this part of the American West, United States in general, as well as Russia and the Soviet Union.

The comparative and transnational turns in American history have included environmental and agricultural histories of the United States and the Great Plains in particular. A decade ago, Sara Gregg wrote: "While most American agro-environmental history to date has been local or regional in its focus, with only a glancing look at parallels elsewhere in the world, the future borders of the field will surely expand to include the manifold interconnections within the global economy."[77] A step in this direction had already been made by Frieda Knobloch, who placed the agricultural colonization of the American West in the broader contexts of European agricultural history as far back as classical Rome. She argued that land, plants, animals, and the indigenous population in the western United

[75] Ian R. Tyrrell, *Transnational Nation: United States History in Global Perspective since 1789* (Basingstoke, UK: Palgrave Macmillan 2007), pp. 2–3. See also Tyrrell, "The United States in World History since the 1750s," in *The Cambridge World History*, vol. 7 *Production, Destruction, and Connection, 1750–Present*, part 1: *Structures, Spaces, and Boundary Making*, eds. J. McNeill and K. Pomeranz (Cambridge, England: Cambridge University Press, 2014), pp. 585–610. See also David Thelen, "The Nation and Beyond: Transnational Perspectives on United States History," *JAH* 86 (1999), 965–75 (and the special issue it introduces); Shelley Fisher Fishkin, "Crossroads of Cultures: The Transnational Turn in American Studies," *American Quarterly* 57 (2005), 17–57.

[76] David M. Wrobel, *Global West, American Frontier: Travel, Empire, and Exceptionalism, from Manifest Destiny to the Great Depression* (Albuquerque, NM: University of New Mexico Press, 2013). See also Kiran Klaus Patel, *The New Deal: A Global History* (Princeton: Princeton University Press, 2016); Daniel T. Rodgers, *Atlantic Crossings: Social Politics in a Progressive Age* (Cambridge, MA: Harvard University Press, 1998); Donald K. Pickens, "Westward Expansion and the End of American Exceptionalism: Sumner, Turner, and Webb," *WHQ* 12 (1981), 409–18.

[77] Sara M. Gregg, "Agro-Environmental History," in *A Companion to American Environmental History*, ed. Douglas Cazanaux Sackman (Malden, MA and Oxford, UK: Wiley-Blackwell, 2010), p. 438.

States were subjected to agricultural development from outside in the name of "civilization."[78] Sterling Evans has recently edited a volume on cross-border connections linking the American West, western Canada, and northern Mexico in the production of agricultural commodities, labor markets, and environmental change.[79] In an important analysis of the contemporary global relevance of the Dust Bowl, sociologist Hannah Holleman located the ecological crisis in the Great Plains in the wider context of "white settler colonialism" around the world. She made passing references to "the Russian plain" and Voeikov's conclusion that erosion was caused by cultivation.[80] As I was completing this book, Kristin L. Hoganson published her stimulating history of *The Heartland*, challenging myths of the American Midwest as "local, insulated, exceptional, isolationist and provincial." Delving into the history of agricultural Champaign County, Illinois (east of the Great Plains), over largely the same period as this book, she found a crossroads of people, commerce, and ideas from around the world: the "Heartland" was part of a global story.[81] This present book aims to make a further contribution to these directions in historical scholarship by focusing on exchanges between the Great Plains region in the United States and an analogous region, the steppes, in Eurasia.

This discussion indicates the importance of placing the agricultural and environmental histories of the Great Plains and the steppes in the global history of settler societies,[82] and the environmental history of European colonialism and imperialism. Several historians, following Alfred Crosby, have shown how European colonialism, imperialism, and settlement transformed the environments and populations of colonized lands in different parts of the world, and changed how Europeans understood human impact on environments.[83] These wider, global, contexts offer perspectives on the

[78] Knobloch, *The Culture of Wilderness.*

[79] Sterling Evans, ed., *Farming across Borders: A Transnational History of the North American West* (College Station, Texas A&M University Press, 2017).

[80] Hannah Holleman, *Dust Bowls of Empire: Imperialism, Environmental Politics, and the Injustice of 'Green' Capitalism* (New Haven, CT: Yale University Press, 2018), pp. 90–2. She focused on English-language publications. Holleman, *Dust Bowls*, p. 90.

[81] Kristin L. Hoganson, *The Heartland: An American History* (New York: Penguin Press, 2019).

[82] See James Belich, *Replenishing the Earth: The Settler Revolution and the Rise of the Anglo-World, 1783–1939* (Oxford: Oxford University Press, 2009). Belich took his story "Beyond the Anglo World" and included the Russian Empire. Belich, *Replenishing*, pp. 505–13.

[83] Alfred W. Crosby, *Ecological Imperialism: The Biological Expansion of Europe, 900–1900* (New York: Cambridge University Press 1986); Grove, *Green Imperialism*; Thomas Dunlap, *Nature and the English Diaspora: Environment and History in the United States, Canada, Australia, and New Zealand* (Cambridge, England: Cambridge University Press, 1999); Libby Robin, "Ecology: A Science of Empire?" in *Ecology and Empire: Environmental History of Settler Societies*, eds. Robin and Tom Griffiths (Edinburgh: Keele University Press, 1997), pp. 63–75; Corey Ross, *Ecology and Power in*

"unexpected" nature of the transfers and influences from the steppes to the Great Plains, from Eurasia to North America, which we will return to in the Conclusion to this book.

The focus of the present book in exploring connections between regions with similar environments and environmental histories in different parts of the globe has been anticipated by two noteworthy studies on the Pacific world. Ian Tyrrell and Edward Melillo analyzed links between California, on the one hand, and Australia and Chile, respectively, on the other. Tyrrell analyzed exchanges of species of trees (eucalyptus and acacia from Australia to California, and Monterey pine the other way), techniques for irrigating land, as well as insects and methods of controlling them. These transfers of plants, techniques, and also pests, took place between two outposts of European global expansion, where settlers were making new lives in environments that were unfamiliar to them, but similar to each other. The settlers were engaged in similar projects, had similar aspirations, and sought to learn from each other.[84] Melillo explored long-term environmental and social connections between Chile and California, which lie in similar latitudes in the southern and northern hemispheres and share Mediterranean-style environments. They resembled each other also in being societies settled by migrants of European origins. The transfers from Chile to California, which radically altered the latter, included new crops such as potatoes, fertilizers, mining technologies, laborers, and ideas. Chile, Melillo argued, was dramatically shaped by systems of servitude, exotic species, and schemes for capitalist development from California. He argued further that the "countless shared external influences that have shaped the histories of Chile and California," which had been little discussed, undermined declarations in both of "exceptionalism."[85] By paying attention to external influences in a region in the United States from another part of the world that have not received due attention from historians, this present book resembles those of Tyrrell and Melillo. It differs in that the main emphasis here is on influences from the steppes to the Great Plains, rather than in both directions.

the Age of Empire: Europe and the Transformation of the Tropical World (Oxford: Oxford University Press, 2017).

[84] Ian Tyrrell, *True Gardens of the Gods: Californian–Australian Environmental Reform, 1860–1930* (Berkeley, CA: University of California Press, 1999).

[85] Melillo, *Strangers on Familiar Soil*, quotation from p. viii. For a discussion of the book, including comments by Tyrrell, see Melillo, "Strangers on Familiar Soil," *H-Environment Roundtable Reviews* 8, 2 (2018), available online at https://networks.h-net.org/melillo-strangers-familiar-soil-roundtable-review-vol-8-no-2-2018, accessed June 13, 2018.

Transnational histories go beyond more traditional studies of international relations and international histories, yet this book also contributes to scholarship on Russian–American relations and perceptions.[86] It consciously builds on and owes a large debt to the scholarship of Kansas historian Norman Saul.[87] Historians of Russian influences and connections in North America have understandably focused on Alaska, which was part of the Russian Empire between 1789 and 1867.[88] The focus here is a long way southeast of Alaska, and points to Russian influences in the Great Plains, a region not normally considered in that context.

Writing a transnational history of the Great Plains and the steppes based on contemporary sources, including accounts by many Americans and Russians who visited the other country's grassland, has resulted in a stress on similarities rather than contrasts between the two countries. Many writers on Russian and Russian–American travel writing have concurred that "confrontation" with "difference" and consciously seeking the other are overriding themes in such works.[89] Such specialists have further argued that attitudes of Russians to America and Americans to Russia, both before and after 1917, have been polarized between sharply positive or negative. In Margarita Marinova's words, they were seen as: "alternative utopian and anti-utopian worlds" to be "enthusiastically emulated or just as passionately rejected."[90] On the other hand, the findings of this book are line with some recent calls for the binary dichotomy of "Russia vs. the West" to be

[86] There has been much interest in mutual relations between Russia and the United States. See Ivan Kurilla and Victoria I. Zhuravleva, eds., *Russian/Soviet Studies in the United States, Amerikanistika in Russia: Mutual Representations in Academic Projects* (Lanham, MD: Lexington Books 2015).

[87] Norman E. Saul, *Concord and Conflict: The United States and Russia, 1867–1914* (Lawrence, KS: University Press of Kansas, 1996); Saul, *War and Revolution: The United States and Russia, 1914–1921* (Lawrence, KS: University Press of Kansas, 2001); Saul, *Friends or Foes? The United States and Soviet Russia, 1921–1941* (Lawrence, KS: University Press of Kansas, 2006). Articles by Saul on specific topics are cited elsewhere in this book.

[88] See Susan Smith-Peter, "Russian America in Russian and American Historiography," *Kritika* 14 (2013), 93–100.

[89] Sara Dickinson, "The Edge of Empire or the Center of the Self: Endpoints and Itineraries in Nineteenth-Century Russian Travel," *RR* 70 (2011), 94; Alexander Etkind, *Tolkovanie puteshestvii: Rossia i Amerika v travelogakh intertekstakh* (Moskva: Novoe Literaturnoe Obozrenie, 2001).

[90] Margarita Marinova, *Transnational Russian-American Travel Writing* (Abingdon, UK: Routledge 2011), p. 152. See also Hans Rogger, "America in the Russian Mind – or Russian Discoveries of America," *PHR* 47 (1978), 27–51; Martin Malia, *Russia under Western Eyes: From the Bronze Horseman to the Lenin Mausoleum* (Cambridge, MA: Harvard University Press, 1999), pp. 287–408; Iu. S. Borisov, A. V. Golubev, M. M. Kudukina, and V. A. Nevezhin, eds., *Rossiia i Zapad: Formirovanie vneshnepoliticheskikh stereotipov v soznanii rossiiskogo obshchestva pervoi poloviny XX veka* (Lewiston, NY: Edwin Mellen Press, 1999).

deconstructed.[91] Martin Malia argued that in the late nineteenth and early twentieth centuries, reforms in the Russian Empire led some people to believe in the possibility of convergence between Russia and the West.[92] The editors of a recent volume on American experiences of Russia argued: "Russians and Americans long coexisted in a tense parity, regarding each other as not-quite-European equals, occasional role models, wartime allies, political antagonists, and desired interlocutors."[93]

Another field of historical study to which this book makes an indirect contribution, and from which the author's research emerged, is the environmental history of the Russian Empire and Soviet Union. "Russian environmental history" has taken off in recent years and there is a growing body of innovative research on the complex and sometimes misunderstood interaction between humans and the rest of the natural world in this large part of the globe.[94] There has been a tendency in global environmental histories for Russia and the Soviet Union to be marginalized, or presented mainly as an example of anthropogenic destruction of the environment, which is only part of the story.[95] From the vantage point of the Great Plains of the United States, however, this book suggests other ways of interpreting the environmental history of Russia: as a source of innovations which others could learn from. A growing direction in "Russian environmental history" is transnational or comparative studies,[96] of which this book is an example, albeit with the main focus on another part of the world.

[91] See S. A. Smith, "Russia under Western Eyes," *Kritika* 1 (2000), 595–6. See also Wendy Bracewell and Alex Drace-Francis, eds., *Under Eastern Eyes: A Comparative Introduction to East European Travel Writing on Europe* (Budapest: Central European University Press, 2008), pp. 118–19.

[92] Malia, *Russia under Western Eyes*, pp. 161–232.

[93] Choi Chatterjee and Beth Holmgren, "Introduction," in *Americans Experience Russia: Encountering the Enigma, 1917 to the Present*, ed. Chatterjee and Holmgren (New York: Routledge, 2012), p. 4.

[94] See Julia Lajus, "Russian Environmental History: A Historiographical Review," in *The Great Convergence: Environmental Histories of BRICS*, eds. S. Ravi Rajan and Lise Sedrez (Oxford: Oxford University Press, 2018), pp. 245–73; Breyfogle, *Eurasian Environments*.

[95] See David Moon, "The Curious Case of the Marginalization or Distortion of Russian and Soviet Environmental History in Global Environmental Histories," *International Review of Environmental History* 3, no. 2 (2017), 31–50.

[96] See, for example, Stephen V. Bittner, "American Roots, French Varietals, Russian Science: A Transnational History of the Great Wine Blight in Late-Tsarist Bessarabia," *Past and Present* 227, no. 1 (2015), pp. 151–79; Bathsheba Demuth, *Floating Coast: An Environmental History of the Bering Strait* (New York: Norton, 2019); Aaron Hale-Dorrell, "The Soviet Union, the United States, and Industrial Agriculture," *Journal of World History* 26, no. 2 (2016), 295–324; Ryan Tucker Jones, *Empire of Extinction: Russians and the North Pacific's Strange Beasts of the Sea, 1741–1867* (New York: Oxford University Press, 2014); Maya Peterson, "US to USSR: American Experts, Irrigation, and Cotton in Soviet Central Asia, 1929–32," *EH* 21 (2016), 442–66; Zeisler-Vralsted, *Rivers, Memory, and Nation-Building*.

This book has claims for originality. In exploring connections between regions with similar environments, but in different parts of world and with different political and economic systems, it presents the first detailed analysis in one place of transfers of plants, agricultural sciences and techniques from the steppes to the Great Plains. The transfers are not unknown to specialists in Great Plains agricultural and environmental history. Some have been referred to in broader accounts, with more attention to the migrants and wheat varieties than the others.[97] Some of the transfers have been analyzed in more detail in specific studies.[98] But, overall, the "Russian roots" of Great Plains agriculture have not received due attention. A further claim to originality is that the analysis is based on extensive research in printed and archival sources produced in both the United States and in the Russian and Soviet states, much of which is held in archives and special collections in those countries. These sources and the scholarly literatures drawn on in this book are in several languages. They are mostly in English and Russian, but also in German – due in part to the prominent role played by Germanic migrants from the steppes to the Great Plains and German as a scientific language – and occasionally in other languages. True transnational histories can only, where necessary, be based on multilingual research in sources produced, and often still stored, in different "nations."[99]

Further, and perhaps not entirely seriously, this book undermines the "first law of geography" proposed by American geographer Waldo Tobler in 1970. Tobler stated that: "everything is related to everything else, but near things are more related than distant things."[100] As we shall see, the environments of the steppes and Great Plains are so similar that people familiar with both can struggle to differentiate between them and that recognition of these close relations in kind but not space was important in the exchanges and transfers between the two grasslands. The Great Plains and the steppes are very distant from each other, but closely related in so many other ways.

[97] See, for example, Hargreaves, *Dry Farming in the Northern Great Plains, 1900–1925*, pp. 3, 84, 290–2, 316–17, 411; Knobloch, *The Culture of Wilderness*, pp. 62, 97, 126–9, 132; Craig Miner, *West of Wichita: Settling the High Plains of Kansas, 1865–1890* (Lawrence: University Press of Kansas, 1986), pp. 80–5, 87-91; Webb, *Great Plains*, p. 373; Worster, *Dust Bowl*, pp. 69, 174–6, 220.

[98] See the specialist works cited in individual chapters of this book.

[99] Since I began research for this book in 2007, the National Agricultural Library in Beltsville, Maryland, has digitized most of the publications of the USDA. Digitization of relevant sources has proceeded more slowly in the Russian Federation.

[100] See Harvey J. Miller, "Tobler's First Law and Spatial Analysis," *AAAG* 94 (2004), 284–9.

As well as stating what this book aims to do, it is important to point out what it does not claim. It does not argue that the transfers and influences from the steppes were the most important factors behind the development of European-style farming in the Great Plains. This book acknowledges the roles of other factors, people, and institutions: the back-breaking toil of the settlers, especially the first wave (including the women) of all ethnic groups; the ways they learned from their own experience; the construction of railroads; and the "expert" advise (some of which drew on experience in the steppes) provided by the USDA, agricultural colleges, and experiment stations. American banks provided loans; American cities provided markets for produce from plains farms; and plains farmers and inhabitants (including Mennonites from the steppes) took risks and showed business acumen. These all contributed to changing the face of the region. In addition, American manufacturers, such as McCormick, International Harvester, and Ford, led the world in building agricultural machinery that assisted the development of agriculture, especially large-scale farming, in the United States and around the world, including the steppes.[101] However, Alan L. Olmstead and Paul W. Rhode have argued forcefully that biological improvements in livestock and crops, including those from the steppes, were more important than mechanization in the development of American agriculture.[102]

Structure of the Book

After this Introduction, the book opens with three chapters exploring contexts. The first presents an overview of the Euro-American agricultural settlement of the Great Plains. Various barriers in the United States to the reception of influences and transfers from the steppes are examined in the second chapter. The third considers the bridges that allowed the barriers to be overcome. The main part of the book has two chapters on each of the transfers: plants; science – soil science; techniques – shelterbelts. The two chapters on plants, on wheat and weeds, are "book ends" around those on

[101] See David B. Danbom, *Sod Busting: How Families Made Farms on the Nineteenth-Century Plains* (Baltimore: Johns Hopkins University Press, 2014), pp. 95–108 and passim; Miner, *West of Wichita*; Alan I. Marcus, *Agricultural Science and the Quest for Legitimacy: Farmers, Agricultural Colleges, and Experiment Stations, 1870–1890* (Ames: Iowa State University Press, 1985); Gordon M. Winder, *The American Reaper: Harvesting Networks and Technology, 1830–1918* (Abingdon, UK: Ashgate, 2012); William Cronon, *Nature's Metropolis: Chicago and the Great West* (New York: W. W. Norton, 1991).

[102] Alan L. Olmstead and Paul W. Rhode, *Creating Abundance: Biological Innovation and American Agricultural Development* (New York: Cambridge University Press, 2008).

sciences and techniques. Each of these chapters, or pairs of chapters, has sections analyzing the plants, science, and techniques in the steppes and then in the Great Plains before the transfers took place. Subsequent sections explore the processes of transfer, including overcoming barriers, and the outcomes. The Conclusion considers the legacies of the "Russian roots" down to the present as well as tying together the argument.

Most chapters begin with examples of historical sources that convey an aspect of the topic. The first chapter opens with mixed reactions by Mennonite migrants from the steppes on their prospects in their new homes in the Great Plains. At the start of Chapter 2 are extracts from General William Tecumseh Sherman's diary of his travels in the steppes of the Russian Empire in 1872. Chapter 3 begins with remarks by Gifford Pinchot, chief of U.S. forestry, when he visited steppe forestry plantations 30 years later. The chapters or pairs of chapters covering the particular transfers open with sources, events or artifacts that illustrate their impact in the Great Plains. Chapter 4 starts at the dedication in a small town in Kansas in 1942 of a statue of a Mennonite settler who had brought wheat from the steppes in the 1870s. At the beginning of Chapter 5 is a USDA publication from 1935 on the "Soils of the United States," in which the soils of the Great Plains were classified using the Russian term "Chernozem." Chapter 7 starts with a Presidential Executive Order issued in July 1934 on planting trees in the Great Plains that drew on Russian and Soviet experience. A cowboy ballad recorded by Roy Rogers and the Sons of the Pioneers in the same year introduces Chapter 9. Finally, the Conclusion begins with a letter from a scientist at the USDA in Washington, DC, to Nikolai Vavilov in Leningrad in late 1933 asking for Soviet help in dealing with the growing problem of soil erosion at the start of the Dust Bowl.

In his book *The Great Plains*, Walter Prescott Webb considered how Anglo-American settlers adapted to the plains environment. After a chapter on "The Plains Indians," he considered "The Spanish Approach to the Great Plains" and "The American Approach to the Great Plains." This book offers another, perhaps unexpected, "Russian Approach to the Great Plains."

PART I

Contexts

CHAPTER 1

Settlement

Mennonite migrants expressed differing views on their prospects when they arrived in the Great Plains from their previous homes in the steppes of the Russian Empire. High winds raising dust storms in Kansas in 1874 prompted one to say that he was "afraid of the future and whether we would make our living here,"[1] despite, or perhaps because of, the fact that such storms were familiar from the steppes. Another new arrival had no such worries. He told Noble Prentis, a journalist from the Topeka *Commonwealth*, that in three years they would transform the "ocean of grass" of the prairies "into an ocean of waving fields of grain, just as we left our Molotschna [in the steppes]." Prentis predicted that Kansas would be "to America what the country of the Black Sea and Sea of Azov is now to Europe – her wheat field."[2]

Mennonites from the steppes were just one group among many Euro-American agricultural settlers in the Great Plains. Over the last four decades of the nineteenth century, due largely to hundreds of thousands of settlers, the population of the central and northern plains grew from around a hundred and forty thousand in 1860 to over three and a quarter million in 1900 (see Table 1.1).[3] The rates of increase in individual states are truly striking. The population of Kansas grew by 174 percent in the 1870s and a further 43 percent in the 1880s. Nebraska saw population increases of 267 and 135 percent in the same two decades. In the 1880s, the numbers of inhabitants of South and North Dakota respectively

[1] Cornelius Krahn, *From the Steppes to the Prairies, 1874–1949* (Newton, KS: Mennonite Publishing Office, 1949), p. 100.

[2] Noble L. Prentis, "The Mennonites in Kansas," *The Commonwealth*, October 15, 1874, reprinted in Krahn, *From the Steppes*, pp. 13–14.

[3] U.S. Census Bureau, "Census of Population and Housing," available online at www.census.gov/prod/www/decennial.html, accessed May 17, 2018. Data for Kansas, Nebraska, South and North Dakota. Few Native Americans were counted in censuses before 1900. U.S. Census Bureau, Censuses of American Indians, available online at www.census.gov/history/www/genealogy/decennial_census_records/censuses_of_american_indians.html, accessed May 17, 2018.

Table 1.1 *Population Increase in the central and northern Great Plains, 1860–1900*

	1860	1900
Kansas	107,206	1,470,495
Nebraska	28,841	1,066,300
Dakota Territory	4,837	
South Dakota		401,570
North Dakota		319,146
Total	140,884	3,257,511

Source: U.S. Census Bureau "Census of Population and Housing"

mushroomed by 256 and 416 percent.[4] The incoming settlers were ethnically diverse, including: white Americans from further east; African–Americans from the south; Hispanics from Mexico, New Mexico, and Texas; French Canadians from the north; and immigrants from Europe, especially Germany, Scandinavia, the Czech lands of the Austro-Hungarian Empire, Britain and Ireland, as well as Germanic peoples, including Mennonites, from the Russian Empire.[5] By their sheer numbers and labor, Euro-American settlers transformed the plains that had been inhabited by smaller populations of Native Americans with their own ways of life into an agricultural region that became a breadbasket for the United States and others parts of the world.[6]

The Euro-American settlement of the Great Plains and development of an agricultural economy were consequences of deliberate policies pursued by the U.S. federal government. The Kansas–Nebraska Act of 1854 established two territories on land it had previously assigned to Native Americans and opened them to outside settlers. In 1860–1, taking advantage of the secession of the southern states, the Republican-dominated Congress advanced

[4] David B. Danbom, *Sod Busting: How Families Made Farms on the Nineteenth-Century Plains* (Baltimore: Johns Hopkins University Press, 2014), p. 65.

[5] See Bruce A. Glasrud and Charles A. Braithwaite, eds., *African Americans on the Great Plains: An Anthology* (Lincoln: University of Nebraska Press, 2009); Karen Hansen, *Encounter on the Great Plains: Scandinavian Settlers and the Dispossession of Dakota Indians, 1890–1930* (New York: Oxford University Press, 2013); Terrence W. Haverluk, "The Changing Geography of U.S. Hispanics, 1850–1990," *Journal of Geography* 96 (1997), 134–45; Frederick C. Luebke, "Ethnic Group Settlement on the Great Plains," *WHQ* 8 (1977), 405–30; D. Aidan McQuillan, *Prevailing Over Time: Ethnic Adjustment on the Kansas Prairies, 1875–1925* (Lincoln: University of Nebraska Press, 1990); William C. Sherman, *Prairie Mosaic: An Ethnic Atlas of Rural North Dakota*, 2nd edition (Fargo: North Dakota State University Press, 2017).

[6] On the development of agriculture in the Great Plains, see Geoff Cunfer, *On the Great Plains: Agriculture and Environment* (College Station: Texas A&M University Press, 2005).

the free settlement of the plains. Kansas was admitted to the Union as a free state in 1861. The Dakota Territory was organized in the same year. It comprised the future states of North and South Dakota and, for a time, much of the future Wyoming and Montana. The federal government enacted laws to make land and transport available for settlers. The Homestead Act of May 1862 authorized settlers, both U.S. citizens and foreign immigrants intending to become citizens, to claim up to 160 acres of "public land" and, after fulfilling certain conditions, paying fees, and waiting five years, to receive title to the land. Also in 1862, Congress enacted the first of several Railroad Acts. They granted public land and loans to companies to build lines across the plains and on to the Pacific coast. The First Transcontinental Railroad, which went through Nebraska, was built between 1863 and 1869. The Atchison, Topeka and Santa Fe Railroad started in Chicago and reached the western border of Kansas in 1873. In time, railroads connected the plains states with the rest of the country. Railroads promoted the agricultural settlement of the plains in three ways. The companies sold the land they had been granted to settlers to cover their costs. The railroads provided transport for settlers to reach the plains. And, once the settlers had crops and livestock for sale, the railroads carried their produce to domestic markets and ports for export.[7]

The U.S. government made expert advice available to support the development of agriculture in the plains and other states. Again in May 1862, the U.S. Department of Agriculture (USDA) was established:

> to acquire and to diffuse among the people of the United States useful information on subjects connected with agriculture, rural development, aquaculture, and human nutrition, in the most general and comprehensive sense of those terms, and to procure, propagate, and distribute among the people new and valuable seeds and plants.[8]

In July 1862, the Morrill Land Grant College Act further supported agriculture by donating public lands to endow "Colleges for the Benefit of Agriculture and the Mechanic Arts."[9] A decade and a half later, in 1877,

[7] For a concise summary, see Danbom, *Sod Busting*, pp. 7–9. The role of the federal government in the American West has received much attention. See for example, Karen R. Merrill, "In Search of the 'Federal Presence' in the American West," *WHQ* 30 (1999), 449–73. See also Richard Edwards, "The New Learning about Homesteading," *GPQ* 38 (2018), 1–23; Richard White, *Railroaded: The Transcontinentals and the Making of Modern America* (New York: Norton, 2011).

[8] "An Act to Establish a Department of Agriculture," May 15, 1862, available online at www.nal.usda.gov/act-establish-department-agriculture, accessed May 15, 2018.

[9] "Morrill Land Grant College Act," July 2, 1862, available online at www.nal.usda.gov/morrill-land-grant-college-act, accessed May 15, 2018.

the Hatch Act provided funds to found agricultural experiment stations attached to the land grant colleges. The stations conducted research and experiments to promote the "agricultural industry," "having due regard to the varying conditions and needs of the respective states and territories."[10]

The settlement of the plains by Euro-Americans and development of agriculture were achieved at the expense of the indigenous population. From the 1860s to the 1890s, the federal government deployed the U.S. Army to fight a succession of wars and military actions against the Plains Indians. By the end of the century, the army and the tide of settlers had driven Native Americans off much of their lands and into reservations or exile, in the process opening up more land for settlement by outsiders.[11] Prior to conquest, the indigenous population had led ways of life that were both settled and mobile. Different peoples had supported themselves in a variety of ways that included growing corn, beans, and squash among other crops in bottom land along rivers, gathering wild food, hunting fur-bearing animals and trading their pelts, and hunting bison in the high plains. Hunting bison from horseback was a relatively recent development. Indigenous peoples began to acquire horses in the sixteenth century from those brought over by Europeans. They acquired firearms from the same source. The Great Plains region before Euro-American conquest and settlement was a dynamic world in which indigenous groups moved around, allying with or defeating other peoples. The Apache, Cheyenne, Comanche, and Sioux peoples moved to the plains from outside after they obtained horses and defeated horticultural peoples, such as the Mandan, Hidatsa, and Arikara, of the river bottoms. Over the eighteenth and first half of the nineteenth century, the Sioux conquered much of the northern plains, until they and their allies were overpowered by the U.S. Army.[12]

[10] A. C. True and V. A. Clark, "The Agricultural Experiment Stations in the United States," *USDA Office of Experiment Stations Bulletin*, no. 80 (1900), 36–7. See also Jeremy Vetter, *Field Life: Science in the American West during the Railroad Era* (Pittsburgh: University of Pittsburgh Press, 2016), pp. 267–330.

[11] For a recent interpretation and analysis, see Gary Clayton Anderson, *Ethnic Cleansing and the Indian: The Crime That Should Haunt America* (Norman: University of Oklahoma Press, 2014), pp. 237–337. For a Native American perspective, see David C. Posthumus, "A Lakota View of Pté Oyáte (Buffalo Nation)," in *Bison and People on the North American Great Plains: A Deep Environmental History*, eds. Geoff Cunfer and Bill Waiser (College Station: Texas A&M University Press, 2016), pp. 278–309.

[12] See Loretta Fowler, "The Great Plains from the Arrival of the Horse to 1885," in *The Cambridge History of the Native Peoples of the Americas*, vol. 1, *North America*, eds. Bruce G. Trigger and E. Washburn Wilcomb, Part 2 (Cambridge: Cambridge University Press, 1996), pp. 1–56; Pekka Hämäläinen, "The Rise and Fall of Plains Indian Horse Cultures," *JAH* 90 (2003), 833–62; Richard White, "The Winning of the West: The Expansion of the Western Sioux in the Eighteenth and Nineteenth Centuries," *JAH* 65 (1978), 319–43.

In recent discussions over how to characterize the actions against Native Americans in the western United States, some scholars have made explicit comparisons with more recent European history, including the Third Reich under Hitler. They have debated the applicability of such terms as "holocaust," "genocide," and "ethnic cleansing."[13] They have pointed to similarities between the military conquest and agricultural settlement of the American West in the nineteenth century and analogous plans for lands to the east of Germany in the twentieth century. From the early 1920s, Hitler and the Nazi party advocated, and during the Second World War pursued by military means, the acquisition of *Lebensraum* (living space) in eastern Europe. The aim was to settle new German agriculturalists in lands to the east from which much of the population, mostly Slavic and Jewish peoples, had been removed. The lands the Nazis designated for German settlement included the fertile steppes of Ukraine. They intended southern Ukraine and the Crimean peninsula to become largely German farming colonies. When the *Wehrmacht* invaded and occupied Ukraine in 1941–3, they found some of the earlier Germanic settlers, including Mennonites, still there, although others had been deported by the Soviet authorities before and after the outbreak of the war. The occupying forces augmented the surviving Germanic settlements in Ukraine with more "ethnic Germans" (*Volksdeutsche*).[14] Historians of Mennonites are now coming to terms with the troubled history of Mennonite communities under German occupation, where some of their members were victims, but others participated in the Nazi's plans, including the Holocaust.[15]

[13] For recent work on this complex subject, see Gary Clayton Anderson, "The Native Peoples of the American West: Genocide or Ethnic Cleansing?" *WHQ* 47 (2016), 407–34, and the response that follows it; Edward B. Westermann, *Hitler's Ostkrieg and the Indian Wars: Comparing Genocide and Conquest* (Norman: University of Oklahoma Press, 2016). On "ethnocide" in the Canadian prairies, see James Daschuk, *Clearing the Plains: Disease, Politics of Starvation, and the Loss of Aboriginal Life* (Regina, SK: University of Regina Press, 2013).

[14] See Richard J. Evans, *The Third Reich at War: How the Nazis Led Germany from Conquest to Disaster* (London: Allen Lane, 2008), pp. 160–1, 171–5, 197. Mennonite communities in southern Ukraine were devastated by the Russian Civil War in 1918–21, famine in 1921–2, and deportations as "kulaks," or rich farmers, in the early 1930s. James Urry, *Mennonites, Politics, and Peoplehood: Europe–Russia–Canada, 1525 to 1980* (Winnipeg: University of Manitoba Press, 2006), pp. 137–40; Benjamin W. Goossen, *Chosen Nation: Mennonites and Germany in a Global Era* (Princeton: Princeton University Press, 2017), pp. 149–53.

[15] See Goossen, *Chosen Nation*, pp. 147–73; Aileen Friesen, "Soviet Mennonites, the Holocaust & Nazism: Part 1," Anabaptist Historians: Bringing the Anabaptist Past into a Digital Century, April 25, 2017, available online at https://anabaptisthistorians.org/2017/04/25/soviet-mennonites-the-holocaust-nazism-part-1/, accessed June 21, 2018; Friesen, "Mennonites and the Holocaust: Soviet Union and Mennonite-Jewish Connections," Anabaptist Historians, March 20, 2018, available online at https://anabaptisthistorians.org/2018/03/20/mennonites-and-the-holocaust-soviet-union-and-mennonite-jewish-connections/, accessed June 21, 2018.

It has been argued that the Nazi policy of acquiring *Lebensraum* in the east and replacing the indigenous population with white settlers was not just similar to, but was partly inspired by, the American doctrine of "Manifest Destiny" that lay behind the conquest and settlement of the West and removal of many Native Americans. This argument is relevant to this book as it is a possible example of a transfer from the Great Plains to the steppes. There is evidence to support the idea that the Nazi policy drew on American actions in the West. In 1928, Hitler spoke with approval about how the Americans had "gunned down the millions of Redskins to a few hundred thousand, and now keep the modest remnant under observation in a cage."[16] In October 1942, during the German invasion of the Soviet Union, Hitler announced that there was "only one duty," which was "to Germanize this country by the immigration of Germans, and to look upon the natives as Redskins."[17] Arguments that Hitler's policy of *Lebensraum* and extermination in the east were modeled directly on the American conquest of the West have attracted critics who have not found sufficient evidence for a strong, causal, connection.[18] While such debates are beyond the scope of this book, the United States' conquest of the Great Plains and displacement of its previous inhabitants in the second half of the nineteenth century created the circumstances in which influences and transfers from the steppes of the Russian Empire and Soviet Union could partly shape the next stage in the region's history.

The settlement of the Great Plains by Euro-Americans and the development of agriculture were achieved also at the expense of the region's distinctive environment.[19] Anthropogenic environmental change did not start with the arrival of this latest wave of settlers. There have been debates concerning the impact of Native Americans on the plains environment before Euro-American settlement. Notions of "Indians" as more ecologically minded than the settlers who replaced them have been advanced and

[16] James Q. Whitmore, *Hitler's American Model: The United States and the Making of Nazi Race Law* (Princeton: Princeton University Press, 2017), pp. 9–10 (quotation from p. 9). For a comparative study making the connection, see Carroll P. Kakel III, *The American West and the Nazi East: A Comparative and Interpretive Perspective* (London: Palgrave Macmillan, 2011).

[17] Goossen, *Chosen Nation*, pp. 147–8.

[18] Jens-Uwe Guettel, "The US Frontier as Rationale for the Nazi East? Settler Colonialism and Genocide in Nazi-Occupied Eastern Europe and the American West," *Journal of Genocide Research* 15 (2013), 401–19; Westermann, *Hitler's Ostkrieg and the Indian Wars.*

[19] On the environmental history of the Great Plains and American West, see James E. Sherow, *The Grasslands of the United States: An Environmental History* (Santa Barbara, CA: ABC-CLIO, 2007); Sara Dant, *Losing Eden: An Environmental History of the American West* (Malden, MA: Wiley Blackwell, 2017).

challenged.[20] In shaping the landscape to suit their purposes, Native Americans contributed to the evolution, and in places creation, of the grassland ecosystem. Plains Indians burned the grasslands to assist them in hunting bison and other large ungulates. They set fire to some areas of land to compel the animals to graze in others, such as river bottoms, where it was easier to hunt them. Plains Indians periodically burned the plains to promote the growth of fresh grasses to provide better fodder for their prey. They also used fire to clear land for crops. Some of the Euro-Americans who ventured into the plains found fire deployed against them to force them out of the ravines or woodland where they were seeking cover or to drive them away altogether. A consequence of setting fires was to promote the growth of some kinds of vegetation over others. Regular burning was favorable to annual plants, in particular grasses, but discouraged woody vegetation. Fire harmed many species of trees and shrubs, particularly young saplings, thus inhibiting the spread of woodland out of river valleys and ravines to the high plains. The bison and other animals the Plains Indians nurtured were also detrimental to tree growth as they grazed on and destroyed young trees. There were further reasons why there were few trees in the Great Plains that were unconnected with human activity, such as the relatively low precipitation and types of soil that were unsuitable for many species of arboreal vegetation, but the use of fire by Plains Indians was a significant factor, especially in the eastern plains.[21] Scholars have also drawn attention to the role of Native Americans, especially once they had acquired horses and firearms, in the decline in the numbers of bison even before white Americans engaged in large-scale hunting in the second half of the nineteenth century. But, it was the Euro-Americans who delivered the coup de grâce to the vast herds of bison that once roamed the plains, driving them to the brink of extinction by the end of the century.[22]

Euro-American settlement and the development of European-style agriculture marked a turning point in the environmental history of the

[20] See Shepard Krech III, *The Ecological Indian: Myth and History* (New York: W. W. Norton. 1999); Michael E. Harkin and David Rich Lewis, eds., *Native Americans and the Environment: Perspectives on the Ecological Indian* (Lincoln: University of Nebraska Press, 2007); Geoff Cunfer, "Overview: The Decline and Fall of the Bison Empire," in Cunfer and Waiser, eds., *Bison and People*, pp. 10–12.

[21] See Daniel I. Axelrod, "Rise of the Grassland Biome, Central North America," *Botanical Review* 51, 2 (1985), 163–201; Krech, *Ecological Indian*, pp. 104–22; Stephen J. Pyne, *Fire: A Brief History* (Seattle: University of Washington Press, 2001), pp. 58–63.

[22] See Dan Flores, *The Natural West: Environmental History in the Great Plains and Rocky Mountains* (Norman: University of Oklahoma Press, 2001), pp. 49–70; Andrew Isenberg, *The Destruction of the Bison: An Environmental History* (New York: Cambridge University Press, 2000); Cunfer and Waiser, eds., *Bison and People*.

Great Plains. As the settlers converted much of the land to farms and ranches, for crops and grazing, they greatly reduced the rich biodiversity of a region once inhabited by such an array of wild life that Dan Flores called it the "American Serengeti."[23] The native flora of the vast areas of grasslands, with tall grasses in the east and short grasses in the west, and the colorful array of wild flowers and other plants, became increasingly rare as land was plowed up, sown with crops introduced from outside, in addition to corn, or subjected to intensive grazing.[24] Geoff Cunfer has termed plowing grassland as "the ecological equivalent of genocide," and the "plow-up of the Great Plains" as "the most important ecological change to emerge out of the shift from Indian to Euro-American land use." "The act of plowing," Cunfer continued, "alters vegetation, animal populations, water dynamics, and soil chemistry and physics in catastrophic ways."[25] Russian scientists have made similar assessments of plowing up the steppes.[26]

From the perspective of the first generation of Euro-American settlers, this transformation of the plains environment entailed back-breaking labor as they engaged in the "arduous, expensive, and time-consuming process" of "sod busting": breaking up the "hard, compact, and tenacious" soil bound by the dense and thickly matted roots of the prairie grasses that had been growing there largely undisturbed for millennia. This was not just very hard work for the settlers, but required particular techniques, special plows, and heavy oxen to pull them. Even with the best equipment and strongest draft animals, the most a farmer could hope for was to break one acre a day.[27] And yet, from around 1870 until about 1930, millions of plains farmers succeeded in plowing up over 100 million acres of grassland. While the main focus of this book is on arable farming, the agricultural economy in the region as a whole was diverse. Throughout the plains region the total proportion of the land that was plowed up amounted to around 30 percent. In the more humid eastern part, around half the land was plowed for crops, while in the more arid western part, only 10–20 percent was plowed up, the rest used for grazing domesticated livestock, in

[23] Dan Flores, *American Serengeti: The Last Big Animals of the Great Plains* (Lawrence: University Press of Kansas, 2016).

[24] See Cunfer, *On the Great Plains*, pp. 16–68; Carolyn Hull Sieg, Curtis H. Flather, and Stephen McCanny, "Recent Biodiversity Patterns in the Great Plains: Implications for Restoration and Management," *Great Plains Research* 9, no. 2 (1999), 277–313.

[25] Cunfer, *On the Great Plains*, pp. 16, 35.

[26] A.A. Chibilev, *Priroda znaet luchshe* (Ekaterinburg: YrO RAN, 1999), p. 173.

[27] Danbom, *Sod Busting*, p. 47.

particular cattle, in contrast to the undomesticated animals that Plains Indians had hunted.[28]

The agricultural settlement of the plains was not a steady progression, but fluctuated between years of advance and retreat that coincided with favorable and unfavorable climatic conditions, especially the amount of rainfall, and in the market for agricultural produce. In good years when the rains came and the markets were good, settlement and agriculture expanded. Plains farmers experienced good years for much of the 1870s and first half of the 1880s, encouraging further settlement and expansion. High demand for wheat during the First World War led to high prices and a further growth in the area of crop land. On the other hand, in bad years when the rains failed or the plains were visited by grasshoppers and other pests that destroyed crops, or when market demand was low, the agricultural settlement of the plains slowed or retreated. Farms were abandoned. Settlers moved away. They went back to their previous homes or on to seek new lives on the west coast. The late 1880s and much of the 1890s witnessed droughts and low prices for farm produce. The more arid western plains were worst affected. Parts of western Kansas lost a quarter of their population between 1888 and 1898. Other areas in the plains saw farm abandonment on a larger scale.[29]

The experiences of one settler family were portrayed in her memoir by Rachel Calof, a Jewish migrant from the Russian Empire. She moved to North Dakota in 1894 with her husband and his extended family. She described how, in the desperate cold of their first winters, three families lived for several months in a sod house measuring twelve by fourteen feet, where they were crammed together with their livestock. Over the following years, as they worked hard to establish a successful farm in the face of the harsh environment, they experienced setbacks when their crops, livestock, buildings, and hard-won achievements were damaged by gophers, hailstorms, and lightning strikes. Fear for her children when they fell ill, with doctors several days' journey away, pervades her memoirs.[30] The arduous

[28] Cunfer, *On the Great Plains*, pp. 8, 16–36. See also Kenneth M. Sylvester and Geoff Cunfer, "An Unremembered Diversity: Mixed Husbandry and the American Grasslands," *AH* 83 (2009), 352–83; Kenneth M. Sylvester, "Ecological Frontiers on the Grasslands of Kansas: Changes in Farm Scale and Crop Diversity," *Journal of Economic History* 69 (2009), 1041–62.

[29] See, for example, Danbom, *Sod Busting*, pp. 66–7, 96–101. See also Kevin Z. Sweeney, *Prelude to the Dust Bowl: Drought in the Nineteenth-Century Southern Plains* (Norman: University of Oklahoma Press, 2016); Richard White, *"It's Your Misfortune and None of My Own": A History of the American West* (Norman: University of Oklahoma Press, 1991), pp. 227–31.

[30] Rachel Calof, *Rachel Calof's Story: Jewish Homesteader on the Northern Plains*, ed. J. Sanford Rikoon (Bloomington: Indiana University Press, 1995).

lives of the pioneers, from Europe and the eastern United States, were captured by writers such as Willa Cather. She drew on her own experiences with her family, who moved to Nebraska from Virginia in the 1880s, in her Great Plains novels.[31]

Seared into the American memory are the hard years of the Dust Bowl in the 1930s, when the plains were hit by extreme drought,[32] which dried out the top soil and allowed it be blown away by high winds. Coinciding with the economic disaster of the Great Depression, many plains farmers could not withstand the double catastrophe. Banks foreclosed on bankrupt farms, farmers and their families abandoned their land and headed off in search of hope. The Dust Bowl conjures up images of the skies blackened by dust storms, abandoned farms, "Okies" leaving the plains, heading west along Route 66 for a promised land in California, and Dorothea Lange's photograph of the "Migrant Mother" anxious about her children. The human story of the Dust Bowl was reinforced by John Steinbeck's epic novel *The Grapes of Wrath*, published in 1939, and John Ford's movie of 1940.

The prevailing narrative of the Dust Bowl has been a declensionist one that challenged the progressive history of American settlement of the Great Plains told by Walter Prescott Webb (whose history was published on the eve of the disaster in 1931). In the declensionist interpretation, the Euro-American agricultural settlers, who had come from outside the plains and were supported by the government and the banks, had a destructive impact on a fragile grassland environment they did not understand. The settlers plowed up too much land, including areas unsuitable for farming, used inappropriate methods to cultivate the soil, and overgrazed other land. The fertile soils yielded bumper harvests in wet years, but the settlers reaped only misery, and choked on the dust, in bad years when the rains failed and the winds blew the top soil. These were the findings of the House of Representative's Great Plains Committee, which submitted its report, "The Future of the Great Plains," in 1937. The Committee concluded: "Nature has established a balance in the Great Plains . . . The white man has disturbed this balance." In contrast to the Plains Indians, who the report's authors believed had lived in "harmony" with "nature,"

[31] Willa Cather, *O Pioneers!* (Oxford: Oxford University Press, 1999) [1st published 1913]; Cather, *My Antonia* (Oxford: Oxford University Press, 2006) [1st published 1918]. Life for family farmers in the plains was still arduous a century later. See Sarah Smarsh, *Heartland: A Memoir of Working Hard and Being Broke in the Richest Country on Earth* (New York: Scribner, 2018).

[32] See Benjamin I. Cook, Richard Seager, and Jason E. Smerdon, "The Worst North American Drought Year of the Last Millennium: 1934," *Geophysical Research Letters* 41, no. 20 (2014), 7298–305.

the settlers' "destructive tendencies" resulted from their "lack of understanding concerning the critical differences between the physical conditions of the Great Plains ... and those ... east of the Mississippi whence they had come."[33] (The notions that Native Americans had lived in "harmony" with the "natural world" and that nature establishes a "balance," were widely held at the time, but have since been challenged.[34])

The version of the story of the Dust Bowl that blamed the plains farmers for their own plight, because they had upset the "balance," was graphically portrayed in the documentary film "The Plow that Broke the Plains" made by Pare Lorentz for the federal government Resettlement Administration in 1936.[35] The narrative has been followed by many historians of the Dust Bowl, including the prominent environmental historian and native of Kansas, Donald Worster. In 1979, following detailed research in the parts of the plains most badly affected, he argued: "Some environmental catastrophes are nature's work, others are the slowly accumulating effects of ignorance or poverty. The Dust Bowl, in contrast, was the inevitable outcome of a culture that deliberately, self-consciously, set itself that task of dominating and exploiting the land for all it was worth." Worster blamed capitalism for the ecological and human disaster.[36]

Not all scholars, and certainly not many plains farmers, have accepted this version. From the 1930s to the 1960s, Kansas historian and plainsman James Malin railed against the view that the farmers were responsible for the Dust Bowl. He pointed out that dust storms were not a new phenomenon, but had occurred before the plains were plowed up. He argued that,

[33] U.S. Great Plains Committee, *The Future of the Great Plains* (Washington, DC: The House of Representatives, 1937), pp. 1–2, 49–50, 63. See also Gilbert F. White, "'The Future of the Great Plains' Re-visited," *GPQ* 6, no. 2 (1986), 84–93.

[34] On Native Americans and "nature," see pp. 44–5, and on changing views of "nature" among ecologists, see Frank N. Egerton, "Changing Concepts of the Balance of Nature," *The Quarterly Review of Biology* 48, no. 2 (1973), 322–50; Jianguo Wu and Orie L. Loucks, "From Balance of Nature to Hierarchical Patch Dynamics: A Paradigm Shift in Ecology," *Quarterly Review of Biology* 70, 4 (1995), 439–66. Earlier in the twentieth century, the ecologist Frederic Clements had proposed a theory of succession of plant communities leading to a stable equilibrium, or climax, based on his field work in the plains of his native Nebraska. Frederic E. Clements, *Plant Succession: An Analysis of the Development of Vegetation* (Washington, DC: The Carnegie Institution, 1916). See also A. G. Tansley, "Obituary Notice: Frederic Edward Clements, 1874–1945," *Journal of Ecology* 34 (1947), 194–6.

[35] Pare Lorentz (1936). *The Plow That Broke the Plains*, U.S. Resettlement Administration, film, available online at www.archive.org/details/PlowThatBrokethePlains1, accessed May 23, 2018; Finis Dunaway, "New Deal Jeremiahs," *EH* 12 (2007), 308–12.

[36] Donald Worster, *Dust Bowl: The Southern Plains in the 1930s*, 25th anniversary edition (New York: Oxford University Press, 2004), pp. 4–5, 106–7, 140–1, 151. For a powerful restatement of this argument, see Hannah Holleman, *Dust Bowls of Empire: Imperialism, Environmental Politics, and the Injustice of "Green" Capitalism* (New Haven, CT: Yale University Press, 2018).

over the longer term, plains farmers had shown an ability to adapt to conditions in the region. He based his arguments on his considerable knowledge of Kansas history and sought to understand the land and environment of the plains by drawing on the natural sciences, in particular ecology.[37] Building on Malin's work, Cunfer presented a "middle ground" between the declensionist and progressive narratives of Great Plains history. Insisting that people are a part of the natural world, Cunfer used Geographic Information Systems to analyze data on land use and the climate in the precise region affected by the Dust Bowl. He concluded that the main cause of the ecological disaster in the plains in the 1930s was the drought. Plowing up land may have tipped the balance in parts of the southern plains, but overall he argued the weather was a more important cause than settler agriculture, careless or otherwise. Echoing Malin, he pointed out that there were dust storms on land that had never been plowed up. Cunfer concluded that plains farmers achieved a "sequence of periods of temporary equilibrium" interrupted by disruptions such as the Dust Bowl that prompted farmers – most of whom did not leave the region – to seek new ways of working with the land.[38] The understandable attention paid to the Dust Bowl years, moreover, has overshadowed the longer-term history of agriculture and the environment in the Great Plains since the advent of Euro-American settlement.[39]

What the years of droughts and dust storms, such as those in the 1890s and 1930s, indicated – and this is important for the argument advanced in this book – was the types of crops and agricultural techniques that were effective in extreme conditions. The bad years also indicated the importance of acquiring an understanding of the plains environment that could inform identifying the most appropriate crops and methods. This is where the Russian experience of settling and cultivating the steppes came in. In

[37] See James C. Malin, "The Adaptation of the Agricultural System to Sub-Humid Environment: Illustrated by the Activities of the Wayne Township Farmers' Club of Edwards County, Kansas, 1886–1893," *AH* 10, 3 (1936), 118–41; Malin, *Winter Wheat in the Golden Belt of Kansas: A Study in Adaption to Subhumid Geographical Environment* (Lawrence: University of Kansas Press, 1944); Malin, "Dust Storms 1850–1900," *Kansas Historical Quarterly* 14 (1946), 129–44, 265–96, 391–413; Malin, *The Grasslands of North America: Prolegoma to Its History* (Lawrence, KS: privately published, 1947). On Malin's work, see also p. 24.

[38] Cunfer, *On the Great Plains*, pp. 3–13, 150–63, 232–40 (quote from p. 236). See also Timothy Egan, *The Worst Hard Time: The Untold Story of Those Who Survived the Great American Dust Bowl* (New York: Houghton Mifflin Harcourt, 2006), pp. 9–10; Kenneth M. Sylvester and Paul W. Rhode, "Making Green Revolutions: Kansas Farms, Recovery, and the New Agriculture, 1918–1981," *AH* 91 (2017), 342–68.

[39] On narratives of the history of the Great Plains and the Dust Bowl, see William Cronon, "A Place for Stories: Nature, History, and Narrative," *JAH* 78 (1992), 1347–76.

many ways, it resembled, and had anticipated, the agricultural settlement of the Great Plains.

In the steppes, agricultural settlers, mostly of European origins, had moved out of the more humid, forested regions of central Russia, the north of today's Ukraine, and parts of central Europe to the grasslands that lay in the south and southeast of the Russian Empire. Backed by the Russian state and its armed forces, settlers pushed the indigenous peoples, who lived from a mixture of pastoralism, nomadic and settled, and some crop cultivation, to the margins or into exile. The settlers who displaced them engaged in extensive livestock husbandry for several decades before gradually plowing up much of the grasslands to cultivate crops, principally grain, in the fertile soil. Like their counterparts in the Great Plains, they lived in dugouts in the early years, broke the sod of the virgin land with heavy plows pulled by draft animals. They experimented with crops from their previous homes as well as varieties acquired locally as they sought those most suitable for the conditions. The process of adaptation to the steppe environment was easiest for settlers in the more moderate conditions near the heartland of Russia in the eighteenth century. It became harder as the frontier of settlement moved southeast and east over the nineteenth century. In the late nineteenth and early twentieth centuries, some settlers in the north today's Kazakhstan found soils that were shallower and less fertile than elsewhere in the steppes. They struggled in the harsher climate with more frequent droughts. Throughout the steppe region, just as in the United States, the Russian and later the Soviet governments supported research to inform advice to farmers. They also largely disregarded the knowledge and experience of the indigenous peoples. The settlers managed, despite the difficulties and recurring droughts, to create a flourishing agricultural economy in those parts of the steppes that were more favorable to crop cultivation. In the late nineteenth and early twentieth centuries, in many years, the Russian Empire was the world's largest grain exporter. Grain from the steppes was exported to other parts of Europe and the Mediterranean world.[40]

[40] See Willard Sunderland, *Taming the Wild Field: Colonization and Empire on the Russian Steppe* (Ithaca, NY: Cornell University Press, 2004); Leonard Friesen, *Rural Revolutions in Southern Ukraine: Peasants, Nobles, and Colonists, 1774–1905* (Cambridge, MA: Harvard Series in Ukrainian Studies, 2009); David Moon, *The Plough that Broke the Steppes: Agriculture and Environment on Russia's Grasslands, 1700–1914* (Oxford: Oxford University Press, 2013); Kelly O'Neill, *Claiming Crimea: A History of Catherine the Great's Southern Empire* (New Haven, CT: Yale University Press, 2017), pp. 164–218; Sarah Cameron, "'People Arrive but the Land Does Not Move': Nomads, Settlers, and the Ecology of the Kazakh Steppe, 1870–1916," in *Eurasian Environments: Nature and Ecology in Imperial Russian and Soviet History*, ed. Nicholas Breyfogle

While there are similarities between the environmental histories of agricultural settlement of the Great Plains and the steppes, there are differences that are important for this book. First, the Russian state had conquered large parts of the Eurasian steppe by the late eighteenth century, but had started to promote the agricultural settlement of the region by European, including Slavic, farmers several decades earlier. This was long before similar processes in the Great Plains that took off only in the 1860s. Thus, there was significant prior Russian experience in the steppes of identifying appropriate crops, techniques, and gaining a deeper understanding of the steppe environment. Second, the Great Plains are orientated from north to south, while the steppes extend from east to west, and much of the steppe region is on a similar latitude to the Canadian Prairies and northern Great Plains. One of the consequences is that the steppe region as a whole has a slightly harsher climate. The climatic conditions in the eastern steppes in southern Siberia and the north of today's Kazakhstan are significantly more severe than in the central and southern plains (see pp. 8–10). The harsher environment of the steppe region, and the limited availability of land well suited for agriculture in adjoining regions (there is no equivalent of the prairies of the Midwest in Eurasia), has meant that a far larger proportion of the steppes has been plowed up than the Great Plains. Over 30 percent of the steppe region of the European part of the Russian Empire had been plowed up by the late 1880s, and over half in the parts with more favorable conditions, for example, the region to the north of the Black Sea where the Mennonites lived. By the end of the twentieth century, 57 percent of the entire Eurasian steppe had been converted to arable land.[41] The harsher conditions in the steppes as a whole and the cultivation of land with marginal conditions for farming has meant that varieties of crops and techniques that worked in average years in the steppe region were capable of withstanding all but the greatest deviations from average conditions in most of the Great Plains. Most

(Pittsburgh, PA: University of Pittsburgh Press, 2018), pp. 43–59. On particular difficulties experienced by some settlers in today's northern Kazakhstan in the late nineteenth and early twentieth centuries, see, for example, TsGA RK, f.318, op.1, d.18, ll.5-ob., 15, 18, 26; f.30, op.1, d.5, l.194. On pastoral nomadism, see N. E. Masanov, *Kochevaia tsivilizatsiia kazakhov: osnovy zhiznedeiatel'nosti nomadnogo obshshestva* (Almaty: Sotsinvest, 1995); Ian W. Campbell, "'The Scourge of Stock Raising': Zhut, Limiting Environments, and the Economic Transformation of the Kazakh Steppe," in *Eurasian Environments*, ed. Breyfogle, pp. 60–74.

[41] David Moon, "The Grasslands of North America and Russia," in *A Companion to Global Environmental History*, eds. J. R. McNeill and Erin Stewart Mauldin (Chichester, UK: Wiley-Blackwell, 2012), pp. 247–62; A. A. Chibilev and O. A. Grosheva, *Ocherki po istorii stepevedeniia* (Ekaterinburg: UrO RAN, 2004), pp. 34–5.

hardy were the crop varieties cultivated in the southern Siberia and northern Kazakhstan. In the late nineteenth and early twentieth centuries, American "plant explorers" appreciated the significance of the more extreme conditions in the eastern steppes and collected hardy plants that grew there (See pp. 95–7).

Thus, when the agricultural settlement of the Great Plains by outsiders developed from the 1860s, the settlers and their advisors in the USDA, agricultural colleges, and experiment stations stood to learn from the prior experience in the steppes of Eurasia of plowing up and cultivating a semi-arid grassland, and one where, for the most part, conditions were harsher than in much of the North American plains. Before this would take place, however, there were a series of barriers that would have to be overcome.

CHAPTER 2

Barriers

Introduction

U.S. Army general William Tecumseh Sherman visited the Russian Empire in 1872. He described his impressions while traveling from Georgia in the Caucasus to Taganrog on the Sea of Azov: "the 'steppe' . . . is as much like our Western plains as possible. I could hardly realize that we were not in Kansas, except when we reached the Cossack villages, composed of straggling rows of single-story huts with thatched roofs." Sherman thought that "in many respects the Cossacks resemble our Indians, [but] I doubt whether they would equal the Indians as enemies." He continued north by train towards Moscow. "For the whole day," he noted, "there was no variation in the face of the country, no more than occurs in our prairies in western Kansas, . . . the soil was black . . ., and very rich." Sherman's comments on the landscape and inhabitants of the steppes reflected his experience in the western United States where, since the end of the American Civil War, he had been in command of the U.S. Army fighting the Native Americans.[1] Sherman was an early example of an American who recognized similarities between the Great Plains of the United States and the steppes of the Russian Empire. But, his comments were confined largely to curiosity at finding a familiar landscape in a foreign land. Some of his remarks, such as his comparison between "Cossacks" and "Indians" and disparaging comments on cossack housing, conveyed a sense of American superiority. It seems not to have occurred to him that there may have been lessons from Russian experience in settling and plowing up their grassland that could be of value to Americans who were embarking on the same processes in the Great Plains. A sense of

[1] William Tecumseh Sherman, "General Sherman in Russia: Extracts from the Diary of W. T. Sherman," *The Century Magazine* 57 (April 1899), 869–71. See also Norman E. Saul, *Concord and Conflict: The United States and Russia, 1867–1914* (Lawrence: University Press of Kansas, 1996), pp. 73–5.

superiority over Russia was the prevailing attitude among Americans at this time and posed a barrier to transfers and influences from the steppes to the Great Plains.

By the time of Sherman's visit, Russians were already aware of similarities between their steppes and the prairies and Great Plains of North America. Back in 1857, Eduard Tsimmerman and a teenaged Prince Mikhail Khilkov visited the American grasslands. In St. Paul, Minnesota, they bought a wagon and horses and traveled around the then territories of Minnesota, Nebraska, and Kansas. They noted that the settlement of these "expanses of steppe" (*stepnye prostranstva*) was underway with the protection of the U.S. Army. Tsimmerman continued: "Advancing inexorably to the west, these detachments of settlers occupied the fertile prairies, which strongly resemble our New Russian steppes, with their lightly rolling relief and small areas of woodland along the sides of streams and gullies."[2] Tsimmerman returned to the United States in 1869–70. He was again struck by the similarities,[3] concluding: "On the Earth is hardly to be found expanses of land that are so similar to each other as our New Russian steppes . . . and the . . . frontier of the United States between the rivers Mississippi and Missouri." He stressed both regions' economic importance in growing grain for the world market and compared the ports of Chicago on Lake Michigan and Odessa on the Black Sea.[4]

Tsimmerman went further and considered what lessons Russians could draw from American experience of settling their grassland. He described how, since his previous visit in 1857, "everything has changed." In 1869, as he traveled west from Omaha, Nebraska, on the new transcontinental railroad, he observed:

> The whole region has been transformed as if by magic: in place of the previous bare steppes, where our horses grazed on the empty space,

[2] Eduard Tsimmerman, *Ocherki Amerikanskogo Sel'skogo Khoziaistva* (Moscow: tip. I. I. Rodzevich, 1897), pp. 29–30. Tsimmerman was born in 1822 into a Russian German family and was a graduate of Moscow University. "Tsimmerman, Eduard Romanovich," *ES* 75, 185–6; Margarita Marinova, *Transnational Russian–American Travel Writing* (New York: Routledge, 2011), pp. 17–18. Khilikov made a career in railways, in the Americas and Russia, and was appointed Russian Minister of Transport in 1895. "Khilkov, Mikhail Ivanovich," *ES* 73, 199. On "New Russia," see p. xxiii.

[3] Eduard R. Tsimmerman, *Puteshestvie po Amerike v 1869–1870 g.*, 2nd edition (Moscow: Grachev, 1871), pp. 214, 220, 222, 226.

[4] Eduard R. Tsimmerman, "Votchinnyi zakon v Amerike i nashi stepi," *Otechestvennye zapiski* 234, 9 (1877), 109–66, quotation from 109. Another Russian visitor to the Great Plains was the Grand Duke Aleksei, younger son of Tsar Alexander II, who went on a buffalo hunt in Nebraska with Buffalo Bill and General Custer. Lee A. Farrow, *Alexis in America: A Russian Grand Duke's Tour, 1871–1872* (Baton Rouge: Louisiana State University Press, 2014).

> [where] ... in the absence of accommodation ..., [we] pitched a tent for the night, now are stretched out rows of fruitful farms, surrounded by fences. The expanses of steppe are covered with the shoots of wheat; herds graze on the meadows; in short, the former, unpopulated, empty steppe has come to life, has been settled, ... and all this thanks to the beneficial influence of the railroad.[5]

In addition to the railroad, he attributed the transformation to the Homestead Act of 1862, which allowed settlers to acquire public land. Tsimmerman advocated such a law in the Russian Empire.[6]

Sherman's and Tsimmerman's accounts were among the first writings by Americans who visited the steppes and Russians who traveled to the prairies and Great Plains. Most, but not all, commented on the parallels between the landscapes. Tsimmerman and other visitors from the Russian Empire soon realized the potential for drawing on American practices, despite the fact that the agricultural settlement of the steppes had begun several decades before that of the Great Plains. Following the revolution of 1917, moreover, the Soviet government looked to American expertise and experience in agriculture and industry. Tsarist and Soviet authorities who drew on American practices were following a long line of "westernizers," who sought to emulate the western world rather than rely on home-grown, Slavic experience.[7] The culture of the United States over this period was quite different, ever more confident and assertive, and less inclined to look to other countries. Thus, even though the Russians had several decades' experience in settling, studying, and cultivating their semi-arid grassland, it was some time before many Americans accepted that they could benefit from this experience.

There were a number of barriers to be bridged before Americans recognized not just similarities between their grassland regions, but that they could learn from the Russians' expertise and experience. One potential barrier was that in the years before the outbreak of the First World War in 1914, the two countries were leading competitors in the growing world market for grain. When the Americans had opened up their grasslands to

[5] Tsimmerman, *Puteshestvie*, pp. 272–3

[6] Tsimmerman, "Votchinnyi zakon," 9. Another Russian later made a similar recommendation. Aleksandr Kol', "Amerikanskaia gomstednaia sistema nadeleniia pereselentsev zemleiu," *Voprosy kolonizatsii* 10 (1912), 1–34; 11 (1912), 121–45.

[7] See Esther Kingston-Mann, *In Search of the True West: Culture, Economics and Problems of Russian Development* (Princeton: Princeton University Press, 1999); Michael David-Fox, *Showcasing the Great Experiment: Cultural Diplomacy and Western Visitors to the Soviet Union, 1921–1941* (New York: Oxford University Press, 2012), pp. 24–7; Robert V. Allen, *Russia Looks at America: The View to 1917* (Washington, DC: Library of Congress, 1988).

cultivation and built the infrastructure to transport grain to the Atlantic ports, their exporters came into competition with Russian traders, who had long been shipping grain grown in the steppes from ports on the Black Sea, such as Odessa, and the Baltic Sea. This rivalry could have led the American and Russian authorities and their business interests to hinder agricultural specialists from the other country learning about their experience in growing grain in their grassland.

There were other barriers, potential and actual, for example, the American sense of superiority over Russian "backwardness." In part, such attitudes were a result of lack of knowledge and long-standing western perceptions of Russian "barbarism."[8] They were also based on specific issues that were well known in the United States in the late nineteenth and early twentieth centuries. Steppe agriculture would not have seemed a good model to follow at this time, since the region was hit by a succession of bad harvests caused in part by recurring droughts. In both the early 1890s and early 1920s, Americans organized famine relief in the steppe region and witnessed the desperate plights of the inhabitants at first hand. In addition, the Russian Empire acquired a reputation for persecuting its Jewish population. Further, the tsarist autocracy was encountering opposition from radicals and revolutionaries, who promoted their cause abroad, including in the United States, where they found some sympathy. All these issues contributed to a growing negative image of Russia that served as a barrier to Americans thinking that they could have anything to learn from Russian experiences.

A practical impediment to Americans learning from Russian experience and expertise was the language barrier that prevented most Americans from reading Russian scientific studies. While many educated Russians had some knowledge of western European languages, few Americans knew Russian or had much opportunity to learn the language, since it was taught in few American schools and universities before the 1940s. Among the small number of Americans who did know Russian were Jewish émigrés from the tsarist empire. A major obstacle to Americans learning from Russian experience was the Russian Revolution of 1917. In October 1917, the Bolsheviks established a Soviet government that was avowedly hostile to the capitalist world and counted the United States among its

[8] See Marshall T. Poe, *"A People Born to Slavery": Russia in Early Modern European Ethnography, 1476–1748* (Ithaca, NY: Cornell University Press, 2000); Larry Wolff, *Inventing Eastern Europe: The Map of Civilization on the Mind of the Enlightenment* (Stanford, CA: Stanford University Press, 1994); Martin Malia, *Russia under Western Eyes: From the Bronze Horseman to the Lenin Mausoleum* (Cambridge, MA: Belknap Press, 1999).

chief adversaries. Due to a deep suspicion of Communism in the United States, moreover, the government declined to establish diplomatic relations with the Soviet Union until 1933.

A persistent barrier in the United States to transfers and influences from the steppes was resistance to changes. There was some opposition inside the United States Department of Agriculture (USDA) to adopting practices from the Russian Empire and Soviet Union, not specifically on account of their origins, but because they challenged established hierarchies and institutions. Influences from the steppes that were adopted by the federal government, moreover, encountered some resistance in the plains states as members of the local population resented intrusion in their lives by Washington, DC. The rest of this chapter will analyze these barriers to transfers and influences from the steppes of the Russian Empire and Soviet Union in the Great Plains of the United States.

Competition in World Grain Market

A potential barrier was that the Russian Empire and the United States were in competition in the growing international market for grain. In the decades before the First World War, they were world's largest grain exporters.[9] In many years, Russian exports outstripped American. According to Russian government data, in 1870, Russian grain exports were valued at 652 million French francs, while American exports were worth 378 million francs.[10] In 1889, the Russian Empire controlled 35.5 percent of the world grain market, compared with a share of 33.1 percent for the United States.[11] In years of poor harvests, Russian exports fell sharply and American producers exported more. In the drought year of 1891, the Russian wheat harvest fell from 81 million hectoliters to 61 million hectoliters. Russian exports fell to 20 million hectoliters, while those from the United States and Canada combined rose to 72 million hectoliters.[12] Competition could have inhibited exchanges of expertise and cooperation between American and Russian agricultural scientists and their respective government departments of agriculture. In both the steppes and the Great Plains, farmers and their advisors were accumulating valuable experience in

[9] See M. E. Falkus, "Russia and the International Wheat Trade, 1861–1914," *Economica*, n.s. 33 (1966), 416–29; C. Knick Harley, "Transportation, the World Wheat Trade, and the Kuznets Cycle, 1850–1913," *Explorations in Economic History* 17, no. 3 (1980), 218–50.

[10] "Ekonomicheskoe obozrenie," *SKhiL* 112 (1873), 3rd pagn., 11.

[11] "Polozhenie Rossii na mezhdunarodnom khlebnom rynke," *SKhiL* 162 (1889), 3rd pagn., 161.

[12] "Iz zagranichnoi literatury: Urozhai pshenitsy v 1891 g.," *SKhiL* 168 (1891), 3rd pagn., 31–2.

growing grain in similar, and challenging, environments that they may have preferred to keep to themselves to give them a competitive advantage.

An arena in which this competition was played out was the world's fairs of the period. Countries exhibited their agricultural and industrial produce and displayed their scientific and cultural achievements. American agricultural exhibits, including grain and farm machinery, impressed Russian visitors, such as scientist Viktor Mochul'skii, who attended the World's Fair in New York in 1853, and future Minister of Agriculture Aleksei Ermolov, who visited the Vienna fair in 1873.[13] A Russian account of the Exposition Universelle in Paris in 1889 (famous for the Eiffel Tower) noted that the Americans were keen to create strong impression. Their agricultural machinery was the most advanced in world and American farmers were "dangerous competition" for European farmers. The writer expressed concern that the Russian exhibit was not a success.[14]

The Russian government tried harder to promote its agricultural produce and achievements at the World's Columbian Exposition in Chicago in 1893. The general commissar of the Russian section reported to Minister of Finance Sergei Witte: "with a feeling of national pride I can say that our grain under expert analysis was shown to be of such high quality... that it was superior not only to all foreign, but even the American [grain]."[15] A Russian observer noted the great interest attracted by their grain exhibits, but was disdainful of the agricultural exhibitions by individual American states.[16] On the other hand, American observers were less than fulsome in their remarks about the displays of crops by its main competitor, when they chose to mention them at all. *The Chicago Tribune*'s reporter was complimentary about the Russian agricultural exhibit of "all the ... products of the empire," but was far more interested in other parts of the Russian displays, such as the furs, lacquer boxes, and "embroideries, weapons, articles of dress and household ornamentation,"

[13] A. I. Khodnev, *Istoriia Imperatorskogo Vol'nogo Ekonomicheskogo Obshchestva s 1765 do 1865 goda* (Spb: tip. Obshchestvennaia pol'za, 1865), pp. 90–1; A. Ermolov, *Sel'skokhoziaistvennoe delo Evropy i Ameriki na venskoi vsemirnoi vystavke 1873 goda i v epokhu ee* (Spb: tip. Panteleevykh, 1875), pp. 610–67.

[14] A. A. Efron, *Torzhestvuiushchaia Frantsiia: Nabroski s parizhskogo vsemirnoi vystavki* (Spb: Avseenko, 1890), pp. 99–100, 140–1.

[15] P. I. Glukhovskii, *Otchet general'nogo kommissara Russkogo otdela Vsemirnoi kolumbovoi vystavki v Chikago* (Spb: Kirshbaum, 1895), pp. 84–6. See also *World's Columbian Exposition 1893 Chicago: Catalogue of the Russian Section* (Spb: Imperial Russian Commission, Ministry of Finance, 1893); S. M. Sokolov, "Rossiia na vsemirnoi vystavke v Chikago v 1893 g.," *Amerikanskii ezhegodnik 1984*: 152–64.

[16] P. Slezkin, "Zametki o minuvshei vystavke v Chikage," *SKhiL* 174 (1893), 1st pagn., 365–77.

than Russian grain.[17] In his *Book of the Fair*, Hubert Bancroft noted that the Russian agricultural exhibit of grain was larger than those of France and Australia, but smaller than Great Britain's. He compared migration to Siberia with the movement of people to the American West. But, he had much, much more to say about the exhibits of American farm produce.[18] The American author of a history of the fair published in 1897–8 opened his account of the agricultural exhibit with a flourish: "When the first agricultural nation of the world [i.e. the United States] organized a world's exposition in the heart of its agricultural region ... it was natural that its agriculturalists should worthily signalize their primordial and fundamental art and bring offerings of its best fruits to celebrate the chief national industry." He presented a full account of the grain, including wheat, displayed by American agriculturalists. He singled out the "magnificent results obtained from the rich prairie soil by the educated skill of Kansas farmers." (But, he omitted to mention that Kansas farmers owed a big debt to wheat introduced from the steppes.) He referred to the Russian agricultural exhibit briefly, but did not comment on its quality.[19]

The competitive edge in the comments by Russians and Americans on the other country's displays of agricultural produce continued at the Exposition Universelle held in Paris in 1900. Sergei Bogdanov praised the Russian grain exhibit, but was less than enthusiastic about the American offering.[20] The American grain exhibit was prepared by the agricultural scientist Mark Carleton, who knew a great deal about Russian agriculture and crops. Nevertheless, he breezily wrote back from Paris: "No other country will compare with ours in the cereal exhibit except France and Canada and possibly Roumania. ... I am much disappointed in Russia's exhibit of grain, though in general her showing is the largest of all foreign countries." He was, however, hurrying to complete his work in Paris so that he could make a second visit to the Russian Empire to collect varieties of crops in the steppes to introduce to the Great Plains.[21]

[17] "Czar Land Treasure: Russia's Display One of Features of the Fair," *Chicago Tribune*, July 24, 1893.

[18] Hubert Howe Bancroft, *The Book of the Fair: An Historical and Descriptive Presentation of the World's Science, Art and Industry, As Viewed through the Columbian Exposition at Chicago in 1893* (New York: Bounty, 1894), pp. 341, 344–62, 372–3.

[19] Rossiter Johnson, *A History of the World's Columbian Exposition Held in Chicago in 1893*, 4 vols. (New York: D. Appleton and Co., 1897–98), vol. 3, pp. 1–14.

[20] S. Bogdanov, "Zemledelie na Parizhskoi vsemirnoi vystavke 1900 goda," *SKhiL* 199, no. 12 (1900), 535–6, 539–41.

[21] NARA CP, RG 54, Finding Aid A1, Entry 58, Division of Vegetable Pathology and Physiology: Correspondence of M. A. Carleton, 1891–1900, Folder M. A. Carleton – 1900, Carleton to A. F. Woods, June 3, 1900.

The two countries' agricultural specialists kept close eyes on their main rival in the world grain market. From the end of the nineteenth century, Russian observers of American agriculture noticed that competition from across the Atlantic was decreasing. This was because of greater domestic demand for farm produce as the American population was increasing, and because the vast numbers of settlers in the plains had already occupied most of the land suitable for cultivation. The prospects for Russian grain exports seemed to be looking good on the eve of the First World War.[22] Following the outbreak of war in 1914, however, Russian agriculture was seriously disrupted as the tsar's armed forces mobilized men and horses from the countryside. Russian railroads prioritized military needs over transporting grain to the ports. And Russian ships, including vessels carrying grain, were prevented from leaving the Black and Baltic Seas by the Ottoman and German navies. Grain exports from the Russian Empire fell from 308.8 million poods (c. 5.6 million U.S. tons) in 1914 to 10.5 million poods (c. 190,000 U.S. tons) in 1915, and 2.1 million poods (38,000 U.S. tons) in 1916.[23] Russian specialists noted how American exports of grain and other foodstuffs to Europe increased sharply to compensate.[24]

American grain exporters had little competition from farmers in the steppes for almost a decade and half. Agriculture in Soviet Russia, including the steppes, experienced serious problems after 1917. The land reform of 1917–18 transferred land to small-scale peasant farmers, whose primary aim was subsistence rather than export. The authorities requisitioned grain from the peasantry during the Russian Civil War (1918–21) alienating many and contributing to a collapse in agricultural production. When drought hit parts of the steppes in 1921, there was a catastrophic famine. Gradual agricultural recovery over the 1920s was followed by the forced collectivization of family farms into large, collective, farms (*kolkhozy*). The policy was enforced most strictly in the fertile steppe region.[25] Parallel to

[22] V. Miuller, "Velichina narodonaseleniia v Soedinennykh Shtatakh Severnoi Ameriki i vyvoz pshenitsy iz nikh," *SKhiL* 168 (Oct. 1891), 95–7; G. Chirkin, "O zadachakh kolonizatsionnoi politiki v Sibiri," *Voprosy kolonizatsii* 8 (1911), 1–37.

[23] George Pavlovsky, *Agricultural Russia on the Eve of the Russian Revolution* (London: Routledge, 1930), pp. 321–4; Hew Strachan, *The First World War* (Oxford: Oxford University Press, 2001), pp. 1091–2.

[24] N. A. Borodin and M. I. Volkov, *Sel'skokhoziaistvennaia Amerika vo vremia voiny: na osnovanii lichnykh vpechatlenii 1917 g.* (Moscow: Izd-vo Narodnoe pravo, 1918), pp. 26–32.

[25] See David Moon, *The Russian Peasantry 1600–1930: The World the Peasants Made* (London and New York: Addison Wesley Longman, 1999), pp. 356–65; R. W. Davies, *The Socialist Offensive: The Collectivisation of Soviet Agriculture, 1929–1930* (London: Macmillan, 1980); Davies, *The Soviet collective Farm, 1929–1930* (London: Macmillan, 1980).

the collective farms on peasant land were state farms (*sovkhozy*) on large, private estates confiscated after the revolution. Soviet economic planners aimed to create large-scale, mechanized farms that drew on the latest organizational principles and farm machinery. To this end, the Soviet government imported American technology and hired American specialists.[26] Collectivization proved a disaster. It was imposed on a reluctant peasantry. Opponents, branded as "kulaks," were dispossessed, exiled, or executed.[27] Famine returned in 1931–3. Worst hit was the steppe region from Ukraine though southern Russia and the North Caucasus to Kazakhstan. Millions died.[28] The Soviet authorities tried to conceal the disaster from their own population and the outside world. They were assisted by *The New York Times* correspondent Walter Duranty, who dismissed reports of famine as exaggerations or propaganda. The famine was not widely reported in the American press.[29]

It is a measure of the Soviet authorities' success in concealing the famine and the consequences of collectivization that in the United States there were renewed concerns over competition from revived exports of wheat and other crops grown in the steppes. In 1931, American soil scientist Curtis Marbut made a positive assessment of the prospects for Soviet grain production and exports. He concluded that the collectivization and mechanization of Soviet agriculture envisaged in the First Five-Year Plan (1928–32) combined with the large area of suitable land in the steppes promised a big increase in "Russian" wheat production. "Assuming nothing interferes with the carrying out of the plan," Marbut asserted that in 1933, "Russia" would be able to export more wheat "than the maximum

[26] See Deborah Fitzgerald, *Every Farm a Factory: The Industrial Ideal in American Agriculture* (New Haven CT: Yale University Press, 2002), pp. 157–83; Jenny Leigh Smith, *Works in Progress: Plans and Realities on Soviet Farms, 1930–1963* (New Haven CT: Yale University Press, 2014), pp. 1–62.

[27] See Lynne Viola, V. P. Danilov, N. A. Ivnitskii, Denis Kozlov, eds., *The War Against the Peasantry, 1927–1930: The Tragedy of the Soviet Countryside* (New Haven CT: Yale University Press, 2005).

[28] For examples of the literature on the causes of the famines, the responsibility of the Soviet government, and its impact, see Robert Conquest, *The Harvest of Sorrow: Soviet Collectivization and the Terror-Famine* (Oxford: Oxford University Press, 1986); R. W. Davies and Stephen G. Wheatcroft, *The Years of Hunger: Soviet Agriculture, 1931–1933* (Basingstoke and New York: Palgrave Macmillan, 2004); Frank Sysyn and Andrij Makuch, eds., "The Ukrainian Famine of 1932–33, the Holodomor," *East/West: Journal of Ukrainian Studies* 2, no. 1 (2015); Sarah Cameron, *The Hungry Steppe: Famine, Violence, and the Making of Soviet Kazakhstan* (Ithaca NY: Cornell University Press, 2018); Niccolò Pianciola, "Famine in the Steppe: The Collectivization of Agriculture and the Kazak Herdsmen 1928–1934," *Cahiers du monde russe* 45, nos. 1–2 (2004), 137–92.

[29] Sally J. *Taylor, Stalin's Apologist, Walter Duranty: The New York Times Man in Moscow* (New York: Oxford University Press, 1990), pp. 210–23.

amount exported ... before the war."[30] Marbut had visited the Soviet Union in 1930 for the International Soil Science Congress and excursion around the country. The delegates were taken to the state farms "Gigant" ("Giant") and "Verbliud" ("Camel") in the steppe region. These were showcase farms equipped with imported American machinery and staffed by American advisors.[31] Not all American specialists were as optimistic as Marbut about the prospects for Soviet agriculture.[32] Nevertheless, in 1933, when President Franklin Roosevelt considered formally recognizing the Soviet Union, it provoked controversy in the plains states. Kansas farmers were concerned that normalizing relations with the Soviet Union could lead to an increase in grain exports that would provide competition for their grain in the world market.[33]

Concerns in the United States over competition from grain grown in the steppes coexisted with widespread perceptions of Russian "backwardness" that the revolution of 1917 served to enhance.

American Perceptions of Russian "Backwardness"

The image of "Russia" in the United States over the late nineteenth and early decades of the twentieth centuries was complex, multi-faceted, and changing. There were "positive" aspects to American perceptions, including the popularity of the classics of Russian literature, for example, the novels of Leo Tolstoy, which were widely read in translation. Tolstoy's ideas on non-violence and communalism also attracted interest. Russian ballet, opera, and music all enjoyed recognition among the American public. Peter Tchaikovsky was acclaimed at the Carnegie Hall in New York in 1891.[34] Nonetheless, a growing and pervasive perception of tsarist

[30] C. F. Marbut, "Russia and the United States in the World's Wheat Market," *GR* 21 no. 1 (1931), 1–21; Marbut, "Agriculture in the United States and Russia: A Comparative Study of Natural Conditions," *GR* 21, no. 4 (1931), 612; See also N. M. Tulaikov, *Sovremennoe polozhenie sel'skogo khoziaistva v Soedinennykh Shtatakh Severnoi Ameriki* (Moscow: Novaia Derevnia, 1923), pp. 5–7.

[31] SHSM MC, Marbut, Curtis Fletcher (1863–1935) Papers, 1852–1963 (C3720) [hereafter Marbut Papers (C3720)], Folder 143, "A visit to Russia" and "Supplement" by C. F. Marbut, Washington, DC, November 1930; Marbut, "Russia and the United States," 1–21. See also pp. 77–9.

[32] For contrasting views, see C. F. Marbut, "Russia and Wheat," [review of V. P. Timoshenko, *Russia as a Producer and Exporter of Wheat*] *GR* 23, no. 1 (1933), 159–60; C. F. Marbut and V. P. Timoshenko, "The Expansion of the Wheat Area in Arid Russia," *GR* 23, no. 3 (1933), 479–83. (Timoshenko, a Russian émigré, was not favorably inclined to the Soviet Union.)

[33] Norman E. Saul, *Friends or Foes? The United States and Soviet Russia, 1921–1941* (Lawrence: University Press of Kansas, 2006), pp. 254–315.

[34] See V. I. Zhuravleva, *Ponimanie Rossii v SShA: Obrazy i mify, 1881–1914* (Moscow: Rossiiskii Gosudarstvennyi Gumanitarnyi Universitet, 2012), pp. 951–72; Steven G. Marks, *How Russia*

and later Soviet Russia in the United States over our period was one of "backwardness." By the 1890s, in Norman Saul's words: "Americans, awash with national pride and new found imperialism, were ... more prone to compare Russia unfavorably with the United States and Western Europe." Negative American perceptions of Russia were fueled by reports of crop failures and famines in the steppe region, which prompted American relief efforts, by news of the persecution of Jews which were repeated by Jewish émigrés in the United States, together with accounts of growing opposition to the tsarist autocracy and harsh treatment of political exiles in Siberia.[35] Throughout the period, moreover, the technological superiority of American agriculture was demonstrated by exports of American farm machinery to the steppes.[36]

American Famine Relief in the Steppes

In 1891, following a serious drought, the harvest failed throughout large parts of the steppe region.[37] The U.S. Consul in Odessa, Thomas E. Heenan, reported to Washington, DC: "I am much afraid that we are on the eve of witnessing one of the greatest, if not the greatest, calamity of modern times." He had heard tales of entire communities starving, but that the authorities were unable to help because they were in debt. At the end of November, the Russian government banned exports of grain. Heenan noted "excitement" in the harbor at Odessa as exporters made "desperate attempts" to ship out as much wheat as possible before the ban came into effect. The consul thought a public appeal for aid in the United States would lead to an objection by the Russian government, but that this should not prevent aid being sent through private sources.[38] Private efforts were indeed made in the United States to organize famine relief for steppe provinces hit by the disaster. Reports by American relief workers

Shaped the Modern World from Art to Anti-Semitism, Ballet to Bolshevism (Princeton, NJ: Princeton University Press, 2003), pp. 109–10, 130–9, 191–3, 201–4.

[35] Saul, *Concord and Conflict*, pp. 189–96, 257, 282, 312 (quotation), 335–64, 395–6; Saul, *Friends or Foes?*, pp. 44–97.

[36] Saul, *Concord and Conflict*, pp. 151, 275–8, 410–1; Saul, *Friends or Foes?*, pp. 108–14, 119–21, 210, 218.

[37] See Richard G. Robbins, *Famine in Russia, 1891–1892: The Imperial Government Responds to a Crisis* (New York: Columbia University Press, 1975).

[38] NARA CP, RG 59, Microcopy no. 459, Despatches from United States Consuls in Odessa, 1834–1906, Roll 5, Vols. 9–10, January 2, 1889–December 28, 1895, Heenan to William F. Wharton, Assistant Secretary of State, November 2, 1891; Heenan to Wharton, November 21, 1891.

contributed to American notions of the "backwardness" of Russian steppe farming. William C. Edgar, a miller from Minneapolis, Minnesota, organized donations of surplus flour from the Great Plains, prairies, and further afield. He noted that people of "those states . . . which had, themselves, at some time in their history, felt the need of help, either through drought, crop failures, grasshoppers or other afflictions, were the first to respond to the call from their Russian brethren, and gave . . . more liberally . . . than any others."[39] In 1892, Edgar traveled to the famine-hit Volga region. He described his efforts to deliver flour to the starving, the people he met, the towns and villages he visited, and the effects of the famine and the typhus epidemic that accompanied it. He made a few references to the landscape.[40] But, he did not note any similarities to his native Minnesota or the plains states. Edgar and most Americans who visited the steppes during famines had little thought of parallels between the landscapes of the steppes and Great Plains that were so common among other American visitors, such as General Sherman, in happier times.

An even greater human catastrophe in the steppe region in 1921–2 increased American perceptions of the "backwardness" of Russian agriculture. A drought, in the aftermath of the revolution, civil war, and forced requisitions of grain from a suspicious peasantry, led to a massive famine. Unable to deal with the disaster, the Soviet government allowed the American Relief Administration (ARA), led by Herbert Hoover, to set up operations. The relief effort lasted for two years, brought nearly 300 Americans to Russia, and probably saved millions of lives. Their efforts could not, however, avert around five million deaths.[41] One of the relief workers, the future Stanford historian Harold Fisher, wrote: "On ruined towns and desolated villages across the bleak, dreary steppes had fallen the heavy pall of black misery, of inert despair. Into this atmosphere of fatalistic hopelessness came representatives of that distant incredible land – America."[42] Fisher expanded on the experiences of the American relief workers:

[39] William C. Edgar, *The Russian Famine of 1891 and 1892* (Minneapolis, MN: Millers and Manufacturers' Insurance, 1893), p. 8.

[40] Edgar, *The Russian Famine*, pp. 45–61; Harold F. Smith, "Bread for the Russians: William C. Edgar and the Relief Campaign of 1892," *Minnesota History Magazine* 42, 2 (1970), 54. See also George S. Queen, "American Relief in the Russian Famine of 1891–1892," *RR* 14 (1955), 140–50; Victoria I. Zhuravleva, "American Corn in Russia: Lessons of the People-to-People Diplomacy and Capitalism," *Journal of Russian American Studies* 1, no. 1 (2017), 24–32.

[41] Bertrand M. Patenaude, *The Big Show in Bololand: The American Relief Expedition to Soviet Russia in the Famine of 1921* (Stanford, CA: Stanford University Press, 2002), pp. 57–8, 196–8, 262–70.

[42] Quoted in Patenaude, *The Big Show in Bololand*, p. v.

> The work in the villages was naturally arduous. The Americans . . . traveled long distances over the most abominable roads in the world in primitive, springless Russian vehicles. They traveled in all weathers; in the beginning through the rain and mud of autumn, and later through snow with the temperature many degrees below zero. Nights were spent in peasant huts which offered nothing in the way of comfort. . . There was the constant strain of working in the midst of suffering . . .[43]

William Shafroth commented on the fertile black earth, but the primitive farming techniques, implicitly contrasting the latter with advanced American farming.[44]

Another historian among the relief workers was Frank Golder. He was born in 1877 in Odessa into a Jewish family, who moved to the United States in 1881. In October 1921, he wrote from Soviet Russia to a colleague at Stanford:

> The famine is bad beyond all imagination, it is the most heart breaking situation that I have ever seen. Millions of people are doomed to die, for little planting is done, the live stock is killed off, and the population is growing weaker. ... One asks in vain where are the healthy men, the beautiful women, the cultural life. It is all gone and in place of it we have starving, ragged, undersized men and women who are thinking of only one thing, where the next piece of bread is coming from.[45]

While traveling down the Volga to Samara he noted: "There is not much scenery to thrill over."[46] He had no thought of comparisons with the United States. On an earlier visit to Russia in 1914, however, Golder had made such comparisons while traveling on the Trans-Siberian Railroad: "Siberia is flat and monotonous, but less so than our Dakotas because of the scrub timber and gulches"; "We seem to have passed out of the flat lands and are now in a rolling country, something like the woodland meadows of Missouri."[47]

One American witness of the famine in the steppe region in 1921–2 who did make brief comparisons with the United States was the Canadian-born correspondent of the *Chicago Daily News*, Frederick Arthur Mackenzie. On his journey to the Volga region, he referred to the: "Stinging cold on the

[43] Harold H. Fisher, *The Famine in Soviet Russia: The Operations of the American Relief Administration* (Stanford, CA: Stanford University Press, 1927), p. 101.
[44] Quoted in Patenaude, *The Big Show in Bololand*, p. 55.
[45] Frank Alfred Golder, *War, Revolution, and Peace in Russia: The Passages of Frank Golder, 1914–1927* (Stanford, CA: Stanford University Press, 1992), p. 92.
[46] Quoted in Patenaude, *The Big Show in Bololand*, p. 58.
[47] Golder, *War, Revolution, and Peace*, pp. xi–xii, 23–5, 92.

great steppes – the prairies of south-east Russia." He described a village showing "every sign of former prosperity . . . It was easy to tell that there were German colonists here, men who had brought to the Volga the same exactness and hard-working qualities that have made so many prosperous communities in the Far West [of America]." "It was in Samara, the Chicago of Russia," he noted, "that I saw things at their worst." He found "lads, gaunt and tall, thin beyond any conception a Westerner can have."[48] As a journalist, it was his job to explain what he saw to his readers in ways that would help them understand. His allusions to the familiar conveyed the full extent of the suffering by making readers think of their home city, its hinterland, inhabitants, and people like themselves. In doing so, however, he can only have deepened perceptions of the "backwardness" of Russian steppe farming.

One group of relief workers from the United States whose responses to the famine in the steppes were more profound than most were members of the American Mennonite Relief organization. Their impressions were particularly striking, since many came from the Great Plains, most were descendants of migrants from the communities they were assisting, and some had themselves been born in the steppes. Two Mennonite relief workers recorded their impressions when they arrived at Chortitza in Soviet Ukraine in late 1921:

> As we visited the Mennonite villages we were made aware of the terrible conditions. The quiet of death hung over the clustered houses like a pall. Not a dog barking, for the Mennonites had eaten their dogs, their cats too had all been consumed. Here and there a cow or a horse was left.[49]

In April 1922, Peter Hiebert and Christian E. Krehbiel of Newton, Kansas, arrived in Halbstadt, in the Molotschna colony in southern Ukraine:

> We went directly from the train to the church, where we came as unannounced messengers from another world, and were heartily welcomed. I was asked to address a fairly large gathering of sad-faced, under-fed

[48] Frederick Arthur Mackenzie, *Russia Before Dawn* (London: Unwin, 1923), pp. 127, 131, 152, 278.

[49] Peter C. Hiebert and Orie O. Miller, eds., *Feeding the Hungry; Russia Famine, 1919–1925: American Mennonite Relief Operations under the Auspices of Mennonite Central Committee* (Scottdale, PA: Mennonite Central Committee, 1929), pp. 67, 75. On relief workers who were born in the steppes, see pp. 340, 390. Hiebert's parents had migrated to Kansas from the steppes in 1876. Katie Funk Wiebe and Richard D. Thiessen (June 2010), "Hiebert, Peter C. (1878–1963)," Global Anabaptist Mennonite Encyclopedia Online, available online at https://bit.ly/2qhOvLc, accessed October 7, 2019; Christian E. Krehbiel was born in Summerfield, Illinois, and moved to Kansas with his family as a child. "Krehbiel, Christian E. (1869–1948)," [Obituary], *Mennonite Weekly Review* (June 10, 1948), 5.

> Mennonites. I had never in my life heard any one pray, "Give us this day our daily bread!" as these people did with tears running down their haggard faces.[50]

Overwhelmed by the horrors of famine, even to Americans accustomed to similar landscapes in the prairies and Great Plains, during the human tragedies of 1891–2 and 1921–2, the steppes seemed to be another world that defied comparison with anything familiar. The recurring famines in the steppes reinforced notions of "Russian backwardness," and contributed to a persistent image in the United States of "starving Russia" in contrast to a "prospering America."[51]

The Persecution of Jews in the Russian Empire

The negative image of Russia in the United States in this period was exacerbated by news of the persecution of its Jewish population. There were extremes of violence during the revolution of 1905. In November the *Los Angeles Herald* reported:

> Thomas E. Heenan, American consul at Odessa, has sent a telegram to the American embassy saying that since Tuesday the bloody attempts upon the Jews have continued and that he estimates the number killed in the thousands. Artillery, he says, has been employed to suppress the rioting and the Jews have fired from windows upon the troops in the streets.

There was more in the article on this theme:

> Other dispatches received from Odessa say that the Cossacks and Infantry fought a regular battle with Jews and revolutionaries ... and estimated the dead at 100 and the wounded at over 2000. Press accounts from Odessa give details of horrible atrocities committed. The tongues of Jews were torn out by the roots, nails were driven in the heads of living persons and others were rolled in spiked barrels, but these reports must be accepted with a large amount of caution.[52]

Such accounts were consistent with other reports that had reached the United States of the mounting repression of Jews in the Russian Empire since the early 1880s. Russian–Jewish émigrés in the United States, for example the forestry scientist Raphael Zon, received news from family members in Russia about the difficulties they encountered. The growing

[50] Hiebert and Miller, *Feeding the Hungry*, p. 200.

[51] Zhuravleva, *Ponimanie Rossii v SShA*, pp. 209–58.

[52] *Los Angeles Herald*, November 6, 1905.

numbers of Russian Jews who settled in the United States publicized their experiences and the plight of Jews who remained in the Russian Empire.[53]

Opposition to the Tsarist Autocracy

American attitudes to Russia were further colored by reports of mounting opposition to the tsarist government from radicals and revolutionaries who wanted to replace the autocratic regime with a liberal or socialist alternative. Some of the regime's opponents resorted to terrorism. The tsarist authorities used the power of the state and its secret police against the revolutionary movement. Many revolutionaries were executed or banished to Siberia. Others fled abroad, where they attacked the tsarist autocracy. An American who contributed to the growing negative image of the tsarist regime was George Kennan. He first visited Russia in 1865, when he traveled to northeastern Siberia with the Russian–American Telegraph Expedition to survey the route of a cable to link the United States and Russia via the Bering Straits. The project was abandoned after a transatlantic cable was laid in 1866. But, Kennan's life-long fascination with Russia had begun. At first an enthusiast for all things Russian, his study of the Siberian exile system in 1885–6 turned him implacably against the tsarist regime for its authoritarianism and harsh treatment of its opponents. Kennan supported exiled revolutionaries, such as the former terrorist Sergei Kravchinskii ("Stepniak"). His articles and public lectures were published in book form as *Siberia and the Exile System* in 1891. In the words of Frederick Travis, Kennan "led a shift in American public opinion of the tsarist government from enthusiastic and uninformed friendliness to hostility."[54] It is a measure of the hostility to the tsarist regime in the United States that there was heated debate in the late 1880s and early 1890s over whether Russian revolutionaries, including terrorists, who had sought refuge should be extradited to face trial in Russia.[55]

An increasingly negative image of tsarist Russia in the United States in the late nineteenth and early twentieth centuries was shaped, therefore, by

[53] See Zhuravleva, *Ponimanie Rossii v SShA*, pp. 90–148; Saul, *Concord and Conflict*, pp. 257, 396, 475. On the Jewish question in Russian–American relations, see Valerii Engel', *"Evreiskii vopros" v russkoamerikanskikh otnosheniiakh: Naprimere pasportnogo' voprosa 1864–1913* (Moscow: Nauka, 1998).

[54] Frederick F. Travis, *George Kennan and the American–Russian Relationship: 1865–1924* (Columbus: Ohio University Press, 1990), p. xiii and passim; Zhuravleva, *Ponimanie Rossii v SShA*, pp. 149–209.

[55] E. L. Nitoburg, *Russkie v SShA: Istoriia i sud'by, 1870–1970: Etnoistoricheskii ocherk* (Moscow: Nauka, 2005), pp. 30–32. In the end they were not extradited.

its growing reputations for persecuting its political opponents, Jewish subjects, and for periodic droughts, bad harvests, and famines that continued after 1917. It may not be altogether surprising, therefore, that Russia was not the first place Americans would look for expertise in agriculture or indeed much else beyond the realms of literature, the arts, and revolutionary ideas. These widespread negative impressions of Russia in the United States provide a context to understanding why most American agricultural specialists were largely ignorant of Russian agricultural sciences at this time.

American Ignorance of Russian Agricultural Sciences before c. 1900

At the end of the nineteenth century, American agricultural scientists were a little surprised by and slightly condescending towards the work of their Russian counterparts. The author of an article published in 1897 in the central journal of the American agricultural experiment stations noted that "agricultural science" was "new" in Russia, but "promising." Above all, the article noted the inaccessibility of Russian research, since it was written in Russian, and because of the "scattered nature" of Russian publications.[56] It was certainly true that most Russian publications on agricultural sciences were in Russian and that they were "scattered," if that implied that there were a number of specialist publications. But, Russian agricultural sciences were certainly not "new." They dated back to the seventeenth century. Serious study of the environment and farming in the steppes began when agricultural settlement of the region took off in the eighteenth century. By the turn of the twentieth century, such studies existed in large quantities, aspects of which, for example, soil science and steppe forestry, were original and relevant to the Great Plains.[57] In Nikolai Vavilov (1887–1943), whose career began at this time, Russia had a crop scientist and geneticist of international importance.[58] Russian agricultural scientists

[56] "Investigation and Research in Russia," *ESR* 9 (1897–8), 201–5.

[57] For a study of Russian agricultural experimental institutions, see Ol'ga Elina, *Ot tsarskikh sadov do sovetskikh polei: istoriia sel'sko-khoziaistvennykh opytnykh uchrezhdenii XVIII-20-e gody XX v.*, 2 vols. (Moscow: Rossiiskaia Akademiia Nauk, 2008). For a monograph based on Russian studies of the steppe environment and agriculture, see David Moon, *The Plough that Broke the Steppes: Agriculture and Environment on Russia's Grasslands, 1700–1914* (Oxford: Oxford University Press, 2013).

[58] For examples of the large literature on Vavilov, see Peter Pringle, *The Murder of Nikolai Vavilov: The Story of Stalin's Persecution of One of the Great Scientists of the Twentieth Century* (New York: Simon and Schuster, 2008); N. P. Goncharov, *Nikolai Ivanovich Vavilov* (Novosibirsk: Izd-vo SO RAN, 2014).

who visited the United States in the early twentieth century, for example, Nikolai Tulaikov, were dismayed that their American counterparts did not know about Russian work. Dmitrii Artsybashev may have been exaggerating, but perhaps only slightly, when he wrote in 1909 that Americans did not know about and did not study "our continent."[59]

The limited knowledge among American agricultural scientists of the work of Russian scholars persisted despite original work of international importance by scientists in Russia in the late nineteenth century. The most famous is Dmitrii Mendeleev, who devised the periodic table of elements in the late 1860s. He faced a long battle to prove his priority, however, in part because he announced his innovation in an article published in Russian.[60] Another important Russian scientific innovation in this period, which was also published in Russian, was the theory of soil formation by Mendeleev's colleague at St. Petersburg University, Vasilii Dokuchaev. As we shall see, it took rather longer for Dokuchaev's theory and its wider implications to become accepted in the United States. This was despite the similarities between the soils of the steppes, where Dokuchaev carried out his field work, and the Great Plains. The limited knowledge of Russian research among many American agricultural scientists was due in part to a language barrier.

Language Barrier

Most Russian research in agricultural sciences was written in Russian for the very good reason that it was intended for other Russian specialists as well as landowners interested in "improving" the ways they cultivated their land.[61] Over the second half of the nineteenth century, Russian scientists in other disciplines published their work in Russian rather than in German

[59] N. M. Tulaikov, "Pochvennye issledovaniia v Soedinennykh Shtatakh," *Pochvovedenie* 10, 4 (1908), 321–2; D. Artsybashev, *Sel'sko-khoziaistvennoe mashinostroenie v Soedinennykh Shatakh Severnoi Ameriki i v Kanade* (Spb: Sel'skii Vestnik, 1909), p. 5.

[60] Michael D. Gordin, "The Table and the Word: Translation, Priority, and the Periodic System of Chemical Elements," *Ab Imperio*, 2013, 3 (2013), 53–82.

[61] See Joseph Bradley, *Voluntary Associations in Tsarist Russia: Science, Patriotism, and Civil Society* (Cambridge, MA: Harvard University Press, 2009), pp. 38–85; Moon, *Plough*, pp. 46–88. The Southern Russian Agricultural Society published a monthly journal in Russian. M. P. Borovskii, *Istoricheskii obzor piatidesiatiletnei deiatel'nosti Imperatorskogo Obshchestva Sel'skogo Khoziaistva Iuzhnoi Rossii s 1828 po 1878 god* (Odessa: P. Frantsov, 1878). An exception was a German-language periodical for settlers in the southern steppes, founded in 1846, which published articles on agriculture. *Unterhaltungsblatt für deutsche Ansiedler im Südlichen Russland.* "Die erste Zeitung für die deutschen Kolonisten in Südrussland," available online at www.hfdr.de/sub/besonderes.htm, accessed March 29, 2018.

as had been common earlier in the century. This was part of a drive to establish Russian as a major scientific language alongside the "triumvirate" of German, French, and English that were the main international languages of science at this time. After 1917, Russian was the main language of publication for Soviet scientists.[62]

The language barrier to exchange of information between Russian and American agricultural scientists was more impenetrable for Americans than their Russian and Soviet counterparts. Throughout our period many, but not all, Russian and Soviet scientists had some knowledge of western European languages, in particular German, French, and English. In the Soviet Union, scientists were required to study at least one foreign language so that they could read international scientific literature. While some struggled, especially with verbal fluency, many could read scientific texts with the aid of dictionaries and communicate at international meetings, sometimes with help from colleagues with better language skills.[63] An exception was pioneering soil scientist Dokuchaev, who had only a limited knowledge of German and French and little English. He had some help with languages from his wife, Anna Egorevna, who was of Scottish descent.[64]

On the other hand, with some important exceptions, few American scientists over our period knew Russian, although some knew western European languages. The most widely studied foreign language in the United States in the late nineteenth and early twentieth centuries was German. For many recent immigrants, it was their native language. But, teaching German in the United States collapsed, never to recover, after the United States joined the Allies fighting against Germany in the First World War.[65] David Fairchild, who headed the USDA's Office of Seed and Plant Introduction for many years from 1898 and traveled extensively,

[62] See Michael D. Gordin, *Scientific Babel: The Language of Science from the Fall of Latin to the Rise of English* (London: Profile Books, 2015), pp. 49, 103, 220–6 and passim.

[63] See Gordin, *Scientific Babel*, pp. 87–93, 222–3. Tulaikov, who knew English well, interpreted for some of his Soviet colleagues at the International Soil Science Congress in the United States in 1927. ARAN, f.582, op.3, 1927, d.454, l.3.

[64] On Dokuchaev's limited language skills, see Bancroft Library, Berkeley, California, The Hilgard family papers [hereafter, Hilgard Papers], Hilgard, E. W., Incoming Letters, Box 23, File: Voeikov, Aleksandr Ivanovich, Voeikov to Hilgard, January 6/18, 1892; E. S. Kul'pin-Gubaidullin, "Vasilii Dokuchaev kak predtecha biosferno-kosmicheskogo istorizma: sud'ba uchenogo i sud'by Rossii," *Obshchestvennye nauki i sovremennost'*, no. 2 (2010), 103–13; I. P. Vtorov, "Pervoe vospriitie ideai V. V. Dokuchaeva mezhdunarodnym nauchnym soobshchestvom," unpublished paper presented to Mezhdunarodnyi nauchnyi seminar "Nauchnoe nasledie V. V. Dokuchaeva," Moscow, May, 30–31, 2016. (I am grateful to Dr. Vtorov for a copy of his paper.)

[65] Gordin, *Scientific Babel*, pp. 180–3.

claimed to speak French, Italian, and German, and also some Malay.[66] A Soviet scientist who attended the International Botanical Congress in Ithaca, New York, in 1926 reported that American scientists sometimes knew German, rarely French, but almost none knew Russian, besides "old émigrés."[67] Many of these Russian-speaking émigrés were Jews who had moved to the United States from the Russian Empire in large numbers from the 1880s (see pp. 105–7). Jewish immigrants comprised many of the 1.2 million people counted in the 1910 U.S. Census who gave "Russia" as their place of birth. Since the first language of many Russian Jews was Yiddish, however, only around 60,000 people listed Russian as their native language.[68]

Russian is considered to be a "difficult" language for native speakers of English by specialists in language learning, because it has "significant linguistic and/or cultural differences from English." The U.S. Foreign Service Institute's School of Language Studies ranks Russian in its third category out of four in degree of difficulty. The School calculates that it takes 1,100 class hours for an English speaker to reach "professional working proficiency" in Russian.[69] In the late nineteenth century and first four decades of the twentieth century, moreover, Americans who had not been born in Russia had little opportunity to study the language. Before 1914, only six American universities, starting with Harvard in 1896 and followed by California at Berkeley and Chicago in 1901, offered courses in Russian. Enrolments were low. Russian language teaching did not become widely available in the United States until the 1940s and 1950s, when interest and need were spurred by the alliance with the Soviet Union in the Second World War and the rivalry, including scientific rivalry, in the ensuing Cold War.[70] Even then, the number of American scientists who

[66] David Fairchild, *The World Was My Garden: Travels of Plant Explorer* (New York: Scribner's Sons, 1938), pp. 180, 194, 300; Daniel Stone, *Food Explorer: The True Adventures of the Globe-Trotting Botanist Who Transformed What America Eats* (New York: Dutton, 2018), pp. 164, 315.

[67] A. F. Lebedev, "V Biuro Upolnomochennykh Vsesoiuznykh S"ezdov po pochvovedeniiu (pis'mo iz Ameriki)," *Biulletini pochvoveda*, nos. 5–7 (1926), 54.

[68] Gordin, *Scientific Babel*, p. 226.

[69] "FSI's Experience with Language Learning," U.S. Department of State, available online at www.state.gov/foreign-language-training/, accessed March 29, 2018. I was taught Russian on an intensive course at the start of the graduate program at the Centre for Russian East European Studies at the University of Birmingham in the UK in the early 1980s. Most of our instructors had learned Russian at the UK Joint Services School for Linguists early in the Cold War. See Geoffrey Elliott and Harold Shukman, *Secret Classrooms: An Untold Story of the Cold War* (London: St Ermin's Press, 2002).

[70] See Albert Parry, *America Learns Russian: A History of the Teaching of the Russian Language in the United States* (Syracuse, NY: Syracuse University Press, 1967); Saul, *Concord and Conflict*, pp. 390–5, 566–7; Saul, *Friends or Foes?*, pp. 186–92, 363–5.

knew Russian was tiny.[71] Prior to the 1940s, attempts to introduce Russian language courses at American universities met with limited success. In 1926, for example, Zon tried to persuade the dean of administration at the University of Minnesota to start teaching Russian at the university. His request was turned down as the dean felt there would not be sufficient demand.[72]

For most of our period, American scientists who wished to know more about Russian studies in their fields often struggled. The American soil scientist Eugene W. Hilgard (1833–1916), was more eager than most of his compatriots to study the work of his Russian counterparts. But, although he was multilingual, he was unable to read Russian. He was born in Germany, moved to the United States with his family as a child, later studied in Germany and Switzerland, and lived for two years in Spain, where he met his wife. Besides English and German, he knew Latin, French, and Spanish.[73] From 1870, he corresponded with several Russian scientists, but relied on their knowledge of western European languages. Gavrill Tanfil'ev wrote to him in German. Jean [Ivan] Vilbourchevitch, who had emigrated from Russia to France, composed his letters in French. Hilgard's longest-standing Russian correspondent, Aleksandr Voeikov, wrote to him in English.[74] In 1902, Voeikov wrote: "It is much to be regretted that you are unable, in California, to use Russian books, for as to the study of soils we are further advanced than any country in Europe. You should study Russian on the Pacific coast ..." Voeikov was probably unaware that Russian teaching had started at Berkeley the previous year. But Hilgard, who was a professor at Berkeley, did not avail himself of the opportunity. In January 1908, he asked if Pavel Ototskii, the editor of the Russian soil science journal (*Pochvovedenie*), could include abstracts of articles in other languages. Ototskii replied that it was a matter of resources and he was already overburdened producing the journal.[75]

[71] Gordin, *Scientific Babel*, p. 218.

[72] MHS, Raphael Zon papers, 1887–1957 [hereafter Zon Papers], Box 5, Folder 7, F. J. Kelly, Dean of Administration, University of Minnesota, to Zon, March 2, 1926.

[73] Frederick Slate, "Biographical memoir of Eugene Woldemar Hilgard,1833–1916," *National Academy of Sciences Biographical Memoirs* 9 (1919), 94–155, available online at www.nasonline.org/publications/biographical-memoirs/memoir-pdfs/hilgard-eugene.pdf, accessed March 27, 2018.

[74] For examples of their letters, see Hilgard Papers, Hilgard, E. W.: Incoming Letters, Box 21, Folder: T-Misc, Gavrill Tanfil'ev to Hilgard, August 1/13, 1893 [in German]; Box 23, Folder: Vilbourchevitch, Jean, Vilbourchevitch to Hilgard, January 7, 1891 [in French]; Folder: Voeikov, Aleksandr Ivanovich, Voeikov to Hilgard, January 10, 1874.

[75] Hilgard Papers, Hilgard, E. W.: Incoming Letters, Box 23, File Voeikov, Voeikov to Hilgard, June 22/July 5, 1902; Hilgard, E.W.: Outgoing Letters, Letterpress copy books, vol. 26, June 1907–July

The language barrier hindered American scientists who attended conferences and congresses in Russia from making the most of the opportunities to learn about Russian scientific work and engage in informal exchanges with Russian scholars. Towards the end of the excursion that followed the International Soil Science Congress in the Soviet Union in 1930, Charles Shaw of the University of California, Berkeley, remarked that he had appreciated the opportunity to spend time with Soviet scientists and other people, but that: "It was hard to learn of you for we do not speak your language."[76] American scientists who wanted to know more about the work of Russian and Soviet scientists generally depended on them to provide abstracts and translations in western European languages. In 1927, the Nebraska-born ecologist Frederic Clements wrote, a little abruptly, to Vavilov:

> I shall appreciate receiving copies of your publications as they appear, and would like also to obtain those of Russian workers . . . in . . . ecology and evolution. The steppe vegetation has so much in common with our grassland formation that I am especially anxious to secure anything in this field. Naturally, however, publications in Russian without summary in some other language are practically of no use to me.[77]

Vavilov would have been entitled to feel slightly irritated, since he spoke several languages and took every chance to improve his English. When he visited the United States in 1930, he sought the help of one of the USDA's administrative staff, Miss Martini, to help him learn "slang," as the only "slang" he knew was "hot dogs." Miss Martini was happy to oblige their distinguished foreign visitor, and prepared what Vavilov referred to as a "splendid text-book on slangs."[78] This rather charming episode has to be seen against a background of tense relations between the capitalist United States and communist Soviet Union in the wake of the Russian Revolution of 1917.

1910, pp. 120–3, Hilgard to Dr. N. Toulaikoff [Tulaikov], January 28, 1908 (Tulaikov and Hilgard corresponded in English); Tulaikov, "Pochvennye issledovaniia," 321–2.

[76] ARAN, f.487, op.1, d.27, Stenogramy vstrech delegatov II-go Mezhdunarodnogo Kongresa pochvovedov s sovetskimi uchenymi, l.120.

[77] TsGANTD Spb, f.318, op.1-1, d.143, l.147, Frederic E. Clements to Vavilov, February 7, 1927.

[78] NARA CP, RG 54, Finding Aid PI-66, Entry 30, Records of the Division of Cereal Crops and Diseases. Foreign Correspondence, 1900–34, Box 8, File Russia, Vavilov, Indio, California, to Harlan, October 13, 1930; Vavilov to Harlan, October 25, 1925. Vavilov spoke English, French, German, Italian, and Persian, in addition to his native Russian. Gary Pau Nabhan, *Where Our Food Comes From: Retracing Nikolay Vavilov's Quest to End Famine* (Washington, DC: Island Press, 2009), p. 126.

American Suspicion of Communism after 1917

The Bolshevik seizure of power and installation of the Soviet government in Russia in October 1917 led to suspicion of communism in the United States and around the capitalist world. It fed on earlier perceptions of Russian "backwardness."[79] From its foundation, the Soviet government provoked a hostile response in the United States. In 1918, the United States sent troops to join the foreign "intervention" that aimed to keep Russia in the First World War and to support the White forces in the Civil War against the Bolsheviks. The intervention failed and American troops were withdrawn in 1920. At the same time, there was a "Red Scare" in the United States, prompting fears of subversion orchestrated by the Communist International (Comintern) from Moscow. At this time, there was widespread opposition in the United States to immigration from Russia.[80] The U.S. government refusal to recognize the Soviet government meant there were no diplomatic relations between the two countries. The tense international situation and lack of formal relations hindered contacts between American and Soviet scientists, and thus posed a further barrier to Americans learning from Russian expertise. In June 1922, Vavilov (who had managed to visit the United States in 1921–2) received a request to assist Americans visit Soviet Russia. He replied that it was difficult for them to do so on account of the absence of American recognition, and because the Soviet authorities would not admit Americans into their country. This was in retaliation for the U.S. government's refusal to allow Communists to visit the United States.[81]

The diplomatic difficulties did not put a complete stop to American scientists building good relations with Vavilov and other Soviet scientists with whom they had mutual interests. There were limits to how far they were permitted to go by the U.S. government. One episode stands out. In 1925–6, Vavilov was planning to travel to Abyssinia (present-day Ethiopia) to collect samples wild grain. The trip was important for his research as he believed Abyssinia to be the home of the ancestors of some cultivated cereals, in particular barley. There was a serious impediment. The Soviet Union did not have diplomatic relations with Abyssinia and so he could

[79] Michael David-Fox emphasized the "superiority-inferiority calculus" in the interaction between western visitors and their Soviet hosts in the interwar years. David-Fox, *Showcasing*, pp. 26–7.

[80] Norman E. Saul, *War and Revolution: The United States and Russia, 1914–1921* (Lawrence: University Press of Kansas, 2001), pp. 309–75, 390–1.

[81] TsGANTD Spb, f.318, op.1-1, d.23, l.173, [Vavilov] to D. Borodin, June 22, 1922. On Vavilov's visit to the United States in 1921–2, see pp. 119–20.

not obtain a visa. He wrote to Harry Harlan, a fellow cereal scientist and friend, asking if the USDA could help. Harlan, who appreciated the scientific importance of Vavilov's planned trip, offered to write to Ras Tafari, the regent and heir to the throne. He drafted a cable and was about to send it when his immediate superior thought it advisable to hold an "informal and confidential consultation with the State Department." As a result, Harlan was informed that: "Such a request from him ... would undoubtedly be accepted as the official request of the United States, which in view of the existing conditions the Department of State would be unwilling to have made." Harlan did not send the cable. (Vavilov visited Abyssinia with help from French scientists.)[82]

Until November 16, 1933, when the United States finally recognized the Soviet Union and established diplomatic relations,[83] American scientists who worked for the federal government, including the USDA, were not permitted to travel there in their official capacities. This created serious problems in 1930, when Marbut, the chief of the U.S. Soil Survey, and Zon, director of a government forest experiment station, were invited to attend the 2nd International Congress of Soil Science in the Soviet Union. After much effort, Marbut took unpaid leave and went as a "special representative of the American Society of Agronomists," which paid for his attendance. The USDA helped by sending him on an official trip to Europe before the Congress. Nevertheless, taking part in the Congress and the excursion that followed cost Marbut $725 in lost salary for the 41 days he was in the Soviet Union.[84] Zon was also eager to attend the congress. The U.S. Forest Service was prepared to facilitate his visit to Europe. However, when he was informed that he could go to the Soviet Union only in his own time and at his own expense, he declined to do so. His

[82] NARA CP, RG 54, Finding Aid PI-66, Entry 30, Records of the Division of Cereal Crops and Diseases. Foreign Correspondence, 1900–34, Box 8, File Russia, Vavilov to Harlan, November 9, 1925; Harlan to Vavilov, January 21, 1926; Vavilov to Harlan, cables, June 26 and 30, 1926; Harlan to H. H. Ras Tafari, cable [draft, not sent], June 30, 1926; Harlan to Vavilov, cable [draft, not sent]; M. A. Taylor, Chief of BPI, memo to Mr. McCall, July 1, 1926; McCall to Harlan, July 2, 1926; Harlan to Vavilov, August 26, 1926; Vavilov to Harlan, April 25, 1927.

[83] "Recognition of the Soviet Union, 1933," Milestones in the History of U.S. Foreign Relations, Office of the Historian, United States Department of State, available online at https://history.state.gov/milestones/1921-1936/ussr, accessed March 27, 2018.

[84] Marbut Papers (C3720), Folder 47 Correspondence, 1930, P. E. Brown, The American Society of Agronomy, to Marbut, March 20, 1930; Marbut to W. Elmer Ekblaw, Clark University, Worster, MA, April 18, 1930; Ekblaw to Marbut, April 21, 1930; Folder 49 Correspondence, 1930, Marbut to Louise and Roy, April 27, 1930; E. N. Meador, Assistant to the Secretary [of Agriculture] to the Secretary, June 25, 1930; Henry G. Knight, Chief of Bureau [of Chemistry and Soils] to Dr. Stockberger, July 1, 1930.

letters from the time show great restraint that probably concealed his bitter disappointment at not being able, for the first time since he had fled tsarist Russia in 1896, to visit the country of his birth.[85]

The lack of diplomatic relations complicated the procedure for the Soviet organizers to invite the American scientists who were able to attend the soil science congress in 1930 as they could not be sent through diplomatic channels. Well in advance, in December 1928, A. G. McCall of the USDA wrote to David Jacobus Hissink, the general secretary of the International Soil Science Society, to suggest the invitations be sent to American institutions or the American Organizing Committee for the previous congress. Some scientists expressed concern about their safety if they traveled to the Soviet Union. Those most worried were émigrés who had left Russia before or after 1917. The Soviet scientists organizing the congress secured reassurances from the Soviet government that the "visa for entering and leaving Russia would give the fullest guarantee that could be desired."[86]

The suspicion worked both ways. The Soviet authorities restricted and controlled visits by its citizens abroad, in particular to capitalist countries. The Soviet delegation to the 1st International Congress of Soil Science in Washington, DC, in 1927, included "two or three commissar type advisers," who maintained surveillance over the Soviet scientists, in particular Konstantin Glinka, who led the Soviet group.[87] Likewise, American visitors to the Soviet Union were subjected to surveillance, and restricted in where they could go.[88] The scientific aims of the excursion around the Soviet Union that followed the Soil Science Congress in 1930, in which Marbut took part, were to study different types of soils and visit

[85] Zon Papers, Box 6, Folder 8, Zon to Ed [Munns], Bureau of Soils, January 3, 1930; "Σ" [possibly chief forester Stuart] to Zon, n.d., received January 8, 1930; Zon to Susanna Paxton, Russian Travel Dept., Open Road Inc., New York, March 22, 1930; Box 7 Folder 1, Zon to E. H. Clapp, Forest Service, April 21, 1930; D. G. Volensky, International Society of Soil Science, Moscow to Zon, May 16, 1930; R. J. Stuart, Forester to Zon, May 8, 1930; Zon to Stuart, May 15, 1930.

[86] ARAN, f.487, op.1, 1929, d.18, l.105, A. G. McCall to Hissink, December 13, 1928; l.165, Hissink to K. K. Gedroitz. August 20, 1929; l.167, Gedroits to Hissink, September 2, 1929; l.168, Zavaritskii to Hissink, September 18, 1929; l.197, Hissink to Iarilov, December 17, 1929 (quotation).

[87] J. S. Joffe, "Russian Contributions to Soil Science," in *Soviet Science: Symposium at the 1951 Philadelphia Meeting of the American Association for the Advancement of Science*, ed. R. G. Christman (Washington, DC: AAAS Publications, 1952), p. 61. Jacob Joffe, an American scientist of Russian–Jewish origin, was assigned to look after the Soviet delegation as he spoke Russian. On Joffe, see F. E. Bear, "Jacob Samuel Joffe (1886–1963)," *SS* 97 (1964), 1–3.

[88] See David-Fox, *Showcasing*; Choi Chatterjee and Beth Holmgren, "Introduction," in *Americans Experience Russia: Encountering the Enigma, 1917 to the Present*, eds. Chatterjee and Holmgren (New York: Palgrave, 2012), pp. 1–11.

experiment stations that were applying soil science to agriculture. The itinerary was carefully prepared, however, to acquaint their foreign guests with the "magnificent program of socialist construction, which is carried out in the land of the Soviets, particularly the socialist reconstruction of agriculture."[89] The organizing committee was concerned also to counter "foreign propaganda" about the impossibility of holding such an event in the Soviet Union.[90] The preparations for the congress and excursion were overseen by the Soviet government, which aimed to ensure their country was shown in the best possible light.[91] They had some success. As well as Marbut's positive assessment of the prospects for Soviet grain production and exports noted earlier, University of California scientist Shaw raised doubts concerning the accuracy of information about the Soviet Union reported in the American press: he thought that about two-thirds was correct, but one-third incorrect.[92]

The awkward international situation and suspicions it engendered continued after U.S. formal recognition of the Soviet Union in 1933 and impeded, but did not prevent, contacts and exchanges between American and Soviet agricultural scientists.

Resistance to Change

A general resistance to change was a barrier to some of the transfers and influences from the steppes in the United States as they would entail alterations to machinery, changes in the ways things were done, and upset existing hierarchies and authorities. When varieties of wheat from the steppes were introduced in the 1870s and afterwards, some plains farmers were reluctant to adopt them in case they failed. Farming in the semi-arid and drought-prone Great Plains was a risky business, especially as many farmers – except the Mennonites who introduced the new crops from the steppes – had little previous experience of such conditions. Moreover, the kernels of the wheat from the steppes were harder than the sorts widely grown at the time. Millers resisted the new wheats from the steppes, because they faced the expense of installing new equipment to grind the harder grains.[93]

The innovation from the steppes that encountered the most and longest resistance was the new Russian soil science. Russian scientists had come up

[89] ARAN, f.487, op.1, d.15, l.79. [90] ARAN, f.487, op.1, d.14, l.117-ob.
[91] ARAN, f.487, op.1, d.12, ll.58, 66. [92] ARAN, f.487, op.1, d.27, l.138 ob.
[93] See James C. Malin, *Winter Wheat in the Golden Belt of Kansas: A Study in Adaption to Subhumid Geographical Environment* (Lawrence: University of Kansas Press, 1944).

with the new way of conceiving, analyzing, and classifying soils during field work in the steppe region in the 1870s–80s. The Russian scientific innovation took study of soils away from the geological and physical conceptions that prevailed among most American scientists at that time. These ideas had underpinned the U.S. Soil Survey that began under Milton Whitney in 1899. Significantly, Whitney and other American scientists who persisted in adhering to these older understandings of soils had gained most of their experience in studying the well-worked soils of the humid and forested eastern United States. They had little experience of the Great Plains, where the environmental conditions and soils differed sharply from those they were accustomed to, but resembled the steppes. Changing the scientific basis of the U.S. Soil Survey would render the existing surveys obsolete and entail starting the survey again. The Russian soil science thus threatened Whitney's authority and the years of work invested in his survey.[94]

When the U.S. government decided to plant shelterbelts of trees in the Great Plains in 1934, on the basis of some Russian and Soviet studies promoted by Zon and others, there was resistance from several American foresters who challenged viability of the project. They enlisted the support of one of the leading Soviet forestry scientists, Georgii Vysotskii, who had long doubts about the viability and benefits of planting shelterbelts in the steppes.[95] The Great Plains Shelterbelt Project provoked opposition among local Republican politicians, newspaper editors, and inhabitants, including some farmers, who resented meddling and interference by east-coast elites, led by Democratic President Franklin Roosevelt, who they felt had no understanding of plains farming.[96]

Conclusion

The issues considered in this chapter posed significant barriers to influences from the steppes taking root in Great Plains agriculture. The barriers were unlikely to be overcome, moreover, until there was widespread appreciation in the United States of the similarities between the two regions, the common challenges facing farmers in both, and that

[94] For an overview, see Roy W. Simonson, *Historical Highlights of Soil Survey and Soil Classification with Emphasis on the United States, 1899–1970* (Wageningen: International Soil Reference and Information Centre, 1989).

[95] H. H. Chapman, "Editorial: The Shelterbelt Tree Planting Project," *JoF* 32 (1934), 801–3; G. N. Vyssotsky, "Shelterbelts in the Steppes of Russia," [translated from Russian] *JoF* 33 (1935), 781–8.

[96] See Craig Miner, *Next Year Country: Dust to Dust in Western Kansas, 1890–1940* (Lawrence: University Press of Kansas, 2006), pp. 262, 281.

Americans could learn from experience and expertise in the Russian Empire and Soviet Union. In contrast to the accounts of their visits to the other country's grassland by Sherman and Tsimmerman (see pp. 54–6), not all Americans who traveled to the steppes noted the resemblance between the two regions. It is not surprising that American relief workers who witnessed human suffering during famines in the steppe region in the early 1890s and early 1920s did not comment on parallels with the American plains, nor that they did not recommend learning from steppe agriculture. Travelers with little experience of the grassland region in their own country were also unlikely to note similarities or recommend learning from their experience. Eugene Schuyler, an American diplomat who helped Sherman when he visited Russia,[97] had traveled across the steppes himself in 1867. He described the wildlife, the flowers, the trees in river valleys, the fertile soil, as well as the gullies. The steppes, for Schuyler, were exotic: "I then thoroughly understood that Asiatic scenery, such as is to be seen from the Caspian to Pekin, really begins with the east shore of the Volga." He saw more "Asiatic" scenery on a later trip across Central Asia. Schuyler had a romanticized view of Russian culture and believed that Russia had a mission to "civilize" Central Asia. He was a native of Ithaca in upstate New York and had little experience of the prairies and Great Plains of his home country.[98] Nowhere does it seem to have occurred to him that there was a landscape similar to the steppes in the United States. For him, the steppes did not remind him of Kansas, but were the gateway to the exotic east.

People whose visits were motivated by political considerations had matters beside landscapes and agriculture uppermost in their minds. Some American Communists who visited the Soviet Union in the 1930s seem to have been blinded to any similarities between the steppes and the American grasslands by ideological fervor and a desire to find a new social order. A striking example was Anna Louise Strong, who spent time on collective farms in the steppe region. Even though she was from Nebraska, her account of Soviet wheat farming contains no comparisons with her home

[97] Saul, *Concord and Conflict*, p. 74.

[98] Eugene Schuyler, "On the Steppe," *Hours at Home* 9,4 (Aug., 1869), 319–29; Schuyler, *Turkestan: Notes of a Journey in Russian Turkistan, Kokand, Bukarha, and Kuldja*, 2 vols. (New York: Scribner, Armstrong & Co., 1876); Patricia Herlihy, "Ab Oriente ad Ulteriorem Orientem: Eugene Schuyler, Russia, and Central Asia," in *Space, Place, and Power in Modern Russia*, ed. Mark Bassin, Christopher Ely, and Melissa K. Stockdale (DeKalb: N. Illinois University Press, 2010), pp. 119–41.

state.[99] Vern Ralph Smith, the Moscow correspondent of the American Communist Party newspaper, *The Daily Worker*, visited a collective farm village in the steppe region. Despite his earlier experience of working in the wheat fields of Kansas, he noted no parallels in his account of the village in the steppes, besides a passing reference to "the great sweeps of the prairie." Instead, he presented an officially approved account of the village's history that focused on the achievements of Soviet agriculture and the "success" of collectivization. Most comparisons he made with the United States were negative. For example, he lavished praise on the scientific achievements of Trofim Lysenko (who opposed Vavilov's genetics) in contrast with "capitalist agronomy."[100]

Zara Witkin, an engineer from California, traveled to the Soviet Union in 1932 "fired by the belief that a noble attempt to refashion human society was taking place there."[101] On his way across the United States to catch his ship to Europe he journeyed: "Over the stupendous Rocky Mountains, across the great western desert, ... Then vast rolling prairies ..." Once he arrived in the Soviet Union, he traveled to the steppes. His account of his trip is full, not of admiration for the noble experiment, however, but of complaints about the incompetence of Intourist (the Soviet travel agency for foreigners), the lack of beds and linen in third-class train cars, half-built or dilapidated new hotels. At the Khar'kov tractor plant he encountered an "air of confusion" and finished tractors left outside to rust. At the "Verbliud" state farm he found deserted fields, tractors with broken headlights, incompetent workers, a foul toilet, and bad food that upset his stomach. He seems to have been too distracted by the "wanton neglect," and by sending furious telegrams to Intourist, to notice whether the terrain resembled the "vast rolling prairies" back home. He was impressed by the dam and hydroelectric power station on the Dnepr River, one of the prestige projects of the First Five-Year Plan, but noted that "much of [it] was of American make."[102]

[99] Anna Louise Strong, *The Soviets Conquer Wheat: The Drama of Collective Farming* (New York: Holt, 1931). See the Encyclopedia of Marxism biographical note, available online at www.marxists.org/glossary/people/s/t.htm#strong-anna-louise, accessed July 10, 2013.

[100] Vern Ralph Smith, *In a Collective Farm Village* (Moscow: Co-operative Society of Foreign Workers in the USSR, 1936), pp. 7–28, 53–67, 97; "Vern Smith is New 'Daily' Correspondent in USSR," *The Daily Worker*, August 24, 1933.

[101] Michael Gelb, ed., "Editor's Introduction," in Zara Witkin, *An American Engineer in Stalin's Russia: The Memoirs of Zara Witkin, 1932-1934* (Berkeley: University of California Press, 1991), p. 1.

[102] Witkin, *An American Engineer*, pp. 36, 53–63, 68.

Witkin's disenchantment on his encounter with the Soviet Union and the sense of American superiority his experiences provoked in him echoed the views of an earlier American visitor to the steppes. In 1907, while secretary of war in President Theodore Roosevelt's administration, William Howard Taft traveled west from Vladivostok along the Trans-Siberian railroad. He observed: "The country is like the Dakotas or Nebraska and will support a population of millions. The opportunities for development, therefore, of Russia toward the Pacific on the one hand are quite like the actual development in the United States towards the Pacific on the other." Russia needed, he believed, "industrious people" to emulate the American experience. He hoped that the development of Siberia would make it "one of the most prosperous and healthily populated parts of the globe" and "bring Russian and American civilization closer and closer together."[103] Thus, Taft's response to recognizing similarities between the steppes and the Great Plains was that the Russians should learn from American experience.

Before many Americans came to appreciate not just the parallels between their grassland regions, but that they could learn from the Russians' longer experience of settling, cultivating, and studying their grasslands, the barriers considered in this chapter needed to be overcome. Some of these barriers contained elements that allowed them to be bridged. The competition in the world grain market that played out at the world's fairs also enabled American and Russian specialists attending the fairs to see each other's exhibits and learn about their agriculture. Competition spurred Russian specialists to study grain production in the prairies and Great Plains and, later, American specialists paid serious attention to steppe agriculture. Competition thus prompted greater knowledge of their chief rival. One of the issues that encouraged negative American impressions of Russia also assisted in bridging the language barrier and American ignorance of Russian agricultural sciences. The large numbers of Jewish immigrants in the United States who had fled oppression in the Russian Empire in the late nineteenth and early twentieth centuries provided a pool of educated people, some with scientific training, who knew Russian.

In aftermath of the Russian Revolution of 1917, the new Soviet state and the capitalist United States were deeply suspicious of each other as their political and economic systems were diametrically opposed, and there were restrictions on travel and contacts between the two countries. But,

[103] Ralph Eldin Minger, "William Howard Taft's Forgotten Visit to Russia," *RR* 22 (1963), 153–4.

the restrictions were not absolute. Some Soviet scientists, including Vavilov and the delegates to the Soil Science Congress in 1927, did visit the United States. Some of their American counterparts, for example, Marbut and other Americans who attended the next Soil Science Congress in 1930, traveled to the Soviet Union. The trips that took place after 1917 built on contacts that had been established before the revolution. Many of the Russians who visited the Great Plains and Americans whose itineraries took in the steppes saw parallels between them and opportunities to exchange experience. Even the resistance to changes could eventually be overcome, if the advantages of transfers from the steppes were too compelling to ignore. It was not only steppe agriculture that experienced serious problems as a result of recurring droughts. The Great Plains experienced the same phenomenon, most urgently during the Dust Bowl in the 1930s, which prompted searches for remedies from regions with similar experiences, such as the steppes. The ways the barriers to transfers and influences from the steppes to the Great Plains were bridged is the subject of the next chapter.

CHAPTER 3

Bridges

Introduction

Gifford Pinchot, Chief of the U.S. Bureau of Forestry, visited the Russian Empire in 1902. On his way south from Moscow, he noticed the landscape change from forest to steppe: "The next day ... found us in a wonderfully fertile agricultural country, spotted with woods but inhabited by paupers, and the next [day], in a region of rich black soil like our own prairies." His hosts took him to forestry plantations in the steppes, where he learned that the first plantation in the region had been established by a German colonist in the 1820s, and that the government began planting trees in the steppes in the 1840s. Pinchot observed that the Russians' techniques, including planting trees and shrubs in particular combinations in "shelter belts," could be appropriate in similar conditions in the "Western United States."[1] Pinchot's active interest in Russian steppe forestry can be contrasted with General Sherman, who on his visit to the steppes in 1872 had just noted similarities with the Great Plains without drawing wider conclusions (see p. 55). Pinchot falls into David Wrobel's category of travel writers who "managed to travel with their eyes and minds wide open, ready to experience *and* understand the peoples and landscapes they traveled in."[2]

People with similar attitudes to Pinchot helped build bridges that overcame the barriers to transfers and influences from the steppes to the Great Plains discussed in the previous chapter. Important roles were played by people who traveled from the Russian Empire, and later the Soviet

[1] Gifford Pinchot, *Breaking New Ground* (Washington, DC: Island Press, 1998) [1st edition 1947], pp. xi–xiii, 214–16 (quote p. 216); LoC, Manuscript Division, Gifford Pinchot Papers (hereafter Pinchot Papers), Box 715, File: "Notes taken by Mr Pinchot on his trip through Russia and Siberia. Prepared by Mr Pinchot 1902," ff.42–65.

[2] David M. Wrobel, *Global West, American Frontier: Travel, Empire, and Exceptionalism, from Manifest Destiny to the Great Depression* (Albuquerque: University of New Mexico Press, 2013), p. 13.

Union, to the prairies and Great Plains of North America, and who went the other way from the United States to the steppes. Some were temporary visitors. From the 1870s, growing numbers of Russian agricultural scientists crossed the Atlantic for field work, to study American agriculture, meet their American counterparts, and participate in scientific meetings.[3] Following Nikolai Vavilov's visit to the United States in 1921–2, other Soviet agricultural scientists followed suit in the interwar years. American agricultural specialists began to visit the steppes around the turn of the twentieth century. Pinchot had been preceded a few years earlier by agricultural explorers Niels Hansen and Mark Carleton. The United States Department of Agriculture (USDA) and state governments in the plains sent explorers to the Eurasian steppes to collect seeds and plants that could be useful in similar environments back home.[4] In addition, American scientists visited the steppes during international meetings, and American specialists were hired as advisors by the Soviet government. In contrast to the American famine relief workers and Communists discussed in the previous chapter, the Americans who traveled to the steppes considered in this chapter took a keen interest in parallels with the Great Plains and possible lessons from Russian experience.

Migrants from the Russian Empire who settled permanently in the United States also helped build bridges. Beginning in the 1870s, groups of farmers left the steppes to relocate in similar environments in the Great Plains. Many were of Germanic origins and included the Mennonites who play a large role in our story. Some of the Russian Jews who fled the Russian Empire for the United States in this period also contributed to building bridges. Since they knew Russian, scientists among them

[3] See G. P. Kuropiatnik, "Russians in the United States: Social, Cultural, and Scientific Contacts in the 1870s," in *Russian–American Dialogue on Cultural Relations 1776–1914*, ed. Norman E. Saul and Richard D. McKinzie (Columbia: University of Missouri Press, 1997), p. 128.

[4] On American "plant exploration" ("bioprospecting") and plant explorers, see David Fairchild, "Our Plant Immigrants: An Account of Some of the Results of the Work of the Office of Seed and Plant Introduction of the Department of Agriculture," *National Geographic Magazine* 17, 4 (1906), 179–201; Courtney Fullilove, *The Profit of the Earth: The Global Seeds of American Agriculture* (Chicago: University of Chicago Press, 2017), pp. 23–66; Howard L. Hyland, "History of U.S. Plant Exploration," *Environmental Review* 4 (1977), 26–33; Knowles A. Ryerson, "History and Significance of the Foreign Plant Introductions Work of the U.S. Dept of Agriculture," *AH* 7 (1933), 110–28; Ryerson, "Plant Introductions," *AH* 50 (1976), 248–57; Allan Stoner and Kim Hummer, "19th and 20th Century Plant Hunters," *HortScience* 42, 2 (2007), 197–99. For a recent biography of David Fairchild, the first head of the USDA's Office of Seed and Plant Introduction, see Daniel Stone, *Food Explorer: The True Adventures of the Globe-Trotting Botanist Who Transformed What America Eats* (New York: Dutton, 2018).

facilitated the transfer of Russian scientific knowledge about the steppes to American scientists. There was a further wave of migration from Russia to the United States by opponents of the Bolsheviks after the revolution of 1917 and Russian Civil War of 1918–21. These migrants, who included more ethnic Russians than previous waves, augmented the numbers of people in the United States with knowledge of Russian and expertise on the steppes.

Growing awareness in both countries of the similarities between the steppes and Great Plains combined with competition between them in the world grain market led first Russians, and later Americans, to study grain farming in the other country's grassland. The Russian and Soviet governments sent specialists to the United States for this purpose. The Americans learned about steppe farming from U.S. Consul Thomas Heenan in the Black Sea port of Odessa and migrants from the Russian Empire. Studying the other country's agriculture spurred further interest. All these factors led to more contacts between the two countries' scientists and government departments of agriculture. In time, recognizing mutual interests and potential benefits, they shared expertise and assisted each other.

In contrast to the previous chapter, that considered barriers to transfers from the steppes to the Great Plains, this chapter explains how bridges were built that allowed those transfers to take place. It will cover efforts by both Americans and Russians to facilitate the transfers of plants and seeds, scientific expertise, and techniques that will be discussed in the chapters that follow. This chapter is structured thematically and so some people, such as American scientists who took an interest in the steppes, and groups, in particular the Germanic and Jewish migrants, will recur several times. In this and in later chapters, the experiences in the steppes will be considered first, since the steppes were settled by farmers and plowed up before the Great Plains.

Recognition of Similarities between the Steppes and Great Plains: Travelers' Accounts

Russian travelers to the prairies and Great Plains and American visitors to the steppes became aware of similarities between the environments and difficulties in cultivating crops in both regions. Both had fertile soils, but semi-arid and windy climates. Encountering familiar environments in foreign lands has not received much attention from specialists on travel writing.[5] Mutual perceptions by Russians and Americans of each other's countries have tended, moreover, to emphasize differences.[6] Nevertheless, evidence for mutual sympathy can be found among some Russians and Americans who visited each other's grasslands. Nikolai Borodin, a Russian engineer who traveled in the United States several times from the 1890s, pointed to "many common features in natural conditions and economic life of the two vast states." He contrasted both the American "provinces" and the Russian steppes with densely populated parts of Europe. Writing in 1915, he referred to a "psychological moment" he and other Russians living in America had experienced:

> As inhabitants of the vastness of Russia, used to the wide-open spaces of the steppes, to boundless distances, organically valuing the great freedom of newly-colonized places and finding satisfaction of the needs of the expansive Russian nature only in these conditions, in America (naturally not in its large cities, but in the provinces) we felt as if we were at home – in the Trans-Volga region, New Russia or Siberia.[7]

Some Americans felt at home in the steppes. Mark Twain visited Odessa – admittedly an urban environment – towards the end of his tour of Europe and the Holy Land in 1867. He commented, with a touch of irony:

> I have not felt so much at home for a long time as I did when . . . I stood in Odessa for the first time. It looked just like an American city; fine, broad streets, . . . low houses . . ., wide, neat, and free from any quaintness of

[5] In a rare example, cultural geographer James Duncan analyzed accounts by nineteenth-century British travelers of the Kandyan Highlands in Ceylon (Sri Lanka), which they thought "looked uncannily like Highland Britain," and they re-created in their writings as "a bit of Britain in the tropics." James Duncan, "Dis-orientation: On the Shock of the Familiar in a Far-Away Place," in *Writes of Passage: Reading Travel Writing*, ed. Duncan and Derek Gregory (London: Routledge, 1999), pp. 151–63.

[6] See, for example, Aleksandr Etkind, *Tolkovanie puteshestvii: Rossiia i Amerika v travelogakh intertekstakh* (Moscow: Novoe literaturnoe obozrenie, 2001); Sara Dickinson, "The Edge of Empire or the Center of the Self: Endpoints and Itineraries in Nineteenth-Century Russian Travel," *RR* 70 (2011), 94. See also Eugene Anschel (ed.), *The American Image of Russia, 1775–1917* (New York: Frederick Ungar, 1974); Anna M. Babey, *Americans in Russia, 1776–1917: A Study of the American Travelers in Russia from the American Revolution to the Russian Revolution* (New York: Comet, 1938); Choi Chatterjee and Beth Holmgren (eds.), *Americans Experience Russia: Encountering the Enigma, 1917 to the Present* (New York: Routledge, 2012); Olga Peters Hasty and Susanne Fusso (eds.), *America through Russian Eyes, 1874–1926* (New Haven, CT: Yale University Press, 1988); Margarita Marinova, *Transnational Russian–American Travel Writing* (New York: Routledge, 2011). Many scholars have analyzed Russian perceptions of America and American perceptions of Russia. None has focused on perceptions of their environments or the steppes and Great Plains in particular.

[7] Nikolai A. Borodin, *Severo-Amerikanskie Soedinennye Shtaty i Rossiia* (Petrograd, 1915), p. ix. See also A. I. Iziumov, "Nikolai Andreevich Borodin," *SShA: Ekonomika, politika, ideologiia*, no. 6 (1991), 48–53. Nikolai Borodin is not to be confused with Dmitrii Borodin, see p. 121.

> architectural ornamentation; locust trees bordering the sidewalks ...; a stirring, business-look about the streets ... and a smothering cloud of dust that was so like a message from our own dear native land that we could hardly refrain from shedding a few grateful tears ... Look up the street or down the street, ... we saw only America!

The illusion was broken only by an Orthodox church with a dome resembling "a turnip turned upside down."[8]

Travelers from Russia in the Great Plains

From the mid nineteenth century, travelers from the Russian Empire and, later, the Soviet Union visited the Great Plains and prairies. Many noted similarities with the steppes. In the summer of 1872, before many Mennonites decided to relocate to the Great Plains, a young man named Bernhard Warkentin took a vacation in the United States with three friends. He was the son of a prosperous miller in the Mennonite colony of Molotschna in today's southern Ukraine. Warkentin wrote regularly to his friend David Goerz back in Molotschna. He described their arrival, after a two-week voyage during which he was sea sick, in Hoboken, New Jersey: "But suddenly the fog was gone and at 7:00 am we saw a beautiful panorama, namely the continent of America." The four friends celebrated their arrival in the first beer hall they found.[9] From New York they traveled via Niagara Falls to the states of Ohio, Indiana, and Illinois. The Mississippi River reminded Warkentin of the Dnepr back home. They visited American Mennonite communities and stayed with Christian Krehbiel in Summerfield, Illinois. (Krehbiel had moved to the United States from Germany in 1851.) While still on vacation, news reached Warkentin that the Russian government was considering ending the Mennonites' exemption from military service. This prompted some to consider leaving Russia. From this point, his letters changed from those of a tourist to those of a scout looking for suitable places to settle. Krehbiel assured him that "only America was the land for the Mennonites." Warkentin wrote that in Illinois they had found farming conditions similar

[8] Mark Twain, *The Innocents Abroad, or The New Pilgrims' Progress* (Ware, UK: Wordsworth, 2010) [1st published 1869], pp. 248–9. See also Norman E. Sol, "'Prostaki za granitsei,' ili kak Amerikantsy i Russkie otkryvali drug druga, 1867–1881," *Amerikanskii ezhegodnik 1993*: 80–95.

[9] MLA, MS 68 Bernhard Warkentin (1847–1908) and Wilhelmina Eisenmayer Warkentin (hereafter MLA, Warkentin Papers), Box 1, Folder 7, Correspondence: Warkentin to [David] Goerz, 1861–1872 (translations), Warkentin to Goerz [June 1872].

to their homes, but they had been told that further west they would find the same conditions as the steppes.[10]

From Summerfield, Illinois, Warkentin and his friends traveled north to Chicago, on to Wisconsin, and then west into Minnesota. From New Ulm in southwestern Minnesota they "made a five-day wagon tour through partially to still completely uninhabited prairie" accompanied by an American land agent. Warkentin commented on the "wonderful" landscape, the "black earth," "magnificent, growing grass," the rivers, lakes, and fish. They then traveled across Iowa to Omaha, Nebraska, and west on the Union Pacific Railroad. Warkentin remarked on the flat landscape, the grain and corn in the fields, the "enormous herds of cattle," and encampments of Pawnee Indians. They went on to Denver, Colorado, which they did not like, before turning back east into Kansas. There they saw "the whole prairie strewn with buffalo skeletons," but only one herd of living "buffalo" in the distance. When they reached Brookville, Saline County, in central Kansas, Warkentin noted that the land had become "better and more inhabited."[11] He returned to Illinois, before resuming his travels northwest into the Dakota Territory. Journeying west from the Red River to the James River, he described the "flat prairie," the wood in the river valleys, black soil two or three feet deep, and grass which was "usually good."[12] He ventured north into the Canadian prairies, where he enquired about the winter weather. After he had got back to Summerfield, Warkentin wrote to Goerz that "winters here as well as up in the North vary, as they do at home in the southern part of Russia."[13] Warkentin made few such explicit comparisons with the steppes. He may not have felt it necessary, since the parallels with the steppes were so striking that they would have been as readily apparent to him and Goerz as they were to the author of this book.

At around the same time, in 1873, the Russian scientist Aleksandr Voeikov visited North America. Drawing on his expertise in climatology,

[10] MLA, Warkentin Papers, Box 1, Folder 7, Warkentin, Summerfield, Illinois, to Goerz, July 29/11 1872. (The double dates are for the Gregorian calendar used in the United States and the Julian used in Russia.) On Krehbiel, see Olin A. Krehbiel (1957) "Krehbiel, Christian (1832–1909)," Global Anabaptist Mennonite Encyclopedia Online, available online at https://gameo.org/index.php?title=Krehbiel,_Christian_(1832-1909), accessed October 8, 2019.

[11] MLA, Warkentin Papers, Box 1, Folder 7, Warkentin, Brookville, Saline Co., KS, to Goerz, August 18/6, 1872.

[12] MLA, Warkentin Papers, Box 1, Folder 7, Warkentin, Glyndon, MN, to Goerz, September 29, 1872.

[13] Cornelius Krahn, "Some Letters of Bernhard Warkentin Pertaining to the Migration of 1873–1875," *Mennonite Quarterly Review* 24, 3 (1950), 251–2.

he made detailed observations of the environment of the Great Plains and Canadian prairies and their suitability for agriculture. In the process, he made direct comparisons with the steppes. Voeikov traveled west by train across the prairies and Great Plains to the Rocky Mountains. Between the Mississippi and the 98th or 100th meridians he observed "stretches of black-earth steppe, very similar to our steppe in the south and east of Russia." Another resemblance was the limited extent of woodland outside river valleys. The part he thought most like "our" steppes was between the latitudes of 36 and 52 degrees north in Illinois, southeast Wisconsin, Iowa, northern Missouri, eastern Kansas and Nebraska, southern and western Minnesota, and the east of the Dakota Territory. He noted more "black-earth steppes" in the Canadian prairies. Voeikov compared the climate and soil of this part of Canada to southwestern Siberia, and Winnipeg to Omsk. He thought the plains south of 36 degrees north, in southern Missouri, Arkansas, Indian Territory (Oklahoma), and northern Texas, where the climate was hotter, differed from the steppes. West of the 100th meridian, moreover, he observed that it was too dry for arable farming without artificial irrigation. In the northern and central plains – the part Voeikov thought most resembled the steppes – he observed much larger areas of unplowed black earth than in Russia, where the plow-up of the steppes had begun much earlier.[14]

Especially eager to visit the prairies and Great Plains was the Russian geo-botanist Andrei Krasnov. Although born and brought up in St. Petersburg, his family were Don Cossacks, who had lived in the steppes for centuries. Krasnov studied natural sciences at St. Petersburg University, where his professors included the soil scientist Vasilii Dokuchaev. Krasnov had participated in expeditions around the Russian Empire and had studied the "Kirgiz" (Kazakh) and Kalmyk steppes. In 1889, he took up a post in the geography department at Khar'kov University in the steppe region.[15] The previous year, he had written to Vladimir Vernadskii: "It is more useful for geography and for Russia to use opportunities to travel to visit America [than Europe]."[16] In 1890, he beseeched Dokuchaev for help

[14] A. I. Voeikov, "Russkii puteshestvennik v Amerike," *Izvestiia Russkogo geograficheskogo obshchestva* 10, no. 1 (1874), 2nd pagn., 59–61.

[15] See P. K., "Andrei Nikolaevich Krasnov (Materialy dlia biografii)," in *Professor Andrei Nikolaevich Krasnov (1862–1914 g.)*, ed. V. I. Taliev (Khar'kov: Tip. M. Kh. Sergeeva, 1916), pp. 5–23. His father, Nikolai, had compiled the *Materialy dlia geografii i statistiki Rossii, sobrannye ofitserami general'nogo shtaba: Zemlia voiska donskogo* (Spb Tip. Departamenta General'nogo Shtaba, 1863).

[16] [A. N. Krasnov], "Iz perepiski A. N. Krasnova. Pis'ma A. N. Krasnova k V. I. Vernadskomu 1888 goda," in Taliev, *Professor Andrei Nikolaevich Krasnov*, pp. 122, 124.

in obtaining permission and funding to attend the International Geological Congress in Washington, DC. He was excited about the post-Congress excursion to America's "prairies, steppes and deserts." Krasnov proposed to give a paper on the "successes" of Russian soil science and studies of "our" steppes (i.e. Dokuchaev's own work). With Dokuchaev's help he received permission and funding.[17] Krasnov was so keen to see the American grasslands that, while sailing across the Atlantic, and in a reversal of the more common depiction of the steppes and plains as an ocean of grass, he described the ocean waves breaking with white crests as "an immense, hilly, black, steppe . . . sprinkled with snow."[18]

At the end of the Congress, Krasnov wrote to Dokuchaev: "Now I am going on the excursion, on which . . . I will see all the characteristic black earth regions from Illinois and Minnesota to Indiana, Kansas and Texas."[19] He decided to leave the excursion, however, as it was too fast for him.[20] Proceeding more slowly, he noted that in America the quantity of rain declined from east to west and, as in "our country" when traveling from the northwest to the southeast, travelers saw the gradual disappearance of forest and its replacement by prairie, prairie by steppe, and then steppe by desert. He felt that the "steppe" region of America was too varied to compare it directly to "our" black earth region. However, Kansas and southern Illinois "recalled our steppes . . . described by [the writer Nikolai] Gogol'" (see Introduction, p. 12). As in "our" steppe region there was woodland only on the steep banks of rivers, beyond which were "level or lightly rolling expanses" where it was possible to "see endless, grassy steppe, with black earth, completely treeless." The part that "most resembles our southern Russian steppe," according to Krasnov, was:

> the region . . . the Americans call the plains, and which start to the [west] of the Missouri [River]. These expanses are strikingly similar to . . . those in southeast Khar'kov province, Ekaterinoslav province, . . . the Don Cossack region and . . . Astrakhan' province. Eastern Kansas and Dakota are the American equivalent of our south, to the extent that . . . it is possible to imagine oneself somewhere on the banks of the Krynka or Elanchik [rivers in the Don Cossack region].

[17] V. V. Dokuchaev, *Sochineniia*, 9 vols. (Moscow–Leningrad: AN SSSR, 1949–61), vol. 8, pp. 498–9.

[18] A. N. Krasnov, "U amerikantsev: Cherez okean – N'iu Iork – Amerikanskie zheleznye dorogi – Vashington – Amerikantsy ne prigotovilis'," *Knizhki nedeli* (November 1891), 161.

[19] Dokuchaev, *Sochineniia*, vol. 8, p. 499. For Krasnov's paper, see A. N. Krassnof, "The 'Black Earth' of the Steppes of Southern Russia," *Bulletin of the Geological Society of America* 3 (1892), 68–81.

[20] Krasnov, "U amerikantsev: Cherez okean," 204.

The resemblance was reinforced when Krasnov met "Russian German" settlers. He also noted differences, but these were human artifacts rather than natural features: individual farmsteads, rather than villages, and the houses, wells, and roads all looked different. After all his comparisons, however, he admitted that the United States was not really a parallel for Russia, as conditions were more favorable, and that Canada was more like Russia.[21] Krasnov later wrote his doctoral dissertation on the "grassy steppes of the Northern hemisphere," including both the Russian and American grasslands.[22]

One of the leading Russian specialists on American agriculture, in particular in the western half of the United States, was Nikolai Tulaikov (1875–1938). A native of Volga region in the steppes, he trained as an agronomist and soil scientist at the Petrovskaia Agricultural Academy in Moscow, and for many years directed agricultural experiment stations near Samara and Saratov on the Volga.[23] Tulaikov first visited the United States in 1908–9. He returned in 1922–3 to write a study of American farming. He drew a direct parallel between the environmental conditions – especially the soil and climate – of the Volga and Don regions in the steppes and the Dakotas and Minnesota in the northern plains. Noting that American farmers were more productive than their Russian counterparts in similar conditions, he attributed this to greater mechanization.[24] Tulaikov's final visit was in 1927 as part of the Soviet delegation to the 1st International Congress of Soil Science in Washington, DC.[25]

The Soviet delegation traveled by train from Moscow to Bremen, from where they sailed across the Atlantic on the liner "George Washington." The soil scientists spent the days on deck preparing for the Congress. In the evenings, to the amusement of their fellow passengers, they sang Russian and Ukrainian folksongs. The Soviet soil scientists, in particular

[21] A. N. Krasnov, "U amerikantsev: 25 dnei v zheleznodorozhnom vagone – Cherez stepi i prerii v stranu chudes – V gostiakh u mormonov – Paiks-pik – Niagra i velikie ozera – Zakliuchenie," *Knizhki nedeli* (December 1891), 63–8, 108, 114–15.

[22] A. N. Krasnov, "Travianye stepi Severnogo polushairiia," *Izvestiia Obshchestva liubitelei estestvoznaniia, antropologii i etnografii* 81, no. 7 (1894), 1–294.

[23] See K. P. Tulaikova, *Ot pakharia do akademika: Ob akademike Nikolae Maksimoviche Tulaikove* (Moscow: Prosveshchenie, 1964).

[24] N. M. Tulaikov, *Sovremennoe polozhenie sel'skogo khoziaistva v Soedinennykh Shtatakh Severnoi Ameriki* (Moscow: Novaia Derevnia, 1923), pp. 16–22. For comparisons he made after his previous trip, see Tulaikov, *Ocherki po sel'skomu khoziaistvu v Soedinennykh Shtatakh* (Moscow: A. A. Levenson, 1912), pp. 93–4.

[25] N. M. Tulaikov, "Neskol'ko vpechatlenii po poezdke v Soedinennye Shtaty i Kanadu," *Nizhnee Povol'zhe*, no. 9 (1927).

their leader Konstantin Glinka, played a prominent role at the Congress. After the formal proceedings, the American scientists, led by the chief of the U.S. Soil Survey, Curtis Marbut, took their guests on a twenty-five-day excursion all over North America.[26] The Soviet scientists thanked Marbut: "This extensive excursion throughout the United States and Canada so well planned and so splendidly executed by yourself and your colleagues gave to the Russian soil scientists a fine opportunity to familiarize themselves with the general character of the soils of North America."[27]

Several of the Soviet scientists noted similarities between the Great Plains and the steppes. Arsenii Iarilov paid particular attention to Kansas. On their way to the agricultural experiment station in Hays, he noted that they came across two villages resembling Russian villages. They turned out to be inhabited by the descendants of Volga Germans who had moved there in the 1870s (See p. 104). Iarilov further noted that the land surrounding their villages was "familiar expanses of steppe." In western Kansas, the short-grass "steppe" reminded him of Siberia.[28] Dmitrii Vilenskii also described similarities:

> In general both the vegetation and the overall landscape of this region [the Great Plains], especially in the extreme west of the state of Kansas, are very reminiscent of our Astrakhan' steppes, and only the presence of prickly pears ... cacti ..., and yucca ..., indicate that you are in another hemisphere.

In Alberta, Canada, and in North Dakota, Vilenskii saw black earth very similar to "our chernozem."[29] Iakov Afanas'ev found soils resembling Russian chernozem in Canada and "southern" black earth in Kansas.[30] The soil scientists' excursion attracted the attention of the press. The *Kansas City Journal* reported: "200 soil experts ... visited Jackson County [Missouri] this week... In the group are ... Russians, from the wheat plains of South Russia."[31]

[26] A. A. Iarilov, "Na kongresse i o kongresse," *Biulleten' pochvoveda*, nos. 5–8 (1927), 120–48. See also Curtis F. Marbut, "The Transcontinental Excursion, under the Auspices of the American Soil Survey Association," *1st International Congress of Soil Science, Proceedings*, 5 (1927), 40–88.

[27] SHSM MC, Marbut, Curtis Fletcher (1863–1935) Papers, 1852–1963 (C3720) [hereafter Marbut Papers (C3720)], Folder 32, The Russian delegation to Marbut, July 18, 1927.

[28] Iarilov, "Na kongresse," 134.

[29] D. Vilenskii, "Osnovnye cherty paspredeleniia pochv i rastitel'nosti SShA i iuzhnoi Kanady," *Pochvovedenie*, nos. 1–2 (1928), 118–19, 129.

[30] Ia. N. Afanas'ev, "O pochvennykh zonakh Severnoi Ameriki (Iz rezul'tatov poezdki na I mezhdunarodnyi kongres pochvoved.)," *Pochvovedenie* no. 3–4 (1928), 185–6.

[31] "Love of the Soil," *Kansas City Journal*, July 5, 1927, p. 4.

Americans in the Steppes

From the late nineteenth century, growing numbers of Americans traveled to the steppes, where many noted similarities with the Great Plains. The observations by the scientist Carleton went much deeper. The purpose of his visits to the Russian Empire in 1898–9 and 1900 was to acquire crop varieties suitable for the Great Plains. To this end, he researched the two regions' soils and climates. He wrote that the "Chernozem" or "black earth" of Russia's "grain region" was "remarkably similar" to the soils of the American prairies and Great Plains. Both contained large amounts of "thoroughly humified organic matter" (humus), "an unusual proportion of phosphoric acid" and "a great amount . . . of lime, potash, and other alkalies," substances that "are just those usually needed by the wheat plant." The structure of the soils in both regions was "very retentive of moisture," which was also conducive to plant growth. "The similarity between the Russian steppes and the Great Plains," he continued, "is fully as great in climate. . ., both regions being emphatically continental and subject to great extremes of temperature and moisture." He provided data showing similarities between the temperatures in the steppes and the central and northern plains, although his figures showed that on average the plains were slightly warmer. His data on "normal annual rainfall" showed more of a contrast, with average rainfall of 15–16 inches in most of the steppe region, while nowhere in the Great Plains, even "as far west as the hundredth meridian," received less than 18 inches on average each year.[32]

The more severe climate in parts of the steppes was noted by other American plant explorers. Niels Hansen, a Danish-born scientist from South Dakota, first visited the steppes in 1897, and returned periodically until the 1930s. He noted the similarities between the Great Plains and the steppes, but also the harsher climate in parts of the latter region.[33] Frank Meyer reached a similar conclusion. In 1913, after traveling around Akmolinsk region in the north of today's Kazakhstan (then considered part of western Siberia), he wrote to David Fairchild at the USDA in

[32] Mark Alfred Carleton, "Russian Cereals Adapted for Cultivation in the United States," *USDA Division of Botany Bulletin*, no. 23 (1900), 7–11; Carleton, "Macaroni Wheats," *USDA Bureau of Plant Industry Bulletin*, no. 3 (1901), 13–15, 18–19. See also Carleton, *The Small Grains* (New York: Macmillan, 1916), pp. 233–6, 243–4.

[33] See Niels Ebbesen Hansen, *The Banebryder (The Trail Breaker): The Travel Records of Niels Ebbesen Hansen 1897–1934*, ed. Helen Hansen Loen (n.p.: privately printed, 2002); Helen Hansen Loen, *With a Brush and Muslin Bag: The Life of Niels Ebbesen Hansen* (Kalamazoo, MI: privately printed, 2003).

Washington, DC, that he "deem[ed] it worth our while to have a correspondent in Western Siberia, where the climate is so uncongenial and so like the North Western Plains of our own country."[34] Meyer's sense of "uncongeniality" may have been conditioned by the milder climate of his native Amsterdam in The Netherlands, from where he had moved to the United States in 1901.[35] It was not only European-born Americans who found conditions in northern Kazakhstan severe and uncongenial. In the summer of 1913, I. N. Stakhovskii, a Russian district agronomist in Turgai region in the Kazakh steppe, requested a transfer to southern Siberia or European Russia on health grounds, as he found it "really difficult to work in the [local] conditions."[36] At the same time, the Turgai regional agronomist reported that settlers in Kustanai (Kostanai) district in the north of Turgai region were in particular need of specialist advice, due to the lightness of the soil, the aridity of the climate, and the prevalence of pests. Demonstration fields of wheat, millet, and corn had all failed that year due to soil exhaustion, after only a few years of cultivation.[37]

The American plant explorers turned the more severe conditions they recognized in parts of the steppes to their advantage in their quest for hardy plants for their northern plains. In 1912, in between trips to the steppes in 1908–9 and 1913, Hansen discussed with Secretary of Agriculture James Wilson their "theory" concerning "getting hardy plants from countries colder and drier than [western] South Dakota."[38] In December 1934, after his final visit to the Soviet Union, he recommended taking marginal land in South Dakota that had been damaged by dust storms out of cultivation and planting it with grasses, including crested wheat grass, that he had collected in the north of present-day Kazakhstan in 1897–8. He suggested this type of grass for his home state, because it grew in harsh conditions in its native environment: alkaline soil, eight inches of rainfall a year, and temperatures ranging from −51 °F in the winter to +106 °F in

34 NARA CP, RG 54, Finding Aid NC-135, Entry 135I, Records of Frank N. Meyer, Plant Explorer, 1902–1908, Box 9, Folder January 16–21 May 21, 1913, Meyer, Tomsk, to Fairchild, February 1, 1913.

35 Isabel Shipley Cunningham, *Frank N. Meyer: Plant Hunter in Asia* (Ames, IO: Iowa State University Press, 1984), pp. 10–15.

36 TsGA RK, f.30, op.1, d.35, l.148. His replacement, Mr. Maslovskii, arrived in November. Ibid. l.254.

37 TsGA RK, f.30, op.1, d.35, ll. I wrote this paragraph on conditions in northern Kazakhstan in the winter of 2018–19 as a visiting professor at Nazarbayev University in that precise region, but in the comfort of a modern, centrally heated apartment.

38 SDSU ASC, UA 53.4, N. E. Hansen papers [hereafter Hansen Papers], Series 2. Helen Hansen Loen Collection, Box 3, Folder 154, Secretary [of Agriculture, James Wilson] to Hansen, Brookings, South Dakota, January 2, 1913.

the summer. Hansen continued: "There are other plants from the driest regions of the Soviet Union which we should study, because their climatic conditions are even more severe than ours."[39]

U.S. Consul Heenan in Odessa on the Black Sea coast, where conditions were rather more congenial than northern Kazakhstan, also noted similarities between the steppes and the Great Plains. He was well equipped to do so, since he had farmed for many years in Minnesota. In March 1890, he described southern Russia as "an immense plain," where the soil was very rich and "goes to produce a bountiful and good quality of grain." He continued: "In its appearance it is much the same as our western prairies but it lacks the attractive qualities which so rapidly peopled the west."[40] Heenan's last remark conveyed a sense of American superiority that colored the views of many of his compatriots.

Heenan assisted Hansen and Carleton during their visits to the steppes. It fell to the consul to arrange for many bales of seeds that arrived at his consulate to be shipped to Washington, DC. Heenan's displeasure may be detected in a letter to the State Department:

> Mr. Hansen, an agent of the Agricultural Department, visited Central Asia last year and I presume it was at his instance that the bales were shipped to me. I am at all times glad to be of service to any of the Departments at Washington, and I take a keen interest in all that pertains to the Department of Agriculture, still I am of the opinion that it would be advisable, in order that I may be in a position to act in an intelligible manner, to previously inform me that I am expected to render such a service as may be necessary. As a matter of fact these bales which I have sent to the Department of Agriculture may not be intended for that Department at all.[41]

The tasks Hansen had imposed on Heenan may explain his exasperation when, in January 1899, another American presented himself with a business card reading: "Mark Alfred Carleton, Agricultural Explorer for the United States of America, Washington, DC." Heenan wrote to the

[39] Hansen Papers, Series 2. Helen Hansen Loen Collection, Box 4, Folder 46, MS: "Improving the Great Plains by Sheltbelt Planting." Notes of meeting with John H. Hatton, Regional Director, Shelterbelt Project, U.S. Forestry Service, December 1, 1934, 1, 7–8. See also Folder 84, MS: Russia as Observed by an Agricultural Explorer: 1934–37, 72–3.

[40] NARA CP, RG 59, Microcopy no. 459, Despatches from United States Consuls in Odessa, 1834–1906, Roll 5, Vols 9–10, January 2, 1889 – December 28, 1895, Consul Heenan to William F. Wharton, Assistant Secretary of State, Washington, DC, March 22, 1890. He mentioned his experience in Minnesota in Heenan to Wharton, January 17, 1891.

[41] NARA CP, RG 59, Microcopy no. 459, Despatches from United States Consuls in Odessa, 1834–1906, Roll 6, Vols. 11–12, January 4, 1896–December 29, 1902, Consul Heenan, Odessa, to Assistant Secretary of State William R. Day, March 10, 1898. In a similar vein, see also Heenan to Day, December 23, 1897; Heenan to Assistant Secretary of State J. B. Moore, May 20, 1898.

State Department that the "present 'explorer' [Carleton] was preceded about one year ago by another explorer [Hansen], and both travelled over the same ground." He had to run his Consulate on a limited budget and evidently resented the money allocated to the "Agricultural Dept."[42] The generous funds assigned to Hansen and Carleton, however, indicate the importance the USDA attached to collecting crops appropriate for the Great Plains in the similar, if slightly harsher, conditions of the steppes.

The soils and climates of the Great Plains and the steppes were of particular importance to the soil scientist Marbut. He was a native of the Ozark Mountains of southern Missouri, "in a transition zone between timber and prairie," and had traveled extensively around the United States and overseas.[43] In the early 1920s, after learning about Russian soil science and soils, Marbut surveyed the soils of the Great Plains, noting similarities between them and the Chernozem of the steppes.[44] His interest in Russian soil science and steppe soils culminated in his participation in 1930 in the 2nd International Congress of Soil Science in the Soviet Union and excursion around the country to see soils in the environments in which they had formed. Marbut arrived in Leningrad on July 19, 1930. Soviet cities were alien to him. He was depressed by their "dilapidated state," dirty sidewalks, absence of "show windows" in the stores, "long bread lines," but abundance of communist propaganda, as he tried to make sense of a foreign land.[45] From Berlin on his way home, he wrote to his daughter about how arduous the excursion had been:

> Our Russian scientific friends were worried and anxious. They did their best, but the rigid hard almost hopeless conditions of Russian life, to one not accustomed to it made anything that they could possibly do ineffective. . . . The extreme heat – dust – dirt – lies – lack of water, except mineral water that had to be got in bottles and never enough in quantity – the poor food – the slovenly way the food was served, the general confusion and lack of systematic handling of any of the ordinary daily affairs combined to make the trip a hard one.

[42] Georgetown University Library, Special Collections Division, Washington, DC, USA, Robert S. Chilton Jr. Papers, Box 1, Folder 22, Thomas E. Heenan, U.S. Consul in Odessa, Russia to Chilton, January 23, 1899.

[43] Elmer H. Johnson, "Prof. Dr. Curtis Fletcher Marbut," in *Life and Work of C. F. Marbut, Soil Scientist: A Memorial Volume*, ed. H. H. Krusekopf ([Columbia, MO]: Soil Science Society of America, [1942]), pp. 29–35.

[44] Curtis F. Marbut, "Soils of the Great Plains," *AAAG* 13 (1923), 41–66.

[45] Marbut Papers (C3720), Folder 143, "A Visit to Russia" by C. F. Marbut, Washington, DC. November, 1930.

Nevertheless, in the same letter he enthused about how much he and the other delegates had got out of the excursion.[46] With characteristic courtesy, he told his hosts that the scientific side of the excursion had been much better prepared than the trip (which he had led) after the previous congress in the United States.[47] Following his visit, he noted that the steppes "correspond to the Northern and Central Great Plains of the United States," and that both were their countries' most important wheat-growing regions. Based on his observations, however, he calculated that "Russia" had a considerably larger area of "first-grade land."[48]

Another American who saw similarities between the two grasslands was Thomas D. Campbell. Born into a farming family in North Dakota, he founded the Campbell Farming Corporation that cultivated wheat on an industrial scale in Montana. The Soviet government invited him to visit to learn from his experience of large-scale, mechanized farming in an environment similar to the steppes.[49] Campbell was initially reluctant to accept due to his "prejudice" towards Soviet policies against private property and anxiety not to be away from his own farms at harvest time. Nevertheless, he traveled to the Soviet Union twice. On the second visit, in July 1930, he was taken to the showcase state farms of "Gigant" and "Verbliud." Campbell was impressed by their size – "Gigant" covered half a million acres, over five times the size of his operation in Montana – and by the "enthusiasm of the directors and zeal of the workers." But, he was appalled by the roads. His car got stuck in mud and on one occasion had to be pulled out by "a group of powerful Russian farmers." He continued: "I must admit I don't feel comfortable in automobiles on country roads at such speed [over 60 miles an hour], when the driver thinks he has to talk and explain with his hands at the same time." His discomfort did not stop him noting parallels with his native plains: "We were shortly on the open prairie, and for many miles passed over excellent soil similar to our great prairies in Kansas and North Dakota." When they arrived at "Gigant": "I was astonished to see the same activity as ... on our own place in Montana – trucks going in all directions, tractors with powerful headlights

[46] Marbut Papers (C3720), Foldler 50, Marbut, Berlin, to Louise Marbut Moomaw, September, 1930.

[47] ARAN, f.487, op.1, d.27, l.159.

[48] C. F. Marbut, "Agriculture in the United States and Russia: A Comparative Study of Natural Conditions," *GR* 21, no. 4 (1931), 609, 612; Marbut, "Russia and the United States in the World's Wheat Market," *GR* 21, 1 (1931), 5–11.

[49] Deborah Fitzgerald, *Every Farm a Factory: The Industrial Ideal in American Agriculture* (New Haven: Yale University Press, 2002), pp. 129–56; Hiram Drache, "Thomas D. Campbell: The Plower of the Plains," *AH* 51 (1977), 78–91.

disking, seeding, plowing, and hauling, and tractor-wagon trains of grain moving to market." He continued to be impressed: "I have raised wheat throughout my life, I have driven through grain fields in all portions of the American continent, but never before had I seen such fields of grain as I saw on that night drive across the Giant ["Gigant"] Farm." Campbell was then taken to "Verbliud," which "compares favorably with any experimental farm that we have in the United States." In general, he noted that "Southern Russia . . . is a semi-arid region and to a great extent similar to our semi-arid region in the West."[50]

When writing about each other's countries, including their respective grasslands, Americans and Russians were "defin[ing] themselves and their place in the world."[51] The travelers most likely to make direct comparisons between their respective grasslands, and in some cases to argue that they could learn from the other's experiences, were natural scientists and agricultural specialists. Some, such as Krasnov and Carleton, had researched the other grassland before they visited and were eager to see it at first hand. Both traveled before the Russian Revolution added a further complexity to mutual perceptions of the two countries. Writing about the period after 1917, Lewis Feuer noted that among American visitors, the "best observers of Soviet reality" were engineers, as they were less ideological than writers, social scientists, and other "intellectuals."[52] This could be applied to agricultural scientists and specialists, whose knowledge could prevail over political concerns. Campbell told Stalin: "I am not a Communist . . . Nevertheless, I am interested in your agricultural development, as I am an agricultural mechanical engineer."[53]

Among the visitors who recognized the striking parallels between the steppes and the Great Plains, it was those from the Russian Empire and Soviet Union who were quicker to realize the potential to learn from the other's experience. This was despite the fact that the agricultural settlement of the steppes preceded that of the Great Plains and so the Russians had more experience. With some exceptions, such as Pinchot, Carleton, Hansen, and Marbut, Americans were slower to recognize that they could

[50] Thomas D. Campbell, *Russia: Market or Menace?* (London: Longman, Green, and Co., 1932), pp. 1, 93, 96–9, 101, 107, 109.

[51] William W. Stowe, "'Property in the Horizon': Landscape and American Travel Writing," in *The Cambridge Companion to American Travel Writing*, ed. Alfred Bendixen and Judith Hamera (New York: Cambridge University Press, 2009), p. 26.

[52] Lewis S. Feuer, "American Travelers to the Soviet Union 1917–32: The Formation of a Component of New Deal Ideology," *American Quarterly* 14, no. 2, part 1 (1962), 141–3.

[53] Campbell, *Russia: Market or Menace?*, pp. 15–17.

learn from experience the Russians had acquired in the steppes. A factor that contributed to more Americans recognizing the value of Russian experience in the steppe region was the growing number of immigrants from the Russian Empire and Soviet Russia.

Immigrants in the United States from the Russian Empire and Soviet Russia

Immigrants in the United States from the Russian Empire and Soviet Russia contributed to building bridges between the steppes and the Great Plains. They brought with them plants appropriate for semi-arid grasslands, experience of farming in such conditions, and facilitated the transmission of agricultural and forestry techniques and sciences. Over the late nineteenth and early twentieth centuries, ever larger numbers of migrants from the Russian Empire settled in the United States. Between 1820 and 1910, people from the Russian Empire comprised 9 percent of all immigrants in the United States.[54] In the five decades to 1870, fewer than 8,000 people made the journey from the Russian Empire, but thereafter the numbers increased sharply. Over 50,000 made their way to the United States in the 1870s. From 1881, the waves turned into floods: over a quarter of million made the trip in the 1880s; more than half a million in the 1890s; and in excess of two and half million between 1900 and 1917. Few were ethnic Russians: only 5 percent in the first decade of the twentieth century. By far the largest group, around 44 percent, were Jews, and nearly 6 percent were Germans from Russia. (Most of the rest were Poles, Lithuanians, and Finns.)[55] A new wave of migrants was sparked by the Bolshevik revolution in October 1917 and victory over their "White" opponents in the Civil War. Up to a million people, including many ethnic Russians, left Russia in this period. Most ended up in other European countries and China. Relatively few settled in the United States, because it adopted stringent immigration policies in 1920. A substantial émigré community of Russians who had left after 1917 did

[54] N. Turchanikov, "Emigratsiia v Sev. Amerikanskie Soedin. Shtaty i novyi Immigratsionnyi zakon Shtatov," *Voprosy kolonizatsii* 9 (1911), 39.

[55] E. L. Nitoburg, *Russkie v SShA: Istoriia i sud'by, 1870–1970: Etnoistoricheskii ocherk* (Moscow: Nauka, 2005), pp. 22–3. See also G. de Vollan, "Emigratsiia v Severo-Amerikanskie Soedinennye Shtaty: Donesenie 1-go sekretaria posol'stva v Vashingtone," *Sbornik konsul'skikh donesenii* 5, 1 (1902), 22–6; A. I. Shcherbatskii, "Russkaia emigratsiia v Soedinennye Shtaty," *Izvestiia Ministerstva inostrannykh del* 6, 1914 (Petrograd: 1915), 117–34. In both the Russian Empire and Soviet Union Jews were considered an ethnic group. Data on the ethnic composition of immigrants in the United States are available only from 1900.

not develop in the United States until the late 1930s, when many fled there to escape the onset of war in the Far East and the threat of war in Europe. Many more arrived in the aftermath of the Second World War. Nevertheless, some Russian émigrés, including some scientists, found their way to the United States in the interwar years.[56]

The migrants who are most important in this book were the Jewish and Germanic peoples. Farmers of Germanic origins, including Mennonites, moved from steppe region to the Great Plains of the United States and Canadian prairies. They relocated to areas with environments similar to their previous homes, partly because the Great Plains and prairies were the frontiers of settlement in the 1870 and 1880s as the U.S. government removed the Native American population.[57] The Mennonites' decision to settle in a familiar environment was also deliberate, and common among agricultural migrants. Inside the Russian Empire, farmers who moved from the European part to Siberia tended to settle in areas with similar natural conditions (forest, forest–steppe, or steppe) to their previous homes.[58] Farmers considering moving to the United States were advised in a Russian newspaper article in 1911 that they should not be put off by an "unknown land." They had "nothing to fear" from the fertility of the soil or the climate as migrants from the Russian Empire had been farming successfully in the plains region for over 25 years.[59]

Between 1873 and 1885, around 18,000 Mennonites from the steppes to the north of Black Sea and the Sea of Azov and in the Crimean peninsula moved to states up and down the Great Plains and to the Canadian prairies. Around 6,000 relocated to Kansas in the 1870s alone.[60] They chose to move, because of changes in the Russian Empire in the 1860s and 1870s. To understand their motives for leaving we need to go back further. Most of the Mennonites who left the steppes for the Great Plains were descendants of people who had lived for centuries in the Vistula delta near the Baltic port city then known as Danzig (now Gdańsk in Poland). In the late eighteenth and early nineteenth centuries, the Russian government invited them to move to the steppes in return for

[56] Marc Raeff, *Russia Abroad: A Cultural History of the Russian Emigration, 1919–1939* (New York: Oxford University Press, 1990), pp. 4–7, 28–9, 202–3.

[57] See Frederick C. Luebke, "Ethnic Group Settlement on the Great Plains." *WHQ* 8 (1977), 405–7; Gary Clayton Anderson, "The Native Peoples of the American West: Genocide or Ethnic Cleansing?" *WHQ* 47 (2016), 407–34.

[58] O. Shanskii, "Na rubezhe pereselenskogo dela," *Voprosy kolonizatsii* 1 (1907), 118, 128.

[59] "Russkie khleboroby v Soedin. Shtatakh," *Russko-Amerikanskii Vestnik: Russian–American Messenger*, no. 2 (1911), 14.

[60] Norman E. Saul, "The Migration of the Russian–Germans to Kansas," *KHQ* 60 (1974), 45–58.

grants of land and privileges. The latter included exemption from military service, which was vital to the Mennonites as they were pacifists, and self-government under the supervision of the Russian authorities through a Guardianship Committee. They retained German as their language of administration and education as well as everyday use.[61] In the early 1870s, plans by the Russian authorities to introduce universal eligibility to military service caused alarm among the Mennonites. Further, the Russian government had recently reformed local government, ending the Mennonites' separate system, and insisted on Russian as the language of administration and education. Many Mennonites objected to these changes. In addition, there were disagreements among Mennonites over religious matters and tensions arising from growing numbers of landless families.[62] In recent years, moreover, they had experienced poor harvests due to droughts and dust storms.[63] (See Map 5: Map of the Mennonite colonies in "New Russia" [S. Ukraine] in 1875.)

Mennonites wishing to leave the Russian Empire considered several destinations. They were encouraged to move to the United States by co-religionists who lived there and by members of their communities who were visiting North America. Prominent among these was Warkentin, whose vacation turned into a scouting expedition (see pp. 89–90). He was joined by a delegation who journeyed to North America to look for possible locations to settle. Warkentin and the delegation traveled all over the Great Plains from the Canadian prairies in the north to Texas in the south. Of particular importance to them were the climate, the soil, and the availability of water. They met President Grant and other members of the governments of the United States and Canada to discuss exemption from military service and other matters. They held meetings with land agents, including those of railroad companies selling land along their lines across the plains. The U.S. and Canadian governments, state and provincial administrations, and railroad companies were eager to attract the

[61] James Urry, *Mennonites, Politics, and Peoplehood: Europe–Russia–Canada, 1525 to 1980* (Winnipeg: University of Manitoba Press, 2006), pp. 17–54, 85–92. Mennonites spoke a dialect of low German, Plautdietsch, but also knew high German. See Reuben Epp, "Plautdietsch: Origins, Development and State of the Mennonite Low German Language," *Journal of Mennonite Studies* 5 (1987), 61–72.

[62] See Saul, "Migration," 41–4; James Urry, "The Russian State, the Mennonite World and the Migration from Russia to North America in the 1870s," *Mennonite Life* 46, no. 1 (1991): 11–16.

[63] See V. Postnikov, "Molochanskie i Khortitskie Nemetskie kolonii: Khoziaistvenno-statisticheskii ocherk," *SKhiL* 140 (1882), 39; N. K. Kalageorgi and B. M. Borisov, *Ekskursiia na reku Molochnuiu: opyt sel'skokhoziaistvennogo i ekonomicheskogo issledovaniia iuzhnorusskikh khoziaistv* (Spb.: V. F. Demakov, 1878), p. 58.

Mennonites as they had a reputation as hardworking and prosperous farmers. Warkentin eventually decided that the winters were too harsh in Canada, the summers too hot in Texas (which had been his first choice), but that the environment in south-central Kansas would best suit his purposes. He did a deal with Atchison, Topeka, and Santa Fe Railway Company to buy land at Halstead, Harvey County, Kansas. Other groups of migrants purchased land from the same company in Harvey County and in adjoining McPherson and Marion counties. The Santa Fe Railway hired Carl Bernhard Schmidt as commissioner for immigration and sent him to Russia to encourage more Mennonites to buy land from the company and move to Kansas.[64] Warkentin and the Mennonites who moved to Kansas played a part in the introduction of crops, in particular wheat, to the Great Plains. Warkentin later worked with Carleton to import wheat from the steppes (See Chapter 4).

The Mennonites were not the only Germanic farmers from the steppes who settled in the Great Plains. Several thousand Germans, both Protestants and Roman Catholics, moved from the steppes to the north of the Black Sea and the east of the Volga River. Their forebears, like those of the Mennonites, had moved there in the late eighteenth and early nineteenth centuries in response to offers of land and privileges from the Russian government. By the 1870s, many of these Russian–Germans were anxious to leave for similar reasons to the Mennonites. They sent scouts to the Americas (north and south). Communities of "Volga Germans" settled around Hays, Ellis County, Kansas, to the west of the Mennonite settlements.[65] The migration by Russian–Germans from the steppes continued. In 1895, Consul Heenan reported from Odessa that a thousand had bought tickets from a Hamburg steamship company to sail from Odessa directly to New York and Norfolk, Virginia, and thence by train to the Dakotas and Manitoba. Heenan wrote: "I am informed that these emigrants are of ample means and from my knowledge of German colonists I have no hesitation in

[64] See MLA, Warkentin Papers, Box 1, Folders 7–8; Cornelius Krahn (ed.), "Some letters of Bernhard Warkentin pertaining to the migration of 1873–1875," *Mennonite Quarterly Review* 24, 3 (1950), 248–63; Kempes Schnell (ed.), "John F. Funk's Land Inspection Trips as Recorded in his Diaries, 1872 and 1873," *Mennonite Quarterly Review* 24, 4 (1950), 295–311; Kempes Schnell, "John F. Funk, 1835–1930, and the Mennonite Migration of 1873–1875," *Mennonite Quarterly Review* 24, 3 (1950), 199–229; KSHS SA, Ms. Coll. 789, Atchison, Topeka, and Santa Fe Railway Company records, Box 308 Files of the Assistant Secretary – New York, Old Correspondence and Papers, Folder 13 Land Department, Sales to Mennonites, 1st party, January 1874 and 2nd party, January 1875.

[65] Saul, "Migration," 50–2; A. B-skii, "Emigratsiia privolzhskikh kolonistov-nemtsov v ameriku," *Saratovskii listok* no. 18 (1888), 1–2.

stating that they are intelligent, sober and industrious."[66] Communities of "Germans from Russia" established themselves throughout the Great Plains. In parts of North Dakota they were the largest ethnic group.[67] Among the immigrants in the plains from the Russian Empire there were also some Slavs (mostly Poles or Ukrainians, rather than Russians), who moved from the empire's western provinces.[68]

The migrants who moved to the Great Plains from the steppes had a great advantage over American settlers from further east and immigrants from elsewhere in Europe. In contrast to these settlers, who came from wetter, forested (or once-forested) environments, those from the steppes already had experience of farming in semi-arid, drought-prone grasslands. They could draw on their existing farming methods and crops. While they experienced initial problems due the unfamiliar American culture and English language, were concerned about prairie fires and grasshoppers (both of which were familiar from the steppes), and about "Indian raids," after a few years most were satisfied with their new lives in their new country.[69]

Far larger numbers of Jewish people moved permanently from the Russian Empire to the United States from the 1860s. They left to escape growing legal restrictions, anti-Semitism, persecution, and violent pogroms. Most originated from the "Pale of Settlement," which stretched from the Baltic to the Black Seas and included the steppes of present-day Ukraine and the port of Odessa. Some Russian–Jewish immigrants came from other parts of the Empire. For example, Raphael Zon was from Simbirsk on the River Volga on the edge of the steppes.[70] The overwhelming majority of Russian–Jewish immigrants made their homes in cities in

[66] NARA CP, RG 59, Microcopy no. 459, Despatches from United States Consuls in Odessa, 1834–1906, Roll 5, Vols. 9–10, January 2, 1889–December 28, 1895, Heenan to William F. Wharton, Assistant Secretary of State, Washington, DC, January 23, 1893.

[67] James R. Shortridge, "The Heart of the Prairie: Culture Areas in the Central and Northern Great Plains," *GPQ* 8 (1988), 206–21; William C. Sherman, *Prairie Mosaic: An Ethnic Atlas of Rural North Dakota*, 2nd edition (Fargo: North Dakota State University Press, 2017), pp. 17–18, 35, 48–51, 68–9, 72, 88–9, 109–10.

[68] "Russkie khleboroby v Soedin. Shtatakh," *Russko-Amerikanskii Vestnik: Russian–American Messenger* no. 2 (1911), 13–17; Sherman, *Prairie Mosaic*, p. 22.

[69] See Dennis D. Engbrecht, "The Settlement of Russian Mennonites in York and Hamilton Counties, Nebraska," *Mennonite Life* 39, no. 2 (1984), 4–10; Melvin Gingerich, "Reactions of the Russian Mennonite Immigrants of the 1870s to the American frontier," *Mennonite Quarterly Review* 34, 2 (1960), 137–146; T. J. Kloberdanz, "Plainsmen of Three Continents: Volga German Adaptation to Steppe, Prairie, and Pampa," in *Ethnicity on the Great Plains*, ed. F. Lubke (Lincoln: University of Nebraska Press, 1980), pp. 54–72.

[70] See Samuel Joseph, "Jewish Immigration to the United States from 1881 to 1910," *Studies in History, Economics and Public Law* 59, no. 4 (1914), 56–69, 98–104, 149–51.

the eastern United States. Most preferred urban life, since this was how many had lived in the Russian Empire. Only a few had experience of farming, because there were restrictions in the Russian Empire on Jews owning and renting land and living in rural areas.[71] Nevertheless, a few thousand Jewish immigrants took up farming. They settled in several states, in particular the Dakotas, from the 1880s. Some joined Jewish agricultural communities, but many left after a few years for urban areas.[72] Among those who persisted was Rachel Calof, who moved from the steppes to North Dakota in 1894 and worked hard to establish a family farm (see p. 47).

Among the Russian–Jewish immigrants in the United States who took up farming were members of a short-lived Jewish agricultural community at Woodbine, New Jersey. They included Jacob Lipman (1874–1939), who relocated to the United States as a teenager with his family from the Russian Baltic province of Kurland (in today's Latvia) in 1888. At Woodbine, he was mentored by another Russian–Jewish immigrant, Hirsch L. Sabsovich. He was an agricultural chemist and prepared Lipman for Rutgers College. Lipman studied agricultural sciences, and went on to make a successful career as professor of soil chemistry and bacteriology and dean of the New Jersey College of Agriculture at Rutgers University and head of the nearby New Jersey Agricultural Experiment Station. In 1916, he founded the American journal *Soil Science*. Lipman acted as a link between American and Soviet soil scientists.[73] Lipman is an example of several Russian–Jewish scientists and scholars in other fields who, with their scientific training, knowledge of the Russian language, and contacts with Russian and Soviet scientists, facilitated the transmission of Russian scientific ideas to the United States. Other such Russian–Jewish émigrés

[71] See Eric Lohr, *Nationalizing the Russian Empire: The Campaign against Enemy Aliens during World War I* (Cambridge: Harvard University Press, 2003), p. 85; Heinz-Dietrich Löwe, *The Tsars and the Jews: reform, reaction, and anti-semitism in imperial Russia, 1772–1917* (Chur and New York: Harwood Academic Publishers, 1993); V. N. Nikitina, *Evrei zemledel'tsy: Istoricheskoe, zakonodatel'noe, administrativnoe i bytovoe polozhenie kolonii so vremeni ikh vozniknoveniia do nashikh dnei, 1807–1887* (Spb: tip gazety "Novosti," 1887).

[72] Violet and Orlando J. Goering, "Jewish Farmers in South Dakota: The Am Olam," *South Dakota History* 12, 4 (1983), 232–47; Janet E. Schulte, "'Proving up and Moving up:' Jewish Homesteading Activity in North Dakota, 1900–1920," *GPQ* 10 (1990), 228–44; Sherman, *Prairie Mosaic*, pp. 19–20, 53–4, 70, 112; Leo Shpall, "Jewish Agricultural Colonies in the United States," *AH* 24 (1950), 120–46. See also Rebecca E. Bender and Kenneth M. Bender, *Still* (Fargo: North Dakota State University Press, 2019).

[73] See Selman Abraham Waksman, *Jacob G. Lipman: Agricultural Scientist, Humanitarian* (New Brunswick, NJ: Rutgers University Press, 1966); David Moon and Edward R. Landa, "The Centenary of the Journal Soil Science: Reflections on the Discipline in the United States and Russia Around a Hundred Years Ago," *SS* 182, no. 6 (2017), 203–15.

who feature in this book are: Issac Rubinow, an economist who wrote studies of Russian steppe farming for the USDA (see pp. 111–13); Peter Fireman, who translated works by Russian scientists into English (see pp. 171, 217); and Zon, who trained as a forestry scientist in the United States (see Chapter 7).

The ethnic Russians who moved to the United States after the Russian Revolution and Civil War included a soil scientist, Konstantin Nikiforov (who spelled his name Constantine Nikiforoff in the United States). His academic career was interrupted by service as an officer in the tsarist army during the First World War and the White army in the Russian Civil War. He arrived in the United States in 1921, but was not able to resume his scientific career until the late 1920s, when he helped familiarize American soil scientists with Russian methods (See pp. 258–61).

Thus, some of the immigrants assisted American specialists learn about Russian steppe agriculture. This was timely, since the growing mutual awareness of the similarities between the steppes and the Great Plains and the competition between the two countries in the world grain market prompted specialists in both countries to study the other's grain production.

Russian and American Studies of the Other's Grassland Agriculture

From the mid nineteenth century, Russians paid close attention to the development of arable farming in the prairies and Great Plains. Towards the end of the century, the Americans began to take an interest in grain production in the steppes.

Russian Studies of American Agriculture

In the mid nineteenth century, as the frontier of Euro-American settlement and grain production moved west across the prairies towards the Great Plains, Russian observers expressed mixed views over whether grain grown in America's grasslands would threaten their export trade. Some pointed to the vast distance to the east-coast ports as a constraint on the Americans' ability to export large quantities of grain.[74] Over the following

[74] See "Smes,'" *ZhMGI* 25 (1847), 4th pagn., 153–7; [V. I. Veshniakov], "O sostoianii sel'skogo khoziaistva v Severo-Amerikanskikh shtatakh," *ZhMGI* 48 (1853), 2nd pagn., 29–58; "Neosnovatel'nost' opasenii neodolimogo sopernichestva v khlebnoi torgovle so storony Soedinennykh Shtatov," *ZhMGI* 53 (1854), 5th pagn., 5–13. On the importance of grain, especially wheat, exports for Russia, see P., "Vyzov produktov russkogo sel'skogo khoziaistva za

decades, Russian specialists watched as settlers plowed up the plains to grow ever larger quantities of grain and sent it by railroad to the east coast, from where some was exported to Europe. In Europe, the surge in American exports caused grain prices to fall and undermined Russian trade.[75] Towards the end of the nineteenth century, the economist Sergei Bulgakov wrote that railroads had "abolished the distance" between the east coast and the center. He considered American agriculture an "all-conquering force" providing ruinous competition for European farmers. He attributed this largely to the extensive, mechanized development of grain cultivation on an abundance of "free land." Peter Struve reached similar conclusions, but noted that, to a lesser extent, agriculture was developing in a similar way in the steppes of New Russia.[76]

Most Russian commentators believed that American plains farming was more advanced than steppe agriculture and that, to compete, Russian farmers would have to learn from American expertise.[77] As far back as 1853, the Free Economic Society (a government-funded institution that sought to improve agriculture) sent the scientist Viktor Mochul'skii to the United States to collect data on American agriculture and the climate. He also brought back seeds of types of grain and other plants.[78] The aspect of American agriculture that attracted most attention, and in which Americans were far ahead, was mechanization.[79] In the late nineteenth and early twentieth centuries, the Russian Ministry of Agriculture sent specialists to

granitsu," *SKhiL* 112 (1873), 1st pagn., 25–53, 107–29, 175–201. For a survey of Russian views of American agriculture, see Robert V. Allen, *Russia Looks at America: The View to 1917* (Washington, DC: Library of Congress, 1988), pp. 114–82.

75 See, for example, V. Samolevskii, "Glavnye faktory khoziaistva v Soedinennykh Shtatakh Ameriki," *SKhiL* 121 (1876), 103–18; Robert Vasil'evich Orbinskii, *O khlebnoi torgovle Soedinennykh Shtatov Severnoi Ameriki* (Spb: Trenke i Fusno, 1880); [M. E. Saltykov-Shchedrin], "Popytki konkurirovat' s Amerikoi i polozhenie nashei khlebnoi torgovli," *Otechestvennye zapiski* no. 6 (1881), 245; A. A. Radtsig, "Statisticheskii ocherk proizvodstva pshenitsy i mezhdunarodnoi torgovli eiu," *SKhiL* 156 (1887), 2nd pagn., 257–96; M. P. Fedorov, *Obzor mezhdunarodnoi khlebnoi torgovli* (Spb: MVD, 1889); S. I. Gulishambarov, "Vsemirnaia torgovlia v XIX v. i uchastie v nei Rossii," *Zapiski imperatorskogo russkogo geograficheskogo obshchestva po otdeleniiu statistiki* 7, no. 3 (1898), 1–230.

76 S. Bulgakov, *Kapitalizm i zemledelie*, 2 vols. (Spb: Tikhonov, 1900), vol. 2, pp. 413–40; P. Struve, *Kriticheskie zametki k voprosu ob ekonomicheskom razvitii Rossii*, part 1 (Spb: Skorokhodov, 1894), pp. 207, 210, 260–6.

77 See [James Caird], "Pis'ma o stepnom khoziaistve v severnoi Amerike," *ZhMGI* 80 (1862), 112 (translator's note); N. Sibirtsev, *Chernozem v raznykh stranakh* (Warsaw: Varshavskii uchebnyi okrug, 1898), p. 42; A. Filipchenko, "Proizvodstvo, vyvoz khleba i blizhaishiia zadachi nashego khoziaistva," *SKhiL* 160 (1889), 2nd pagn., 137–55.

78 A. I. Khodnev, *Istoriia Imperatorskogo Vol'nogo Ekonomicheskogo Obshchestva s 1765 do 1865 goda* (Spb: tip. Obshchestvennaia pol'za, 1865), pp. 90–2, 156.

79 See, for example, A. Mikhel', "Zametki ob Amerikanskikh topchakakh, molotilkakh i veialkakh," *SKhiL* 94, 1 (1867), 2nd pagn., 25–40; V. Cherniaev, "Ocherk istorii usovershevstvovaniia zhatvennykh i senokosil'nykh mashin v Rossii," *SKhiL* no. 11 (1870), 130.

examine American farm machinery. Dmitrii Artsybashev was impressed and urged the use of American machinery in the steppes. Russians had been importing agricultural machines from the United States since at least the 1870s. The main port of entry was Odessa, where some American companies had representatives.[80] The Russian government also took an interest in the elevators used to store grain along American railroads.[81] In the 1890s, it looked to American agricultural experiment stations as a model when it established a national network in Russia.[82] Tulaikov, who was sent to the United States by the Russian government in 1908, wrote a detailed study of American agriculture, agricultural experiment stations, education, and the USDA.[83]

Such was the interest in learning from American agriculture that Russian provincial and central government bodies established permanent agricultural agencies in the United States. In 1907, the provincial council (zemstvo) in Ekaterinoslav[84] in the steppes set up an agency in Minneapolis, Minnesota. The zemstvo chose Minneapolis as it was a major center for milling and shipping grain grown in the prairies and Great Plains, and because the climate and soils in the plains west of the city resembled the steppes. A journalist in Rostov-on-Don (east of Ekaterinoslav) went into more detail, explaining that northwest Nebraska and southeast Wyoming were similar to Ekaterinoslav province, the Dakotas resembled Khar'kov and Poltava provinces, northwest Nebraska – the Don region, and northwest Kansas and northeast Colorado – Kherson province.[85] The Ekaterinoslav agency was headed by Iosef Rozen (Joseph Rozen). He was a Russian Jew who had moved to the United States in 1903, graduated from Michigan agricultural college, and carried out studies of agriculture in Kansas, Nebraska, and South and North Dakota.[86] The Russian central

[80] See D. Artsybashev, *Sel'sko-khoziaistvennoe mashinostroenie v Soedinennykh Shatakh Severnoi Ameriki i v Kanade* (Spb: Sel'skii Vestnik, 1909); S. Panaev, "Ocherk razvitiia proizvodstva sel'sko-khoziaistvennikh mashin i orudii v Severno-Amerikanskikh Soedinennikh Shtatakh," *SkhiL* 199, no. 10 (1900), 171–206.

[81] "Iz zagranichnoi literatury: Amerikanskie khlebnye elevatory," *SKhiL* 142 (1883), 216–20; A. L. Rafalovich, *Zernokhranilishcha v Soedinennykh Shtatakh Severnoi Ameriki* (Spb: Tip. V. F. Kirshbauma, 1903).

[82] PFA RAN, f.184, op.1, d.137; d.138; d.144, l.5.

[83] Tulaikov, *Ocherki*, p. 94 and passim.

[84] Ekaterinoslav is today's Ukrainian city of Dnipro. From 1926 to 2016, it was known as Dnepropetrovsk.

[85] F. Gotval'd, "Ob uchrezhdenii zemskoi sel'sko-khoziaistvennoi agentury v Amerike," *Iugo-vostochnyi khoziain* (November 15, 1907), 1.

[86] I. B. Rozen, *Otchet starshego agenta Ekaterinoslavskogo zemstva v Soed. Shtatakh Sev. Ameriki za iiul' – oktiabr' 1908 goda, Izvestiia zemskoi sel'sko-khoziaistvennoi agentury v Soedinennykh Shtatakh*, no. 6 (Ekaterinoslav: Gubernskoe zemstvo, 1908), p. 6. See also A. G. Komsha and L. P. Sokal'skii, "Ob uchrezhdenii russkoi sel'sko-khoziaistvennoi agentury v Soed. Shtatakh Severnoi Ameriki,"

government department responsible for agriculture followed the Ekaterinoslav zemstvo's example in 1909, when it set up an agricultural agency in St. Louis, Missouri, under Kryshtoforvich and Aleksandr Kol'.[87] Both agencies sent back detailed reports on American practices and developments.[88] The central government department took them very seriously and circulated copies to agronomists around the Russian Empire.[89] In 1921–2, on behalf of the Soviet government, Vavilov set up a bureau in New York City to facilitate purchases and exchanges of plant material and scientific literature (see pp. 119–20).

American Studies of Russian Agriculture

In contrast to the Russian interest in American farming in the prairies and plains, many Americans seem not to have taken Russian agriculture very seriously. To take just one anecdotal example, in 1883, Ezra Butler McCagg, a resident of Chicago, spent a few weeks in the Russian Empire. While cruising down the River Volga near Saratov, he commented that the fertile plain contained "land of exceeding richness," where farmers were growing grain, but had no machinery, nor, apparently, any skilled labor. He advocated (to the readers of his travel account) the use of reaping machines in the steppes.[90] McCagg's disparaging attitude to steppe farming, which was before the famine of 1891–2, contributed further to impressions of the "backwardness" of Russian agriculture (see pp. 63–70).

Nevertheless, competition in the world market from grain grown in the steppes was prompting some Americans to pay more attention to Russian agriculture. U.S. Consul Heenan in Odessa reported that, over the 1880s and 1890s, there was a "marked increase" in the quantities and

Zapiski imperatorskogo obshchestva sel'skogo khoziaistva iuzhnoi Rossii nos. 2–3 (1906), 2nd pagn., 1–20; Dana G. Dalrymple, "Joseph A. Rosen and Early Russian Studies of American Agriculture," *AH* 38, no. 3 (1964), 157–60.

[87] *Obzor deiatel'nosti sel'sko-khoziaistvennogo agentsva v Severo-Amerikanskikh Soedinennykh Shtatakh s 1 iiulia 1909 po 1-e ianvaria 1912 g.* (Spb: Glavnoe Upravlenie Z. i Z., 1913); Aleksandr Kol', "Organizatsiia tsental'nogo upravleniia po zemledeliiu v Soedinennykh Shtatakh Severnoi Ameriki," *SKhiL* 246 (1914), 60–87. I could not find Kryshtoforvich's first name or initials.

[88] See Allen, *Russia Looks at America*, pp. 152–60.

[89] For example, in December 1913, the government agronomist in Turgai region, in present-day Kazakhstan, received 25 copies of the latest report and distributed them among district agronomists, experimental fields, agricultural societies, and the regional office of the Resettlement Administration. TsGA RK, f.30, op.1, d.35, ll.286–9.

[90] Ezra Butler McCagg, *Six Weeks of Vacation in 1883* (Chicago: McDonnell Brothers, 1884), pp. 12–15.

value of exports of grain, especially wheat, from the Black Sea port, but that exports were periodically hit by bad harvests caused by drought. Grain exports from the steppes were destined mainly for the Mediterranean region and western Europe, including Great Britain. Heenan described the condition of agriculture in Odessa's steppe hinterland. The region was "essentially" an "agricultural country," with fertile soil ideal for growing high-quality grain. He was less impressed with the transport infrastructure, which he thought inefficient and wasteful, and the some of the agricultural techniques. He described peasant farming as "primitive," but noted that large proprietors were buying American harvesters, binders, and reapers, and that local farm machinery was much improved. Heenan raised the issue of competition. In 1887, he reported: "The condition of the Russian grain trade is such as to cause the future of our farming population much anxiety." Three years later, he wrote: "Russian cereals, wheat especially, are the finest in the world."[91] Heenan used his own salary to buy the Russian publications and meteorological observations on which he based his reports. Only when he complained in 1896 was he assigned $25 a year for this purpose.[92] This allocation, worth about $750 in 2019,[93] suggests that the U.S. government was starting to take Russian grain exports more seriously.

To find out more about Russian wheat farming and exports, in 1907, the USDA hired the economist Issac M. Rubinow. He was born in a Jewish family in 1875 in Grodno in the west of the Russian Empire (in today's Belarus). In 1893, he moved to the United States, where he studied for degrees at Columbia University, first in medicine, and then political science.[94] With his knowledge of Russian, he could read the extensive collection of Russian agricultural literature that the Library of Congress had received in official exchanges (see pp. 118–19). In his reports for

[91] NARA CP, RG 59, Microcopy no. 459, Despatches from United States Consuls in Odessa, 1834–1906, Roll 5, Vols. 9–10, January 2, 1889–December 28, 1895, Consul Heenan to William F. Wharton, Assistant Secretary of State, Washington, DC, July 25, 1889; Heenan to Wharton, March 22, 1890; Roll 6, Vols. 11–12, January 4, 1896–December 29, 1902, Heenan to William W. Rockhill, Assistant Secretary of State, Washington, DC, December 28, 1896, "The Export Trade of Odessa."

[92] NARA CP, RG 59, Microcopy no. 459, Despatches from United States Consuls in Odessa, 1834–1906, Roll 6, Vols. 11–12, January 4, 1896–December 29, 1902, Consul Heenan, Odessa, to William W. Rockhill, Assistant Secretary of State, Washington, DC, December 24, 1896; Rockhill to Heenan, January 28, 1897.

[93] See, CPI Inflation Calculator, available online at www.officialdata.org/us/inflation/1895?amount=25, accessed June 28, 2019.

[94] "Isaac Max Rubinow Papers, Collection Number 5616: Biographical Note," available online at http://rmc.library.cornell.edu/EAD/htmldocs/KCL05616.html, accessed February 14, 2018.

the USDA, he made direct comparisons with the United States. He noted that the climate in the Russian wheat belt – the steppes – was slightly harsher than the American wheat-growing region. The difference in climates influenced the types of wheat farmers grew. Due to the severe winters in the steppes, especially to the east, many farmers preferred spring-sown wheat. On the other hand, in the milder winters of much of the Great Plains, farmers sowed more winter wheat. Despite the harsher conditions, Rubinow noted that steppe farmers were sowing larger quantities of wheat, and that the wheat belt was expanding into the North Caucasus, western Siberia, and Central Asia, including present-day Kazakhstan. Between 1896 and 1904, the area sown with wheat in the Russian Empire increased by 22.8 percent, compared with a growth of 18.3 percent in the United States. The Russian wheat belt, moreover, was larger than the American. On the other hand, Russian yields were lower, which Rubinow attributed "primitive techniques," and production was hit by periodic crop failures due to droughts. Nevertheless, Rubinow noted that Russian wheat production was increasing much faster than American, and had overtaken its main rival in 1904. Other changes threatened American exports. Russian farmers were using more machinery, including imported American equipment, and the expansion of education was ending the "ignorance" of the peasant population. Rubinow argued that unless there was a major redistribution of land to the peasants, who were likely to grow rye for consumption, rather than wheat for sale, "a further increase [in Russian wheat production] may be expected."[95] (The land redistribution and the consequences Rubinow foresaw took place after the Russian Revolution in 1917–18.)

Rubinow went on to argue that the relatively poor organization of the Russian wheat trade, together with the "primitive" production methods, and a shortage of elevators to clean and sort grain, meant that Russian wheat was of poorer quality and commanded lower prices than American wheat. Russian transport infrastructure, in particular railroads and port facilities, was less developed than in the United States. He estimated that it was probably more expensive to transport grain to western Europe from Russia than from across the Atlantic. But, because changes were under way, Russian exporters could have the advantage in the future.[96] While

[95] I. M. Rubinow, "Russia's Wheat Surplus; Conditions Under Which it is Produced," *USDA Bureau of Statistics Bulletin* no. 42 (1906), 9–10, 15, 18–29, 51, 60–2, 98–102; Rubinow, "Russian Wheat and Wheat Flour in European Markets," *USDA Bureau of Statistics Bulletin* no. 66 (1908), 9–10.

[96] I. M. Rubinow, "Russia's Wheat Trade," *USDA Bureau of Statistics Bulletin* no. 65 (1908).

Rubinow's general conclusions stressed the negative features of "the competitive ability of our main competitor in the international wheat trade," his data showed Russia's wheat exports surging past those of the United States.[97] The USDA's chief of statistics wrote that Rubinow's study "can not fail to be of service to wheat growers, buyers, millers, and exporters in this country."[98]

The Russian and American studies of their rival's agriculture in its grasslands were motivated by a desire to learn more about the competition. In doing so, however, they made specialists aware of the potential of learning from each other. While the Russians had long been prepared to do this, Americans took longer to realize the possibilities. One of the first was the crop scientist Carleton. After his visit to the steppes in 1898–9 to collect grain varieties for the Great Plains, he made precise parallels between the soils and climates of parts of the steppes and the Great Plains to support his recommendations where particular crops might be best suited. For example, he felt the Volga region resembled the northwestern Great Plains, while the milder climate of the North Caucasus and the Crimea was similar to Kansas.[99]

Despite the competition between them, both Russian and American agricultural specialists and government departments of agriculture assisted visitors from the other country who came to study their agriculture. This continued after the Russian Revolution of 1917, including the period prior to 1933 when the two countries had no diplomatic relations, and despite the deep suspicion on both sides. This mutual assistance enabled specialists in both countries improve their understanding of grain farming in semi-arid grasslands, and contributed to greater contacts between Russian and American agricultural scientists.

Contacts between Scientists

Contacts developed between scientists with interests in grassland environments and agriculture in the United States and the Russian Empire. Towards the end of the nineteenth century, Russian scientists, such as Krasnov, started to attend international meetings in the United States. Russian and American scientists met at conferences and congresses in

[97] Rubinow, "Russian Wheat and Wheat Flour," 12–13, 98–9.

[98] Rubinow, "Russian Wheat and Wheat Flour," 2; Rubinow, "Russia's Wheat Trade," 3. See also Edward T. Peters, "Russian Cereal Crops: Area and Production by Governments and Provinces," *USDA Bureau of Statistics Bulletin* no. 84 (1911).

[99] Carleton, "Russian Cereals," 14–18, 28–32.

Europe and at the world's fairs in various countries.[100] With a few exceptions, meetings between American and Russian scientists came to a temporary halt on the outbreak of the First World War in 1914, but resumed with the end of the Russian Civil War in 1921.[101] There will be more detailed discussion of contacts between scientists in the chapters that follow. A few examples here, covering the main areas of influences from the steppes – crops, soil science, and steppe forestry – will serve to introduce them.

The American agricultural explorers in the Russian Empire met Russian scientists. Hansen visited agricultural institutes, schools, model farms, experiment stations, and botanical gardens, where he became acquainted with, amongst others, Vasilii Robertovich Vil'iams (Williams) in Moscow and Robert Regel' in St. Petersburg. Hansen and Vil'iams first met in 1893 at the World's Columbian Exposition in Chicago, where Vil'iams was one of the Russian delegates.[102] In 1898–9, Carleton met several Russian agricultural scientists, for example, V. A. Bertensen in Odessa and Vil'iams in Moscow.[103] Meyer made the acquaintance of several Russian scientists, including the botanist Vladimir Komarov, Vil'iams, Tulaikov, and the Russian pioneer of plant selection, Ivan Michurin.[104]

[100] American agricultural scientists already had extensive contacts with scientists in many European countries by the late nineteenth century. See Mark R. Finlay, "Transnational Exchanges of Agricultural Scientific Thought from the Morrill Act through the Hatch Act," in *Science as Service: Establishing and Reformulating Land-Grant Universities, 1865–1930*, ed. Alan I. Marcus (Tuscaloosa: University of Alabama Press, 2015), pp. 33–60. (Finlay mentioned only one contact, a Baltic German, in the Russian Empire.)

[101] On scientific contacts between the new Soviet state and the United States, see E. D. Lebedkina, "Mezhdunarodnye sviazi sovetskikh uchenykh v 1917–1924 gg.," *Voprosy istorii* no. 2 (1971), 44–54; L. S. Nikol'skaia, "K istorii razvitiia sovetsko-amerikanskikh nauchnykh i kul'turnykh kontaktakh (1924–1933 gg.)," *Amerikanskii ezhegodnik 1973*: 208–33.

[102] Hansen, *The Banebryder*, [9], [11], [12], [15], [18], [47–8]; Hansen Papers, Series 2. Helen Hansen Loen Collection, Box 4, Folder 174, p. 244. Vil'iams/Williams (1863–1939) was a Russian agronomist and soil scientist. His views on maintaining soil structure through crop rotations were contentious, but he was politically asute and found favor with both the tsarist and Soviet governments. His father was an American engineer who moved to Russia in the mid nineteenth century. See N. P. Panov and E. V. Kulakov, "Vasilii Robertovich Vil'iams," *Vestnik AN SSSR* no. 6 (1984), 101–8; David Joravsky, *The Lysenko Affair* (Cambridge, MA: Harvard University Press, 1970), pp. 295–6.

[103] Carleton, "Russian Cereals," 18, 21.

[104] NARA CP, RG 54, Finding Aid NC-135, Entry 135I, Records of Frank N. Meyer, Plant Explorer, 1902–1908, Box 5, Folder 4259g, Meyer, Spb, to Fairchild, USDA, December 30, 1909; Meyer, Moscow, to Fairchild, January 4, 1910; Zon to Fairchild, January 13, 1910; Box 7, Folder [large, orange, no number], Meyer, Samara, to Fairchild, November 10, 1911; Meyer, Rostov on Don, to Fairchild, December 6, 1911; Box 9, Folder January 16– May 21, 1913, Meyer, Tomsk, to Fairchild, January 31, 1913.

Such meetings in Russia deepened contacts between crop scientists working for their government departments. In 1893, Frederick Corville of the Division of Botany of the USDA wrote to Pavel Kostychev of the Russian Department of Agriculture requesting seeds of alfalfa grown in Turkestan. Kostychev duly sent 25 pounds of the seed.[105] Over the following years, American and Russian government scientists regularly corresponded about exchanges of seeds, publications, and visits. In 1911, for example, Fairchild of the USDA's Bureau of Plant Industry wrote to his counterpart, Regel', director of the Bureau of Applied Botany in St. Petersburg, asking for seeds of the types of alfalfa that Hansen had collected in the steppes and of specific varieties of barley. Fairchild offered other seeds or payment. Regel' sent Fairchild the seeds he had to hand and requested other samples in exchange.[106]

Similar correspondence continued down to 1917. The tone of the letters in that difficult year reveals the warmth of the relationship that had developed. Fairchild wrote to Regel' in June 1917 against the backdrop of the war (in which the two countries were by then allies) and unfolding revolution in Petrograd (the name for St. Petersburg between 1914 and 1924). Fairchild reported on the help they were providing to Nikolai Borodin, who was visiting the United States on behalf of the Russian provisional government. Fairchild had

> assured [Borodin] of our desire to do everything possible to assist him in his mission to our country. If there is any way that occurs to you in which we can be of assistance to you in this tremendous emergency, I should be very glad indeed if you would let me know. Any message you might wish to send to the agriculturalists of America I shall be very glad indeed to get, although I realize how difficult it is for you to get any mail through to this country at this time.

Fairchild thanked Regel' for some clover seeds he had managed to send and offered other seeds in return. Regel' got a letter through to Fairchild in December 1917, a few weeks after the Bolshevik seizure of power. He thanked him for seeds of early corn varieties, which they planned to test at experiment stations in the steppe region. Showing no little restraint, Regel' continued:

[105] NARA CP, RG 54, Finding Aid NC-135, Entry 26L, Division of Botany, Correspondence with Foreign Botanists and others, 1895–1929, Box 3, File, Russia, P. Kostytscheff [Kostychev] to Corville, May 12, 1894.

[106] RGIA, f.382, op.9, 1911–17, d.290, l.3, Fairchild to Regel', March 13, 1911; l.2, Fairchild to Regel', December 8, 1911; l.7, Chief of Bureau of Applied Botany to Fairchild, February 27, 1912. See also ll.5, 7, 9, 11, 12, 15–16, 20, 23, 27.

> As to your request for sending you some quantity of seeds of red clover. . . , we are extremely sorry to say, that on account of the great disorganization and disorder which are broken out in our country, we are unable to do this. However, we can assure you, that we shall avail ourselves of the first opportunity to send you our publications as well as other materials. Thanking you and the other Agricultural and Scientific Establishments which are regularly forwarding us their papers and other materials. . .[107]

Due to the "disorganization and disorder," which worsened as revolution turned into civil war, contacts were largely halted until 1921. In that year, Vavilov, who had succeeded Regel', visited the United States on behalf of the Soviet government. Carleton was unable to meet him as he was working for the United Fruit Company in Panama. It is evidence of the esteem with which Carleton held Vavilov that he wrote to express his regret, but hoped they could continue their correspondence and exchange scientific literature on wheat and other crops.[108]

Close contacts also developed between some Russian and American scientists interested in soils. Aleksandr Voeikov, who visited North America in 1873–4, met Eugene W. Hilgard at the University of Michigan, where Hilgard worked briefly. They corresponded until 1908, discussing studies of the soils of Russia and North America, including black earth of their grasslands. Hilgard took a keen interest in Russian soil science and wanted to learn more.[109] When Krasnov attended the Geological Congress in Washington, DC, in 1890, Hilgard was one of the commentators on his paper.[110] Tulaikov spent a year in the United States in 1908–9 to study the utilization of salty soils in the arid west. He chose to be based at the University of California, Berkeley, to meet Hilgard, who had moved there in 1875. Tulaikov traveled extensively in the United States during his year there and met some of the leading soil scientists, including Milton

[107] RGIA, f.382, op.9, 1911–17, d.290, l.61, Fairchild to Regel', June 27, 1917; l.60 Chief of Bureau of Applied Botany to Fairchild, December 1/14, 1917. On Borodin's mission in 1917, see N. A. Borodin and M. I. Volkov, *Sel'skokhoziaistvennaia Amerika vo vremia voĭny: na osnovanii lichnykh vpechatleñii 1917 g.* (Moscow: Izd-vo Narodnoe pravo, 1918), p. 5.

[108] TsGANTD Spb, f.318, op.1-1, d.6, l.42, Carleton to Vavilov, November 21, 1921.

[109] Bancroft Library, Berkeley, California, The Hilgard family papers [hereafter, Hilgard Papers], Hilgard, E. W.: Incoming Letters, Box 23, File: Voeikov, Aleksandr Ivanovich. Voeikov's first letter to Hilgard was dated January 10, 1874, and sent to Hilgard from Washington, DC; Hilgard, E. W.: Outgoing Letters, Letterpress copy books, vol. 26, June 1907–July 1910, pp. 108–10, Hilgard to "Prof. Alex. Woeikof," St. Petersburg, January 2, 1908. For Hilgard's time at the University of Michigan, see "Eugene Woldemar Hilgard," available online at http://umhistory.dc.umich.edu/history/Faculty_History/H/Hilgard,_Eugene_Woldemar.html, accessed February 8, 2018.

[110] Krassnof, "The 'Black Earth'," 80.

Whitney and Marbut at the Bureau of Soils in Washington, DC.[111] Contacts between soil scientists continued after 1917. In May 1926, the Russian–Jewish émigré Lipman traveled to Moscow to make preparations for the visit by the delegation of Soviet soil scientists to the 1st International Congress in Washington, DC, the following year.[112]

While visitors and émigrés from the Russian Empire and Soviet Union helped facilitate contacts between Russian/Soviet and American scientists, there were tensions between émigrés of different waves. In 1930, for example, Zon, a Jewish immigrant from tsarist Russia, suspected Constantine Nikiforoff, a White Russian refugee from Soviet Russia, of initiating (unfounded) rumors that the Russians did not wish Americans attend the soil science congress in the Soviet Union in that year.[113]

Back in 1902, the reason Pinchot, the chief of U.S. forestry, visited the Russian steppes (see p. 85) was because he had been invited by Nikolai Nesterov. The two men had studied together at the École nationale forestière in Nancy, France, in 1889–90.[114] Zon, who was Pinchot's protégé, maintained contacts with Russian and Soviet forestry scientists and read their work. Zon was a combative and ambitious character. His relations with scientists in the country of his birth, and not just Nikoforoff, were not always cordial. Zon may have drawn too closely on the work of Russian forestry scientists in his own publications. In 1935, during controversy over the American Shelterbelt Project (see Chapter 8), in the American *Journal of Forestry*, Soviet forestry scientist Georgii Vysotskii accused Zon of adapting his ideas concerning the influence of forests on humidity without attribution. He was referring to an article by Zon in *Science* in 1913. In reply, Zon stated that he had sent Vysotskii a copy of his article at the time and he had not objected then. However, in unpublished correspondence with the editor of the *Journal of Forestry* (*JoF*), Zon admitted that Vysotskii "may have some justifiable grievance," but that he had mentioned his name "more specifically" in later articles. In the political climate in the Soviet Union in the 1930s, Vysotskii may have

[111] RGIA, f.398, op. 70, 1907, d.24719; f.426, op.1, 1908, d.659; N. M. Tulaikov, "Pochvennye issledovaniia v Soedinennykh Shtatakh," *Pochvovedenie* 10, no. 4 (1908), 293–322. On his visit to the Bureau of Soils, see p. 222–3 .

[112] ARAN, f.487, op.1, 1927, d.8, l.2 and passim, see also dd.5, 10.

[113] MHS, P1237, Raphael Zon papers, 1887–1957 [hereafter Zon Papers], Box 7, Folder 1, Zon to Prof. L. G. Romell, April 17, 1930. Both Zon and Nikiforoff were based at the University of Minnesota in St. Paul, but I found no reference in their papers to them ever contacting each other.

[114] Pinchot, *Breaking New Ground*, pp. xi–xiii, 214–16; Char Miller, *Gifford Pinchot and the Making of Modern Environmentalism* (Washington, DC: Island Press, 2001), pp. 83–8. On Nesterov, see G. P. Eitingen, "Zhizn' i trudy N. S. Nesterov," *Les i step'* no. 6 (1951), 81–3.

felt it expedient to attack a "capitalist" specialist. He was not placated by Zon's admission, and in a private letter to Arsenii Iarilov wrote that Zon had "admitted to his plagiarism, of course without using that term."[115] While Zon may have been using his knowledge of Russian forestry science to advance his career, nevertheless, he contributed to an increasing awareness of Russian studies among American foresters.

Building on such contacts, from the late nineteenth century there was growing mutual interest among some American and Russian agricultural scientists in reading each other's work. Requesting and offering their latest publications was a recurring theme in their correspondence, and exchanging work was routine when they met. For the Chicago world's fair in 1893, the Russian government published English translations of work on agriculture and agricultural sciences, including soil science,[116] some of which came to the attention of American scientists. There were formal government exchanges of scientific literature. In the 1860s, the U.S. and Russian governments agreed to send each other copies of official and scientific publications. The agreement included a direct exchange between the Russian Ministry of State Domains (whose responsibilities included agriculture) and the USDA.[117] In 1925, the Soviet government reached agreement with the Library of Congress on similar exchanges. Over the following few years, the Soviet authorities sent in excess of 50,000 publications on various subjects, including agricultural sciences, to the United States, and received around three times that number in return.[118] Thus, specialists in both countries had access to scientific publications produced in the other country. There are examples in this book of Americans using the Russian materials in the Library of Congress. Rubinow based his studies of Russian wheat farming and exports on the Russian publications in the Library (see pp. 111–13).

[115] G. N. Vyssotsky, "Shelterbelts in the Steppes of Russia," *JoF* 33 (1935), 781, 788 (Zon's reply is on the latter page); Zon Papers, Box 9, Folder 3, Herbert Smith, Journal of Forestry, to Zon, July 5, 1935, Zon to Smith, July 9, 1935; ARAN, f.582, op.4, d.27a, l.42, Vysotskii to Iarilov, October 27, 1935. Vysotskii continued that Zon had not admitted "everything" and had kept quiet about the "big" matter. He did not specify what it was, however, so we are left in ignorance.

[116] RGIA, f.91, op.2, 1892, d.66, l.34; *The Industries of Russia*, vol. 3, *Agriculture and Forestry*, ed. John Martin Crawford (Spb: The Department of Agriculture Ministry of Crown Domains, 1893). On the arrangements with Crawford, the U.S. Consul General in St. Petersburg, for editing the English translation, see RGIA, f.399, op.1, 1892, d.238, ll.26–7. For the publications on soil science, see Chapters 5 and 6.

[117] Norman E. Saul, *Concord and Conflict: the United States and Russia, 1867–1914* (Lawrence: University Press of Kansas, 1996), pp. 133–6.

[118] Nikol'skaia, "K istorii razvitiia sovetsko-amerikanskikh nauchnykh i kul'turnykh kontaktakh," 223.

In the early 1930s, an official of the U.S. Forest Service sent a team of translators to the Library to read Russian publications on steppe forestry and shelterbelts (see p. 321).

There were also exchanges of literature between research institutes in the two countries' grasslands. To take just one example: In 1917 Aleksandr Kol', the director of the Kuban' Agricultural Experiment Station in Ekaterinodar (today's Krasnodar) in the North Caucasus, wrote to J. H. Worst, the director of the North Dakota Agricultural Experiment Station in Fargo. He asked for copies of their scientific bulletins and sent him his station's in anticipation. Henry Luke Bolley, who had visited Russia as a plant explorer in 1903, responded favorably.[119] When Vavilov visited the Fargo station in 1921, he asked Bolley for copies of its publications as well as crop samples. Bolley dispatched what he had requested directly to Petrograd. The warmth of their acquaintance was conveyed in Bolley's accompanying letter: "I wish to thank you personally for your visit and the large amount of valuable scientific information you . . . were able to bring us." Vavilov wrote back, thanking Bolley for his kindness, and adding: "We shall remember for a long time our stay with you in Fargo."[120]

The main purpose of Vavilov's visit to the United States in 1921 was to secure agreement with the American authorities to ship large quantities of seed grain to Soviet Russia because of a catastrophic shortfall in the harvest, which preceded a massive famine. The Soviet government wanted seeds of crop varieties, such as types of wheat, that had earlier been introduced from the steppes to the United States (see Chapter 4). Vavilov was also instructed to collect American literature on agricultural sciences and meet American scientists to learn about their recent work. He traveled all over the United States in 1921–2, visiting several locations in the Great Plains in addition to Fargo. He set up a Russian Bureau of Applied Botany in New York City to arrange shipments of seed and American publications back to Soviet Russia.[121] Vavilov spent a total of $14,674.95 on publications, seed, and expenses for the New York bureau.[122] This was an

[119] NDSUA, Mss 18, Henry Luke Bolley Papers [hereafter Bolley Papers], Box 51, Folder 11, Bolley to Z. B. Libenson, Intourist, New York, May 20, 1939; Alexander Kol to J. H. Worst [n.d. 1917]; Bolley to Kol, September 15, 1917.

[120] Bolley Papers, Box 51, Folder 12, Vavilov to Bolley, October 21, 1921; Bolley to Vavilov, October 26, 1921; Vavilov to Bolley, November 4, 1921.

[121] N. P. Goncharov, *Nikolai Ivanovich Vavilov* (Novosibirsk: Izd-vo SO RAN, 2014), pp. 42–4; D. N. Borodin, "The Russian Bureau of Applied Botany," *Science*, NS, 55, 1414 (February 3, 1922), 129.

[122] TsGANTD Spb, f.318, Op.1, d.10, ll.1–2. Vavilov's own expenses were in addition to this sum.

enormous sum of money, over $200,000 at 2019 prices,[123] for the new Soviet government after several years of civil war and desperate hardship. It is testament to the value the Soviet government attached to seed and scientific advice from the United States, despite the deep suspicion between the new communist regime and its ideological opponent.

At the same time, there was a remarkable initiative among a group of American scientists, who set up the "American Committee to Aid Russian Scientists with Scientific Literature." They published a notice in *Science* in June 1922 explaining that Russian scientists had been cut off from western European and American scientific literature since 1914 and were appealing for help. The Committee asked for donations of recent literature. They arranged for its distribution through Herbert Hoover's American Relief Administration (ARA), which was directing famine relief in Soviet Russia. The Committee justified its appeal:

> The assistance that American scientists can give to the Russian scientists who are in distress, besides being a good Samaritan act, will be a real contribution to the progress of science. It may also be the means of re-establishing the normal exchange of scientific results between Russian and American scientists, and will be a fine manifestation of the cooperation of men in science throughout the world.[124]

The appeal was directed at all fields of the natural sciences, including agricultural and environmental sciences. The committee's chairman was Vernon Kellogg, professor of entomology at Stanford University who had worked in Russia with the ARA.[125]

One of the driving forces behind the initiative was Zon,[126] who was keen to revive contacts between American and Russian scholars. On June

[123] See, CPI Inflation Calculator, available online at www.officialdata.org/us/inflation/1921?amount=14674, accessed June 28, 2019.

[124] Vernon Kellogg, L. O. Howard, David White, Raphael Zon, "American Committee to Aid Russian Scientists with Scientific Literature," *Science*, NS, 55, no. 1434 (June 23, 1922), 667–8. On the ARA, see Bertrand M. Patenaude, *The Big Show in Bololand: The American Relief Expedition to Soviet Russia in the Famine of 1921* (Stanford, CA: Stanford University Press, 2002). See also Zon Papers, Box 4, Folder 6, Circular letter from American Committee to Aid Russian Scientists with Scientific Literature, June 19, 1922.

[125] C. E. McClung, "Biographical Memoir of Vernon Lyman Kellogg, 1867–1937," *National Academy of Sciences of the United States of America: Biographical Memoirs*, vol. XX, presented to the Academy at the Autumn Meeting, 1938, available online at www.nasonline.org/publications/kellogg-vernon.pdf, accessed March 8, 2016; Patenaude, *The Big Show in Bololand*, p. 65.

[126] Zon Papers, Box 4, Folder 2, Zon to Vernon Kellogg, November 13, 1920; Folder 3, Robert B. Sosman to Vernon Kellogg, March 1, 1921; Folder 4, Zon to D. Borodin, November 1, 1921; Zon to V. V. Tchikoff, November 3, 1921; Zon to Barrington Moore, November 21, 1921; Zon to D. Borodin, November 28, 1921.

17, 1922, he wrote to Dmitrii Borodin (the head of Vavilov's Bureau of Applied Botany in New York) that the Russian Academy of Sciences had agreed to receive, catalog, and distribute the American scientific literature. He hoped: "It may be the forerunner of [a] closer relationship between the men of science in this country and abroad."[127] On June 21, he shared with Vavilov his hope that "within a few months there will be continuous flow of American scientific literature to Russia." He continued, revealing perhaps his main motive: "A number of us who are fortunate enough to read Russian are, of course, anxious to have some of the recent Russian publications, and I wonder whether you could send in exchange Russian publications through the same channels [the ARA]."[128] Zon's request was apposite, since new Soviet regulations virtually prohibited sending literature abroad. Some Russian scientists had the same idea and asked Borodin if they could send literature via the ARA office in Petrograd in return for the publications "American scientific workers" were sending them "quite liberally."[129] Zon kept in touch with Russian forestry scientists, for example Mikhail Tkachenko, to check that they were receiving the American publications, and to request specific Russian publications on forestry science in exchange.[130]

The American Committee to Aid Russian Scientists shipped 445 cases of scientific literature weighing 28,000 lbs. They received hundreds of letters of gratitude from Russian scientists and scientific institutions.[131] While this episode was extraordinary, it was part of longer-term exchanges of scientific literature between the United States and Russia that dated back to the 1860s and continued throughout our period, regardless of the state of relations between the two governments. However, there was more interest among Russian and Soviet scientists in western scholarly literature than among American scientists in Russian and Soviet work. There were exceptions,

[127] Zon Papers, Box 4, Folder 6, Zon to D. Borodin, June 17, 1922. On Dmitrii Borodin, see N. I. Vavilov, *Nauchnoe nasledie v pis'makh: mezhdunarodnaia perepiska*, vol. 1 *Petrogradskii Period, 1921–1927*, ed. V. A. Dragavtsev (Moscow: Nauka, 1994), p. 461. Dmitrii Borodin is not to be confused with Nikolai Borodin, see p. 88.

[128] TsGANTD Spb, f.318, op.1-1, d.23, l.46, Zon to Vavilov, June 21, 1922. After meeting Vavilov in the United States in late 1921, Zon wrote that Vavilov had "made an exceedingly fine impression upon me." Zon Papers, Box 4, Folder 4, Zon to Prof. Nicholas P. Makaroff, November 16, 1921.

[129] TsGANTD Spb, f.318, op.1-1, d.23, l.167, Bogdanoff-Katkoff and Grossman to Borodin, June 1, 1922.

[130] Zon Papers, Box 4, Folder 7, M. Tkachenko to Zon, February 2, 1923; D. Borodin to Zon, February 9, 1923; Zon to D. Borodin, February 15, 1923; Zon to Tkachenko, March 24, 1923.

[131] Harold H. Fisher, *The Famine in Soviet Russia: The Operations of the American Relief Administration* (Stanford, CA: Stanford University Press, 1927), p. 468.

such as American scientists of Russian origin, for example Zon, and Americans, including Bolley, Carleton, and others, who had visited Russia and had close contacts with their Russian counterparts. Even if they wished to read the Russian publication, most Americans could not do so since few, with the notable exceptions of Jewish émigrés, could read Russian.

Thus, by the early twentieth century there were growing contacts between American and Russian scientists with mutual interests in the environment of semi-arid grasslands and the challenges of developing agriculture in such conditions.

Building Bridges: Conclusions

Bridges were built that enabled first Russians and then Americans to recognize not just parallels between the steppes and the Great Plains, but that both could learn from the other's experiences. Russians, for the most part, were the first and the more willing to learn from the other country. As well as using American agricultural experiment stations as a model for a Russian network (see p. 109), on the eve of the First World War, the Russian government began to build grain elevators on the American pattern.[132] Russians investigated and imported American agricultural machinery, both before and after 1917. The tractor factories constructed during the First Five-Year Plan (1928–32) were planned and built by American companies.[133] In the late 1920s, the Soviet government looked to the large-scale farms of the northern Great Plains as templates for its gigantic, mechanized state farms in the steppe region, and hired Americans such as Campbell to advise them.[134] Although the main direction in exchanges of crop varieties was from the steppes to the Great Plains, an American crop that had made its way to the steppes was corn (*kukuruza*), which was grown with some success in today's Moldova and Ukraine and elsewhere in the steppe region, long before Nikita Khrushchev's obsession

[132] *Statisticheskii obzor deiatel'nosti zernokhranilishch za khlebnuiu kampaniiu 1913–1914 gg.* (Petrograd: Izd. Gos. Banka, 1915).

[133] Kendall E. Bailes, "The American Connection: Ideology and the Transfer of American Technology to the Soviet Union, 1917–1941," *Comparative Studies in Society and History* 23 (1981), 421–48; George S. Queen, "The McCormick Harvesting Machine Company in Russia," *RR* 23 (1964), 164–81; Fred V. Carstensen, *American Enterprise in Foreign Markets: Studies of Singer and International Harvester in Imperial Russia* (Chapel Hill: University of North Carolina Press, 1984), pp. 107–234. See also, pp. 5, 34.

[134] See Fitzgerald, *Every Farm a Factory*. On Campbell, see pp. 99–100. The tsarist government took an interest in bonanza farms in North Dakota in the 1890s. "Iz zagranichnoi literatury: Znamenitie pshenichnye fermy v severnoi Dakote v Soed. Shtatakh Sev. Amerike," *SkhiL* 173 (1893), 144–50. See pp. 99–100.

with the crop in the 1950s.[135] Most attention in this book will be paid not to these American influences and technology transfers in Russia and the Soviet Union, which have already been the subject of much research, but to those that went the other way. Americans were slower to learn from Russian experience in the steppes, but once they did so, it led to a series of transfers and influences.

Growing numbers of Americans visited the steppes and were struck by similarities with the Great Plains. Most important were visitors with scientific expertise and practical experience. Some deliberately visited the steppes to study the parallels and look for models to learn from. The plant explorer Mark Carleton solicited support from the USDA to travel around the steppes in search of crop varieties to introduce to the Great Plains on account of parallels between the two regions. Curtis Marbut was so eager to attend the Soil Science Congress and excursion in the Soviet Union in 1930, and see the Chernozem in the steppes, that he took unpaid leave from his government post for almost six weeks (See p. 77).

A major contribution to building bridges between the steppes and Great Plains was played by immigrants from the Russian Empire. Two groups stand out: the Germanic farmers and Russian Jews who moved to North America from the 1870s. The Germanic farmers, including Mennonites, brought their experience of farming in semi-arid grasslands and some of their crops. Carleton traveled to the steppes in search of more varieties of grain after he had seen the success of the Mennonites' wheat in Kansas. The Jewish immigrants who left the Russian Empire to escape persecution brought with them their knowledge of the Russian language, which was rare in the United States at that time. Some had, or acquired in their new homes, relevant scientific and specialist training. Jewish émigrés facilitated the transfer of Russian scientific and practical expertise to the United States by studying and translating Russian-language specialist literature. In addition to the cases discussed in earlier this chapter, Marbut's papers contain a translation of an article by the soil scientist Konstantin Gedroits made by Selman Waksman, who emigrated in 1910 from Odessa to the United States, where he trained as a soil scientist and microbiologist at Rutgers College under Jacob Lipman.[136] Lipman – who had made the same

[135] Benzin, V. M., *Kul'tura kukuruzy* (St Petersburg: tip. Kirshbauma, 1912); Victoria I. Zhuravleva, "American Corn in Russia: Lessons of the People-to-People Diplomacy and Capitalism," *Journal of Russian American Studies* 1, no. 1 (2017), 23–45.

[136] MSU DGML SCA, Ozarkiana Collection: Curtis F. Marbut Collection, Collection Number: M014, Folder, Work Related – Papers, Studies, and Theses, folder 2, K. K. Gedroiz, "The

journey earlier – assisted the dissemination of Russian/Soviet scientists' research in the United States by publishing translations in his journal *Soil Science*. In 1926, he traveled to Moscow to arrange the visit by Soviet scientists to the soil science congress in the United States the following year (see p. 117).

Greater contacts and mutual interest assisted in bridging barriers to exchanges of information. In both countries, specialist studies of the other's grain production were prompted initially by the competition between them in the world grain market. Such studies spurred further interest in learning more about the other country's experience. This led to greater contacts between scientists and other specialists who, once they found they were working on similar subjects, were eager to exchange ideas and results. Their common concerns assisted in overcoming, to some extent, the language barrier that hindered most Americans in reading Russian publications. In part this was due to Russian scientists' efforts to bring their work to international audiences by providing summaries or translations in other languages. Very important was a book on the Russian innovation in soil science published in German by Konstantin Glinka in 1914. There were other such cases. In response to a request from Hilgard in 1908 (see p. 74), the editor of the Russian soil science journal published abstracts in western European languages.[137] To take another example, in 1926, L. E. Melchers of the Kansas State Agricultural College thanked Nikolai Vavilov for a copy of his "Studies on the Origins of the Cultivated Plants." He was particularly appreciative of the English-language summary.[138]

Overcoming the language barrier was due also in part to American scientists and institutions arranging translations. By the late 1920s, the USDA had in-house translators to make the Russian publications it received accessible.[139] A few, keen, American scientists were inspired to study Russian, in spite of the limited opportunities for formal instruction

Absorbing Capacity of the Soil and Soil Zoolitic Bases," trans. S. A. Waksman, 1923. On Waksman, see Peter Pringle, *Experiment Eleven: Deceit and Betrayal in the Discovery of the Cure for Tuberculosis* (London: Bloomsbury, 2012), pp. 12–13.

[137] Hilgard wrote to Pavel Ototskii to thank him. Hilgard Papers, Hilgard, E. W.: Outgoing Letters, Box 3, Hilgard to Professor P. Ototzky [Ototskii], May 11, 1911.

[138] TsGANTD Spb, f.318, op.1-1, d.90, l.8, Melchers to Vavilov, May 26, 1926.

[139] NARA CP, RG 54, Finding Aid PI-66, Entry 30, Records of the Division of Cereal Crops and Diseases. Foreign Correspondence, 1900–34, Box 8, File Russia, G. K. Meister, Saratov Experiment Station, to USDA, March 4, 1927 (trans. Elizabeth Jodidi, senior translator). For an example, concerning translating a book by Vavilov, see TsGANTD Spb, f.318, op.1-1, d.1192, l.46, J. A. Clark, Senior Agronomist, Wheat Investigations, USDA, to Vavilov, January 2, 1936.

in the United States before the 1940s. Among the first was Carleton, who seems to have taught himself. On his second visit to the Russian Empire in 1900, he wrote that he would "soon speak the Russian as much as the German I think. I have now letters to several men in the North Caucasus who speak only Russian, and expect to buy wheat from them."[140] In late 1929, the soil scientist Thomas Rice asked his Russian émigré colleague, Nikiforoff, for advice on studying Russian. The following year, possibly in preparation for the soil science congress in the Soviet Union, the USDA graduate school offered a Russian language course, which Rice attended. He included a few words, in not-entirely-correct Russian, in a letter to Nikiforoff a few months later.[141] In the winter of 1930, after attending the congress in the Soviet Union, Marbut started to teach himself Russian. Two years later, he wrote to his daughter that he was studying the language from 6:30 to 7:30 every morning, adding: "Am now reading Zakharoff's text book on soils and get along with it fairly well. I can read from one to two pages per day. The dictionary has to be used a lot of course but the vocabulary is gradually sticking."[142]

The greater contacts and growing knowledge of Russian work among American specialists enabled some to overcome their ignorance of Russian agricultural sciences and perceptions of Russian "backwardness." The importance both sides attached to contacts meant that they continued after the Russian Revolution of 1917, in spite of the suspicion between their governments.

In time, some Americans came to realize that their Russian counterparts had crop varieties and had produced scientific work that was of great interest to them. In December 1909, the American plant explorer Frank Meyer wrote from St. Petersburg to David Fairchild at the USDA that he had been:

> having talks with various officials in the Central Agricultural Dept. here. There are hundreds of them, just like with us and one could go around for weeks gathering information. Curiously enough, notwithstanding such great differences in race and language, the Russians are developing

[140] NARA CP, RG 54, Finding Aid A1, Entry 58, Division of Vegetable Pathology and Physiology: Correspondence of M. A. Carleton, 1891–1900, Folder M. A. Carleton – 1900, Carleton to Galloway, August 8, 1900.

[141] UMA, Collection Number uarc 173, Department of Soils Records, 1911–1990s, Correspondence, Box 2, Folder 3, Nikiforoff, C. C., 1928–1930, Rice to Nikiforoff, August 17, 1929; Folder 4: Nikiforoff, C. C. 1928–31, Rice to Nikiforoff, December 5, 1929; Rice to Nikiforoff, May 28, 1930.

[142] Louise Marbut Moomaw, "Curtis Fletcher Marbut," in *Life and Work of C. F. Marbut*, p. 25; Marbut Papers (C3720), Folder 57, Marbut to Louise, Roy, and Family, December 12, 1932.

> their agriculture much along the same lines as we do and they too face those numerous problems, that a nation possessing much territory, has to front.[143]

The plant explorers Carleton, Hansen, and Meyer all recognized that in the similar, if slightly harsher, conditions of the steppes, Russian farmers and plant breeders were cultivating varieties of crops and other plants that would be useful in the Great Plains. The participation by Soviet scientists at the soil science in the United States in 1927 and by American scientists at the next congress in the Soviet Union in 1930 confirmed to Marbut and other American scientists the innovative nature of Russian soil science and its practical significance for surveying and mapping soils. Many years later, the soil scientist Jacob Joffe (another Russian–Jewish émigré who studied at Rutgers), wrote about "flares of brilliancy" against a back drop of poverty, famine, and ignorance in tsarist Russia.[144] The Russian experience in planting trees in the steppes that Pinchot saw on his visit in 1902 (see p. 85) proved useful in the 1930s, when Pinchot's Russian-born protégé, Zon, played a key role in drawing on this experience to advocate planting shelterbelts in the Great Plains to combat the Dust Bowl.

Thus, the bridges that were built assisted in overcoming, at least in part, barriers to transfers and influences from the steppes to the Great Plains discussed in the previous chapter. The following chapters will analyze how plants, both "useful" crops and harmful "weeds," scientific expertise in studying soils, and the scientific basis and practical techniques for planting shelterbelts made their way from the steppes of the Russian Empire and Soviet Union to the Great Plains of the United States.

[143] NARA CP, RG 54, Finding Aid NC-135, Entry 135I, Records of Frank N. Meyer, Plant Explorer, 1902–1908, Box 5, Folder 4259g, Meyer, Spb, to Fairchild, December 9, 1909.

[144] J. S. Joffe, "Russian Contributions to Soil Science," in *Soviet Science: Symposium at the 1951 Philadelphia Meeting of the American Association for the Advancement of Science*, ed. R. G. Christman (Washington, DC: AAAS Publications, 1952), pp. 57–8. On Joffe, see *American Men of Science: A Biographical Directory*, ed. Jacques Cattell, 10th edition, *The Physical and Biological Sciences, F–K* (Tempe, AZ: The Jacques Cattell Press Inc., 1960), p. 2006.

PART II

Transfers

CHAPTER 4

Wheat

Introduction

On September 10, 1942, in the Athletic Park in Newton, Harvey County, Kansas, the Junior Chamber of Commerce dedicated an imposing, if plain, seventeen-foot statute of a Mennonite settler from the steppes. The inscription reads: "Commemorating Entry into Kansas from Russia of Turkey Red Hard Wheat by Mennonites 1874." The dedication was threefold: to Bernhard Warkentin, a Mennonite who made his home in Newton, "for bringing hard winter wheat to the United States from the Crimea, Russia"; to the Santa Fe Railway for transporting the wheat and pioneers to the Great Plains; and to "the Mennonite people who . . . toiled to turn a prairie into the richest wheat land in America" as well as "people of all national origins" for producing wheat in the state. The statue was funded by local citizens and farmers and by the New Deal Works Progress Administration (WPA).[1] Turkey Red was the name given in the United States to a hard red winter wheat introduced from the steppes, where it was known as Krymka (Crimean). It was resistant to drought, frost, disease, and pests, and was very productive in the central plains.[2]

There was an upsurge of interest in Kansas in the story around this time. In July 1941, local radio stations broadcast a dramatization of a little Mennonite girl called Anna Barkman selecting wheat seeds for her family to take to the United States from their home in the Crimea:

[1] MLA, MS 68 Bernhard Warkentin (1847–1908) and Wilhelmina Eisenmayer Warkentin (hereafter MLA, Warkentin Papers), Box 5, Folder 20, Dedication, Memorial: "Commemorating Entry into Kansas from Russia of Turkey Red Hard Wheat by Mennonites 1874," September 10, 1942; Image of Mennonite Settler Statue, National Register of Historic Places, available online at https://npgallery.nps.gov/AssetDetail?assetID=b5f792c2-daad-4c3b-88c9-c0784738 0b20, accessed October 11, 2017; Information about the statue is given on a plaque by its base. I visited on May 29, 2017.

[2] For an illustration, see Mark Alfred Carleton, "Hard Wheats Winning Their Way," *Yearbook of the United States Department of Agriculture 1914* (Washington, DC: USDA, 1915), facing p. 400.

ANNA: *(there is the sound of a seed dropping into a tin can)* Fifty-three, fifty-four, nope, color's not right – you can't come with us to America *(pause)*, fifty-five.

MOTHER: *(comes into scene)* And how are you getting along, Anna?

ANNA: All right, mother. But it's pretty slow work.

MOTHER: I know, dear, but your father wants only the best seeds to come along with us.

It was worth it, since, as the radio broadcast explained: "Kansas wheat leads the world."[3]

The story had been collected a few years earlier by WPA Federal Writers Project. In April 1874, in a village in "the Crimea in Russia," eight-year old Anna Barkman had been told by her father to select two gallons of the "largest, hardest, and most perfect grains of wheat." She spent a week choosing 259,862 grains, for which he treated her with a handful of nuts. The Barkmans and twenty-three other Mennonite families packed their belongings, including the seeds, and made their way by train and steamship across Europe and the Atlantic Ocean to New York and then on to Marion County, Kansas. In the fall after they arrived, they planted the seed and harvested their first crop the following June. American farmers noticed that the Mennonites' wheat produced more and better flour than theirs, and farmers came from all over the state to buy their seed. "[I]n a short time, Turkey Red wheat became the most prominent wheat in the state, and gained for Kansas the reputation for the best wheat in the Union."[4] The main source of the story was a book published in 1927 by Wichita journalist and historian Bliss Isely.[5] He had researched it with help from David Richert and Herbert Schmidt of Bethel College, a Mennonite institution in North Newton. They were told the story by Anna Barkman Wohlgemuth, then aged sixty one, in Hoffnungstal, near Gnadenau, in neighboring Marion County.[6]

[3] MLA, Warkentin Papers, Box 1, Folder 15, Radio Address over the Kansas State Network of the Mutual Broadcasting System on July 6, 1941: Wheat, ff.3–6. It was broadcast over four stations in Kansas on July 6 (KTSW, Emporia; KFBI, Wichita; KSAL, Salina; KFGB, Great Bend) and repeated on July 7 on WHB, Kansas City, MO.

[4] WSUL SCUA, Federal Writers' Project, Kansas, Box 10, File 2 Reno County. Sturgeon, Katherine, 1936, [Topic] Racial Groups: The Mennonites, ff.1–2.

[5] Bliss Isely, *Early Days in Kansas* (Topeka: Kansas State Teachers Association, c. 1927), pp. 41–6.

[6] MLA, Warkentin papers, Box 2, Folder 22, *The Bethel Collegian: Weekly Student Publication*, Newton, Kansas, May 25, 1927; Raymond F. Wiebe, *Hillsboro, Kansas: The City on the Prairie* (Hillsboro, KS: Multi Business Press, 1985), p. 140. Anna Barkman was born in the Molotschna colony in the Russian Empire in 1866. She died in Gnadenau, Marion County, Kansas, in 1929. "Anna Barkman (1866–1929)" available online at www.ancestry.co.uk/genealogy/records/anna-barkman-24-68294k, accessed November 7, 2019.

The WPA Guide Book to Kansas, published in 1939, explained how Mennonite immigrants had brought bushels of Turkey Red wheat in their baggage. It continued: "Up to that time attempts to grow wheat on the Plains had not been successful, but the Russian grain was perfectly adapted to these conditions." The guide book pointed also to Warkentin's role in encouraging Mennonites to move to Kansas and in establishing a successful milling company to grind their wheat.[7]

Versions of the story of Mennonites bringing Turkey Red wheat to Kansas from the steppes, with or without Warkentin's encouragement, in 1874 or 1873, have been told, and retold, many times. The first direct reference I have found was in November 1900 in an unpublished letter from Warkentin to Mark Carleton. Warkentin wrote that "hard winter wheat" from the Crimea "originally was brought here to Kansas by the Mennonites, which [sic] settled in 1873 [sic] in Marion County."[8] Carleton elaborated in 1914: "Each family brought over a bushel or more of Crimean wheat for seed, and from this seed was grown the first crop of Kansas hard winter wheat." He added that Warkentin had been "chiefly instrumental in introducing the Turkey wheat."[9] Carleton was invested in the story, since he was a cereal scientist for the United States Department of Agriculture (USDA), had worked with Warkentin to test the wheat, and had visited the steppes twice, in 1898–9 and 1900, to collect varieties of winter and spring wheat suitable for the Great Plains.

The story, which spread from several sources, came to be widely accepted and was reinforced in 1974. The Kansas State Legislature proclaimed 1974 "the year of centennial celebration of the introduction of Turkey red hard winter wheat [by Mennonites from Russia] . . . and the beginning of the great wheat industry in Kansas."[10] Kansas Mennonites took pride in their forebears' contribution.[11] A pageant was held at

[7] Harold C. Evans, *Kansas: A Guide to the Sunflower State* (New York: The Viking Press, 1939), pp. 57, 66–7, 262–3.

[8] MLA, Warkentin Papers, Carleton, Warkentin, Galloway correspondence concerning introduction of Turkey Red wheat, Folder 19, Warketin to Carleton, November 9, 1900, available online at http://mla.bethelks.edu/archives/ms_68/, accessed October 12, 2019. Copies of the correspondence were sent to Bethel College from the National Archives in Washington, DC, in 1945, available online at http://mla.bethelks.edu/archives/ms_68/, Folder 18, T. R. Shellenberg to Cornelius Krahn, November 8, 1944, Folder 19, Shellenberg to Krahn, December 26, 1944, accessed October 15, 2019.

[9] Carleton, "Hard Wheats Winning Their Way," 399.

[10] "Wheat Centennial Proclaimed," *Mennonite Life* 28, no. 2 (June, 1973), 46.

[11] See, for example, John W. Schmidt, "Turkey Wheat: A Mennonite Contribution to Great Plains Agriculture," *Mennonite Life* 29, no. 4 (1974), 67–9.

Warkentin's first Kansas farm, in Halstead, just west of Newton.[12] The U.S. Postal Service issued a commemorative stamp depicting a train steaming through fields of wheat.[13] The USDA released a new strain of wheat, Centurk, to mark the anniversary. Two Kansas agronomists wrote that the Turkey wheat "answered for all time the critics who doubted the future of wheat on the Plains."[14] The story has continued to gather momentum and has been repeated by Mennonites interested in their history.[15] Analyzing similar stories about Mennonite immigrants bringing seeds to the Canadian prairies, Susie Fisher has pointed to the importance in Mennonite culture of "historical writings, physical evidence, memories, and legends of seed transfer and intergenerational propagation."[16]

Right from the start, the story attracted skeptics. In the Kansas State Board of Agriculture's report for 1905–6, Hugh P. Coultis wrote that although it was "commonly believed" that the "hard Russian or 'Turkey' winter wheat was unknown in Kansas before the advent of the Mennonite settlers" in the 1870s, "they were not the first who raised it." Without providing any evidence, he wrote that before they arrived "there was a small colony of French settlers in Marion County who were raising the hard winter wheat, although not to any great extent." The first party of Mennonites, he continued, brought with them no more than twenty or thirty bushels of seed. He noted that the first wheat from the Crimea that was sold to the general public was imported by Warkentin in 1885 or 1886. Coultis did accept, however, that the Mennonites were "largely instrumental" in introducing the variety and were the "greatest factor in its early development."[17]

Kansas historians James Malin and Norman Saul also raised objections, although neither doubted that the Mennonites played some part. In the 1940s, Malin mined Kansas newspapers from the 1870s for reports on Mennonites bringing wheat with them from Russia and sowing it in Kansas. He found references to "Russian sheep, Russian oats, Russian

[12] Margaret Olwine, "In the Cradle of Kansas Turkey Red Wheat," *Star: Sunday Magazine of the Kansas City Star*, June 30, 1974, 10–15.

[13] Lynn Batdorf (2009), "10-cent Wheat Fields & Train," Arago: People, Postage & the Post, available online at https://arago.si.edu/category_2037915.html, accessed October 16, 2017.

[14] K. S. Quisenberry and L. P. Reitz, "Turkey Wheat: The Cornerstone of an Empire," *AH* 48 (1974), 98. See also Robert G. Dunbar, "Turkey Wheat," *AH* 48 (1974), 111–14.

[15] See, for example, Wiebe, *Hillsboro*, pp. 139–40; Glen Ediger, *Leave No Threshing Stone Unturned* (North Newton, KS: Privately published, 2012), pp. 122–32.

[16] Susie Fisher, "(Trans)planting Manitoba's West Reserve: Mennonites, Myth, and Narratives of Place," *Journal of Mennonite Studies* 35 (2017), 144.

[17] Hugh P. Coultis, "The Introduction and Development of Hard Red Winter Wheat in Kansas," *Fifteenth Biennial Report of the Kansas State Board of Agriculture, 1905–1906* (Topeka, KS: 1907), 945–8.

threshers, Russian ovens," to the immigrants growing wheat and purchasing reapers, but no mention that they brought any wheat seed with them. He cast doubt on the Anna Barkman story by reporting evidence that her group had purchased American seed after they arrived.[18] In the 1970s and 1980s, Saul suggested that the story was "largely a myth," but had "some factual basis." Saul was unable to find any "public references to the Mennonites bringing Turkey Red" to Kansas in 1874 until the "early 1900s." Contradictory stories had followed until the 1920s, when Isely's book established the Anna Barkman story. In the 1970s, Saul presented major objections to the story that the Mennonites were primarily responsible for introducing the wheat variety. First, it would have been logistically impossible for them to bring sufficient seed to plant much of the 200,000 acres they cultivated in their early years in Kansas. Second, they had not grown hard red winter wheat in the steppes, but soft and hard spring wheats known as Girka and Arnautka. Third, he suggested that they had converted a hard spring wheat, such as Arnautka, into a winter wheat by sowing it in the fall to take advantage of the hardy Russian wheat and milder Kansas winters. Fourth, when the Mennonites arrived in Kansas, the railroad company supplied them with local seed wheat.[19] Malin and Saul both recognized that unraveling the story entailed researching the types of wheat grown in the steppes around the time of the migration. Malin looked at investigations by Carleton and other USDA cereal specialists. He noted that hard red winter wheat was widespread across southeastern Europe and had been introduced to the United States many times. This cast doubt on the "reasonableness of the Mennonite claim to a virtual monopoly" on the introduction of this type of wheat into Kansas. "Whether true or false," he continued, "provincialism and the wisdom born of hindsight have established a tradition which will long resist any substantial modification."[20]

Recently, Courtney Fullilove has also cast doubt on the primary importance of the Mennonite immigrants, Warkentin, and Carleton. She argued that it "compressed a history of multigenerational agricultural labor and

[18] James C. Malin, *Winter Wheat in the Golden Belt of Kansas: A Study in Adaption to Subhumid Geographical Environment* (Lawrence: University of Kansas Press, 1944), 165-6. For an assessment of Malin's interpretation, see Bonnie Lynn-Sherow, "Beyond Winter Wheat: The USDA Extension Service and Kansas Wheat Production in the Twentieth Century," *Kansas History* 23, nos. 1–2 (2000), 100–11.

[19] Norman E. Saul, "Myth and History: Turkey Red Wheat and the 'Kansas Miracle'," *Heritage of the Great Plains* 22, no. 3 (1989), 1–4; Saul, "The Migration of the Russian–Germans to Kansas," *KHQ* 60 (1974), 56, 60–1.

[20] Malin, *Winter Wheat*, pp. 166–7, 278 n. 25.

biological exchange," and underplayed the longer history of wheat cultivation in the southern steppes before the Mennonites moved there from the late eighteenth century, as well as their experience in the steppes over the following decades. Drawing on materials in English, she outlined this longer history.[21] Her work gave a partial answer to Saul's observation that the "Russian chapter of Turkey Red has yet to be written and the sources are not easily obtainable."[22] This chapter will endeavor to present the "Russian chapter" by drawing on Russian sources and to analyze the objections raised by Malin, Saul, and Fullilove.

To broaden the focus, over the nineteenth and early twentieth centuries, farmers around the world expanded crop cultivation into regions, such as the North American Great Plains and Eurasian steppes, with continental and semi-arid climates, and did so at the expense of the indigenous populations. The Mennonites' arrival in Kansas in 1873–4, for example, coincided with the relocation of the last of the Kanza people to Indian Territory (present-day Oklahoma).[23] Around the globe, migrant farmers sought varieties of familiar crops that could cope with the more extreme climatic conditions, as well as diseases and pests, in continental interiors. They did so with the help of private plant breeders and government agricultural scientists. Together, they improved existing types of wheat by selecting the best strains. From the 1870s, they created hybrids suitable for the harsher conditions. And "agricultural explorers," such as Carleton, traveled the world, including the Russian Empire, in search of hardy and productive varieties.[24] In the Great Plains, farmers, scientists, and government agencies benefitted from prior Russian experience, since large areas of the steppes had already been plowed up to grow grain. It was significant also that conditions in the steppes were slightly harsher than in much of the American plains.

Although Turkey Red hard winter wheat has assumed much of the glory, the origins of other types of wheat grown in the American plains can be

[21] Courtney Fullilove, *The Profit of the Earth: The Global Seeds of American Agriculture* (Chicago: The University of Chicago Press, 2017), pp. 101, 111–19.

[22] Saul, "Myth and History," 12, n. 25.

[23] Ronald D. Parks, *The Darkest Period: The Kanza Indians and Their Last Homeland, 1846–1873* (Norman: University of Oklahoma Press, 2014).

[24] A. L. Olmstead and P. W. Rhode, "Biological Innovation in American Wheat Production," in *Industrializing Organisms: Introducing Evolutionary History*, ed. S. R. Schrepfer and P. Scranton (London: Routledge, 2004), pp. 62–5; J. Allen Clark, "Improvement in Wheat," *Yearbook of Agriculture 1936* (Washington, DC: USDA, 1936), pp. 207-302; M. K. Bennett, "World Wheat Crops, 1885–1932," *Wheat Studies* 9 (April 7, 1933), 239–74; Bennett and H. C. Farnsworth, "World Wheat Acreage, Yields, and Climates," *Wheat Studies* 13, (March 6, 1937), 265–308.

traced to the steppes. These included durum wheat, a very hard spring wheat, varieties of which were also introduced by Carleton. Canadian agricultural scientists created a hybrid hard red spring wheat, known as Marquis, for their prairies. One of Marquis's parents came from the steppe region. The wheat varieties were part of a larger story of plant introductions from the Eurasian steppes to the American plains. Other arrivals included more varieties of grains, fodder crops, in particular alfalfa and crested wheat grass, and hardy trees and shrubs. In the bigger history of American bioprospecting in Eurasia, in addition to Carleton, other significant figures were Niels Hansen, Frank Meyer, and Henry Bolley, whom we encountered in the previous chapter. Hansen traveled extensively on several visits between 1894 and 1934 and collected many plants, including wheat, fodder crops, and trees. He worked for the USDA and at the state agricultural college and experiment station in Brookings, South Dakota. Meyer made four trips to Russia and China between 1908 and 1918. He sent back fodder crops, trees, wheat, and other grains. Bolley, who worked at the state agricultural college and experiment station in Fargo, North Dakota, visited Russia in 1903, and collected varieties of flax as well as wheat.[25]

To trace precisely which varieties of wheat were introduced we need to compare the types cultivated in the Great Plains with those grown in the steppes. This is quite complicated. There were no standard systems for classifying wheat at this time and, to confuse matters, Russians and Americans described different varieties in different ways. Further, there were numerous sorts of wheat, which varied from place to place, and many names that were not used consistently. Similar names were used for different types, and different names for similar types. In 1897, Carleton explained that "great confusion ... exists in regard to the naming of varieties," which was "increased by the complexity of languages." He added that "Beloturka and Kubanka of Russia are the same probably. Beloturka is also applied to one or two other very different sorts from

[25] See Niels E. Hansen, "Fifty Years Work as Agricultural Explorer and Plant Breeder," *Transactions of the Iowa State Horticultural Society* 79 (1944), 28–49; Helen Hansen Loen, *With a Brush and Muslin Bag: The Life of Niels Ebbesen Hansen* (Kalamazoo, MI: privately printed, 2003); Isabel Shipley Cunningham, *Frank N. Meyer: Plant Hunter in Asia* (Ames: Iowa State University Press, 1984); NDSUA, ND, Mss 18, Henry Luke Bolley Papers (hereafter Bolley Papers), Box 51, Folder 8: Flax – Agreement between the North Dakota Agricultural Experiment Station and the Bureau of Plant Industry [USDA]; Folder 11: General Correspondence – Russia, 1897-1932, Bolley to Z. B. Libenson, Intourist, New York, May 20, 1939; Box 52, Folder 2: Flax: Record of Samples Collected in Russia in 1903; Bolley to A. J. Pieters, USDA, Washington, DC, July 15, 1904; "Record of Samples of Flax Seed collected by H. L. Bolley in Russia during Season of 1903;" "List of small samples of miscellaneous kinds of seeds."

Siberia."[26] For the same reasons, it is difficult to identify accurately the types of wheat that were already being cultivated in the prairies and Great Plains on the eve of the Mennonites' arrival. In 1857, John Hancock Klippart of the Ohio State Board of Agriculture published a compendium of existing knowledge of wheat varieties. One of his correspondents, D. Gregory of Delaware County, Ohio, summarized the dilemma: "The kinds of wheat formerly raised had local names, and the same variety in different sections was not infrequently known by different names."[27] As if the task of identifying which types of wheat were grown where was not difficult enough, USDA scientists did not devise a comprehensive system for classifying crop varieties until the 1920s, when they consolidated various schemes used in the United States and around the world.[28]

Nevertheless, the main distinctions that need to be born in mind here, and will assist in identifying the types of wheat introduced from the steppes to the Great Plains, are between: winter and spring wheats; hard and soft wheats; red and white wheats; and awned (or bearded) and smooth wheats.[29] Winter wheats were sown in the fall, while spring wheats were, logically, sown in the spring. Winter wheats were grown in regions with milder climates where the shoots were likely to survive the winters. In the steppes of the Russian Empire, they were grown near the moderating influence of the Black and Azov seas, while in the United States they thrived in the central and southern plains. Spring-sown wheats were more common in regions with harsher climates, where crops planted in the fall were liable to be killed by cold winters or spring frosts. Thus, spring wheats were grown in the north and east of the steppe region, far from the seas, and in the north and west of the Great Plains. The distinction between winter and spring wheats was not absolute, as varieties sown in the spring in one location could be planted in the fall in others with milder winters.

Hard and soft wheats had different markets. Russian traders exported the very hard durum wheats to the Mediterranean world, where they were used to make pasta. Soft wheats were sold to western Europe for bread flour. It is, unfortunately, more convoluted. The terms "hard" and "soft"

[26] NARA CP, RG 54, Finding Aid A1, Entry 58, Division of Vegetable Pathology and Physiology: Correspondence of M. A. Carleton, 1891–1900, Folder M. A. Carleton – 1897, Chief of Division [Galloway] to Mr. John Hoffer, Manager, Paxton and Stelton [?] Flouring Mills Co., Harrisburg, PA, June 10, 1897. (The letter is in Carleton's handwriting.)

[27] John Hancock Klippart, *The Wheat Plant: Its Origin, Culture, Growth, Development, Composition, Varieties, Disease* (Cincinnati: Moore, Wilstach, Keys, & Company, 1860), pp. 479, 486.

[28] See Fullilove, *Profit*, pp. 135–6.

[29] See Jessica Barnes, "Separating the Wheat from the Chaff: The Social Worlds of Wheat," *Environment and Society* 7 (2016), 91.

were used differently in Russia and the United States. In Russia, the term "hard" was used only for the very hard durum wheats, and Crimean (Turkey Red) wheat was considered "soft." In the United States, however, this type of wheat was described as "hard."[30] To cut through these complications, I will rely on the system employed by Carleton, who understood the different varieties in both Russia and the United States. He distinguished between *three* groups: "soft bread wheats"; "hard bread wheats," in which he included Turkey Red; and very hard durum wheats.[31] The color of the grains was used in the names of some types of wheat, such as Turkey Red and a type of durum wheat known as Beloturka (White Turk). The final distinction was between types with grains that had "awns" or spikelets growing from the grains, which assisted in the dispersal of seeds, and were known as awned or bearded wheat, and smooth types that did not.[32] Turkey Red was an awned, hard bread, winter wheat, while durum wheats also had awns, were very hard, sown in the spring, and many were white.

This chapter also considers how barriers were overcome to the successful introduction of wheat from the steppes. Since Russia and the United States were competitors in the world grain market, it would not have been surprising if the Russian authorities and farmers had been reluctant to share their most productive and hardy crop varieties with their chief rival. The recurring crop failures in the steppes, such as those of 1891 and 1921, were well known in the United States and could have reduced interest in acquiring crop varieties from a region with such a checkered agricultural history. American ignorance of Russian crops had been overcome since the 1870s by the Mennonite migrants who imported some to the Great Plains. But, USDA scientists had to overcome the language barrier to read Russian specialist literature and travel around Russia as "plant explorers." A final, and significant, barrier was resistance to change among farmers reluctant

[30] See Malin, *Winter Wheat*, p. 162; Mark Alfred Carleton, "Russian Cereals Adapted for Cultivation in the United States," *USDA Division of Botany Bulletin*, no. 23 (1900), 13–15, 34–5; P. Slezkin, "Zametki o minuvshei vystavke v Chikage," *SKhiL* 174 (1893), 376; F. Keppen, "O polevodstve v Tavricheskoi gubernii i o vrednykh na nego vliianiiakh," *ZhMGI* 83 (1863), 98; M. E. Falkus, "Russia and the International Wheat Trade, 1861–1914," *Economica*, NS 33 (1966), 416–29, 418, n.3; "Pshenitsa," in S. M. Bogdanov, *Illiustrirovannyi sel'skokhoziaistvennyi slovar'* (Kiev: Barskii, 1891–[95]), pp. 1081–2.

[31] Mark Alfred Carleton, "Improvements in Wheat Culture," *Yearbook of the United States Department of Agriculture 1896* (Washington, DC: USDA, 1897), p. 492; Carleton, *The Small Grains* (New York: Macmillan, 1916), pp. 27–86.

[32] See Rivka Elbaum, Liron Zaltzman, Ingo Burgert, and Peter Fratzl, "The Role of Wheat Awns in the Seed Dispersal Unit," *Science* 316, no. 582 (2007), 884–6.

risk sowing new crops, and among millers opposed to the expense of retooling their mills to grind the harder Russian wheats.

The next section takes us to the steppes to investigate the types of wheat and other cereals farmers, in particular but not solely Mennonites, grew there. The following section examines the crops, including types of wheat, grown by Native American and Euro-American farmers in the plains before the arrival of immigrants from the steppes. At the heart of the chapter is analysis of the roles in the introduction of wheat varieties from the steppes played by immigrants, especially Mennonites, plant explorers, and scientists from the USDA, in particular Carleton, contacts between American and Russian agricultural scientists, and the international grain trade.

Wheat and Other Cereals in the Steppes

By the second half of the nineteenth century, cultivation of wheat and other grains in the steppes was a major commercial operation. In the 1890s, over a third of all wheat harvested in the steppes went for export. It was a precarious enterprise as yields fluctuated sharply from year to year on account of recurring droughts, disease, and pests.[33]

The Mennonites who moved to North America during and after the 1870s came from the steppes of the Crimean peninsula (Krym) and the region north of the Sea of Azov in present-day Ukraine. (See Map 5: Map of the Mennonite colonies in "New Russia" [S. Ukraine] in 1875.) The Molotschna colony had been founded north of the Sea of Azov at the start of the nineteenth century by Mennonite immigrants from the vicinity of the Baltic port of Danzig (now Gdańsk). In the steppes, they partly displaced Tatar and Nogai peoples, who lived mainly by raising livestock, but also grew some crops, including wheat.[34] The Mennonites adapted to the local conditions, including relatively low rainfall, periodic droughts, high winds, but also fertile soil. Under the firm leadership of Johann Cornies (1789–1848), they planted fruit trees, mulberry trees for silkworms, and forest trees to shelter their homes and fields. Grazing large herds of sheep and cattle was an important part of their economy. They also plowed up land to grow grain. From the mid 1830s, they used

[33] See John Martin Crawford (ed.), *The Industries of Russia*, vol. 3, *Agriculture and Forestry*, trans. Crawford (Spb: Department of Agriculture, Ministry of Crown Domains, 1893), pp. 93–107.

[34] See Fullilove, *Profit*, pp. 111–19; David Moon, *The Plough that Broke the Steppes: Agriculture and Environment on Russia's Grasslands, 1700–1914* (Oxford: Oxford University Press, 2013), p. 17.

four-field rotation and techniques to conserve scarce moisture in the soil, such as keeping the fallow field clean of vegetation, known as "black fallow" (*chernyi par*, *schwartze Brache*), plowing deeply, and harrowing their fields to reduce evaporation. Their methods worked and they became successful farmers. This was due partly to their hard work, but also to their privileged status: they had more land on better terms than their Slavic, Tatar, and Nogai neighbors. Mennonite wheat production in the steppes developed rapidly from the 1840s as farmers plowed up pasture land and sowed wheat for two years rather than one in their rotation.[35]

A serious objection to the story that the Mennonites introduced the hard red winter wheat (Turkey Red/Krymka) to Kansas is Saul's view that they had not grown it in the steppes. He stated that they had cultivated mostly Girka, a soft spring wheat, and Arnautka, a very hard spring durum wheat (See p. 133). Saul's claim can be investigated by consulting Russian sources. In his annual reports for 1808 and 1811, the governor of Tauride province – which included the Crimean peninsula and the steppes north of the Sea of Azov where the Mennonites lived – wrote that red winter wheat was grown throughout most of the province and comprised around a fifth of all grain sown.[36] Nevertheless, it may not have been very widely sown by Mennonites in Molotschna. In 1837, an official reported that they were starting to sow a red wheat from the Crimean peninsula as well as their customary Arnautka. An attraction of the Crimean wheat was that it was in high demand in the nearby ports of Berdiansk and Mariupol'.[37] Henry Danby Seymour, a British politician who visited the region in the early 1850s, noted the high quality of the "Krymka" wheat produced by the "wealthy German colonies on the Moloshna [sic]" and exported from Berdiansk. It commanded a higher price than other types of wheat. English millers appreciated its quality and started to import large quantities in the early 1850s.[38]

Nevertheless, between the 1840s and early 1860s, Mennonite farmers in Molotschna did not cultivate large quantities of red winter wheat. Visiting Russian and German agricultural specialists wrote that Girka and Arnautka

[35] David Moon, "Cultivating the Steppe: The Origins of Mennonite Farming Practices in the Russian Empire," *Journal of Mennonite Studies* 35 (2017), 251–78.

[36] RGIA, f.1281, 1804-10, op.11, d.131, ll.203ob.-208; op.11, 1804-11, d.132, ll.9-14.

[37] RGIA, f.383, op.29, 1837-8, d.609, ll.37ob.-8.

[38] Henry Danby Seymour, *Russia on the Black Sea and Sea of Azof: Being a Narrative of Travels in the Crimea and Bordering Provinces* (London: John Murray, 1855), pp. 315, 319–20.

were indeed the main types of wheat they grew.[39] They had moved away from growing the winter wheat, because in the steppes north of the Sea of Azov it suffered from winter killing. Instead, they planted spring wheats. On the Crimean peninsula, farmers continued to sow the red winter wheat, as the winters were milder and they valued the wheat's greater resistance to drought as well as the high demand it enjoyed.[40] Farmers north of the Sea of Azov resumed growing the hard red winter wheat in the 1860s. This was due, in part, to the Crimean War (1854–6). During the war, the Russian authorities ordered Mennonites who lived north of the Crimean peninsula to transport supplies, troops, casualties, and prisoners of war to and from the war zone. Mennonites involved in the transports saw large expanses of sparsely populated steppe in the north of the peninsula. After the war, some Mennonites founded new settlements on this land to escape land shortages in their former communities and disputes over religious matters. They adapted their farming methods to the slightly different conditions, including the milder winters, and started to sow hard red winter wheat. Among the Mennonites who moved to the Crimean peninsula and cultivated the Krymka wheat was a group from Molotschna led by Jakob Wiebe, who became known as the Krimmer (Crimean) brethren. Their numbers included the Barkman family.[41]

From the 1860s, winter wheat spread from the peninsula to the Mennonite communities to the north of the Sea of Azov. Farmers sowed it in place of spring wheat that had recently been suffering from disease. They learned to protect the young shoots of the winter wheat from spring frosts by harrowing the fields to cover them with soil. Visitors to Molotschna noted the Mennonites' shift to winter wheat. Vladimir Postnikov, who stopped by in 1881, wrote that the "Germans" had been

[39] Iu. Vitte, "O sel'skom khoziaistve v Khersonskoi, Tavricheskoi i Ekaterinskoi guberniiakh," *ZhMGI* 8 (1844), 105; F. Ķeppen, "Sel'skokhoziaistvennoe polozhenie Tavricheskoi gubernii v 1863 g.," *ZhMGI* 86 (1864), 478; K. Bunitskii, "Voprosy i otvety o vozdelyvanii pshenitsy: stat'ia vtoraia," in *Sbornik statei o sel'skom khoziaistve iuga Rossii*, ed. I. Palimpsestov (Odessa: P. Frantsov, 1868), pp. 311–12; Alexander Petzholdt, *Reise im westlichen und südlichen europäischen Russland im Jahre 1855* (Leipzig: Hermann Fries, 1864), p. 165.

[40] Keppen, "O polevodstve," 98; Iu. Vitte, "Ocherk khlebopashestva na Krymskom poluostrove," *SKhiL* 8 (1870), 241–6; David G. Rempel, "The Mennonite Colonies in New Russia: A Study of their Settlement and Economic Development from 1789 to 1914" (PhD dissertation, Stanford University, 1933), p. 247.

[41] James Urry and Lawrence Klippenstein, "Mennonites and the Crimean War, 1854–1856," *Journal of Mennonite Studies* 7 (1989), 9–32; David V. Wiebe, *Grace Meadow: The Story of Gnadenau and Its First Elder, Marion County, Kansas* (Hillsboro, KS: Mennonite Brethren Publishing House, 1967), pp. 27–9; Heinrich Goerz, *Mennonite Settlements in Crimea*, trans. John B. Toews (Winnipeg: CMBC Publications and the Manitoba Mennonite Historical Society, 1992), pp. 6–8, 24–6, 38.

cultivating winter wheat successfully "for over ten years." He later noted that the winter wheat was mainly "Krymka."[42] N. K. Kalageorgi and B. M. Borisov, who visited Molotschna in July 1874 – the precise time of the migration to the United States – wrote that Mennonite farmers' best crop was "red wheat" ("*krasnaia pshenitsa*").[43] In the late nineteenth century, cultivation of hard red winter wheat spread throughout the steppes north of the Black Sea and Sea of Azov and further east into the North Caucasus. In this large region it had become the predominant type of wheat by the turn of the twentieth century.[44]

Thus, we know from several contemporary Russian sources that Mennonite farmers in the steppes were cultivating Krymka – a hard red winter wheat – in the 1870s, at the time of the Mennonite migration to the United States, and in the following decades, when immigrants imported wheat from their former homes, and USDA plant explorers, including Carleton, visited Mennonite communities in the steppes to collect more of this type of wheat. This was the variety that turned up in Kansas, where it became known as Turkey Red. "Krymka wheat and Turkey wheat were effectively synonyms," remarked Fullilove after discussing samples with a scientist at the Vavilov Research Institute of Plant Industry in St. Petersburg in 2012.[45] Mennonite farmers in the steppes had switched to hard red winter wheat in the 1860s, because of its productivity and resistance to disease and drought: the precise qualities that made it so successful in the Great Plains.

Throughout the steppe region, wheat was the main grain and occupied between a third and half of all arable land.[46] The predominance of wheat in the steppes contrasted with the rye and oats that prevailed in the forested heartland of Russia around Moscow, where rainfall was more

[42] V. E. Postnikov, "Molochanskie i Khortitskie Nemetskie kolonii," *SKhiL* 139 (1882), 80; 140 (1882), 36–9; V. E. Postnikov, *Iuzhno-Russkoe krest'ianskoe khoziaistvo* (Moscow: I. N. Kushnerev, 1891), p. 178.

[43] N. K. Kalageorgi and B.M. Borisov, *Ekskursiia na reku Molochnuiu: opyt sel'skokhoziaistvennogo i ekonomicheskogo issledovaniia iuzhnorusskikh khoziaistv* (Spb: V. F. Demakov, 1878), pp. 70, 100–2. For similar accounts in American sources, see Carleton, "Hard Wheats," 399; Rempel, "The Mennonite Colonies," 247; Malin, *Winter Wheat*, 166.

[44] M. Sofronov, "Pshenitsa," *Polnaia entsiklopediia russkogo sel'skogo khoziaistva*, 12 vols. (Spb: Devrien, 1900–12), vol. 8, p. 6; Carleton, "Hard Wheats," map on 400. See also Bennett and Farnsworth, "World Wheat Acreage," 269.

[45] Fullilove, *Profit*, p. 17.

[46] Unless otherwise stated, the following discussion is based on a volume prepared by the Russian department of agriculture for the World's Columbian Exposition in Chicago in 1893, i.e. information that was accessible in the United States from the 1890s. Crawford, *The Industries of Russia*, vol. 3, *Agriculture and Forestry*, pp. 93–107.

abundant, but the soil less fertile. When, from the seventeenth and eighteenth centuries, settlers moved to the steppes, they adapted their crops and farming methods to the different conditions. Steppe farmers grew various types of wheat that suited the local conditions and met market demand. The names used cannot be treated as "scientific" terms for specific varieties, but can serve as general guides. Among the winter wheats grown in the west of the steppe region, with its milder climate, were soft varieties, such as Sandomirka, as well as the hard red wheats known as Krymka. The spring wheats cultivated throughout the steppes, especially in the harsher climate to the east, included soft types, such as Girka, which varied from red to yellow in color and produced good bread flour. The most common very hard durum spring wheats were Arnautka, Beloturka ("White Turk") and Krasnoturka ("Red Turk"), which were excellent for pasta, but could also be ground into flour for bread. (The last named is *not* to be confused with the quite different, and less hard red winter wheat called Krymka in Russia and "Turkey Red" in the United States.) These spring-sown durum wheats were the main types in the Volga basin and further east in southern Siberia and the north of present-day Kazakhstan. Nineteenth-century Russian specialists thought soft winter wheats, such as Sandomirka, had come from Polish lands to the west, while Girka and Arnautka had been grown in the steppes for centuries.[47] In the 1920s, Nikolai Vavilov stated that hard red winter wheat was a Crimean wheat of ancient lineage.[48]

Steppe farmers cultivated other grains. Rye occupied up to a fifth of the arable land. They also grew millet, as well as barley and corn (*kukuruza*). The last two were more common in the west of the region. The expansion of crop cultivation in the steppes over the nineteenth century, fueled in part by demand for exports, squeezed the livestock husbandry that had been so important for centuries, among the both the indigenous inhabitants, some of whom were nomadic pastoralists, and the early generations of settlers.

In the late nineteenth and early twentieth centuries, agricultural settlers moved in growing numbers to the steppes of southern Siberia and northern Kazakhstan, where they plowed up Kazakh pasture land. This was the latest phase in the outside, agricultural settlement of the steppe region as a whole, and directly paralleled the similar process taking place in the

[47] See Moon, *Plough*, pp. 16–21; A. Skal'kovskii, *Opyt statisticheskogo opisaniia Novorossiiskogo kraia*, 2 parts (Odessa: Frantsov and Nitche, 1850–3), part 2, pp. 78–80.

[48] Fullilove, *Profit*, pp. 100, 119.

northern Great Plains at the same time.[49] In the steppes, as well as the plains, the government offered agricultural advice to settlers. In Kustanai (Kostanai) district, Turgai region, in Kazakhstan, the authorities set up an experiment field in 1907 to prepare recommendations for settlers who were struggling with the severe local conditions. In 1911, the field's director reported that the climate was sharply continental and dry. Temperatures ranged from −44°C in the winter to +37°C in the summer, the frost-free period was only 120–125 days, 291 mm (11.5 inches) of precipitation fell each year, much of it around harvest time, and droughts and high winds were common. There were areas of fertile soil, but some land was too salty for crops due to the low rainfall. The agronomists tested a range of crops, including Krymka winter wheat, spring wheats, other grains, fodder grasses such as alfalfa, various crop rotations, methods such as "black fallow" to conserve moisture, and tried to grow trees. The director concluded that "spring hard wheat" was the main crop for the region.[50] A few years earlier, in 1898, the Turgai regional agronomist reported that the main type of wheat sown, by settlers as well as Kazakhs, was spring wheat, and named one of the main varieties as Kubanka.[51] This brief excursion into northern Kazakhstan is directly relevant as USDA plant explorer Carleton traveled to the nearby city of Orenburg in 1898 to collect wheat varieties that could grow in such harsh conditions.

The tests at the experiment field in Turgai region were part of wider trials by Russian scientists throughout the steppes to select the most appropriate crops for local conditions. They tested varieties available inside the Russian Empire and introductions from other countries. Many of the varieties from milder conditions abroad, including the United States, struggled to survive the steppe winters.[52] After 1917, building on work begun before the revolution, Soviet crop scientists, including Vavilov, sought out and tested a wide range of crop varieties, and produced new, more productive, and hardier sorts through selection and hybridization. The new varieties included some derived from hard red winter wheat from

[49] On settlement in present-day Kazakhstan, in comparison with the northern Great Plains, see Steven Sabol, *"The Touch of Civilization": Comparing American and Russian Internal Colonization* (Boulder: University Press of Colorado, 2017), pp. 171–204; and on Siberia, see Donald W. Treadgold, *The Great Siberian Migration: Government and Peasant in Resettlement from Emancipation to the First World War* (Princeton: Princeton University Press, 1957).

[50] TsGA RK, f.30, op.1, d.5, ll.16–34, 92-96, 98ob.–104ob., 106–114ob.

[51] TsGA RK, f.318, op.1, d.48, ll.5ob., 17ob-18, 57ob.–58, 141, 143.

[52] N. Vasil'ev, "Botanicheskie raznovidnosti i sorta khlebnykh rastenii v Rossii," *SKhiL* 214 (1905), 3–47.

the Crimea.[53] In the 1920s, Vavilov devised a theory of the origins of cultivated plants that traced wheat to part of the South Caucasus adjoining Turkey.[54] This was more specific than other contemporary crop scientists. Carleton, for example, noted in 1916 simply that wheat's origins lay in "Western Asia," i.e. the famed "fertile crescent," in the region's prehistory.[55] More recently, scientists have largely confirmed Vavilov's theory and placed the genesis of cultivated wheat in southeastern Turkey.[56] From eastern Turkey, wheat had only a short hop over the Caucasus Mountains to reach the steppes of Eurasia. Wheat faced a rather longer journey before it could reach North America. With the exception of corn, all the cereals cultivated in the New World were introduced from other continents.[57]

Wheat and Other Cereals in the Great Plains to the 1870s

When the Mennonite Warkentin arrived in the New World from another continent in 1872 and traveled around the prairies and Great Plains, his impressions of American wheat were mixed. He wrote from Summerfield, Illinois, to his friend David Goerz in the steppes: "In general the wheat in America is not the best. In Pennsylvania, Ohio and Indiana we found the winter wheat to be very poor due to a dry winter." On the other hand: "Here in Summerfield the wheat is very good." A few weeks later, after visiting the state fair in Minneapolis and traveling around Minnesota and the Dakota Territory, he reported that he had seen some "arnautka," a "Russian summer [i.e. spring] wheat."[58]

When Warkentin and other Mennonites moved to Kansas in the 1870s, they found a state being settled by migrants from further east in the United States and others, like themselves, from parts of Europe. For many early

[53] N. I. Vavilov, *Polevye kul'tury iugo-vostoka* (Petrograd: Red-izdat. Kom. Narodnogo Kommissariata Zemledeliia, 1922); V. V. Talanov, *Raiony sortov iarovoi i ozimoi pshenitsy SSSR i ikh kachestvo (prilozhenie 32-e k Trudov po Prikladnoi Botanike, Genetike i Selektsii)* (Leningrad, 1928). On the selection of varieties of wheat and corn at the Odessa experimental farm in 1912, see DAOO, f.22, op.1, 1912, d.481.

[54] See N. I. Vavilov, "Aziia: istochnik vidov," *Nauka i zhizn'*, 1 (1968), 97–9; Kim E. Hummer and James Hancock, "Vavilovian Centers of Plant Diversity: Implications and Impacts," *Horticultural Science* 50 (2015), 780–3.

[55] Carleton, *Small Grains*, p. 32.

[56] See, for example, Ehud Weiss and Daniel Zohary, "The Neolithic Southwest Asian Founder Crops: Their Biology and Archaeobotany," *Current Anthropology* 52, S4 (2011), 237–54; Fullilove, *Profit*, pp. 1–5.

[57] See Carleton, *The Small Grains*.

[58] MLA, Warkentin Papers, Box 1, Folder 7, Warkentin to Goerz, Summerfield, Illinois, July 29/11 1872; Warkentin to Goerz, September 29, 1872.

settlers, but not the Mennonites, the treeless grasslands with their fertile soil, continental and semi-arid climate, periodic droughts, and high winds were unfamiliar. Many who tried to farm struggled as the crops, implements, and techniques they accustomed to were not suitable in the plains environment. Their difficulties increased the further west they went, because precipitation declined from east to west. Farmers had to deal with crop diseases, especially rust (fungi that attack wheat), and pests, such as chinch bugs, hessian flies, and grasshoppers. The Native Americans whom the settlers displaced, such as the Kanza, had cultivated corn, beans, and squash, as well as prairie potatoes, melons, and other crops in gardens around their villages in the river valleys. In what became western Kansas, nomadic peoples, for example the Cheyenne, had hunted bison and traded bison products for corn. After 1830, Native Americans from further east were relocated to the Kansas Territory. They brought their own crops and farming practices, and some they had adopted from Euro-American farmers. Most inhabitants of Kansas in the mid nineteenth century, Native American and Euro-American alike, raised corn as their main crop as it was easier to grow and higher yielding than other cereals. Corn remained the state's main crop through 1914. Kansas farmers also grew oats, barley, rye, millet, alfalfa, sorghum, and flax, all of which were brought from elsewhere.[59]

Over the same period, Euro-American farmers gradually replaced the corn that had been grown in the region by generations of Native Americans with wheat as it was more lucrative and had ready markets at home and overseas. Farmers sought varieties of wheat and cultivation techniques suitable for conditions in the plains.[60] The first wheat in Kansas had been grown in 1839, paradoxically by Native Americans, at the Shawnee Methodist Mission in present-day Fairfield (a suburb of Kansas City, Kansas). They cultivated around 100 acres of soft winter wheat. Other Native Americans who had been relocated to Kansas grew some wheat, and Euro-American settlers planted more. Between 1872 and 1900, the area

[59] See Homer E. Socolofsky, "The Agricultural Heritage," in *The Rise of the Wheat State: A History of Kansas Agriculture, 1861–1986*, ed. George E. Ham and Robin Higham (Manhattan, KS: Sunflower University Press, 1987), pp. 19–27; E. G. Heyne, "The Development of Wheat in Kansas," in ibid., pp. 41–56; Frank G. Bieberly, "Other Crops in the Wheat State," in ibid., pp. 57–71; Malin, *Winter Wheat*, pp. 3–5, 39, 54–6.

[60] See Geoff Cunfer and Fridolin Krausmann, "Adaptation on an Agricultural Frontier: Socio-Ecological Profiles of Great Plains Settlement, 1870–1940," *Journal of Interdisciplinary History* 46 (2016), 355–92; Jeremy Atack, Fred Bateman, and William N. Parker, "Northern Agriculture and the Westward Movement," in *The Cambridge Economic History of the United States*, vol. 2, ed. Stanley L. Engerman and Robert E. Gallman (Cambridge, England: Cambridge University Press, 2000), pp. 285–328.

under wheat in Kansas increased over forty-fold from 68,000 to around 3,000,000 acres.[61] The early Euro-American settlers tried various types of wheat that they had brought with them, purchased from further east, or had been introduced from abroad. Initially, spring wheat predominated, but over the 1870s and 1880s, it was replaced by winter wheat. Among the varieties of soft winter wheat were Early May and Fultz, or Bluestem, which were valued as they matured early. In the 1870s, T. C. Henry, the "wheat king," grew soft winter wheat east of Abilene, about fifty miles north of the Mennonite settlements in Marion County. Farmers growing wheat struggled with droughts, heavy rains and floods, frosts in the late spring and early fall, disease, and pests.[62]

If wheat production was to take off in Kansas and elsewhere in the central and southern plains, farmers needed varieties that matured early so that they could harvest them before they were struck by heat waves, droughts, pests, and diseases later in the summer. Thus, the most suitable were varieties of fall-sown winter wheat that were sufficiently hardy to survive the winter and spring frosts, and would ripen before the height of the summer. Further north in the plains, where winters were too harsh for winter wheat, farmers needed spring-sown varieties that matured in time to avoid, or were sufficiently hardy to survive, early frosts and diseases.[63] David Fairchild, who was in charge of seed and plant introductions at the USDA, recalled that at the end of the nineteenth century settlers in the central and northern plains "were appealing almost daily to the Department of Agriculture for information as to what crops could be grown. . . . [They] demanded hardier [and more drought-resistant] and plants than were found in the eastern States."[64]

Several historians, including Malin and Fullilove, have suggested that Kansas farmers were growing hard red winter wheat before the Mennonites arrived with their varieties of this type in the mid 1870s.[65] These suggestions are hard to pin down, because of the lack of precision in classifying crop

[61] Socolofsky, "Agricultural Heritage," 20. For different figures on acreages, but the same trend, see Malin, *Winter Wheat*, p. 44.

[62] Malin, *Winter Wheat*, pp. 7–58, 66–79, 96–101; Heyne, "Development," 42.

[63] See Malin, *Winter Wheat*, 16–22; Heyne, "Development," 42; Fullilove, *Profit*, pp. 101–2; James L. Colwell, "American Wheat Varieties: Our History in Microcosm," *Social Science Journal* 16 (October 1979), 71–3; Alan L. Olmstead and Paul W. Rhode, *Creating Abundance: Biological Innovation and American Agricultural Development* (New York: Cambridge University Press, 2008), pp. 17–18.

[64] David Fairchild, *The World Was My Garden: Travels of Plant Explorer* (New York: Scribner's Sons, 1938), p. 114.

[65] See, for example, Malin, *Winter Wheat*, p. 162; Colwell, "American Wheat Varieties," 74; Fullilove, *Profit*, pp. 101, 135–6.

varieties at this time. As we have seen, similar names were used for quite different varieties and different names for similar ones. There is information about types of wheat grown in Ohio (four states east of Kansas) in the 1850s in Klippart's compendium published in Cincinnati in 1860. This enables us to investigate whether a type of wheat similar to that introduced by the Mennonites was already grown in the prairies. Klippart made tantalizing references to wheat with names such as "Turkey large red" and "Turkey." But, were these similar to the "bearded," or "awned," hard red winter wheat that we know was introduced later from the steppes? According to Klippart, "Turkey large red" was "bearded," but it was "tinged," rather than red, and he did not mention whether it was a spring or winter wheat. Klippart's "Turkey" wheat was smooth, and thus lacked the "beard" of the wheat from the steppes. He discussed it in a section on "smooth white winter wheats," i.e. types that lacked all but one of the characteristics (winter) of the variety with a similar name from the steppes.[66]

Klippart also presented information on red winter wheats grown in Ohio before 1860. They included "bearded red winter wheat," "red chaff winter wheat," "China," "Mediterranean red bearded wheat," "Cretan," and "Egyptian." Some were resistant to pests and disease, and both hardier and earlier ripening than other types of wheat. Some encountered opposition from millers, presumably because the kernels were too hard for their millstones. All these features are similar to the wheat introduced later from the steppes. On the other hand, cultivation of some had declined or ceased in Ohio by the 1850s, because they had fallen victim to pests and been replaced by soft white wheats.[67] Klippart did not explain where these red winter wheats came from. The names suggest that some originated in the eastern Mediterranean, not too far from the Crimea and the steppes to the north of the Black and Azov seas. However, in 1921, a USDA crop scientist described some of them as "soft" or "semihard,"[68] and thus different from the hard wheat from the steppes. It seems unlikely, therefore, that the types of red winter wheat grown in Ohio in the first half of the nineteenth century were the forebears of the Turkey Red wheat grown later in Kansas. Klippart's information suggests that hard red awned winter wheats similar to those introduced from the steppes from the 1870s were not present, or not in large quantities, before the Mennonites arrived.

66 Klippart, *The Wheat Plant*, pp. 498, 531, 547.
67 Klippart, *The Wheat Plant*, pp. 484–6, 502, 509, 512–25.
68 C. E. Leighty, "Varieties of Winter Wheat Adapted to the Eastern United States," *USDA Farmers' Bulletin*, 1168 (1921), 18.

The other type of wheat introduced later from the steppes, very hard, awned, spring, including durum, varieties – one of which was called "Black Sea wheat" – were cultivated in Ohio in the 1850s. But, these differed sharply from Turkey Red winter wheat[69]

The next two sections will explore how the hard red winter wheat, known as Krymka (Crimean) in the steppes reached the Great Plains, where it acquired the name Turkey Red. They will also examine the introduction other types of wheat, in particular spring durum varieties (similar to those already in the plains).

Mennonite Immigrants

This section aims to cut through the stories that have been repeated many times about the Mennonite immigrants bringing Turkey Red wheat with them in their baggage in 1873–4, and to consider their contribution to the origins and growth of the hard red winter wheat industry in the central plains by drawing on the available, contemporary sources.

It is important to clarify which types of wheat Mennonite farmers were growing in the steppes before they left and which they were growing in the Great Plains after they arrived. As we saw earlier, contemporary Russian sources show that, contrary to Saul's objection, some Mennonites were growing hard red winter wheat in the steppes on the eve of the migration to the Great Plains. Saul's suggestion that, after they arrived in the United States, they converted a hard spring wheat, such as Arnautka, into a winter wheat by sowing it in the fall can also be countered. The Turkey Red wheat that Mennonites grew in Kansas could not have been a durum wheat sown in the fall. Durum wheats were much harder than Turkey Red and had "yellowish white" rather than red grains. I have demonstrated elsewhere from analysis of contemporary Russian and American sources, and taking into account the different systems for classifying wheat, that the hard red winter wheat cultivated in Kansas from the 1870s was the same as that some Mennonites were growing on the steppes around this time.[70]

In 1873 and 1874, Mennonites preparing to leave the steppes sold their farms and some of their belongings, packed up what they could take with

[69] Klippart, *The Wheat Plant*, pp. 484, 489.

[70] David Moon, "In the Russians' Steppes: The Introduction of Russian Wheat on the Great Plains of the United States of America," *Journal of Global History* 3 (2008), 215–18. Fullilove referred to the conversion of spring varieties of "Black Sea wheat" to winter wheat, *Profit*, p. 101. However, it is likely these were also durum wheats (see Klippart, *The Wheat Plant*, pp. 497–8) and so could not be the source of "Turkey Red."

them, and embarked on the long journey across Europe and the Atlantic Ocean to the United States, and on by train to the Great Plains and prairies. Among the migrants were between twenty and thirty-five families of the Krimmer brethren, led by Jacob Wiebe, including the Barkman family with their daughter Anna. They left their village of Annafeld, near Simferopol', on the Crimean peninsula and moved to Kansas, where they established a new village of Gnadenau ("Grace Meadow") in Marion County.[71] Another group of immigrants were from the Alexanderwohl congregation of Molotschna. They planned initially to move to Nebraska, but decided to settle instead not far from Gnadenau, in Marion County.[72]

Warkentin and other Mennonites already in the United States set up a Board of Guardians to assist the new arrivals. Warkentin traveled to New York to meet some of them off the ships. In July 1874, he wrote to Goerz (who had arrived a few months earlier and was staying in Summerfield, Illinois): "Some 400 members from the Alexanderwohl congregation arrived here yesterday on the steamship, Hammonia." He wrote also that the Krimmer brethren were on their way from Liverpool. Two months later, Warkentin wrote from his new home in Halstead, Kansas: "I . . . arrived safely yesterday at 11 pm. with 11 passenger cars, 102 families making a total of 557 people." He found accommodation for them until the railroad company finished building temporary houses.[73] The immigrants endured a difficult first winter. Some lived in railroad freight cars, where they suffered from the cold, and relied on loans. Warkentin and the Board of Guardians helped by buying flour and other supplies.[74]

Is there any evidence in the contemporary sources that the Mennonite immigrants brought with them from the steppes hard red winter wheat seed and, thus, confirm the story that gained currency from the early twentieth century? In particular, are there any sources from the 1870s that

[71] For accounts from Mennonite sources of the Krimmer brethran's move to Kansas, see Harold S. Bender (1957) "Krimmer Mennonite Brethren," Global Anabaptist Mennonite Encyclopedia Online, available online at https://bit.ly/2NcY7jq, accessed November 1, 2017; Wiebe, *Grace Meadow*, pp. 30–5; C. F. Plett, *The Story of the Krimmer Mennonite Brethren Church* (Winnipeg, MB and Hillsboro, KS: Kindred Press 1985), pp. 57–69.

[72] See Cornelius Krahn (ed.), *From the Steppes to the Prairies, 1874–1949* (Newton, KS: Mennonite Publishing Office, 1949), pp. 7–8, 14, 27–49, 98–101; Kempes Schnell, "John F Funk, 1835–1930, and the Mennonite migration of 1873–1875," *Mennonite Quarterly Review* 24, no. 3 (1950), 199.

[73] MLA, Warkentin Papers, Box 1, Folder 9, Warkentin to Goerz, July 20, 1874; Warkentin to Goerz, July 21, 1874; Warkentin to Goerz, September 24, 1874.

[74] MLA, Warkentin Papers, Box 1, Folder 9, Warkentin to "Mennonite Board of Guardians," Summerfield, IL, January 7, 1875 and January 10, 1875; Warkentin to Goerz, January 26, 1875.

support the story Anna Barkman told in the 1920s that her group from the Crimea, led by Wiebe, brought seed of this type of wheat to Gnadenau, Marion County, Kansas?

In the 1870s, Warkentin wrote frequent, detailed letters to Goerz about the migration. At no point in any of the surviving letters, however, did he write anything about immigrants bringing wheat seed with them. Nor did he advise them to do so.[75] He did offer other advice on what they should bring. In March 1873, when Goerz was planning his move, Warkentin wrote to him: "do not bring any unnecessary articles along." While he made no mention of wheat seed, he did recommend bringing "the good featherbeds, the sheepskins, and ... clothes."[76] Goerz, who joined Warkentin in Halstead in 1875, wrote a pamphlet evaluating Kansas for settlement in February of that year. He mentioned that among the recent immigrants, "expert wheat raisers ... from South Russia say that they are well satisfied with the nature of their lands in Kansas." He referred specifically to Wiebe and their settlement of Gnadenau, in Marion County, but made no reference to them or any other immigrants bringing wheat seed.[77]

In late 1874 and early 1875, the *Marion County Record* published articles about the Mennonites' arrival. In January 1875, a journalist visited Gnadenau. He described the temporary sod houses they had built for the winter, their "old-fashioned German time piece[s] reaching from the ceiling to floor," and their trunks, which they used as seats (and later stories say contained wheat seed). He mentioned that "they all seem to have a liberal share of currency," and concluded: "we believe that they will be a blessing to Marion County." But, there was no mention of any wheat they had brought with them.[78] Malin made a more comprehensive search through the local press for the first few years after the Mennonites arrived and found no reference to them bringing wheat seed (see pp. 132–3).

Other contemporary accounts of the Mennonites in their early years in Kansas mentioned that they were growing wheat, including winter wheat, but not that they had brought the seed from the steppes. In 1878, C. B.

[75] MLA, Warkentin Papers, Box 1, Folders 7–9; Cornelius Krahn (ed.), "Some letters of Bernhard Warkentin pertaining to the migration of 1873–1875," *Mennonite Quarterly Review* 24, no. 3 (1950): 248–63.

[76] Krahn, "Some letters of Bernhard Warkentin," 255.

[77] Melvin Gingerich, "An 1875 Mennonite evaluation of Kansas," *Mennonite Quarterly Review* 28, no. 4 (1954), 307–9.

[78] "Russian Mennonite Emigration," *Marion County Record*, Saturday, November 21, 1874; "Among the Mennonites: Their Houses and Habits. A Visit to Gnadenau," *Marion County Record*, Saturday, January 16, 1875.

Schmidt, the immigration commissioner for the Santa Fe Railway, wrote a pamphlet to encourage more settlers from Russia and Germany to come to Kansas. He portrayed the lives of the recent immigrants in glowing terms, pointing out how they found the land more fertile and yields higher than in their previous homes. He wrote about large acreages sown by Mennonites with winter wheat: 857,000 in 1877 and 1,243,000 in 1878. But, he did not write that they brought the seed with them, nor did he advise prospective immigrants to bring any. He did refer to other items they had shipped across the Atlantic, such as bulky threshing stones and agricultural implements, but only to emphasize that they had proved obsolete in the New World.[79]

In 1878, a Russian traveler, I. Dement'ev, visited several Mennonite settlements in Kansas, including Gnadenau, nearby Hoffnungstal and Alexanderwohl, and Halstead. He described the people he met, their comparisons of their new lives in Kansas with their previous lives in the steppes, their houses and farms, how much land they had and how much they had paid for it, their farm implements and machinery, and their livestock. He also described their crops, including winter wheat as well as corn, oats, rye, and watermelons. He wrote that one farmer, Iakov Smit of Emmatal, Marion County, had sowed 190 acres with winter wheat for the harvest in 1878 (i.e. he had sown it in the fall of 1877). Dement'ev wrote also about the Mennonites' yields and the prices they got for their harvests. He explained how they had brought from Russia their carts, plows, and threshing stones, and also trunks, which were painted with bouquets of improbable red flowers. Nowhere did he write that they had brought wheat seed with them in the trunks. If they had done so, and if they told him about it, he did not see fit to mention it.[80]

The journalist Prentis went to see the Mennonites in Harvey and Marion counties in the spring of 1882, seven years after his previous visit. One of the farmers, Heinrich Richert, took him for drive, and Prentis remarked on the "immense fields of the greenest wheat." He was impressed: "If anyone has not yet made up his mind as to the possibilities of Mennonite agriculture, I commend a visit to the Mennonite settlements."[81] If their success was due to seeds they had brought with them,

[79] C. B. Schmidt, "Kansas Mennonite Settlements, 1877," *Mennonite Life* 25, no. 2 (1970), 51–8, 65–80.

[80] I. Dement'ev, "Poezdka v amerikanskie poseleniia russkikh menonitov v shtate Kanzas. (Iz puteshestviia po Amerike v 1878 g.," *Ustoi* 1, no. 11 (1882), 117–34.

[81] Noble L. Prentis, "A Day with the Mennonites," *Atchison Champion*, May 8, 1882, reprinted in Prentis, *Kansas Miscellanies*, 2nd edition (Topeka: Kansas Publishing Company, 1889), pp. 171, 173–4, 178, 182.

Prentis did not include it in his article. Thus, Schmidt, Dement'ev, and Prentis all described how the Mennonite immigrants were growing winter wheat, but none was specific about the type (hardness, color, bearded, or not), or where they got it from.

The closest I have found to contemporary evidence that Mennonite immigrants from the steppes brought seeds of wheat and other crops with them is from Nebraska. On June 25, 1877, Aganetha and Heinrich Ratzlaff wrote from their new home in Blumenort, near Jansen, Nebraska, to Aganetha's brother, Johan Janzen, who had settled in Blumenhof, near Steinbach, in Manitoba, Canada. They wrote that elder Abraham Friesen "has already seeded only with Russian varieties of wheat, rye and barley." But, he may not have planted very much. The Ratzlaffs wrote that they had broken the sod on only six acres and did not mention that Friesen had broken a larger area. The Ratzlaffs were in touch with Mennonites in Kansas, from where Heinrich Enns had written they would have only a small crop. Three years later, also from Jansen, Nebraska, Mrs. Jacob Enns wrote that they planned to plant another seven acres of wheat later that year (i.e. fall-sown winter wheat).[82] Thus, it may be that some immigrants did bring Russian wheat seed with them, as several recalled much later, and sowed it on a few acres.[83]

Despite his earlier skepticism, in 2006, Saul accepted that Mennonite immigrants from the Crimea brought some Turkey Red wheat, or something similar, in their baggage and "probably" planted it at Gnadenau. However, he noted that they would have struggled to grind large quantities of the wheat, because the millstones then in use in Kansas were inadequate for such hard grains, but that they could have ground small quantities for domestic use.[84] This is consistent with his earlier point that it would have been logistically impractical for them to have brought enough seed to plant

[82] Delbert Plett (ed.), *Pioneers and Pilgrims. The Mennonite Kleine Gemeinde in Manitoba, Nebraska and Kansas, 1874 to 1882* (Steinbach, Manitoba: D. F. P. Publications, 1990), pp. 80–2, 88–9. Blumenort, in Jefferson county, Nebraska, was founded by immigrants from Molotschna in 1874. Cornelius Krahn (1957) "Jansen Kleine Gemeinde Mennonite Church (Jansen, Nebraska, USA)," Global Anabaptist Mennonite Encyclopedia Online, available online at https://bit.ly/2NedCYf, accessed November 3, 2017; Krahn and Richard D. Thiessen (May 2011) "Friesen, Abraham L. (1831–1917)," Global Anabaptist Mennonite Encyclopedia Online, available online at https://bit.ly/32dut1L, accessed November 3, 2017. Jansen, Nebraska, is about 140 miles north of Hillsboro, Marion County, Kansas.

[83] See, for example, Alberta Pantle, "Settlement of the Krimmer Mennonite Brethren at Gnadenau, Marion County," *KHQ* 13, no. 5 (1945), 269.

[84] Norman E. Saul, "Bernhardt Warkentin and the Making of the Wheat State," in *John Brown to Bob Dole: Movers and Shakers in Kansas History*, ed. Virgil W. Dean (Lawrence: University Press of Kansas, 2006), 109–10.

the areas some Mennonite farmers were sowing with wheat in their early years in Kansas (see p. 133).

To test the plausibility of the story that Mennonite immigrants brought wheat seed with them and laid the groundwork for the Kansas winter wheat industry we can conduct a hypothetical investigation into how much wheat they would have needed and how much they could have carried in their baggage. Warkentin noted in 1875 that each family needed "no more than 20 bushels of wheat, 20 bushels of oats and barley, and 2 to 3 bushels of corn."[85] In the early twentieth century, the USDA advised farmers that the usual rate of seeding winter wheat in the Great Plains was between 30 and 60 pounds an acre, but that good crops had been raised from as little as 15 pounds an acre.[86] One bushel of wheat weighed about 60 pounds.[87]

Anna Barkman recalled in the 1920s that she had selected two gallons of seed to take from the Crimea to the United States. Two gallons was one quarter of a bushel or around 15 pounds.[88] This was less than other estimates. In 1914, Carleton wrote that each family brought a bushel or more with them, which was consistent with Coultis's statement that the first group of families brought 20 or 30 bushels in total. These amounts were far short of what Warkentin considered a family's requirements, and were sufficient for each family to have sown only one or two acres. But, it would certainly have been possible for the immigrants to bring these quantities of seed with them from the steppes by ship across the Atlantic and then by train to Kansas. One of the shipping companies permitted each adult a free baggage allowance of 20 cubic feet. This was the equivalent of approximately 128 gallons (dry): i.e. many times the two gallons Anna recalled and easily sufficient for the slightly larger amounts mentioned by Carleton and Coultis. One of the railroad companies allowed each adult 150 pounds of baggage free of charge. A quarter of a bushel of wheat would have taken up only 10 percent of this allowance.[89] If twenty-four families of Krimmer brethren each brought one bushel,

[85] MLA, Warkentin Papers, Box 1, Folder 9, Warkentin to Goerz, February 24, 1875.

[86] E. C. Chilcott and John S. Cole, "Growing Winter Wheat on the Great Plains," *USDA Farmers' Bulletin*, no. 895 (1917), 5.

[87] For the weight of a bushel of wheat, see the Grain and Weight Conversion website available online at http://grainweightconversion.com, accessed November 7, 2019.

[88] To convert gallons to bushels, see Kyle's Converter, available online at www.kylesconverter.com/volume/gallons-(u.s.-dry)-to-bushels-(u.s.-dry-level), accessed March 3, 2018.

[89] For the baggage allowances, see Georg Leibbrandt, "The Emigration of the German Mennonites from Russia to the United States and Canada, 1873–1880. II," *Mennonite Quarterly Review*, 7, 1 (1933), 30. According to Warkentin, another line allowed 200 pounds per passenger. Krahn, "Some letters of Bernhard Warkentin," 255. To convert gallons (dry) to cubic feet, see Metric Conversions available online at https://bit.ly/2K1oFix, accessed November 7, 2019.

it would have weighed a total of 1,440 pounds, or 0.72 U.S. (short) tons. This was a small part of the 10-ton capacity of the freight cars used on American railroads around 1870.[90] Although we do not know how many freight cars the Krimmer brethren had for their journey from New York to Kansas, we do know that the larger Alexanderwohl group had five car loads of baggage.[91] This would have been sufficient to transport at least 50 tons, and thus plenty of wheat seed, if they had brought any with them.

Therefore, it would have been possible for the immigrants to have brought from the steppes the amounts of seed mentioned in different sources as part of the free baggage allowances offered by the shipping and railroad companies. Since some of the Mennonites were prosperous, they could have paid for additional baggage. And, we know from several contemporary sources that they shipped bulky items, including featherbeds, threshing stones, plows, carts, and ceiling-high clocks across the ocean and then on to Kansas. C. B. Schmidt reported that the immigrants had also brought with them seeds of apricot, mulberry, and wild olive trees and shrubs.[92] Should they have chosen to do so, therefore, the immigrants could well have brought a few bushels of wheat seed. So, although there seems to be no contemporary evidence to confirm Anna Barkman's story that immigrants brought wheat seed to Kansas, it was certainly plausible. It would also have been possible for the group who settled in Blumenort, Nebraska, to have brought sufficient seed for the few acres they sowed with Russian crops in their first years. However, in their early years in the Great Plains, the immigrants and people who met them and wrote about them did not think it important enough to mention that they had brought wheat seed from the steppes. Some of the immigrants recalled, or claimed to have recalled, having done so only several decades later. We will return to possible reasons for these silences in the conclusion to this chapter.

For Iakov Smit to have sown 190 acres in Marion County, Kansas, with winter wheat in the fall of 1877, as Dement'ev reported, would have required 95 bushels of seed, if he had sown at 30 pounds/half a bushel an acre. This would have weighed nearly three tons: far in excess of what he could plausibly have brought over from the steppes. However, if he had brought a smaller quantity, reaped good harvests for the previous three years, set aside a significant part as seed, and bought more from his

[90] See Richard White, *Railroaded: The Transcontinentals and the Making of Modern America* (New York: Norton, 2011), p. 152.

[91] MLA, Warkentin Papers, Box 1, Folder 9, Warkentin to Goerz, October 12, 1874.

[92] Schmidt, "Kansas Mennonite Settlements," 60–1.

neighbors, he could hypothetically have built up a large supply of seed. Average yields of wheat in Kansas in the 1870s were around 15 bushels an acre and some Mennonites obtained higher yields.[93] Nevertheless, it seems improbable that Smit sowed 190 acres with seed solely from imported stocks by 1877, unless more had been shipped over. It is even less likely that seed for the 1,243,000 acres C. B. Schmidt reported Mennonites had sowed with winter wheat in 1878 derived from crops grown from seed brought over in 1874: this would have required approximately 18,645 tons of seed, enough to fill nearly two thousand railroad cars.

If they did not bring all this seed with them, then in the first few years the newly arrived Mennonites must have purchased most of the seed they needed in the United States. There is evidence that this is precisely what they did, and that Warkentin assisted them. On October 12, 1874, he wrote to Goerz that he was on his way to Missouri and Iowa "where I am to purchase grain and potatoes for our Russian brethren." On February 24, 1875, he wrote that he was arranging loans for families to buy seed wheat. In March, the Board of Guardians decided to settle each immigrant family on 40 acres of land and provide them with implements, cattle, and seed.[94] Several decades later, Jacob Wiebe recalled that after they arrived, they purchased American winter wheat seed, which they planted in the fall of 1874 and reaped a bountiful harvest the following year.[95] This is significant for the Anna Barkman story as Wiebe was the head of the Krimmer brethren from the Crimea who were reputed to have brought hard red winter wheat with them to Gnadenau, Marion County.

There are hints in the contemporary sources and scholarly literature about the types of seed wheat they purchased in the United States and where it came from. The seed Warkentin and the Board of Guardians bought for the immigrants in February and March 1875 must, given the time of year, have been spring wheat to sow over the following weeks. The grain Warkentin bought in Missouri and Iowa the previous October was probably for them to eat, but if it included seed, then it is likely that it was

[93] Leah J. Tsoodle and Christine A. Wilson, "125 Years of Farmland Values in Kansas: 1870–1997," Kansas State University, September 2001, available online at https://bit.ly/2NOkPgW, accessed November 6, 2017. Among Mennonite farmers, Dement'ev reported average yields of 22 bushels an acre. Schmidt recorded average yields in different localities between 8 and 25 bushels an acre. Both noted that some farmers obained higher yields. Dement'ev, "Poezdka," 115; Schmidt, "Kansas Mennonite Settlements," 51–8. For similar yields a few decades later, see Chilcott and Cole, "Growing Winter Wheat," 9.

[94] MLA, Warkentin Papers, Box 1, Folder 9, Warkentin to Goerz, October 12, 1874; Warkentin to Goerz, February 24, 1875; Schnell, "John F. Funk, 1835–1930," 222.

[95] Glenn D. Bradley, *The Story of the Santa Fe* (Boston: Richard G. Badger, 1920), 118–24.

also spring wheat as it would have been too late to sow winter wheat: in the early twentieth century the USDA considered it "undesirable" to sow winter wheat in the central plains after October 15.[96] Thus, some of the wheat grown by Mennonites in the mid 1870s is likely to have been spring wheat of the types already cultivated in the region (see p. 146).[97]

What of the winter wheat we know from contemporary sources that Mennonites were sowing in Kansas in the 1870s? Historians have found it difficult to pin down precisely what sort this was and where it had come from. As we have seen, small quantities may indeed have come from the steppes. Some, however, may have come from Illinois. When he first arrived in the United States in 1872, Warkentin stayed in an existing Mennonite community in Summerfield, Illinois. His friend Goerz stayed in Summerfield after he arrived. In late 1873, Warkentin relocated to Halstead, Kansas, where he was joined later by Goerz and other Mennonites from Summerfield.[98] Writing in 1878, C. B. Schmidt hinted that they brought winter wheat with them: "The Illinois farmers find that the winter wheat can more easily be raised in Kansas than in Illinois." He continued that "careful farmers," for example Daniel Haury, found that the yields were as good as those in southern Illinois.[99] Haury had arrived in Illinois from Bavaria in 1850, before moving to Halstead in 1875, where he died in 1937. His obituary in a Mennonite newspaper stated that he was "one of first farmers in central Kansas to plant the Turkey Red Wheat."[100] The Mennonite historian David Haury (probably a descendant), mentioned a suggestion that the hard winter wheat came from Illinois, but felt it was inconclusive "in the absence of . . . documentation that it was really hard wheat."[101] His source was Carleton, who wrote in 1914 about claims that earlier Mennonite migrants who moved to Illinois before 1870 may have introduced "Turkey wheat." He presented no evidence, however.[102] And, the Mennonites in Summerfield came from southern Germany,[103] and so were unlikely to have introduced wheat from

[96] Chilcott and Cole, "Growing Winter Wheat," 5.
[97] See Saul, "Myth and History," 4; Colwell, "American Wheat Varieties," 74.
[98] Krahn, "Some letters of Bernhard Warkentin," 250–1; Saul, "Bernhardt Warkentin," 107–8.
[99] Schmidt, "Kansas Mennonite Settlements," 56.
[100] "Haury, Daniel S. (1845–1937)," Mennonite Weekly Review (June 1937), available online at https://bit.ly/2Cp4ycZ, accessed November 7, 2019.
[101] David A. Haury, "Bernhard Warkentin: A Mennonite Benefactor," *Mennonite Quarterly Review* 49, no. 3 (1975), 194.
[102] Carleton, "Hard Wheats," 404.
[103] Harry F. Weber, *Centennial History of the Mennonites of Illinois 1829–1929* (Goshen, IN: The Mennonite Historical Society, 1931), p. 48.

the steppes of the Russian Empire. There are other stories that the hard winter wheat was introduced to Kansas by French settlers, but they also lack "documentation."[104] There is evidence, moreover, that a soft winter wheat, known as Red Genesee, was harvested around Newton, Kansas, in 1875.[105] It is possible, that for a while after they arrived, Mennonite farmers were sowing similar types of soft winter wheat to other farmers in the region. They may also have acquired some of the red winter wheats that we know from Klippart's book were grown in Ohio into the 1860s, but may not have been "hard" wheat. From this point in the story, however, we have more precise sources.

For the cultivation of the hard red winter wheat some Mennonites had grown in the steppes to develop in the Great Plains two things were necessary: first, the shipment of substantial quantities of seed from the steppes and, second, the retooling of local mills to grind the hard wheat. A key role in both was played by Warkentin. From the late 1870s, Warkentin was shipping wheat from the steppes to Halstead, Kansas. On December 4, 1879, the *Kansan*, published in nearby Newton, reported: "Mr. Warkentin is buying Odessa wheat for seed and paying $1.00 a bushel for it last week." A few months later the paper reported that "B. Warkentin is preparing to build a large granary."[106] This was probably not the first shipment. Malin found a reference to "Odessa wheat" on sale in the Newton *Kansan* in November 1877.[107] We cannot be certain that this was hard red winter wheat, as the papers provided no more detail, and the names used for types of wheat were fluid at this time. "Odessa" could have referred simply to the Black Sea port it was shipped from. We do know that Warkentin's business was growing quickly. In 1878, C. B. Schmidt wrote that Halstead had become a "major center for the sale of wheat, drawing more and more farmers to the town."[108] While it can only be speculation, Warkentin's imports may have been the source of the winter wheat Iakov Smit sowed on 190 acres in nearby Marion County at this time (See p. 151).

The first reference to "Turkey" wheat Malin found in the local press was in the *Marion County Record* in June 1881. A farmer stated that he had been growing it for four years (i.e. he had first sown it in the fall of 1877),

[104] Haury, "Bernhard Warkentin," 194; Quisenberry and Reitz, "Turkey Wheat," 103.
[105] Colwell, "American Wheat Varieties," 74.
[106] MLA, Warkentin Papers, Box 2, Folder 21, Notes on articles in local newspapers concerning Warkentin, 1875–1900, *Kansan*, Thursday, December 4, 1879, 2; *Kansan*, Thursday, July 22, 1880, 3.
[107] Malin, *Winter Wheat*, p. 170. [108] Schmidt, "Kansas Mennonite Settlements," 55.

and that it was "superior to all the other" varieties. This report came after three drought years. Over the following few years more farmers grew "Turkey" wheat in Marion, Harvey, McPherson, and Reno counties: the counties where the Mennonites had settled. From 1881, the local press reported that it was the Mennonites who had introduced the wheat from Russia. Cultivation of the wheat spread to surrounding counties and to non-Mennonite farmers as it gained a reputation for resilience in the face of droughts, harsh winters, pests, and disease, and for yielding better harvests than other varieties.[109]

More supplies of seed were needed if cultivation of the "Turkey" wheat was to increase further. Warkentin visited his former home in the steppes in the summer of 1885. We know from stamps in his American passport that he was issued with a Russian visa in Vienna on May 31/June 12, 1885, crossed the Russian frontier shortly afterwards, and departed from Odessa on August 23/September 4.[110] No letters or other contemporary documents about his activities in Russia in 1885 seem to have survived. But, several historians have written, on the basis of later sources, that Warkentin made arrangements to ship "several thousand bushels" of wheat. Haury noted: "This action made the Turkey Red wheat available in quantity for the first time and was thus a very significant step in the history of Kansas agriculture."[111] Warkentin returned to the steppes in 1894, for around two weeks, leaving Odessa on April 5, 1894. He arranged with his brother-in-law, Johannes Wiebe in Molotschna, to ship "a lot of fine and clean ... genuine Crimean hard winter wheat" to Kansas. Malin found evidence that a shipment arrived in December 1895.[112] Thus, possibly from as early as the late 1870s, but almost certainly from the mid 1880s, Warkentin was arranging shipments of substantial volumes of hard red winter wheat seed from the steppes to Kansas, where it was gaining a reputation for hardiness and productivity.

Warkentin also played a key role in retooling the mills in Kansas and elsewhere in the Great Plains to grind the hard wheat, which was essential

[109] Malin, *Winter Wheat*, pp. 171–8.

[110] MLA, Warkentin Papers, Box 4, Folder 13, Bernhard Warkentin, U.S. Passport, issued March 7, 1885. (The earlier dates are for the Julian calendar then in use in the Russian Empire; the later dates for the Gregorian calendar used in much of the rest of Europe and the United States.)

[111] Haury, "Bernhard Warkentin," 194; Saul, "Myth and History," 7. Malin noted the lack of contemporary documentation, but thought it "reasonable" that he might have imported some wheat in 1886. Malin, *Winter Wheat*, pp. 204–5.

[112] MLA, Warkentin Papers, Box 4, Folder 13 Bernhard Warkentin, U.S. Passport, issued January 29, 1894; Warkentin Papers, http://mla.bethelks.edu/archives/ms_68/, accessed March 1, 2018, Folder 19, Warkentin to Carleton, May 31, 1900; Malin, *Winter Wheat*, 204–5.

for large-scale production of Turkey Red wheat. Warkentin was familiar with mills as his father, Bernhard Aron Warkentin, ran a prosperous mill in Molotschna. He learned about American milling technology from his future father-in-law, Conrad Eisenmayer, who owned a mill in Summerfield, Illinois. Eisenmayer was testing steel rollers to replace millstones at this time. After he moved to Halstead, Kansas, Warkentin built a water-powered mill equipped with grindstones. He married Wilhelmina Eisenmayer in the summer of 1875. She moved to Halstead with her family, and Warkentin went into business with his father-in-law. The following year, they built a larger, steam-powered mill near the railroad in Halstead. Their milling business grew, and in 1884 they tested steel rollers. It was at this point, in 1885, that Warkentin visited the steppes and seems to have ordered substantial shipments of hard red winter wheat. On his return home, he bought the Monarch Steam Mill in Newton, where he established the Newton Milling and Elevator Company, and relocated his family to a grand house he had built nearby. He reequipped his new mill with steel rollers. A decade later, the Newton mill had twelve double sets of steel rollers and was turning out 400 barrels of flour a day. The Halstead mill was producing similar quantities. Both mills combined were supplying flour ground from Turkey Red wheat to cities around the United States and for export. In 1900, Warkentin expanded his operations by buying a mill at Blackwell, Oklahoma (ninety miles south of Newton). He played a further role in retooling the Kansas mills. In 1888, he was involved in setting up the Kansas Millers' Association and served as its first president. The Association promoted the cultivation of Turkey Red wheat throughout the state so that millers could minimize their costs by equipping their mills to grind mostly this sort of wheat. By arranging shipments of wheat from the steppes and playing a leading role in retooling the mills to grind it, Warkentin thus played a key part in the development of the hard red winter wheat industry in the Great Plains.[113] He played a further role through his association with Carleton, a cereal specialist who worked in Kansas for the USDA.

[113] See Saul, "Bernhard Warkentin," 105–10; MLA, Warkentin Papers, Box 1, Folder 9, Warkentin to Goerz, January 21, 1874; Schmidt, "Kansas Mennonite Settlements," 55; Malin, *Winter Wheat*, pp. 188–97; Charles W. Deyne, "Flour Milling in Kansas," in *The Rise of the Wheat State: A History of Kansas Agriculture, 1861–1986*, ed. George E. Ham and Robin Higham (Manhattan, KS: Sunflower University Press, 1987), pp. 159–64; Cornelius Krahn and Richard D. Thiessen (June 2007) "Warkentin, Bernhard (1847–1908)," Global Anabaptist Mennonite Encyclopedia Online, available online at https://bit.ly/2NietY9, accessed November 8, 2017.

USDA Plant Explorers

An important contribution to the introduction of wheat varieties from the steppes in the Great Plains was made by "plant explorers" sent to the Russian Empire by the USDA. The most significant was Mark Alfred Carleton (1866–1925). We can trace his path from his childhood on a farm in Kansas to the steppes. He was born in Ohio, but moved with his family to Cloud County, in north-central Kansas in 1876. This was north of where the Mennonites were settling at that time. On his family's farm he witnessed the difficulties in cultivating wheat caused by crop diseases such as rust. He studied at Kansas State Agricultural College in Manhattan. One of his classmates was David Fairchild, who went on to work for the USDA in Washington, DC. After graduating with a BS in 1887, Carleton taught natural history for two years at Garfield University in Wichita, just south of the Mennonite settlements. While taking students on a field trip across the Great Plains and the West, he became interested in the association of certain plants with particular soils and climates. He returned to Manhattan to study for a master's degree in botany and horticulture and work at the new Kansas State Agricultural Experiment Station.[114] Carleton was interested in international comparative studies. In November 1893, he wrote to B. T. Galloway, chief of the Division of Vegetable Pathology and Physiology at the USDA, asking for funds to travel to Australia to study wheat diseases. Galloway turned him down since, at that time, the USDA had no power to send people overseas, and also because there was much work "to be done here."[115]

Since the mid 1870s, the Kansas State Agricultural College had been testing crop varieties and cultivation methods appropriate to the region. After its establishment in 1887, the state Agricultural Experiment Station carried out further work. The scientists, who were soon joined by Carleton, acquired varieties from experiment stations in other states and local farmers, including Mennonites. Among those they tested were "Russian" and "Turkey" wheat. In the early years, they gave mixed results. But, this

[114] See Thomas D. Isern, "Wheat Explorer the World Over: Mark Carleton of Kansas," *Kansas History* 23, nos. 1–2 (2000), 12–25; Gary M. Paulsen, "The Agronomic Legacy of Mark A. Carleton," *Journal of Natural Resources. Life Science Education* 30 (2001), 120–1; Theodore Holm, "Mark Alfred Carleton," *Botanical Gazette* 80, no. 1 (1925), 111–13. Fairchild recalled Carleton as his class mate in his memoir. Fairchild, *World*, pp. 12, 115.

[115] NARA CP, RG 54, Finding Aid A1, Entry 58, Division of Vegetable Pathology and Physiology: Correspondence of M. A. Carleton, 1891–1900, Folder M.A. Carleton – 1893, f.3, Carleton, to B.T. Galloway, Chief of Division, Washington, November 1, 1893; Galloway to Carleton, November 9, 1893.

changed from the early 1890s with the onset of drought. Turkey wheat consistently demonstrated its ability to withstand dry springs, cold springs, late frosts, and severe winters.[116] The tests thus replicated the experiences of local farmers who had grown it since the late 1870s. The Experiment Station reported in 1894 that "Turkey ... is held in high esteem in many parts of the state, and it may properly be classed with our most productive wheats."[117]

In 1894, Carleton was appointed to the Division of Vegetable Physiology and Pathology at the USDA in Washington, DC, but returned regularly to Kansas and Nebraska to continue experiments on crop diseases.[118] He was aware that successful cultivation of wheat and other small grains in the Great Plains required varieties that could cope with the local conditions. Many varieties that had been cultivated for decades in the United States, especially further east, were not well adapted to the plains. It was at this time that Carleton developed his scientific interest in wheat varieties from Russia, in particular the steppes, including durum wheats as well as hard red winter wheats, such as Turkey. He noted that Russian wheats resisted rust "quite well."[119] Writing the USDA Yearbook for 1896, Carleton recommended extending the cultivation of hard wheat in the United States and further trials of "the finest class of bread wheats in the world," including "Turkey." He noted that hard wheats were grown in southeastern Russia under "conditions very similar to ... our own wheat belt."[120] In May 1897, he reported to the USDA from Manhattan, Kansas: "Some of the varieties [of winter wheat], especially the Russian and two or three of the Italian, are excellent. ... Three or four of the Russian sorts are going prove just the thing for this country being extremely hardy and of good quality."[121]

Carleton interest in wheat from the steppes stemmed in part from acquaintance with some of the Kansas Mennonites during the dry

[116] Malin, *Winter Wheats*, pp. 179–87.
[117] C. C. Georgeson, F.C. Burtis, and D.H. Otis, "Experiments with Wheat," *Experiment Station of the Kansas Agricultural College Bulletin*, no. 47 (1894), 16.
[118] Isern, "Wheat Explorer," 15.
[119] *ESR* 5 (1893–4), 497, 881–2; 6 (1894–5), 224–6; 8 (1896–7), 594; 11 (1899–1900), 942–4; NARA CP, RG 54, Finding Aid PI-66, Entry 29, Letters sent by M. A. Carleton; Carleton to F. L. Scribner, November 28, 1897; Carleton to W. Frank Crowley, November 29, 1897; Carleton to J. E. Rickenbaker, December 18, 1897; Carleton to D. G. Fairchild, March 30, 1898; Carleton to A. F. Worde, June 27, 1898; Carleton, "Improvements in Wheat Culture," 492, 496–7.
[120] Carleton, "Improvements in Wheat Culture," 493.
[121] NARA CP, RG 54, Finding Aid A1, Entry 58, Division of Vegetable Pathology and Physiology: Correspondence of M. A. Carleton, 1891–1900, Folder M. A. Carleton – 1897, Carleton to Galloway, May 23, 1897.

1890s. In October 1894, Carleton reported from Newton, Kansas, that he was "investigating the methods of wheat culture . . . among the Mennonite Russians." Two years later, he wrote that he was impressed by the way the Mennonite farmers had "continued to have good harvests" when the wheat crop had failed throughout much of the Great Plains due to the drought.[122] Carleton and Warkentin first met in 1896 and corresponded about the Mennonite immigrants and their wheat.[123]

Carleton embarked on a serious study of conditions in Russia. In 1895, he arranged to be sent samples of wheat soils from Russia. He corresponded with Milton Whitney, the chief of the USDA's Division of Soils, about them.[124] Carleton was keen enough to learn to read Russian to study specialist literature. (He may have found out that Russian scientists in the steppes were carrying out similar experiments, and that hard red winter wheats, including "Crimean," were performing well in harsh conditions.[125]) In March 1898, he wrote to Fairchild at the USDA that he had made a "fairly close study of Russian climate, geography, and agriculture and my conclusion . . . is that Russian and Siberian wheats are the wheats most admirably adapted to the Great Plains." Carleton was pleased to receive samples gathered by Hansen, who had traveled to Russia as a "plant explorer" the previous year.[126] Carleton was eager to travel to the steppes himself, and was delighted when Fairchild sent him a telegram on June 24, 1898, suggesting such a trip. Carleton wrote back: "A trip to Russia is what I have been wishing to make above all others. . ." He aimed to select varieties of wheat "best adapted to our central wheat region" in the

[122] NARA CP, RG 54, Finding Aid PI-66, Entry 1, General correspondence of the chief, 1900–1908, box 7, Burleson-Carleton, File Carleton, Carleton to Mr. Galloway, October 20, 1894; Mark A. Carleton, "Successful Wheat Growing in Semiarid Districts," *Yearbook of the United States Department of Agriculture 1900* (Washington, DC: USDA, 1901), p. 531.

[123] MLA, Warkentin Papers, Box 2, Folder 21, Letter from Carleton to Warkentin, c. August, 1899, from *Weekly Kansan*, August 25, 1899 (typed copy).

[124] NARA CP, RG 54, Finding Aid A1, Entry 58 Division of Vegetable Pathology and Physiology: Correspondence of M. A. Carleton, 1891–1900, Folder M. A. Carleton – 1895, Carleton to Mr. B. T. Galloway, Chief Division of Vegetable Physiology and Pathology, November 1, 1895; RG 54, Finding Aid PI-66, Entry 29, Letters sent by M. A. Carleton, 1897–98, Carleton to Milton Whitney, November 28, 1897; RG 54, Finding Aid PI-66, Entry 180 Records of the Bureau of Chemistry and Soils, General Records – Letters received, 1894–1901, Box 2, General correspondence, C–E 1894–96, M. A. Carleton to Milton Whitney, April 30, 1898. See also p. 220.

[125] See, for example review of *Odesskoe uezdnoe zemskoe sobranie XXVI ocherednoi sessii*, *SKhiL* 181 (1891), 17; S. Kizenkov, "Sel'skokhoziaistvennye voprosy v russkoi pechati za 1897 g.," *SkhiL* 188 (1898), 694.

[126] NARA CP, RG 54, Finding Aid PI-66, Entry 29, Letters sent by M. A. Carleton, 1897–98, ff.76–77, Carleton to D. G. Fairchild, March 30, 1898; f.97, Carleton to Fairchild, May 11, 1898; Fairchild, *World*, p. 115. See also pp. 97–8.

Great Plains taking into account "resistance to drouth [sic], cold and disease, early maturity etc." He was confident that the "knowledge to be gained regarding ... relations of climate and soil to wheat diseases, adaptations of varieties etc. ought to be considerable" and that "the expedition [would] be of great utility to American agriculture."[127] Fairchild could propose the trip to his former classmate from the Kansas Agricultural College as he had recently been appointed head of a new Section of Seed and Plant Introduction at the USDA. He recalled later that Carleton had been sent to Russia as a "Special Agent" with "all possible cooperation on our part."[128]

Carleton was soon on his way. He sailed from New York for Britain in July 1898 and was overseas until February 1899. He visited various locations in Europe before making his way to Romania, where he boarded a ship at the Black Sea port of Constanța bound for Odessa, arriving on August 22. Carleton spent several months traveling around the Russia Empire. He went north to Kiev, on to Moscow, then east to Samara on the Volga. He ventured further southeast, deeper into the steppes, to the city of Orenburg, near the modern border with Kazakhstan. Taking advantage of the Russian railroads, he then visited Moscow and St. Petersburg, before returning to the steppes at Saratov on the Volga and Rostov-on-Don. The seemingly inexhaustible Carleton then traveled through the steppes of the North Caucasus over the mountains to Georgia, before sailing across the Black Sea back to Odessa. En route Carleton met Russian agricultural specialists, but his main concern was "agricultural exploration." He collected:

> twenty-three varieties of cereals [seven ... of wheat], one of buckwheat, and two of forage plants ... and sample packets of seeds of other cereals and grasses for breeding purposes... In addition..., arrangements were made for the importation of others fully as valuable.

He recorded the soil and climate of the areas where he collected the varieties, and indicated the parts of the United States he thought they would be suited to.

Carleton's main interest on this trip was not hard red winter wheats, such as Crimean/Turkey Red, because his visit coincided with a drought in the region where they were grown. Instead, he focused on hard spring durum wheats. He assigned first place among those he collected to

[127] NARA CP, RG 54, Finding Aid PI-66, Entry 29 Letters sent by M. A. Carleton, 1897–98, Carleton to Fairchild, June 27, 1898; Carleton to A. F. Woode, June 27, 1898.

[128] Fairchild, *World*, p. 115.

Kubanka. He noted that it was one of the "most prized" varieties in Russia, where it was cultivated in the steppes in the Volga region, north of the Sea of Azov, the North Caucasus, and parts of the "Kirghiz Steppes and Turkestan" (today's Kazakhstan and Central Asia). Carleton believed that "varieties brought from Russia are all the better adapted for cultivation in ... [the United States], as they are already used to even more rigorous conditions of climate." Following the example of Hansen, who had traveled around the steppes collecting plants for the plains the previous year, Carleton chose seeds of Kubanka wheat from part of steppes with a particularly harsh climate and less fertile soil: "the Turghai territory, in the Kirghiz Steppes," in the north of present-day Kazakhstan. He acquired seed grown by a "Mr. Gnyezdilov ... about 40 miles southeast of Orenburg" (on the way to the modern city of Aktobe). Carleton noted the local conditions: 15 inches or less rainfall a year; short, intensely hot, summers; and soil "much grayer than the usual black earth." Russian agronomists considered these conditions extreme (see pp. 96, 142–3). Carleton believed that his Kubanka would be well adapted to the drier, hotter, southern Great Plains and more arid lands further west. He also collected varieties of "common wheat," which he believed suitable for the northern plains.[129]

On his return to the United States, Carleton asked Warkentin for help in leasing a "10 acre tract" of land with "good, typical, black, wheat soil" near Newton, Kansas, for "field experiments" with "a large number of new Russian varieties." Warkentin obliged and had the winter wheat planted in the fall of 1899. In May 1900, Warkentin wrote to Carleton that some of the wheat was doing well and had "remarkably large heads."[130] Carleton arranged for a scientist from the Kansas State Agricultural College, D. B. Swingle, to oversee the experiments and keep him informed. In June 1900, Carleton reported: "In the field experimental work further proof was

[129] NARA CP, RG 54, Finding Aid A1, Entry 58 Division of Vegetable Pathology and Physiology: Correspondence of M. A. Carleton, 1891–1900, Folder M.A. Carleton – 1898, Carleton, London, to A. F. Woods, Washington, DC, July 23, 1898; RG 54, Finding Aid PI-66, Entry 29, Letters sent by M. A. Carleton, 1897–98, vol. 1, Carleton to Fairchild, June 27, 1898; Carleton, "Russian Cereals," 2 (Carleton's route), 8–9, 12–43 (quotations); Isern, "Wheat Explorer," 18-19. It is not clear if Carleton travelled to Turgai region himself. According to his itinerary, he did not visit it, but spent three days in Orenburg, from where the railroad southeast was not completed until 1906. See "Tashkentskaia zheleznaia doroga," *ES*, 11a (1907), 742. On Hansen's travels, see pp. 96–7, 162.

[130] MLA, Warkentin Papers, Box 2, Folder 21, Letter from Carleton to Warkentin, c. August, 1899, from *Weekly Kansan*, August 25, 1899 (typed copy); *Weekly Kansan*, September 1, 1899; Warkentin Papers, https://mla.bethelks.edu/archives/ms_68/folder_20/, accessed March 1, 2018, Warkentin to Carleton, May 5, 1900.

obtained of the admirable adaption of Russian hardy red winter wheats to the conditions of our semi-arid Plains."[131]

Since March 1900, however, Carleton had been in Paris, France, where the USDA had sent him to prepare the exhibit of American grain at the Exposition Universelle.[132] (See pp. 60–1) While in Paris, the USDA authorized him to make a two-month visit to the Russian Empire. He had ambitious plans for his second Russian trip: "I find it will be cheap and easy travelling in Siberia, and I will go probably over half way across the continent." Moreover, "I intend to bag seeds of every kind of importance from every quarter that I can rake [?] them in."[133] From Paris, he corresponded with Warkentin to arrange to visit his former home at Molotschna in the steppes. Warkentin sent him letters of introduction to his brother, Gerhard, and brother-in-law, Jacob Enns, and wrote to his relatives asking them to assist Carleton when he arrived. Warkentin advised Carleton on the best route from Odessa to the "German Colonies along the Malotchna river." In return, he asked Carleton to ship him "a lot of strictly clean hard winter wheat (Turkey), the same variety we are raising in Kansas." Carleton reported to the USDA that Warkentin had asked him to get 500 bushels of "hard winter Crimean (Turkey) wheat for seed," and that he intended to obtain 50–60 bushels for the Department. Once his work in Paris was completed, Carleton set out to study "the Crimea as a wheat region" as "an exact parallel with Kansas and Nebraska."[134]

[131] NARA CP, RG 54, Finding Aid PI-66, Entry 164, Paris Exposition 1900, vol. 2, ff.194–5, Carleton to A. S. Hitchcock, State Agricultural College, Manhattan, Kansas, January 6, 1900; Finding Aid A1, Entry 58, Division of Vegetable Pathology and Physiology: Correspondence of M. A. Carleton, 1891–1900, Folder M. A. Carleton – 1900, Carleton, "Memorandum on field experiments with cereals to be done by Mr. D. B. Swingle at Halsted, Kansas," n.d.; Carleton to Galloway, May 21, 1900; Carleton "Memorandum relative to the work of the year ending June 30, 1900"; Carleton to Galloway, June 18, 1900.

[132] NARA CP, RG 54, Finding Aid A1, Entry 58, Division of Vegetable Pathology and Physiology: Correspondence of M. A. Carleton, 1891–1900, Folder M. A. Carleton – 1899, Carleton to Galloway, January 30, 1899; Folder M. A. Carleton – 1900, Carleton, on board USMS New York bound for Paris, to A. F. Woods, USDA, March 7, 1900; Carleton, Paris, to Galloway, March 18, 1900.

[133] NARA CP, RG 54, Finding Aid A1, Entry 58, Division of Vegetable Pathology and Physiology: Correspondence of M. A. Carleton, 1891–1900, Folder M. A. Carleton – 1900, Carleton to Galloway, May 3, 1900; Carleton to A. Woods, June 3, 1890.

[134] MLA, Warkentin Papers, https://mla.bethelks.edu/archives/ms_68/folder_20/, accessed March 1, 2018, Warkentin to Carleton, May 5, 1900, May 31 1900, July 2, 1900; Carleton to Galloway, June 18, 1900; NARA CP, RG 54, Finding Aid A1, Entry 58, Division of Vegetable Pathology and Physiology: Correspondence of M. A. Carleton, 1891–1900, Folder M. A. Carleton – 1900, Carleton to Galloway, May 21, 1900; Carleton to Galloway, June 18, 1900.

Carleton left Paris in late June for Odessa,[135] and then on to the Mennonite settlements at Molotschna. With Enns' assistance, Carleton arranged for some winter wheat to be shipped to Warkentin in Kansas.[136] The next leg of his journey took him to Rostov-on-Don, from where on August 8 he reported to the USDA that he was traveling to the North Caucasus to purchase more wheat. He then traveled northeast to the Volga region, and west to Khar'kov province in present-day Ukraine. He sought hardier varieties of winter wheat, similar to Turkey, that could be grown further north and west in the Great Plains, where the climate was colder and drier than Kansas. In the North Caucasus, he collected a variety that was more drought resistant and, he believed, suitable for western Nebraska and eastern Colorado. In Khar'kov province, which "possesses a climate nearly or quite as severe as South Dakota," he obtained "Kharkov Winter Wheat."[137] In addition, he collected hard spring durum wheat, Arnautka, which was widely grown in the Volga region and elsewhere in southern and eastern Russia.[138]

While he did not make it "half way across the continent," Carleton traveled extensively around the steppes for three months rather than the two he was authorized. On September 13, in Budapest on his way back, he wrote to Galloway at the USDA:

> My work in Russia has been very satisfactory indeed. For Russia it was an enormous amount of work to accomplish in the time given. It has been also very difficult. Everything has been obtained (and more too) that I went for, and I have a large number of pictures, and still more information of value.

Carleton returned to Washington, DC, on September 25, 1900.[139]

[135] NARA CP, RG 54, Finding Aid A1, Entry 58, Division of Vegetable Pathology and Physiology: Correspondence of M. A. Carleton, 1891–1900, Folder M. A. Carleton – 1900, Carleton to Galloway, June 18, 1900; Finding Aid PI-66, Entry 164, Paris Exposition 1900, vol. 2, f.354, Carleton to Professor J. H. Gore, Columbia University, October 30, 1900.

[136] MLA, Warkentin Papers, https://mla.bethelks.edu/archives/ms_68/folder_20/, accessed March 1, 2018, Warkentin to Carleton, November 9, 1900; Carleton to Warkentin, November 13, 1900, January 12, 1901.

[137] NARA CP, RG 54, Finding Aid A1, Entry 58, Division of Vegetable Pathology and Physiology: Correspondence of M. A. Carleton, 1891–1900, Folder M. A. Carleton – 1900; Carleton to Galloway, August 8, 1900; Carleton, "Successful," 535–6. See also E. Cherniaev, "Zametki o iuzhnorusskikh pshenitakh," *SKhiL* 118 (1875), 81–2.

[138] MLA, Warkentin Papers, https://mla.bethelks.edu/archives/ms_68/folder_20/, accessed March 1, 2018, Carleton to Warkentin, November 16, 1900.

[139] NARA CP, RG 54, Finding Aid A1, Entry 58, Division of Vegetable Pathology and Physiology: Correspondence of M. A. Carleton, 1891–1900, Folder M. A. Carleton – 1900, Carleton to Galloway, September 13, 1900; Carleton to Galloway, Telegram from Odessa, received in Washington, DC, September 10, 1900.

The wheat Carleton arranged to be shipped to Warkentin in Kansas was in transit. In September, the shipping agent in Odessa notified Warkentin that 36 bags of wheat weighing 200 puds (around 7,200 pounds) were on their way.[140] Each bag contained around 3.5 bushels, thus the total shipment was about 126 bushels (less than the 500 Warkentin had requested). In November 1900, Warkentin and the Millers' and Grain Dealers' associations in Topeka, Kansas, invited subscriptions for a far larger shipment of winter wheat. They advertised it as "the hard winter wheat variety, known as the Russian Turkey, [which] is the best adapted for Kansas soil and climate, and in quality for flour equal to any grown in the world." They explained that this was same as the wheat that had been brought over from "South Russia" in 1873. Further:

> Through the assistance of Mr. M. A. Carleton, cerealist of the U.S. Department of Agriculture . . . , who was sent to Europe and Asia last summer to select and gather suitable varieties of grain for the U.S., we are advised that the best hard winter wheat suitable for Kansas, should be obtained from Central Crimea, where he found a colony of Mennonites, who make it a specialty of raising pure, clean, hard Turkey wheat . . . Mr. Carleton thinks it is the proper time to import a lot of this wheat for next year's seeding.

They sought help from Congressman Justin De Witt Bowerstock for an exemption from duty for the shipment, and obtained preferential rates for transporting the grain to Kansas.[141] In the summer of 1901, 4,364 sacks of winter wheat, which had been shipped by Bernhard Enns from Odessa, arrived in Kansas City, Missouri. It weighed over 450 (short) tons and amounted to 15,274 bushels: sufficient to sow between 15,000 and 30,000, or more, acres.[142]

Carleton was keen to return to the experimental work on the plot Warkentin had leased for him in Kansas. In August, while still in Russia, he had written to Galloway: "I have dozens of things to talk with you about on my return. Am more and more anxious to establish a permanent cereal Experiment Station in Kansas, where there is now already a good

[140] MLA, Warkentin Papers, Box 5, Folder 20, William Jacob Owen, Steam Ship Agents, Odessa, to Bernhard Warkention, Newton, Kansas, September 12/30 1900; William Jacob Owen, Steam Ship Agents, Odessa, to Warkentin, September 22/4 [October] 1900.

[141] MLA, Warkentin Papers, Box 5, Folder 20, Seed Wheat Circular, Topeka, Kansas, November 26, 1900; Seed Wheat Circular no. 2, Topeka, Kansas, January 8, 1901; Box 2, Folder 21, *Weekly Kansan*, November 23, 1900.

[142] MLA, Warkentin Papers, Box 5. Folder 20, R. F. Stevenson and C. H. Brokaw [?], Kansas City, Missouri, to Bernhard Warkentin, Newton, Kansas, July 12, 1901; Power of Attorney by Bernhard Warkentin appointing H. B. Irwin of New York City to receive shipment of wheat from Odessa, July 15, 1901. See also Malin, *Winter Wheat*, pp. 205–6.

foundation started. Am excellently situated now to accomplish interesting things." In September, while still on his way home, he requested authorization and funding to go to Kansas as soon as he got back.[143] The timing suggests he wanted to oversee sowing the winter wheat in the fall. Tests conducted under Carleton's supervision in Kansas and other plains states in the early 1900s on the Crimean/Turkey and Khar'kov varieties of hard red winter wheat demonstrated again their resistance to drought and rust and their high yields. They tested the varieties in rotations with other grains, legumes, and fodder crops such as alfalfa.[144] Carleton cooperated with Hansen in testing winter wheat, which Hansen had introduced before Carleton, at the South Dakota Agricultural Experiment Station in Brookings.[145]

Carleton vigorously pushed the commercial cultivation of the wheat varieties he brought back from steppes and their use in the American food industry.[146] Promoting the hard red winter wheats, such as Turkey, was not too difficult as they were already becoming established. Farmers from Kansas and other plains states wrote to the USDA with requests for "Turkey Red wheat," "winter wheat," "rust proof wheat." The USDA ordered more wheat from suppliers in St. Petersburg, and several locations in the steppes, including Odessa, Rostov-on-Don, the North Caucasus, and Tsaritsyn on the Volga.[147]

Carleton faced a harder task in promoting the very hard spring durum wheats he had also brought back. Durum wheats were little known and not widely grown in the United States at this time. The USDA had imported some Arnautka in 1864 (and had planted some in its grounds near 14th

[143] NARA CP, RG 54, Finding Aid A1, Entry 58, Division of Vegetable Pathology and Physiology: Correspondence of M. A. Carleton, 1891–1900, Folder M. A. Carleton – 1900, Carleton to Galloway, August 8, 1900; Carleton to Galloway, September 13, 1900.

[144] Mark A. Carleton, "Wheat Improvement in Kansas," *Thirteenth Biennial Report of the Kansas Board of Agriculture for the Years 1901 and 1902* (Topeka, KS: Kansas Department of Agriculture, 1903), pp. 514–15; Carleton, "Lessons from the Grain-Rust Epidemic of 1904," *USDA Farmers' Bulletin*, no. 219 (1905), 14–15; NARA CP, RG 54, Finding Aid PI-66, Entry 1, General Correspondence of the Office of the Chief, 1900–8, Box No. 7, Carleton, "Nature and Importance of Cooperative Grain Experiments Conducted in South Dakota since the year 1900," [August 15, 1907]; Carleton to Chief of the Bureau, June 15, 1907.

[145] NARA CP, RG 54, Finding Aid PI-66, Entry 1, General Correspondence of the Office of the Chief, 1900–8, Box 7, Carleton, "Nature and Importance of Cooperative Grain Experiments Conducted in South Dakota since the year 1900," [August 15, 1907].

[146] Carleton, "Successful"; Carleton, "Russian Cereals."

[147] NARA CP, RG 54, Finding Aid PI-66, Entry 1, General Correspondence of the office of the Chief, 1900–8, Box No. 7, Various letters from the Bureau to Carleton, August 1901; Entry 30, Records of the Division of Cereal Crops and Diseases. Foreign Correspondence, 1900–34, Box 8, File Russia, B. T. Galloway, Director, Office of Plant Industry, to Mr. V. E. Grached, Samenhandlung [seed merchants], St. Petersburg, June 4, 1901.

Street in Washington, DC). More had been brought over by immigrants from Russia to North Dakota, where Warkentin saw some in 1872 (see p. 144). But, it did not become commercially established on a large scale.[148] There was opposition to the durum wheats. They were known, disparagingly, as "macaroni wheats," because they were used for pasta in the Mediterranean world, and some considered them unsuitable for bread flour. There were doubts, moreover, about their suitability for American conditions. In August 1901, the USDA had to deny press reports that test sowings of durum wheat had been killed by drought. Serious resistance came from the millers of Minneapolis, because they would have to invest in new machinery to grind them as the grain was harder than Turkey varieties. Carleton worked to win them over. In 1903, for example, he spoke at the millers' national convention. The trade paper, the *Northwestern Miller*, conducted a prolonged campaign against durum wheat. Its ferocity prompted Carleton's superior to write: "Probably the best course to pursue would be to let the editor go his own way and hang himself on his own yard."[149] Carleton took part in work by the USDA to improve the yield and milling quality of the very hard spring wheats by selection and breeding to make them more suitable for the northern plains.[150] He requested, and received, further samples of durum wheat from the steppes in 1911 from Meyer, who was on an USDA expedition.[151]

Carleton argued for the cultivation of durum wheats, because tests had shown that they were resistant to heat, drought, and rust. He asserted that they would allow the expansion of grain production into more arid parts of the Great Plains as they grew in similar conditions in the steppes. He noted that the normal yearly rainfall along the 100th meridian in the Great Plains, where little wheat was grown at that time, was nearly three inches

[148] J. Allen Clark, John H. Martin, and Carleton R. Ball, "Classification of American Wheat Varieties," *USDA Bulletin*, no. 1074 (1922), 183, 187; Carleton, "Russian Cereals," 14–15; J. Allen Clark and B. B. Bayless, "Classification of Wheat Varieties Grown in 1939," *USDA Technical Bulletin*, no. 795 (1939), 127. See also Mary W. M. Hargreaves, "The Durum Wheat Controversy," *AH* 42 (1968), 211–29.

[149] NARA CP, RG 54, Finding Aid PI-66, Entry 1, General Correspondence of the Office of the Chief, 1900–8, Box No. 7, Carleton to Galloway, August 3, 1901; September 11, 1902, September 18, 1903, Chief of Bureau to Carleton, November 9, 1904.

[150] NARA CP, RG 54, Finding Aid A1, Entry 9, Project records relating to crop investigations, 1885–1913, 1 box, Folder "Improvement of Hard Spring Wheat."

[151] NARA CP, RG 54, Finding Aid NC-135, Entry 135I, Records of Frank N. Meyer, Plant Explorer, 1902–8, Box 7 Correspondence 1911–12, 1917, Accounts 1913–15, Folder 11, Meyer, Omsk, to Fairchild, USDA, October 26, 1911; Meyer, Samara, to Fairchild, November 10, 1911. Meyer asked Fairchild to bring them to Carleton's attention as he had "asked me specially . . . to be on the look out for hardy, . . . durum wheats." (Emphasis in the original.)

more than in the steppes of the Volga region, which produced the finest "macaroni wheat" in the world. He noted also direct parallels between the black earth of the western Great Plains and the steppes: both were alkaline and rich in minerals, such as lime, and contained a lot of humus. Carleton further argued that there were markets in southern Europe, where he believed American durum wheat could compete with Russian, and also at home, where it could be used to produce macaroni. He claimed that, mixed with red wheat, durum wheat could be used to make flour for bread and other products. Such was his zeal for his durum wheat, that he advocated its use in muffins, breakfast cereals, and had recipes printed for macaroni cheese.[152]

Carleton initially thought durum wheats would be well suited to the hotter, more arid southern plains, for example in Texas, but tests showed they were more appropriate for the northern plains. Gradually, as news spread of the durum wheat's resistance to drought and disease, farmers overcame their suspicions and sowed larger quantities. From 1901 to 1903 alone, production increased from 100,000 to 6,000,000 bushels. Dealers bought the wheat and, as markets developed at home and abroad, millers adapted their machinery. In 1918, the *Northwestern Miller* (presumably with a new editor) reported that durum wheat was selling at a premium in Minneapolis and the Great Lakes port of Duluth.[153]

Contacts between American and Russian/Soviet Agricultural Scientists

Another route for the introduction of crop varieties from the steppes to the Great Plains was contacts between American and Russian/Soviet agricultural scientists, which facilitated exchanges of seeds for testing in each other's experiment stations and improving strains through selection and hybridization. Such contacts began in the early 1890s and persisted through the 1940s (See pp. 113–22).

Carleton and other American plant explorers and crop scientists became acquainted with Russian scientists on their expeditions in Russia and at world's fairs (see pp. 113–14). While Carleton was at the Exposition Universelle in Paris in 1900, the USDA asked him to acquire varieties

[152] Carleton, "Wheat Improvement in Kansas," 505–14; Mark Alfred Carleton, "Macaroni Wheats," *USDA Bureau of Plant Industry Bulletin*, no. 3 (1901); Carleton and Joseph S. Chamberlain, "The Commercial Status of Durum Wheat," *USDA Bureau of Plant Industry Bulletin*, no. 70 (1904).

[153] Carleton and Chamberlain, *Commercial Status*, p. 9; Carleton, "Hard Wheats," 405–19.

from several countries, including: "Russia: Hardy winter wheats, macaroni or durum wheats, emmer and buckwheats from extreme north and east Russia, ... and everything obtainable from Turkestan."[154] American agricultural scientists also learned about their Russian counterparts' work from the scientific literature. For example, an article by P. Melikov on wheat in southern Russia was abstracted in the central journal of American agricultural experiment stations in 1901. The author of the abstract, Peter Fireman, was a Jewish immigrant from the Russian Empire, who had scientific training as well as knowledge of Russian.[155] Unless Russian articles were translated or abstracted in English (or other western European languages) they remained inaccessible to most American specialists due to the language barrier. An exception was Carleton who learned Russian in the 1890s (see p. 125). The bibliography of his book, *The Small Grains*, listed sixteen works in Russian out of a total around 500 items.[156] When he offered a copy to Vavilov, he pointed out "how much attention I have given to Russian publications," but admitted he found it "difficult" to read Russian.[157]

Some American and Russian crop scientists developed quite strong and productive connections. The volume of correspondence in archives on both sides of the Atlantic is so large that only a few examples can be referred to here. Before 1917, an important Russian contact for American crop scientists was Vasilii (Basil) Benzin. He studied for a master's degree at the University of Minnesota in St. Paul in 1910–11, before returning to Russia to work for the Ministry of Agriculture.[158] In July 1912, Professor C. P. Bull, of the Minnesota agricultural experiment station, wrote to him: "My dear Basil ... I would be very glad indeed to receive samples of wheat and other cereals that you mention in your letter that you are collecting from all over Russia. ... a few hundred seeds each would be plenty for us [for trials]." Benzin replied, after he returned from his trip around "Asiatic Russia," that he had collected plenty of different seeds and proposed to

[154] NACP, RG 54, Finding Aid PI-66, Entry 164, Paris Exposition 1900, ff.332–36, Carleton to Charles R. Dodge, Director of Agriculture, U.S. Paris Exposition Commission, October 19, 1900; ff.365–9, Galloway, Director of Plant Industry, to W. P. Wilson, U.S. Commission to the Paris Exposition, November 6, 1900.

[155] P. Melikov, "Investigation of the Wheat of Southern Russia," *ESR* 13 (1901–2), 451. On Fireman, see p. 107 and pp. 217–18.

[156] Carleton, *The Small Grains*, pp. 639–85.

[157] TsGANTD Spb, f.318, op.1-1, d.6, l.42, Carleton to Vavilov, November 21, 1921.

[158] V. Benzin, "Izuchenie zasukhoustoichivykh ras sel'sko-khoziaistvennykh rastenii," *SKhiL* 241 (1913), 318; Robert V. Allen, *Russia Looks at America: The View to 1917* (Washington, DC: Library of Congress, 1988), pp. 174, 281.

send them "as soon as possible." It is an indication of their familiar relationship that he added: "I certainly have had quite a time in those wild country, but I am glad to be back. I wasn't sick by malaria and bite by spyders and shnakes [sic]." The USDA sent Benzin samples of American wheat varieties, including some that had been introduced from Russia, such as Crimean and Kharkov winter wheat.[159] There were many more such exchanges concerning crop samples, scientific literature, and visits to the other country (See pp. 115–16).

The main point of contact in Russia for American crop scientists was the Bureau of Applied Botany of the Ministry of Agriculture in St. Petersburg headed by the botanist Robert Regel'. He was succeeded as director after his death in 1920 by Vavilov. Vavilov led its successor organization in Soviet Russia, which was later renamed the Institute of Plant Industry (and now bears Vavilov's name). Vavilov's scientific expertise covered crops, their origins, plant breeding, selection, and genetics. His contacts and travels spanned the world, including North America, as he assembled a global collection of seeds for his famous seedbank.[160] In 1921, Vavilov established an office of the Bureau of Applied Botany in New York, under Dmitrii Borodin (See p. 172). Following Vavilov's visit to the United States, the USDA sent him extensive collections of crop samples, in anticipation of receiving similar in exchange.[161] Vavilov sent them copies of his publications, including his study of crops of the steppes of southeastern Russia.[162]

Scientists at several agricultural experiment stations and colleges in the Great Plains developed good relationships with Vavilov after his visit in 1921–2. Louis Jorgenson, of the Langdon Substation of the North Dakota Agricultural Experiment Station, sent Vavilov samples of varieties he had

[159] RGIA, f.382, op.9, 1911–17, d.290, ll.9–10, C. P. Bull, Department of Agriculture, Experiment Station, University of Minnesota, to Benzin, July 30, 1912; l.11, C. R. Ball, Bureau of Plant Industry, USDA, to Benzin, September 10, 1912; l.12 [Benzin] to Professor C. P. Bull, University Farm, St. Paul, MN, December, 18, 1912; l.20, Fairchild, to Benzin, August 28, 1912.

[160] See Anastasia A. Fedotova and Nikolay P. Goncharov, "Robert Regel: Otdel prikladnoi botaniki, 1917–18 gg.," *Istoriko-biologicheskie issledovaniia* 7, 4 (2015), 90–123; Igor G. Loskutov, *Vavilov and His Institute: A History of the World Collection of Plant Genetic Resources in Russia* (Rome: International Plant Genetic Resources Institute, 1999). See also Peter Pringle, *The Murder of Nikolai Vavilov: The Story of Stalin's Persecution of One of the Great Scientists of the Twentieth Century* (NewYork: Simon and Schuster, 2008).

[161] NARA CP, RG 54, Finding Aid PI-66, Entry 30, Records of the Division of Cereal Crops and Diseases. Foreign Correspondence, 1900–34, Box 8, File Russia, Cerealist in charge [USDA] to Vavilov, c/o D. N. Borodin, Russian Bureau of Applied Botany, New York, November 15, 1921; Agronomist in charge Western Wheat Investigations to Vavilov, c/o Borodin, November 16, 1921.

[162] TsGANDT, Spb, f.318, op.1-1, d.128, l.128, C. R. Ball, BPI, USDA to Vavilov, Bureau of Applied Botany, Petrograd, November 3, 1922. On the book, see n. 53 this chapter.

requested. In exchange, he asked for seeds of several Russian varieties, in particular winter wheats, spring wheats, other grains and grasses. He reminded Vavilov that the climatic conditions at Landgon were "very severe" with "very cold climatic conditions" and "rainfall ... not ever-abundant."[163] A close relationship developed between Vavilov and John H. Parker of the Kansas State Agricultural College in Manhattan. Following Vavilov's visit in 1921, Parker sent him a large selection of seeds of wheat and oat varieties, including some introduced from Russia. Parker asked him for publications in return, and aimed to establish a longer-term connection.[164] Parker was engaged in research into rust-resistant varieties of winter wheat that had originally come from Russia.[165] In 1924, he asked Vavilov for samples of early-ripening winter wheat varieties. Vavilov responded by inviting Parker to visit experiment stations in the Soviet Union. Parker, like Carleton a generation earlier, was eager to go:

> In view of the fact that Kansas wheats are now almost entirely from Russia it seems to me highly desirable that I learn something about them in their native land and perchance collect some still more valuable than those we have, such as Kharkov, "Turkey Red", Kanred and most recently "Blackhull."

(The last two had been selected from existing strains of Turkey/Crimean wheat.) He asked Vavilov for information to help him persuade the Kansas state government and the federal government to authorize his trip. But, he had to decline as the state of Kansas was unable to approve or finance a visit to Soviet agricultural experiment stations.[166]

The USDA obtained samples of the latest Soviet strains of wheat in 1928. The relationship benefitted both parties. Through the intermediary of J. W. Pincus of Amtorg (the Soviet trading corporation in the United States), the USDA was offered several new varieties of winter

[163] TsGANTD Spb, f.318, op.1-1, d.6, l.85, Louis Jorgenson, North Dakota Agricultural Experiment Station, Langdon Substation to D. N. Borodin, Russian Bureau of Applied Botany, New York, December 27, 1921.

[164] TsGANTD Spb, f.318, op.1-1, d.21, ll.1–4, Parker to Vavilov, January 5, 1922.

[165] See Leo Edward Melchers, and John L. Parker, "Rust Resistance in Winter-Wheat Varieties," *USDA Bulletin*, no. 146 (1922).

[166] TsGANTD Spb, f.318, op.1-1, d.49, l.83, Parker to Vavilov, March 10, 1924; d.88, l.40, Vavilov to Parker, November 11, 1925; d.90, l.37, Parker to Vavilov, April 13, 1926; ll.43–5, Parker to Vavilov December 26, 1925. Parker's later attempts to visit Russia were also unsuccessful, but the two scientists kept in touch, e.g. TsGANTD Spb, f.318, op.1-1, d.698, l.26, Vavilov to Parker, March 10, 1934. See also E. G. Heyne, "The Development of Wheat in Kansas," in *The Rise of the Wheat State: A History of Kansas Agriculture, 1861–1986*, ed. George E. Ham and Robin Higham (Manhattan, KS: Sunflower University Press, 1987), pp. 45–6.

wheat from experiment stations in the steppe region. They included Nova-Krimka (New Crimean), Kooperatorka, Durabel, and Ukrainka. The Soviet side proposed also to send scientific publications on the new varieties, and offered to translate one of them into English.[167] After some discussion, the USDA accepted the offer, requesting 25 or 50 pounds of the new varieties.[168] Pincus reported that the seeds and publications had been dispatched as requested and, in addition, that 20 lbs each of Durabel and Ukrainka had been sent to L. E. Call at the Kansas State Agricultural College and Bolley at the Agricultural College in Fargo, North Dakota. Pincus later wrote that he had received requests for seeds from agricultural experiment stations in Fort Collins, Colorado, and St. Paul, Minnesota.[169]

The Soviet offers of new varieties and scientific literature were not purely altruistic. They were, in part, in return for crop samples and studies the USDA had sent them earlier and, in part, in anticipation of future requests. In this case, the requests came right away. Pincus asked for samples of 2–3 pounds of "leading, promising varieties of cereals and legumes," for the Soviet experiment stations that had supplied the varieties to the USDA. In addition, the Department of Agriculture of the Crimean Autonomous Republic wished to receive information on the "present distribution of Crimean wheats in the United States and Canada," as well as historical and statistical information.[170] C. E. Leighty of the USDA offered samples of "promising varieties of cereals" and enclosed a USDA bulletin on "Crimean Wheat." He referred Pincus to Carleton's book, *The Small Grains*, and to the Ottawa experiment farm for information on Crimean wheats in Canada. At the same time, Leighty wrote to Parker in Manhattan, Kansas, asking for samples of Kanred and other varieties to send to Amtorg. Leighty

[167] NARA CP, RG 54, Finding Aid PI-66, Entry 30, Records of the Division of Cereal Crops and Diseases. Foreign Correspondence, 1900–34, Box 8, File Russia, J. W. Pincus, Seed Department, Amtorg Trading Corporation, 165 Broadway, NY to Dr. Carleton R. Ball, Senior Agronomist, Cereal Investigation, USDA, March 6, 1928; Pincus to Ball, May 12, 1928; Pincus to Ball, May 23, 1928; Pincus to Ball, June 6, 1928.

[168] NARA CP, RG 54, Finding Aid PI-66, Entry 30, Records of the Division of Cereal Crops and Diseases. Foreign Correspondence, 1900–34, Box 8, File Russia, J. A. Clark, Agronomist in Charge, Western Wheat Investigations, USDA, Bureau of Plant Industry, to Dr. C. R. Ball, Memorandum, May 16, 1928; Ball to Pincus, May 18, 1928; M. A. McCall, Senior Agronomist Acting in Charge, BPI, USDA to Pincus, June 8, 1928.

[169] NARA CP, RG 54, Finding Aid PI-66, Entry 30, Records of the Division of Cereal Crops and Diseases. Foreign Correspondence, 1900–34, Box 8, File Russia, Pincus to McCall, June 15, 1928; Pincus to McCall, July 11, 1928; Pincus to McCall, July 17, 1928.

[170] NARA CP, RG 54, Finding Aid PI-66, Entry 30, Records of the Division of Cereal Crops and Diseases. Foreign Correspondence, 1900–34, Box 8, File Russia, Pincus to McCall, July 17, 1928.

explained to Pincus that they were planning to test the samples of Novo-Krimka in comparison with "our own sorts." This prompted a request from Pincus for: "a careful examination of the analysis of the Russian Wheat Seed sent to you. If possible, we would like to have not only the purity and germination, but also report on its grading, and also if possible a chemical analysis showing the percentage of protein."[171]

While this may give the impression that the Soviet authorities were keen for the USDA to do their work for them, the USDA benefitted by obtaining new strains from the Soviet Union for its own experimental work. Some, such as Ukrainka, went into commercial production. Both sides did not always get all they wanted. For example, a commercial seed grower, R. M. Woodruff of Pratt, Kansas (about ninety miles southwest of Newton), had bought some seed of the Soviet variety Kooperatorka in 1927. But, Woodruff declined to share the results of his tests when they showed it to be of very high quality, with very high protein content, and more productive than all other varieties. He planted his seed to increase his supplies, and sold them to Kansas farmers under the name "Kooperatka."[172]

The International Grain Trade

Varieties of wheat from the steppes also arrived in North America through the international grain trade. In one significant case, an imported variety of wheat was used to create a hybrid that was ideal for the northern plains of the United States and the Canadian prairies. This was the hard red spring wheat known as Marquis. It was created in the 1890s by plant breeders, most likely at an experiment station in Indian Head, Saskatchewan (about 120 miles north of the border with North Dakota). It was a cross of Hard Red Calcutta wheat from India and Red Fife wheat. The latter had been imported in 1842 by Mr. David Fife of Peterborough, Ontario, from Glasgow, Scotland. Its origins were not immediately apparent. Carleton believed it had come from "Russia." Subsequent investigations indicated

[171] NARA CP, RG 54, Finding Aid PI-66, Entry 30, Records of the Division of Cereal Crops and Diseases. Foreign Correspondence, 1900–34, Box 8, File Russia, C. E. Leighty, Senior Agronomist in Charge, Eastern Wheat Investigations, USDA to Amtorg, July 19, 1928; C. E. Leighty, Cereal Investigations, USDA, to Dr. J. H. Parker, Kansas State Agricultural College, Manhatten, KS, July 19, 1928; Pincus to McCall, July 27, 1928.

[172] NARA CP, RG 54, Finding Aid PI-66, Entry 30, Records of the Division of Cereal Crops and Diseases. Foreign Correspondence, 1900–34, Box 8, File Russia, J.W. Pincus, Seed Dept, Amtorg to Dr. M. A. McCall, BPI, USDA, August 31, 1928; Clark and Bayless, "Classification of Wheat Varieties," 98–9.

that it had reached Glasgow from the Baltic port of Danzig, from where grain had been exported from the territory of present-day Poland and Ukraine. The wheat itself was likely to have originated in Galicia, where it was known as Halychanka, on the edge of the steppes. (Galicia, then part of the Hapsburg Empire, is now in western Ukraine.) Cultivation of Marquis in the Canadian prairies began in 1907. The USDA introduced it to the northern plains in 1912–13. It quickly established itself as one of the most successful varieties of wheat in the region for several decades.[173]

Marquis was used to create more hybrids. One was Tenmarq, which was selected from crosses between Marquis and a Crimean/Turkey wheat by Parker at the Kansas Agricultural Experiment Station in 1921. It was distributed in 1932 and grown successfully in Kansas, Oklahoma, and the Texas panhandle.[174] The success of Marquis wheat, and the hybrids in which it was crossed with other strains, including Crimean wheat, further indicates the suitability of varieties derived from crops from the steppe region that found their way to the grasslands of North America by various routes including, last but not least, the international grain trade.

Conclusion

The similarities in the environmental conditions were essential for the success of the varieties of wheat and other crops that were introduced from the steppes to the Great Plains. The importance of the similar conditions was recognized by immigrants, especially the Mennonites, who chose to move to a region with familiar, and in some cases slightly better, conditions for agriculture. It was also acknowledged by American agricultural explorers and crop scientists, such as Hansen, Carleton, and others. They deliberately traveled to a region that resembled the Great Plains, and consciously went to parts with more extreme climates, such as the north of present-day Kazakhstan, in search of hardy varieties of wheat and other crops that could withstand the full range of fluctuations in conditions in the Great Plains.

The barriers to Americans learning from steppe farming and introducing crop varieties from the region that were outlined in the introduction to

[173] J. W. Morrison, "Marquis Wheat: Triumph of Scientific Endeavor," *AH* 34 (1960), 182–8; Clark and Bayless, "Classification of Wheat Varieties," 70–4; Carleton, "Russian Cereals," 7; Carleton, *The Small Grains*, p. 161 (Carleton used the term "Russia" to include present-day Ukraine). See also Stephan Symko, "From a Single Seed: Tracing the Marquis Wheat Success Story in Canada to its Roots in the Ukraine," Agriculture and Agri-Food Canada, 1999, available online at www5.agr.gc.ca/resources/prod/doc/publications/marquis/wheat-ble_e.pdf, accessed November 1, 2018.

[174] J. Allen Clark and K. S. Quisenberry, "Hard Red Winter Wheat Varieties," *USDA Farmers' Bulletin*, no. 1806 (1938), 9–10.

this chapter were overcome in several ways. The realization among American crop scientists that they could only benefit from importing crop varieties from the steppes alleviated concerns inside the USDA or state governments about cooperating with their chief rival in the world grain market. These scientists were well aware, moreover, of the cycles of wet and dry years that afflicted both the steppes and the Great Plains. They recognized that the recurring droughts and crop failures in the steppes were short-term phenomena, which they were familiar with in North America, and not evidence for long-term problems that could have discouraged them from importing crops from Russia. For example, Carleton did not visit the Mennonite settlements in the steppes to obtain varieties of grain on his first trip in 1898–9, because they were suffering from drought, but returned to visit them in 1900. American scientists were eager to resume contacts with their Russian counterparts, after the hiatus brought about the Russian Revolution and Civil War of 1917–21, in spite of the famine raging in the steppe region at that time. While Carleton learned Russian to enable him to read Russian publications and interact with people on his visits, he was unusual among his compatriots in doing so. Most relied on translations and Russian scientists' knowledge of English or other western European languages.

Russian and Soviet crop scientists also realized that similarities between the environments of the steppes and Great Plains were a basis for collaboration, regardless of competition in the world grain trade before 1914 and the serious political impediments after 1917. In March 1927, for example, Professor G. K. Meister of the experiment station at Saratov on the Volga wrote to the USDA requesting assistance to visit experiment stations in the United States, in particular in Kansas. After introducing himself and his interest in genetics and plant breeding, he continued: "It occurs to me that scientific connection in plant breeding which might be established between Russian and American investigators may prove to be useful for both, since we have identical climatic conditions and many varieties of wheats on which Americans are working were imported from Russia." He added: "I am 53 years old, do not care for politics and scientific problems exclusively attract me to the United States." Leighty was familiar with Meister's work as Vavilov had drawn it to his attention. He replied:

> We have been working on some of the same crops and even on the same problems ... and I would consider it an unusual privilege to be able to discuss these matters with you. I feel sure that personal contacts ... with American scientists will be found mutually helpful. Many of our agricultural problems are the same as you have in Russia. We even grow the same

> varieties of certain crops to a considerable extent. More than a third of our wheat acreage is sown to varieties of Russian origin.[175]

The USDA assisted Meister and wrote letters of introduction for him, including to Parker in Manhattan, Kansas.[176] After his visit, Meister thanked Leighty for his help. From the Russian Agricultural Agency in New York, Dmitrii Borodin wrote: "It has always been a source of great satisfaction to know that the Russian colleagues are taken such wonderful care of by the [USDA], and I sincerely trust that it will always prove mutually advantageous."[177]

Probably the greatest barrier to the successful introduction of crop varieties from the steppes was resistance to change among farmers and millers. Plains farmers, however, were convinced during the dry years of the 1890s, when they saw how the introduced wheat weathered the drought, while their usual crops did not. Opposition from millers facing the expense of retooling their mills to grind the harder wheats from the steppes was eventually overcome, due in part to the political skills of Warkentin and Carleton. Their success in expanding the cultivation of the harder wheats left millers little choice but to adapt their equipment.

The introduction of wheat varieties from the steppes was of enormous and enduring significance for wheat production in the Great Plains and American agriculture more broadly.[178] The importance can be gauged from the large acreages sown with these varieties and selections and hybrids derived from them. In 1916, Carleton proclaimed: "About or nearly half the total wheat production of North America" was of the hard varieties introduced from Russia.[179] USDA data for 1919 shows their extent. In Kansas and Nebraska around 82 percent of the wheat area was

[175] NARA CP, RG 54, Finding Aid PI-66, Entry 30, Records of the Division of Cereal Crops and Diseases. Foreign Correspondence, 1900–34, Box 8, File Russia, Professor G. K. Meister, Saratov Experiment Station to USDA, March 4, 1927 (translated by Elizabeth Jodidi, Senior translator, BPI); C. E. Leighty, Agronomist in Charge Eastern Wheat Investigations to Meister, April 7, 1927.

[176] NARA CP, RG 54, Finding Aid PI-66, Entry 30, Records of the Division of Cereal Crops and Diseases. Foreign Correspondence, 1900–34, Box 8, File Russia, Leighty to Dr. W. A. Taylor, Chief, Bureau of Plant Industry, April 7, 1927; Carleton R. Ball, Senior Agronomist in Charge, BPI to Leighty, May 11, 1927; Leightly to Professor J. H. Parker, Agricultural Experiment Station, Manhattan, KS, October 17, 1927.

[177] NARA CP, RG 54, Finding Aid PI-66, Entry 30, Records of the Division of Cereal Crops and Diseases. Foreign Correspondence, 1900–34, Box 8, File Russia, Meister to Leighty, December 7, 1927; D. N. Borodin, Russian Agal Agency in America, 136 Liberty St., NY. to Dr. C. R. Ball, Office of Cereal Crops and Diseases, USDA, September 20, 1927.

[178] See Alan L. Olmstead and Paul W. Rhode, *Creating Abundance: Biological Innovation and American Agricultural Development* (New York: Cambridge University Press, 2008), pp. 17–63.

[179] Carleton, *The Small Grains*, p. 166.

sown with Turkey, i.e. Crimean, varieties, and in Oklahoma and Colorado, the proportion was roughly two-thirds. Durum wheats had found their niche. In North and South Dakota they were sown in around 29 and 17 percent respectively of the total areas under wheat. In neighboring Wyoming 24 percent of the wheat area was sown with durum varieties. In the United States as whole, Turkey was the most widespread variety, sown on over twenty one million acres, comprising almost 30 percent of the total acreage under all wheat. Durum wheats were grown in around 6 percent of the nation's wheat fields. After Turkey, the most popular variety was Marquis, with around 16 percent of the wheat area. It was the most widespread type of wheat in the Dakotas.[180] Marquis was sown in about 60 percent of the total area under *spring* wheat in the entire United States in 1920.[181]

Wheat varieties from the steppes continued to be widely grown in the Great Plains during the interwar years. In 1920, the Kansas Board of Agriculture proudly boasted: "The state is preeminent as a producer of hard winter wheat, and with the decrease in production of wheat of this type in Russia, Turkey and Hungary, due to war conditions, it is believed that Kansas is now the largest producer of hard winter wheat in the world."[182] The Chief of the Bureau of Plant Industry of the USDA reported in 1926 that, following the introduction of the "valuable" "Turkey, Kharkof, and other hard winter wheats" annual wheat production had increased by 10 million bushels.[183] By 1939, Turkey wheats were sown in twenty-six states on nearly thirteen million acres, comprising almost 20 percent of the total area under wheat in the United States. This was less than in 1919, but in part because selections and hybrids derived from Turkey wheats were sown in large amounts. Blackhull and Kanred were sown on around eight million and over a million and a half acres respectively. The area of land on which Tenmarq was cultivated had reached over three and half million acres by 1939.[184] Turkey wheat continued to feature in the pedigrees of hybrids grown in the Great Plains later in the twentieth century, including "Newton," which was named for the city where

[180] Clark and Bayliss, "Classification," 144–7, 208–18.
[181] NARA CP, RG 54, Finding Aid A1, Entry 51, Division of Plant Exploration and Introduction: Reference Files, 1914–51, Report of the Chief of the Bureau of Plant Industry 1920, p. 9.
[182] "Wheat in Kansas," *Report of the Kansas State Board of Agriculture for the Quarter Ending September 1920*, 155, (1920), 10.
[183] NARA CP, RG 54, Finding Aid A1, Entry 51, Division of Plant Exploration and Introduction: Reference Files, 1914–51, Report of the Chief of the Bureau of Plant Industry 1926, 3.
[184] Clark and Bayless, "Classification of Wheat Varieties," 100–1, 103, 105–7. See also Heyne, "Development of Wheat," 45–6.

Warkentin made his home and the statue of the Mennonite settler was dedicated in 1942.[185]

In the northern plains, the acreages sown with durum wheats, in particular Kubanka, grew over the 1920s. Farmers chose durum wheats over other varieties, because of their greater resistance to drought and disease and higher productivity.[186] When Kubanka succumbed to rust, one of the durum varieties that replaced it was Pentad, which had been selected from the samples Bolley brought back from the steppes to the North Dakota Agricultural Experiment Station in 1903.[187] Experiment stations in the northern plains produced hybrids of durum wheats. One, created in North Dakota in 1943, was named for Carleton.[188] Marquis wheat, which had been so successful in the northern plains for several decades, succumbed to rust in the mid 1930s. One of its main replacements was Thatcher, which had been bred specifically as a rust-resistant variety by the experiment station in St. Paul, Minnesota. Its pedigree included Marquis and Kanred, amongst others, and so could trace its pedigree to the steppes.[189]

This chapter has demonstrated various routes by which varieties of wheat reached the Great Plains from the steppes: the grain trade; exchanges between American and Russian scientists; plant explorers such as Carleton; and Mennonite immigrants. In introducing and establishing hard red winter wheat, Turkey Red, decisive roles were played by Carleton and Warkentin. They organized the shipment and distribution of large quantities of seed around the turn of the twentieth century, persuaded millers to retool to grind the harder kernels, and by testing them in experiment plots, showed their suitability for the central plains. Hansen, who had first introduced hard red winter wheat from the steppes and

[185] See Gary M. Paulsen, "International Contributions to the Improvement and Marketing of Kansas Wheat," *Kansas State University Agricultural Experiment Station and Cooperative Extension Service* 126, February 2000, available online at https://krex.k-state.edu/dspace/handle/2097/16391, accessed October 17, 2019; Dana G. Dalrymple, "Changes in Wheat Varieties and Yields in the United States, 1919–1984," *AH* 62 (1988), 20–36; Heyne, "Development of Wheat," 49–56.

[186] NARA CP, RG 54, Finding Aid A1, Entry 51, Division of Plant Exploration and Introduction: Reference Files, 1914-1951, Report of the Chief of the Bureau of Plant Industry 1920, p. 10; Report of the Chief of the Bureau of Plant Industry 1921, 3; Report of the Chief of the Bureau of Plant Industry 1926, 3.

[187] Clark and Bayless, "Classification of Wheat Varieties," 125–9; Bolley Papers, Bolley to Pieters, July 15, 1904; "List of small samples of miscellaneous kinds of seeds."

[188] Paulsen, "The Agronomic Legacy," 121.

[189] Clark and Bayless, "Classification of Wheat Varieties," 70–4; W. C. Coffey and M. A. McCall, "Thatcher Wheat," *University of Minnesota Agricultural Experiment Station Bulletin*, no. 325 (1936).

cooperated with Carleton in testing it, felt his role had been understated and sought to remedy this in the 1930s and 1940s.[190] As an aside, both Warkentin and Carleton came to untimely ends. Warkentin fell victim to a shooting incident in 1908 on a train from Damascus to Beirut (in the vicinity of the "fertile crescent") while on vacation to the Holy Land with his wife.[191] Carleton died of heart disease made worse by malaria in Peru in 1925, where he was trying to make a living following professional, financial, and family problems.[192]

While Warkentin's role in the origins and development of the hard red winter wheat industry in the central plains is well documented, how much importance can be attached other Mennonite immigrants from the steppes? The available contemporary evidence suggested only that the migrants certainly could have brought quantities of wheat seed with them, and some probably did, but that they did not think to mention it or attach much importance to it until several decades after they arrived. In contrast to the repeated retellings of story from early 1900s, little attention was paid to story before the turn of the twentieth century. Even the Kansas journalist Noble Prentis, who visited the Mennonites in the 1870s and 1880s and had a keen eye for detail and interesting anecdotes, made no reference to Mennonite immigrants bringing wheat with them from the steppes in his history of Kansas published in 1899. He mentioned the Mennonite migration, and that they brought with them Russian apricot and mulberry trees, and also German thrift, industry, and belief in education, but he made no mention of winter wheat seed. He discussed agriculture in his history, referred to wheat on eight separate pages, but not winter wheat from the steppes.[193]

Another acute observer of Mennonite agriculture in the Great Plains, Carleton, did not refer to the immigrants bringing winter wheat seed with them from the steppes in an article for the USDA Yearbook for 1900.

[190] See SDSU ASC, UA 53.4, N. E. Hansen papers, Series 2. Helen Hansen Loen Collection, Box 3, Folder 15, "Recent Developments in Agriculture and Horticulture in Ammerica [sic] and Russia (Extracts from an address – 1935)," p. 1; Box 4, Folder 41, MS: "Hard Wheats from Russia. Help in the Battle for Bread"; Hansen, "Fifty Years," 35–6. Hansen went so far as to annotate the copy of Fairchild's memoir, *The World Was My Garden: Travels of Plant Explorer*, in the South Dakota State University Library, to emphasize his role. His notes are on p. 125 of the library's copy. I am grateful to Kevin D. Kephart, Vice President for Research and Economic Development, SDSU, for drawing this to my attention.

[191] Krahn and Thiessen, "Warkentin, Bernhard (1847–1908)."

[192] Paulsen, "The Agronomic Legacy," 123.

[193] Noble L. Prentis, *A History of Kansas* (Topeka, KS: Caroline Prentis, 1899), pp. 146–7, 92, 151, 153, 158, 169, 192, 227, 250.

He did write that they and other immigrants from the steppes brought their agricultural methods:

> Mention has been made of the success that has attended the practices of the Russian settlers in various localities of the Great Plains. These people have simply followed the methods they learned in their native country. In the southern and eastern wheat districts of Russia the people have contended with extremes of climate even more severe than ours for long periods of time. It should not be surprising, therefore, if they have learned to get the best results possible under adverse conditions.

Among the practices he referred to were "black fallow," i.e. keeping the fallow field clear of vegetation, which was particularly associated with Mennonites, deep plowing, methods and timing of planting seed, and particular crop rotations.[194] Carleton seems to have heard the story about the immigrant families bringing seed with them for the first time in November 1900, in a letter from Warkentin, but did not mention it in his writings until 1914 (see p. 131).

Another way to look at this issue is to ask: Why did the story of immigrants bringing winter wheat seed with them take off after 1900, gather momentum, and attract further attention in the late 1930s and early 1940s? The stories took off at exactly the time when Warkentin and Carleton imported large quantities of Crimean hard red winter wheat seed following the latter's visit in 1900 to the Mennonite communities in the steppes. It was from this time that the importance of this particular type of wheat became fully apparent. This could have been the point when Mennonites realized that the seeds their families may have brought with them in the 1870s were actually extremely important, more so than the clocks, featherbeds, threshing stones, and other items they had attached more significance to when they arrived. Concerning the story of the Krimmer brethren, led by Jacob Wiebe, which included Anna Barkman and her family, the ever-critical Malin argued:

> According to recent Mennonite historians it was this colony of twenty-four families ... that is credited with the introduction of Turkey hard winter wheat, each family of whom had brought about a peck [1/4 bushel] of it, planting it in the fall of 1874 and harvesting it in 1875. Such a story of exclusive credit is possible, but scarcely seems to meet the first test of historical criticism, that of reasonableness. *It would seem that, if the Wiebe group brought a remarkable new wheat, ... he did not realize its significance, not yet being blessed with sufficient wisdom of hindsight* [my emphasis DM].[195]

[194] Carleton, "Successful," 538–42. And see pp. 18–19.
[195] Malin, *Winter Wheat*, pp. 166, 167, 250.

In a volume published in 1949 to mark the seventy-fifth anniversary of the migration, Cornelius Krahn, a local Mennonite historian and director of the Bethel College library, wrote:

> One seed transplanted from the steppes to the prairies grew and multiplied *far beyond any expectation* [my emphasis DM]. This little kernel of wheat that fell into the ground has conquered the prairie and made it the bread basket of the nation. The seed that crossed the ocean in the handmade chests . . . was the hard winter wheat grown natively along the coasts of the Black Sea and Sea of Azov.[196]

There may be further reasons why the story took hold after 1900, about a generation after the migration, and then persisted. Mennonite historian Royden Loewen detected an increase in yearning for their former homes in the steppes in letters from female migrants in North America to their families in the Russia Empire at the turn of the twentieth century. He saw in this growing longing an attempt to reconnect with their old homeland at a time of change in their new communities on the other side of the Atlantic.[197] Nostalgia for their previous homes in the steppes also informs the sophisticated argument made by Susie Fisher for the persistence of stories about Mennonites transferring seeds from the steppes to their new, American, homes:

> We ought to read these stories *not* for the facts of settlement, . . . but rather for the experience of settlement: a reckoning with unfamiliar space. In doing so, we notice that the longing to plant gardens with seeds from the ancestral home represents an attempt at refashioning what was perceived to be an *ahistorical space* into a *settled place*.[198]

It is possible that the story attracted further attention in the late 1930s and early 1940s, featuring in the WPA guidebook to Kansas and the radio show dramatizing Anna Barkman selecting all those seeds, because it served as a comfort to farmers who had experienced the depression, drought, and dust storms of the "dirty thirties." It allowed them to look back to the endurance of the early settlers as something to cling to in difficult times. By 1942, when the statue of the Mennonite settler was dedicated in the Athletic Park in Newton, Kansas, plains farming was emerging from the crisis as the rains returned and the Second World War

[196] Krahn, *From the Steppes*, p. 10.
[197] Royden Loewen, *Hidden Worlds: Revisiting the Mennonite Migrants of the 1870s* (Winnipeg: The University of Manitoba Press, 2001), pp. 63–5.
[198] Fisher, "(Trans)planting Manitoba's West Reserve," 131.

increased demand and prices for agricultural produce.[199] The statue perhaps represented a new confidence and hope in what they could achieve once more in emulation of the immigrants from the steppes. In a similar vein, writing about Western movies of the Great Depression and Second World War era, Richard Aquila argued that they "transported audiences to a mythic West, a land of hope and redemption, . . . [evoking] simpler days in a magical place," a place where: "Mythic cowboy heroes could be counted on to right every wrong and overcome any obstacle."[200]

It could be argued that the stories, and the persistence of the stories, about Mennonite immigrants, including Anna Barkman, bringing seeds of Turkey Red wheat with them from the steppes in the mid 1870s are more important as part of the cultural heritage and self-perception of these communities than as historical evidence for how this particular type of wheat reached the United States. The early settlers probably did bring some wheat seed with them, and its success likely spurred further imports, but in the wider story of the introduction from the steppes of varieties of wheat, including the other types discussed in this chapter, appropriate to the environmental conditions in the plains, due regard needs to be given to the full range of factors and other actors considered in this chapter.

Ultimately, the persistence of the Warkentin–Mennonite immigrants–Carleton story to explain the introduction of "Turkey Red" wheat, despite objections by several historians, has been a successful exercise in branding. We know from Klippart's compendium of knowledge on wheat grown in the United States published in Cincinnati, Ohio, in 1860 that there were varieties with similar names and some similar characteristics already in cultivation in states east of Kansas before the Mennonites arrived. These varieties seem, however, to have had softer kernels than the Crimean/Turkey wheat. Further, since the Crimean hard red winter wheat was only starting to come into wider cultivation in the steppe region in the mid nineteenth century, it is unlikely that this specific type of wheat could have made its way across the Atlantic through the international grain trade in time to be recorded in Klippart's book. When the immigrants from the steppes purchased American wheat in their first years in their new homes, they may well have planted some of these other red winter wheats that were

[199] On the end of the Dust Bowl, see Geoff Cunfer, "The Dust Bowl," EH.Net Encyclopedia, ed. Robert Whaples, August 18, 2004, available online at http://eh.net/encyclopedia/the-dust-bowl/, accessed October 30, 2017. R. Douglas Hurt, *The Big Empty: The Great Plains in the Twentieth Century* (Tuscon: University of Arizona Press, 2011), pp. 166–8.

[200] Richard Aquila, *The Sagebrush Trail: Western Movies and Twentieth-Century America* (Tucson: University of Arizona Press, 2015), pp. 118–19.

already in the United States. Since these types were already falling out of favor among farmers further east, however, the Mennonites may not have cultivated them to any great extent or for very long. Nor is it likely that these red winter wheats suddenly acquired the attributes of hardiness in the face of pests, disease, and drought that made the imported Crimean hard red winter wheats so successful in the 1880s and 1890s. The imported Crimean wheat stood out to such an extent in the drought years of the 1890s that it came to the attention of the USDA in the person of Carleton. Meanwhile, Warkentin had been shipping over new supplies, marketing them, retooling his mills to grind the harder grains, encouraging other millers to do the same. Carleton built on this work, drew on Warkentin's contacts with family members in the steppes, and assisted him import more "clean" Crimean wheat and promote it in Kansas and other plains states.

Warkentin and Carleton thus established the successful brand of "Turkey Red" at the very time that USDA crop scientists, Carleton included, were devising formal, scientific systems for classifying wheat and other crop varieties. These allowed scientists, officials of the federal and state governments, the seed industry, and farmers to have a clearer idea of precisely what varieties they were dealing with. Their counterparts in the Russian Empire and Soviet Russia were doing the same thing at the same time. And American and Russian/Soviet crop scientists were in regular contact. Thus, the days when similar names were used for different varieties and different names for similar types were long gone. The greater degree of certainty contributed to the success of the brands "Turkey Red" and "Krymka" (Crimean) wheat in both the United States and the Soviet Union. Over subsequent decades, Mennonites in Kansas and elsewhere and others interested in the story, amateur and professional specialists in crops and history alike, have repeated versions of the story many times, keeping it alive. They even erected a statue to it in an athletic park in a small town in south-central Kansas. In doing all these things, they have ensured the brand's, and its history's, continued prominence. Some historians have attached the word "myth" to the Warkentin–Mennonite immigrants–Carleton story of the introduction of the wheat from the steppes. But, the myth has some grounding in evidence that has withstood the scrutiny of skeptical historians, who have unearthed contemporary sources to back up at least parts of it, and of agricultural scientists, whose methods allow them to establish pedigrees of particular types, or cultivars, of crops.

The story of the wheat from the steppes, promoted to such good effect in the Great Plains, also attracted the interest of soil scientists. They recognized the importance of identifying appropriate crop varieties for the conditions in the plains. In the early twentieth century, American soil scientists George Coffey and Thomas Rice noted that wheat varieties introduced from the steppes prospered in the Great Plains, precisely because the soil and climatic conditions were appropriate for them. On the other hand, tests at agricultural experiment stations in North Carolina and Pennsylvania showed that such crops did not do well in the quite different conditions in the eastern United States. Coffey concluded: "the varieties . . . which are standards in the black prairie sections of the country are not adapted to the light-colored soils of the timbered sections."[201] Russian agricultural specialists had discovered back in the 1840s that crops from the eastern United States, including types of winter wheat (not the hard varieties introduced later from the steppes), yielded mixed results and some failed altogether when tested in the harsher conditions of the steppe region.[202] The importance of soils for agriculture spurred Russian soil scientists to devise an innovative way of understanding, studying, and classifying soils from the 1870s.

Coffey and Rice, as we shall see, were among the first American soil scientists to recognize the significance of the innovative Russian work on soils. They recognized also the similarities between soils of the two countries in general, and of their grasslands in particular. A few decades later, in a talk to the Congress of American–Soviet Friendship in 1943, the Chief of the U.S. Soil Survey, Charles Kellogg, noted the achievement of Russian soil scientists since the 1870s, and the value of their work for understanding American soils that resembled those in Russia. He continued that similarities between the United States and Soviet Union went beyond soils to crops and livestock:

> Many of the most important crops had their origin in Soviet territory. Through long experience, selection, and scientific breeding superior strains are developed for the different regions. Through exchange of results and materials, the experience in each country becomes available to the other. The mutual benefits have already been great, even with scientific work in this field comparatively young.[203]

[201] George Nelson Coffey and Thomas D. Rice, "Reconnaissance Soil Survey of Western Kansas, United States," *Field Operations of the Bureau of Soils 1910* (Washington, DC: USDA, 1912), pp. 93–4; G. N. Coffey, "Value of the Field Study of Soils," *Proceedings of the American Society of Agronomy* 1 (1907–9), 171.

[202] RGIA, f.515, op.12, 1846, d.1572, ll.17–19, 27-ob, 146–8, 170–20b., 180–1, 183–5, 203–7, 221–3; RGIA, f.515, op.12, 1849, d.1712, ll.20-ob., 32-ob., 48–9, 64–6.

[203] NAL, Special Collections, Manuscript Collection 91, The Charles Edwin Kellogg Papers, Box 5, Folder 83, 47. "Soviet soil technology and agriculture," 1944 (paper presented at Congress of American–Soviet Friendship, New York City, November 7, 1943), 26.

By the 1940s, thanks largely to Kellogg's predecessor, Curtis Marbut, soil scientists at the U.S. Soil Survey had adopted a new science of soils that drew heavily on the Russian innovation. The transfer of Russian soil science from the steppes to the Great Plains, however, proved to be a more lengthy and complex process than the transfer of the crops that grew in the similar soils.

CHAPTER 5

Soil Science I

Introduction

In publications of the U.S. Department of Agriculture (USDA) in the mid 1930s, the soils of much of the Great Plains were classified as "Chernozems." (See Map 6: Map of the distribution of the great soil groups, Curtis Marbut, 1935.) This was the Russian scientific term for the black earth that also covered parts of the steppes.[1] The author of the essay on the "Soils of the United States" in the USDA's *Altas of American Agriculture* explained: "Soils are the product of the environmental conditions under which they have developed or are developing." The term "Chernozem" and the concept of soils based on their development, or formation, were taken from pioneering work by Russian scientists led by Vasilii Dokuchaev (1846–1903). They had devised a new understanding of soils during field work in the steppe region of the Russian Empire in the 1870s–80s. The author of the American text was Curtis F. Marbut (1863–1935), who was Chief of the USDA's Soil Survey Division from 1913 almost until his death. Under his direction, the U.S. Soil Survey adopted a concept of soils and system of classification based on Russian soil science.[2]

This took some time, however, because Marbut encountered resistance to his "Russian Revolution" in U.S. government soil science. Especially hostile was his boss, chief of all government soil investigations, Milton Whitney (1860–1927), who had established the U.S. Soil Survey in the late 1890s.[3] The scientific basis and classification system of Whitney's soil survey were quite different from the innovative Russian soil science, which he unequivocally rejected. Dokuchaev's theory of soil formation,

[1] The terms "Chernozem" and "black earth" (a literal translation) will be used interchangeably.

[2] C. F. Marbut, "Soils of the United States," in *Atlas of American Agriculture: Physical Basis including Land Relief, Climate, Soils, and Natural Vegetation of the United States*, ed. O. E. Baker, Part III (Washington, DC: USDA, 1935), pp. 11, 17, 72. On Marbut, see pp. 231–3.

[3] The unit underwent several changes in its name, see p. 203.

or genesis, had profound implications for the U.S. Soil Survey. In the words of Roy Simonson:

> Theories of soil genesis, the concept of soil and the classification of soils are closely related ... How an individual thinks soils were formed largely governs his concept of them. The concept in turn affects the thoughts and activities of every soil scientist. Theories of genesis are thus of primary importance in soil science.[4]

Changing the Soil Survey's scientific basis and classification system would require redoing all the surveys that had been carried out, which partly explains Whitney's attitude. His stubborn resistance held up for over two decades the adoption by the U.S. Soil Survey of an understanding of soils that was gaining currency among soil scientists around the world, and remains the basis of modern soil science. Whitney's obstruction delayed formal recognition of similarities between the soils of the Great Plains and the steppes. Whitney had little experience of such soils as he had spent much of his life and career in the eastern United States, where the soils, and the humid, forested environment in which they had formed, were quite different.[5]

The new understanding soils had great practical significance. The pioneering Russian work in the 1870s coincided with the expansion of crop cultivation in the steppes and in the North American plains, thus a new concept of soils was "the order of the day."[6] The new soil science would be of value to agricultural scientists in devising appropriate advice for farmers in both regions.[7] It would be very important in the United States in the 1930s, when the USDA urgently needed a well-founded theory of soils to underpin methods to deal with wind erosion during the Dust Bowl in the Great Plains. Whitney had impeded the USDA in gaining such an understanding in good time.[8]

Marbut disseminated the Russian innovation inside the USDA. In 1928, significantly the year after Whitney's death, he delivered a course of lectures to the USDA Graduate School. He explained how soils formed

[4] Roy W. Simonson, "The United-States Soil Survey: Contributions to Soil Science and Its Application," *Geoderma* 48, nos. 1–2 (1991), 8. Simonson worked for the Survey.

[5] On Whitney, see p. 202.

[6] Jean Boulaine, "Le Contrepoint et le Cortege de Dokuchaev: Quelques Contemporains du Fondateur de la Pedologie Genetique," *Pedologie* 34 (1984), 9–10.

[7] On the importance of making soil science research accessible to farmers, see Jan Arend, "'Simple, Clear, and Easily Understood by the Farmer...': On Expert–Layman Communication in American Soil Science, 1920s–50s," *History of Science* 57 (2019): 324–345.

[8] See Paul S. Sutter, *Let Us Now Praise Famous Gullies: Providence Canyon and the Soils of the South* (Athens: University of Georgia Press, 2015), pp. 37–50.

as a result of several forces, including temperature, rainfall, natural vegetation, geological formations, and topography. Marbut continued: "That is fundamentally the character of the work done in Russia in the late [18]70's and 80's ..., and on the basis of which the science of pedology [soil science] as it is now understood was built."[9] He described how by studying the black soil of the continental interior, i.e. the steppes, and correlating its distribution to "rainfall, temperature and natural vegetation," the Russians had worked out "the relation of soil character to ... the physical environment in which the soil developed."[10] Marbut stressed the concept of the "soil profile," which comprised a series of layers, or "horizons," from the "parent material," or rock, at the bottom to two layers of "true soil" above it (see Figure 5.1). The horizons, designated A, B, and C from top to bottom, revealed the process of soil formation from the parent material as a result of "forces," especially climate and natural vegetation.[11] Different types of soils formed in different environmental conditions and had different profiles. Drawing on "the work of the Russians," Marbut discussed the profiles of the "Chernozems" that had formed in America's plains as a result of their relatively low rainfall and grassy vegetation.[12] The "soil profile" made up of "horizons" was at the heart of Dokuchaev's soil science. A Soviet scientist visiting the United States in 1926 noted that Americans referred to Russian soil science as "profile soil science."[13]

At the 1st International Congress of Soil Science in Washington, DC, in 1927, Marbut singled out the contributions of the "Russian" (Soviet) delegation. He wrote that the "brochures [they] ... presented to ... the Congress ... [constitute] the first complete statement in English of Russian accomplishments in this field, except the book published in German by Glinka more than ten years ago and now for the first time available in English."[14] Konstantin Glinka (1867–1927), who headed the "Russian"

[9] Curtis Fletcher Marbut, *Soils: Their Genesis and* Classification. *A Memorial Volume of Lectures Given in the Graduate School of the United States Department of Agriculture in 1928* (n.p.: Soil Science Society of America, 1951), pp. 17–18. This volume contains lectures 1–10, which covered theoretical issues. Lectures 11–29, which covered regions of the United States, including the prairies and Great Plains, can be found in MSU DGML SCA, Ozarkiana Collection: Curtis F. Marbut Collection, Collection Number: M014, Box 2, Folders [13]–[16].

[10] Marbut, *Soils*, p. 25. [11] Marbut, *Soils*, pp. 26–7. [12] Marbut, *Soils*, pp. 125–30.

[13] A. F. Lebedev, "V Biuro Upolnomochennykh Vsesoiuznykh S"ezdov po pochvovedeniiu (pis'mo iz Ameriki)," *Biulletini pochvoveda*, nos. 5–7 (1926), 54.

[14] C. F. Marbut, "Geography at the First International Congress of Soil Science," *GR* 17, 4 (1927), 662. The Soviet delegation brought a set of thirteen "brochures" in English on Russian soil science published by the Academy of Sciences, see ARAN, f.487, op.1, 1926, d.5, ll.3, 32; Roy W. Simonson, "Early Teaching in USA of Dokuchaiev Factors of Soil Formation," *Soil Science Society of America Journal* 61 (1997), 12. The Library of Congress has two sets under S590.A6.

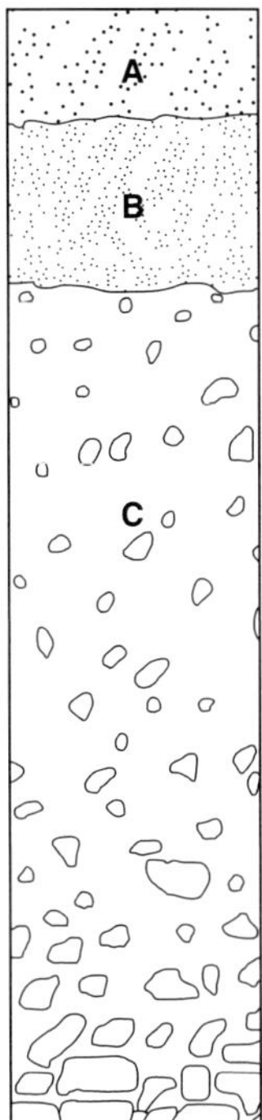

Figure 5.1 Soil profile showing horizons A (top soil), B (subsoil or transition layer), C (parent rock)

Adapted by permission from Springer from William Mahaney and Krishnamani Ramanathan, "Understanding Geophagy in Animals: Standard Procedures for Sampling Soils," *Journal of Chemical Ecology* 29 (2003) 1503–23, doi:10.1023/A:1024263627606, figure 2, p. 1513.

delegation, was a former student and colleague of Dokuchaev. At the Congress's opening ceremony, Glinka was welcomed with "loud applause." The English translation (by Marbut) of his "book ... in German" was on display with a note explaining its huge influence on American soil science.[15] The Congress was followed by a month-long excursion around North America to see different soils. Particular attention was paid to, in Marbut's words, the "groups of Steppe, or as we call them in this country, Great Plains, soils." During the excursion, American and

Marbut had a full set: SHSM MC, Marbut, Curtis Fletcher (1863–1935) Papers, 1852–1963 (C3720) (hereafter Marbut Papers [C3720]), Folder 169, Periodicals and Pamphlets, 1927. Some of the brochures are cited later.

[15] A. A. Iarilov, "Na kongresse i o kongresse," *Biulleteni pochvoveda*, nos. 5–8 (1927), 127, 129.

Soviet scientists debated the similarities and differences between their respective "Chernozems."[16]

Thus, by the late 1920s, under Marbut's leadership, U.S. government soil scientists were using Russian terms for soils of the Great Plains (and for other types of soils in other regions), employing a classification system strongly influenced by the Russian concept of soils based on their formation, analyzing soil "horizons" in "profiles," and citing the importance of work by Russian scientists in the steppes in the late nineteenth century.

Some American scholars have challenged the innovative nature of Dokuchaev's soil science. In 1947, Kansas ecological historian James Malin (who also challenged the Russian origins of Turkey Red wheat) argued that German-born and educated American soil scientist Eugene W. Hilgard (1833–1916) "anticipated much of the modern theories which are credited to the Russian school of V. V. Dokuchaev." Malin admitted that Hilgard had emphasized geological origins of soils, rather than the broader range of soil-forming factors in the Russian theory, but that "also he struck out on new lines." Hilgard grasped "quickly the importance of climate," as he had experience of quite different environments in the United States. Malin paid attention to Hilgard's work, because of what he saw as "the distortions by enthusiasts of the newer Russian–American school of soil scientists of the United States Department of Agriculture." He claimed that Marbut had ignored Hilgard's contribution, but that "[o]utside the Federal governmental publications," Hilgard had been given "substantial recognition" by American soil scientists Jacob Joffe and Hans Jenny.[17] In the 1950s, Malin had a lengthy correspondence with Jenny (a professor in Hilgard's former department at the University of California, Berkeley). Malin insisted on Hilgard's achievement, "the errors of the Russian doctrinaire formula," and Marbut's variant of it.[18] Thus, this pair of chapters will consider whether there were also American roots for the new understanding of soils in the United States.

[16] Marbut, "Geography," 663.

[17] James C. Malin, *The Grasslands of North America: Prolegoma to Its History* (Lawrence, KS: privately published, 1947), pp. 48–50. On Hilgard, see p. 74.

[18] KSHS SA, Ms. Coll. 183, James C. Malin Collection (hereafter Malin Collection), Microfilm MF 6304–8, Letters written by Malin, Malin to Jenny, January 18, 1951. Malin sent Jenny a copy of his 1947 book, *The Grasslands of North America*. Malin Collection, Box 9, Folder 5, correspondence from 'J', Jenny to Malin, October 30, 1947.

Aspects of the reception of Russian soil science in the United States have been explained before.[19] These two chapters, drawing on both Russian and American sources, highlight the role played by studying the soils of the steppes and the Great Plains. The Russian scientists devised their theory while investigating soils in the steppe region in the 1870s–80s. Marbut convinced himself of the importance of the Russian work by testing it in the Great Plains between 1917 and 1922. In the late 1920s, the USDA appointed a Russian émigré soil scientist, Constantine Nikiforoff, to lead a survey of the Chernozem of the Northern Great Plains and to share his Russian experience with American soil survey workers.

This chapter surveys first Russian and then American soil science in the late nineteenth and early twentieth centuries. It contrasts the approaches in both countries, despite similarities between the soils of the steppes and Great Plains, before turning to a discussion of the limited reception of Russian soil science among American scientists before 1914. The following chapter examines the dissemination of Russian soil science in the United States from 1914, Marbut's field work to test the Russian methodology in the Great Plains, and his leadership of the "Russian Revolution" in U.S. government soil science.

Russian Soil Science in the Late Nineteenth and Early Twentieth Centuries

In the late 1870s and early 1880s, Vasilii Dokuchaev, a scientist from St. Petersburg University, led a series of expeditions to study the fertile black earth (Chernozem) of the steppe and forest steppe regions of the Russian Empire. In 1883, he published his path-breaking monograph, *The Russian Black Earth* (*Russkii chernozem*).[20] Much of the work was funded by the

[19] See, for example, Jan Arend, "Russian Science in Translation: How *pochvovedenie* was brought to the West, c. 1875–1945," *Kritika* 18 (2017), 683–708; Arend, *Russlands Bodenkunde in der Welt. Eine ost-westliche Transfergeschichte 1880–1945* (Göttingen: Vandenhoeck und Ruprecht, 2017); Roy W. Simonson, *Historical Highlights of Soil Survey and Soil Classification with Emphasis on the United States, 1899–1970* (Wageningen, The Netherlands: International Soil Reference and Information Centre, 1989); T. R. Paton and G. S. Humphreys, "A Critical Evaluation of the Zonalistic Foundations of Soil Science in the United States. Part I: The Beginning of Soil Classification," *Geoderma* 139, nos. 3–4 (2007), 257–67; G. V. Dobrovol'skii, *Lektsii po istorii pochvovedeniia* (Moscow: Izd-vo MGU, 2010), pp. 132–63; I. A. Krupenikov, *Istoriia pochvovedeniia: ot vremeni ego zarozhdeniia do nashikh dnei* (Moscow: Nauka, 1981), pp. 192–57.

[20] V. V. Dokuchaev, *Russkii chernozem* (Spb: Imperatorskoe Vol'noe Ekonomicheskoe Obshchestvo, 1883). On Dokuchaev, see David Moon, *The Plough that Broke the Steppe: Agriculture and Environment on Russia's Grasslands, 1700–1914* (Oxford: Oxford University Press, 2013), pp. 53–5.

Free Economic Society, a government-backed organization that supported research in agriculture, following recent poor harvests caused by droughts.[21] The expeditions covered a region extending from central Russia in the north to the coasts of the Black and Azov seas and the Caucasus Mountains in the south, from today's Moldova in the west to the Ural River in the east. At regular stops, they analyzed cross sections, or profiles, of soil several feet deep. They devised soil augers or drills (*pochvennye bury*) to bore sample cross sections, and observed that they comprised a series of layers, or horizons. The scientists measured the depths of the horizons within the profiles, recorded the textures of the top soil (or plowing layer) and the subsoil, and noted the underlying geology. They designated the different "horizons" by Roman letters: A for the top soil; B for the subsoil or transition layer; and C for the parent rock at the bottom (see Figure 5.1). They calculated the amounts of organic matter (humus) and water in the horizons, and sent samples back to St. Petersburg for chemical analysis. The scientists also gathered data about the localities where they collected their soil profiles: the climate; flora and fauna; the relief or topography; and tried to estimate the age of the soil by analyzing earth under ancient burial mounds (*kurgany*) of known age.[22]

Traveling around the region, the researchers observed gradual changes in relief, parent rock, vegetation, and climate, and corresponding changes in the color, depth, humus, and water content of the soil.[23] They noticed that different kinds of soil overlay similar parent rock, and similar soils occurred over different underlying geology. They weighed up whether the geology, vegetation, or climate played the leading role in the formation of the black earth. Dokuchaev concluded – and this was his innovative theory of soil formation – that the "special character of soil" is "the result of an extremely complicated interaction between the local climate, plant and animal organisms, the composition and structure of the parent rock, the relief of the locality, and finally the age of the land."[24] Thus, the parent rock in the C horizon was transformed into the soil in the A horizon, via the transitional B horizon, by the five soil-forming factors: climate; organic matter; parent rock; relief; and time. Dokuchaev recognized that the relative importance of the factors varied from case to case. By the end of the nineteenth century, Russian soil scientists had developed a theory and practice for

[21] I. Krupenikov and L. Krupenikov, *Puteshestviia i ekspeditsii V.V. Dokuchaeva* (Moscow: Gos. izdat-vo geograficheskoi literatury, 1949), p. 31.
[22] Dokuchaev, *Russkii chernozem.* [23] For example, Dokuchaev, *Russkii chernozem*, pp. 161–2.
[24] Dokuchaev, *Russkii chernozem*, p. iii, see also pp. 177–8, 218–24, 255–7, 307.

studying soils as "natural-historical bodies" that had formed over time in the environmental conditions in which they were found. They called their new subject "genetic soil science" (genetic from genesis, i.e. origins, not the science of genetics). Although their new science was based primarily on Dokuchaev's theory, the Russians were aware of work by other European and American scientists, including Hilgard in Berkeley, California.[25]

The Russian scientists were assisted in developing their theory by the geography of the Russian Empire. In the empire's vast plains there is a succession of east–west belts with fairly uniform climatic conditions, in particular temperature and precipitation. The uniform climates in each belt give rise to fairly uniform vegetation inside them. The most northerly belt is the Arctic tundra, which gives way further south to forest, forest steppe, grassy steppe, and finally deserts in the south. There are no high mountain ranges inside the Eurasian plain to distort this pattern with rain shadows, as mountains are to be found mostly along the southern and southeastern peripheries. The Ural Mountains, especially in the south, are quite low and have limited influence on rainfall. Therefore, when the Russian soil scientists observed uniform types of soils inside each of the belts, such as the Chernozem of much of the steppes, it did not require a massive leap of imagination to work out the connection between the soils, vegetation, climate, and topography.[26]

Dokuchaev and his followers institutionalized their new discipline in the Russian Empire. The implication of their work was that, since soils were not simply eroded rocks, soil science was a separate branch of science from geology. They received support from the Russian government, which recognized the importance of the new science to support agriculture that was such an important part of the economy. The first chair in soil science was created at the Novaia Aleksandriia Institute of Agriculture, near Lublin in Russian-ruled Poland, in 1894. The first scientific journal devoted to soil science, *Pochvovedenie*, was founded in 1899. Dokuchaev trained the first generation of "genetic soil scientists," including Nikolai Sibirtsev (1860–1900) and Glinka, who advanced and promoted the new science.[27]

[25] For a concise essay on Russian soil science in the late 1890s by one of Dokuchaev's students, see P. V. Ototskii, "Pochva," in *ES* 24a (1898), 768–775.

[26] See David Moon, "Scientific Innovation in the Russian Empire: The Case of Genetic Soil Science," in *Science and Empire in Eastern Europe: Imperial Russia and the Habsburg Monarchy in the 19th Century*, ed. Jan Arend (Munich: Vandenhoeck and Ruprecht, 2020), pp. 305–20.

[27] See Dobrovol'skii, *Lektsii*, pp. 88–109; L. A. Varfolomeev, "Pamiat' cherez stoletie (k 150 letiiu so dnia rozhdeniia i 110-letiiu letiiu so dnia smerti Nikolaia Mikhailovicha Sibirtseva)," *Pochvovedenie*, no. 11 (2010), 1394–1400; S. V. Zonn, *Konstantin Dmitrievich Glinka, 1867–1927* (Moscow: Nauka, 1993).

The new soil science had immediate practical applications in the Russian Empire.[28] It was used to produce maps of different types of soils to assist the government manage its lands and support agriculture. The scientists classified soils into broad, overarching groups, for example, the Chernozem of the steppes, Chestnut (*kashtanovye*) soils in the more arid steppes, and Podzols in forested regions. The broad soil groups were initially classified on the basis of the environmental conditions in which they were located, i.e. the soil-forming factors in particular regions. The broad soil groups contained larger numbers of subordinate types. The decision to start with broad soil groups was deliberate to identify their extent and to facilitate the preparation of soil maps of large areas, such as the entire European part of the Russian Empire, as well as larger-scale provincial soil maps.[29]

The provincial authorities also saw practical value in the new soil science. In the 1880s, the provincial councils (*zemstva*) of Nizhnii Novgorod (which straddled the forest steppe and steppe regions east of Moscow) and Poltava (in the steppes of present-day Ukraine) commissioned Dokuchaev to carry out scientific surveys of their provinces, including the types and fertility of the soil, to assess the land for taxation. Dokuchaev assembled teams of soil scientists, agronomists, botanists, chemists, and meteorologists. Sibirtsev and Andrei Krasnov were members of the team in Nizhnii Novgorod and Glinka in Poltava. Dokuchaev asserted that scientific studies of soils should underlie any assessment of land values, but needed to be complemented by research into social and economic conditions. The approach became known as the Nizhnii Novgorod method for land evaluation.[30]

Dokuchaev received a further assignment after the drought, crop failure, and famine in the steppes in 1891–2. The Forestry Department of the Ministry of State Domains charged him with organizing a "Special Expedition" of scientists to the steppes to investigate ways of averting future disasters. They planted trees to prevent erosion and assist in conserving

[28] See A. A. Yarilov [sic], *Brief Review of the Progress of Applied Soil Science in USSR* (Leningrad: Publishing Office of the Academy, 1927).

[29] Dokuchaev revised an existing soil map of the European part of the Russian Empire on these principles in 1879. V. V. Dokuchaev, *Kartografiia Russkikh pochv: Ob"iasnitel'nyi tekst k pochvennoi karte Evropeiskoi Rossii* (Spb: Ministerstva Gosudarstvennykh Imushchestv, 1879). See also L. I. Prasolov, *Cartography of Soils* (Leningrad: Publishing Office of the Academy, 1927).

[30] Catherine Evtuhov, "The Roots of Dokuchaev's Scientific Contributions: Cadastral Soil Mapping and Agro-Environmental Issues," in *Footprints in the Soil: People and Ideas in Soil History*, ed. B. P. Warkentin (Amsterdam: Elsevier, 2006), pp. 134–9; Krupenikov and Krupenikov, *Puteshestviia i ekspeditsii*, pp. 56–80. On Krasnov, see pp. 91–3.

scarce moisture, including shelterbelts to protect fields from the drying influence of the wind and retain snow in the winter. Chemical analysis of steppe soils, however, revealed a layer of alkaline salts in the lower part of the B horizon that was harmful to some tree species. Further experiments were conducted to work out the most appropriate crops, cultivation methods, and balance between different land uses (arable, meadow, pasture, forest, water). Their plans were based on their soil science: since soils formed in the local environmental conditions, then the most appropriate ways to cultivate them were those that best mirrored the "natural environment." Their proposals were not implemented on a wide scale due to constraints of funding and private property, and lack of will by the authorities. Nevertheless, by the 1890s, Russian soil scientists had devised a theory and practice for sustainable agriculture in the steppe region, four decades before the Dust Bowl in the similar environment of the Great Plains.[31]

The Russian government applied the new soil science to locate land suitable for settlement by migrants from overcrowded central regions of the empire. In 1908, to assist the Resettlement Administration identify such land, Glinka was put in charge of expeditions to survey soils in Siberia, Central Asia, and the Russian far east.[32] In the summer of 1912, for example, Glinka led an expedition to survey the soils in western Siberia along the route of the Trans-Siberian railroad from Tiumen', in the forest region, to Omsk, in the steppes, around 400 miles to the southeast, near the modern border with Kazakhstan. They cut soil profiles and analyzed the horizons in the context of the relief, hydrology, climate, vegetation, and geology of the localities. Their reports included descriptions of the color and texture of the soils and the results of chemical analysis of samples.[33]

The Russian soil scientists developed their system for classifying soils. In the 1890s, Sibirtsev introduced the notion of "zonal soils." It was based on the geographical distribution of the main groups of soils that had developed in different zones as a result of physical and geographical factors, i.e. the soil-forming factors. The zonal soils were the main groups, for example Chernozem, and were considered "mature" soils. Sibirtsev also identified

[31] See V.V. Dokuchaev, *Nashi Stepi prezhde i teper'* (Spb: E. Evdokimov, 1892); David Moon, "The Environmental History of the Russian Steppes: Vasilii Dokuchaev and the Harvest Failure of 1891," *Transactions of the Royal Historical Society*, 6th series, 15 (2005), 149–74; Moon, "The Grasslands of North America and Russia," in *A Companion to Global Environmental History*, ed. J. R. McNeill and Erin Stewart Mauldin (Chichester: Wiley-Blackwell, 2012), pp. 247–62.

[32] Zonn, *Konstantin Dmitrievich Glinka*, p. 18.

[33] K. D. Glinka, K. P. Gorshenin, V. V. Stratonovich, and A. A. Iakovlev, "Pochvy vdol' linii Tiumen'-Omskoi zheleznoi dorogi," *Trudy Dokuchaevskogo Pochvennogo Komiteta* 1 (Spb, 1914).

"intrazonal soils" that had formed as a result of particular local factors in different zones, for example the soils of river floodplains. His third type was "azonal" soils, which were still forming, i.e. were "immature." Dokuchaev adopted the zonal principle in his final classification system.[34] When Glinka summarized the work of his expeditions, he stated that they had delineated the boundaries of the main "soil zones," including the Chernozem, of Siberia and Central Asia.[35] The system of "zonal soils" could be applied around the world. Dokuchaev, however, had almost a nationalistic pride in his beloved Chernozem, which he called the "tsar of soils." In 1883, he stated that it was the "native . . . and incomparable wealth of Russia," the "*result of a remarkably happy and very complicated complex of a whole series of physical conditions*!" Even after millions of years, he concluded, Chernozem would never form in the climate of western Europe, mountainous parts of Asia, or tropical regions. He did concede that it was possible that the "steppes of Siberia, the Missouri and Mississippi" (i.e. the Great Plains) could complete with "our black earth region."[36]

Two of Dokuchaev's former students, Krasnov and Sibirtsev, took an interest in the American parallels. Krasnov studied grasslands around the northern hemisphere following his visit to the United States in 1890 to attend the Geological Congress and explore the prairies and Great Plains (see pp. 92–3). Sibirtsev discussed the parallels in a lecture on "chernozem in various countries" published in 1898. He anticipated Marbut in noting similarities between the "chernozem" of the "moderately-arid" states, such as Nebraska, Dakota (sic), Montana, Kansas, and Texas, in the Great Plains, and parts of the steppes of his homeland. He realized the significance of the Rocky Mountains' rain shadow for the geography of environmental regions in North America (see pp. 201, 230–1).[37] In 1899, Dokuchaev wrote that, following the work of Sibirtsev, Krasnov, and the American scientist Hilgard, major soil zones could be identified around the northern hemisphere, including the black-earth steppes and prairies of

[34] N. M. Sibirtsev, "Ob osnovaniiakh geneticheskoi klassifikatsii pochv," in N. M. Sibirtsev, *Izbrannye sochineniia*, 2 vols. (Moscow: Gos. Izdat. Sel-khoz. lit, 1953), vol. 2, pp. 271–93 (first published in 1895); V. V. Dokuchaev, "Novaishaia Klassifikatsiia pochv," (table at end of article) *Pochvovedenie* 2, no. 2 (1900); E. S. Migunova, "Rol' N M Sibirtseva v stanovlenii pochvovedeniia," *Visnyk Khakivs'koho natsional'noho agrarnogo universytetu. Seriia: Hruntoznavstvo, ahrokhimiia, zemlerobstvo, lisove hosodarstvo* 1 (2016), 40–54.

[35] K. G. [Konstantin Glinka], "Khronika: III Vserossiiskii S'ezd [sic] russkikh pochvovedov," *Russkii pochvoved*, nos. 1–3 (1922), 35–6.

[36] Dokuchaev, *Russkii chernozem*, pp. 351–2 (emphasis in original); Dokuchaev, *K ucheniiu o zonakh prirody. Gorizontal'nye i vertikal'nye pochvennye zony* (Spb, 1899), pp. 8–9.

[37] N. M. Sibirtsev, *Chernozem v raznykh stranakh* (Warsaw: Varshavskii uchebnii okrug, 1898), pp. 42–3.

Hungary, Russia, Asia, and America. In characteristic language, he described them as the "most successful creation of Zeus or Jupiter."[38]

This summary of origins, theory, practice, and application of Russian genetic soil science presents the pioneering Russian work that Marbut referred to several decades later and was the basis of his overhaul of the U.S. Soil Survey after 1914. The Russians' multi-faceted approach to surveying soils' geological, chemical, mechanical, and physical characteristics in the environmental conditions in which had formed can be contrasted sharply with the narrower approach devised by Whitney for the U.S. Soil Survey that began in 1899, two decades after the Russian innovation in soil science. The Russian system of classifying soils first into broad, overarching groups that could be applied around the world can also be contrasted with Whitney's scheme that began by identifying local soil types and then trying, with difficulty, to correlate them into larger groups to allow the mapping of soils of larger areas.

Soil Science in the United States in the Late Nineteenth and Early Twentieth Centuries

It was some time before American soil scientists applied concepts from the new Russian soil science and classified the soils of the Great Plains as "Chernozems." Regardless of the Russian innovation in understanding soils in their own right, when the U.S. Soil Survey began in 1899 and for some time afterwards, most American scientists still considered studying soils a part of geology.[39] The authors of two important works on soils in the 1890s adhered closely to the bedrock. In his 357-page monograph, *A Treatise on Rocks, Rock-Weathering, and Soils*, George Perkins Merrill devoted only the final 40 pages to soils, most of which concerned geological matters, apart from a brief discussion of the effects of "plant and animal life." He mentioned the "black earth of Russia," in a footnote, as "a local phase of the loess," but did note that its color was a result of organic matter.[40] Nathaniel Southgate Shaler's "The Origin and Nature of

[38] Dokuchaev, *K ucheniiu o zonakh prirody*, pp. 8–9. See also Jonathan Oldfield and Denis J. B. Shaw, *The Development of Russian Environmental Thought: Scientific and Geographical Perspectives on the Natural Environment* (London: Routledge, 2016), pp. 61–3.

[39] David Rice Gardner, *The National Cooperative Soil Survey of the United States* (Washington, DC: USDA, Natural Resources Conservation Service, Resource Economics and Social Sciences Division, 1998) ("Thesis presented by David Rice Gardner to the Graduate School of Public Administration, May 1957, Harvard University"), p. 7.

[40] George Perkins Merrill, *A Treatise on Rocks, Rock-Weathering and Soils* (New York: The Macmillan company, 1897), pp. 318, 381 and passim.

Soils," was mostly geological, but did take some tentative steps in other directions. He considered soils to be "decayed vegetable matter ... [m]ingled with ... more or less distinct fragments of a stony nature." He used a cross section, with "Bed Rock" at the bottom, and layers of "Sub Soil," "True Soil," and "Forest Mould" above, to illustrate the effect of tree roots on soil formation. In explaining the characteristics of "prairie soils," he took in account of the level topography and "moderate" rainfall.[41] There is no evidence that he knew about Dokuchaev's work, which paid far more attention to vegetation, topography, climate, and analysis of cross sections.[42] Like many of their American contemporaries, Shaler and Merrill spent most of their careers in the eastern and northeastern United States, where there are not marked differences in soils and environments, especially rainfall, and where soils have been altered by several centuries of farming. Shaler was from Newport, in northeastern Kentucky, and his career was divided between his home state and Harvard University.[43] Merrill, who was from Maine, worked in different locations in the east coast, including Washington, DC.[44]

An outlier among American scientists working on soils at this time was Hilgard. Although also a geologist by training, he emphasized climate in explaining differences between soils in the humid eastern United States and the arid West. He was also an outlier among his American contemporaries in his geographical experience. He developed his ideas on climate in soil formation over a career spent in humid Mississippi and arid California (with a short interlude in Michigan). Hilgard was therefore familiar with sharply contrasting environments, in particular levels of precipitation, and quite different types of soil throughout the continental United States.[45] Soil scientists in the Great Plains were sometimes frustrated by the limited experience of scientists in the eastern United States. In 1909, Frederick J. Alway of the Nebraska Agricultural Experiment Station wrote to Hilgard about a meeting on soil classification in Omaha,

41 Nathaniel Southgate Shaler, "The Origin and Nature of Soils," *U.S. Dept. of the Interior, Geological Survey, Annual Report* 12 (1890–1), 223, 238, 270, 323.

42 Later scholars did not, however, distinguish Shaler's work from other contemporary American studies of soils. See J. S. Joffe, "Soil Profile Studies I: Soil As an Independent Body and Soil Morphology," *SS* 28 (1929), 42; Gardner, *National Cooperative Soil Survey*, pp. 19–20; Malin, *Grasslands*, p. 200. Joffe was a strong advocate of Russian soil science and Malin of Hilgard's work.

43 Nathaniel Southgate Shaler, *The Autobiography of Nathaniel Southgate Shaler 1841–1906* (Boston, New York: Houghton Mifflin Company, 1909).

44 Waldemar Lindgren, "Biographical Memoir George Perkins Merrill 1854–1929," *National Academy of Sciences: Biographical Memoirs* 17 (1935), available online at www.nasonline.org/publications/biographical-memoirs/memoir-pdfs/merrill-george.pdf), accessed March 6, 2018.

45 On Hilgard, see p. 74.

Nebraska. Professor Homer J. Wheeler of Rhode Island seems particularly to have annoyed him: "The climatic classification did not seem to mean much to him. I called his attention to the necessity of taking this into consideration. Men who have had nothing to do with the arid and semi-arid soils do not seem able to appreciate the importance of the climatic classification."[46] Hilgard and Alway, and later Marbut, were more like their Russian counterparts in the breadth of their geographical experience that had exposed them to different types of soils that had formed in a variety of climatic and other environmental conditions.

It should be pointed out that, in contrast to their Russian counterparts, American soil scientists were hindered by the geography of their country in recognizing the connections between soils and the environments in which they had formed. The high Rocky Mountains that extend from north to south in western North America create a rain shadow in the middle of the continent, where the low rainfall is conducive to grassy vegetation rather than trees. This is why the Great Plains are orientated from north to south, parallel to the Rockies, in contrast to the east–west orientation of the steppes. In North America rainfall increases from west to east, while temperature rises from north to south. The resulting checkerboard pattern in environmental conditions, rather than the uniform east–west belts in the Eurasian plain, made it harder for American scientists to recognize the causal link between different types of soils and the vegetation, climate, and topography in different regions.[47]

A major concern among American specialists in the late nineteenth century was the relative importance of chemical and physical analysis in determining soils' productivity. By the 1880s, attempts to find definite relationships between chemical properties of soils and their fertility had proved inconclusive. While Hilgard argued that both types of analysis still were needed, other scientists asserted that physical analysis was more important. One of the staunchest advocates of physical analysis was Whitney. In the early 1890s, based on field work in Maryland, he argued that the productivity of crops and the adaptability of different crops to different soils were determined by the supply of moisture in the soil. This, he argued, was controlled by the space between soil particles, and could be measured by physical analysis of the texture of soils, in

[46] Bancroft Library, Berkeley, California, The Hilgard family papers (c. 1848–1945), BANC MSS C-B 972 (hereafter, Hilgard Papers), Hilgard, E. W.: Incoming Letters, Box 4, Alway to Hilgard, December 15, 1909.

[47] See Curtis F. Marbut, "Introduction," in Jacob Joffe, *Pedology*, 2nd edition (New Brunswick, NJ: Pedology Publications, 1949), pp. ix–xii. See also pp. 412–14.

particular the sizes of soil particles. He disregarded the significance for plant growth of mineral nutrients in the soil, which could be determined by chemical analysis, and ignored the role of chemical fertilizers in providing plant nutrition. Thus, for Whitney, soil texture, measured by physical analysis of particle size, was the key to analyzing soils.[48] Whitney's limited understanding of soils reflected his geographical experience. A native of eastern Maryland, he was educated in chemistry at, but did not graduate from, Johns Hopkins University in Baltimore. He worked at agricultural experiment stations in North Carolina, under Charles Dabney, Connecticut, and South Carolina, before returning to Maryland to work at the station at College Park.[49] Unlike many American scientists, such as Dabney and Hilgard, Whitney's education did not include time in Europe.[50] Macy Lapham, who worked under Whitney for the Soil Survey, wrote that he "seldom ventured far from Washington."[51] With the exception of a few visits elsewhere in the United States, Whitney's working life was spent near the eastern seaboard and his experience was restricted largely to the well-watered and long-worked soils of this region.

Whitney restated his beliefs about soils in a publication co-authored with Franklin K. Cameron in 1903:

> It appears further that practically all soils contain sufficient plant food for good crop yield, that this supply will be indefinitely maintained, and that this actual yield of plants adapted to the soil depends mainly, under favorable climatic conditions [plentiful and well-distributed rain], upon the cultural methods and suitable crop rotation, ... and that a chemical analysis of a soil, ... will in itself give no indication of the fertility of this soil or of the probable yield of a crop, and it seems probable that this can only be determined, if at all, by physical methods, as it lies in the domain of soil physics.[52]

[48] See Gardner, *National Cooperative Soil Survey*, pp. 18–29; Sutter, *Let Us Now Praise Famous Gullies*, pp. 38–40; Douglas Helms, Anne B. W. Effland, and Steven E. Phillips, "Founding the USDA's Division of Agricultural Soils: Charles Dabney, Milton Whitney, and the State Experiment Stations," in *Profiles in the History of the U.S. Soil Survey*, ed. Helms, Anne B. W. Effland, and Patricia J. Durana (Ames: Iowa State University Press, 2002), p. 10.

[49] See Douglas Helms, "Early Leaders of the Soil Survey," in *Profiles*, pp. 20–2; D. S. Fanning and M. C. B. Fanning, "Milton Whitney: Soil Survey Pioneer," *Soil Horizons* 42 (2001), 83–89.

[50] See Connie L. Lester (2017), "Charles W. Dabney Jr.," *Tennessee Encyclopedia*, available online at http://tennesseeencyclopedia.net/entry.php?rec=348, accessed April 24, 2017.

[51] Macy H. Lapham, *Crisscross Trails: Narrative of a Soil Surveyor* (Berkeley, CA: W. E. Berg, 1949), p. 155.

[52] Milton Whitney and F. K. Cameron, "The Chemistry of the Soil as Related to Crop Production," *USDA Bureau of Soils Bulletin*, no. 22 (1903), 64.

Whitney's views, and this publication in particular, attracted sharp criticism from some American and foreign scientists, including Hilgard and the Russian scholar Nikolai Tulaikov.[53]

Whitney and Cameron's controversial views had official status since they were published as bulletin no. 22 of the Bureau of Soils of the USDA. In 1894, Assistant Secretary of Agriculture Dabney had appointed Whitney head of the USDA's new Division of Agricultural Soils.[54] Whitney remained chief of all U.S. government soil investigations until shortly before his death in 1927. The soils unit underwent various reorganizations: in 1897 it was renamed the Division of Soils; it was upgraded to a Bureau of Soils in 1901; and in 1927, at the time of Whitney's retirement, it became part of a larger Bureau of Chemistry and Soils. Over the thirty-three years Whitney was in charge, in large measure due to his political acumen and support from successive secretaries of agriculture, the soils unit was awarded increasing funds by the federal government that enabled it to grow from a staff of 10 to 218, and considerably expand its operations.[55]

Among Whitney's undeniable achievements was establishing the U.S. Soil Survey. It began modestly in four states in 1899, but extended across the United States as Whitney received more funding. The Division, later Bureau, of Soils in Washington, DC, cooperated with states in surveying and mapping soils. Much of the field work was carried out by the staff of state agricultural experiment stations, land-grant colleges, and state universities. At end of the survey's first decade, the Bureau set up a system of supervision and inspection of field work.[56] The Soil Survey's methodology was based on Whitney's understanding of soils. The author of an official history noted in 1928: "When the soil survey work was begun it was the assumption that the determination of the texture of the surface six-inch layer and the geology of the underlying material constituted the

[53] See Lapham, *Crisscross Trails*, pp. 78–9; N. M. Tulaikov, Review of Milton Whitney and F. K. Cameron, "The Chemistry of the Soil as Related to Crop Production," *Pochvovedenie*, no. 4 (1909), 348–51; R. Amundson, "Philosophical Developments in Pedology in the United States: Eugene Hilgard and Milton Whitney," in *Footprints in the Soil: People and Ideas in Soil History*, ed. B. P. Warkentin (Amsterdam: Elsevier, 2006), pp. 149–65; Sutter, *Let Us Now Praise Famous Gullies*, pp. 41–5. On Tulaikov, see p. 93.

[54] Helms et al., "Founding," 1–18.

[55] See Gustavus A. Weber, *The Bureau of Chemistry and Soils: Its History Activities and Organization* (Baltimore, MD: The Johns Hopkins Press, 1928), pp. 1–2, 85, 87–8 and passim; Gardner, *National Cooperative Soil Survey*, pp. 38, 76.

[56] Carleton R. Ball, *Federal, State, and Local Administration Relationships in Agriculture*, 2 vols. (Berkeley: University of California Press, 1938), vol. 1, pp. 203–4; Gardner, *National Cooperative Soil Survey*, pp. 29, 207–8; Weber, *Bureau*, pp. 39, 89; Helms, "Early leaders," 23.

proper method of attack."[57] At first, Whitney prioritized analysis of specific soils in particular localities. There was no preconceived plan to define broad, overarching soil groups. Nor did Whitney pay serious attention to classification systems in the scientific literature or used in soil surveys in other countries. The U.S. Soil Survey started at the bottom with local types of soils and then, gradually and with difficulty, tried to correlate them into broader groups across wider areas.[58] In 1918, Whitney justified his decision to embark on the survey without an overall system of soil classification:

> The agricultural situation in this country made it necessary to begin by attempting to produce a detailed soil map at once... [T]he soil investigator ... could not obtain the information he needed [regarding ... a workable scheme of soil grouping or classification] from any other source, since no systematic studies of soils in the field, either in detail or in a broad way, had ever been attempted in any country *readily accessible to Americans* [my emphasis DM].[59]

Whitney's approach was thus the exact opposite of what Dokuchaev and his colleagues had done in Russia since the 1870s (which Whitney claimed not to know about). The Russians had started by defining a group of Chernozem soils in terms of multiple soil-forming factors while carrying out sample surveys of cross sections (profiles) revealing different layers (horizons) several feet deep over a large geographical area. They then defined and located other broad soil groups on the basis of different soil-forming factors and further surveys of soil profiles. By the 1890s, before the start of the U.S. Soil Survey, the Russian scientists were identifying, from first principles, similar groups of soils not only in the vast Russian Empire, but around the globe, including Chernozem in the Great Plains of the United States.

In the first four years of the U.S. Soil Survey (1899–1903), Whitney instructed survey workers to identify and classify soils in the same area into types with similar textures at the surface. Texture referred to the size of particles of soil (clay, silt, and sand), and the proportions of different-sized particles in soil samples. The "soil types" were given binomial names: the locality or a geological criterion and then the texture, for example, "Jordan sandy loam" and "Triassic stony loam." After a few years, on further

[57] Weber, *Bureau*, pp. 90–1.
[58] See Ball, *Federal*, p. 204; Gardner, *National Cooperative Soil Survey*, p. 40.
[59] Milton Whitney, *Field Operations of the Bureau of Soils, 1918* (Washington, DC: USDA, 1924), p. 3.

instructions from Whitney, survey workers looked for additional soil characteristics, such as color, organic content, alkalinity, soil structure, nature of subsoil, parent material, and others. Soil surveyors started to sample soils to greater depths, down to three feet in humid regions and six feet in arid. As variations inside "soil types" were recognized, subdivisions, known as "soil phases," were added. In 1903, a higher level of classification, "soil series" was established. "Soil series" included several "soil types" in different localities deemed to be similar, initially in terms of their geologies, but with different textures. For example, by 1906, the "Miami" soil series included sixteen "soil types" in glaciated regions. A third and higher level of classification, the "soil province," was added in 1906 in an attempt to organize systematically the increasing numbers of "soil types" and "soil series." The large geographical provinces were assumed to be homogeneous in their geology and geomorphology. It was expected, or perhaps hoped, that "soil series" would be unique to particular provinces. By 1912, there were thirteen provinces, seven to the east of the Great Plains and six (termed "regions") to the west, including the Great Plains.

In principle, the three levels of classification, with "soil types" (and "phases") at the bottom, "soil series" in the middle, and "soil provinces" at the top, was a neat system. In practice, it simply did not fit the varieties of soils defined by the Survey's criteria. The tasks of correlating similar "soil types" in different localities to avoid duplication and then grouping them into "soil series" turned out to be colossal, if not unachievable. In spite of Whitney's instructions not to create unnecessary new "types" and "series," the numbers multiplied. By 1912, there were 1,650 "soil types" grouped into 534 "soil series." The attempt to identify homogeneous soil provinces was hampered by diverse topographies, climates, vegetations, and geologies inside them. Further, the decision to define "provinces" in mainly geological terms was inconsistent with the attention survey workers were instructed to pay to other soil characteristics. The soil provinces abandoned in 1916.[60]

The results of local surveys of different soil types were represented on maps of individual counties. The maps got more complicated as more types were added. A 1915 soil map of Montgomery County, in southeastern Kansas, based on a survey in 1913 by F. V. Emerson and C. S. Waldron under Hugh H. Bennett, showed twenty-five different soil types and phases, from "Bates very fine sand" to "Osage clay," intermingled in

[60] See Gardner, *National Cooperative Soil Survey*, pp. 40–50, 81; Ball, *Federal*, p. 204; Simonson, *Historical Highlights*, pp. 22, 45.

curious swirls across the county's 650 square miles.[61] A map of a single Kansas county, using Whitney's system, was far more complicated than a soil map of the entire world prepared by Glinka in 1906 using the Russian system of classifying soils into broad groups. On Glinka's map, Montgomery County is in the east of a wide belt of Chernozem extending from Texas to the Canadian prairies. Glinka's map showed similar soils in other parts of the world with similar environmental conditions, for example the steppes of the Russian Empire and pampas of Argentina.[62] In the U.S. Soil Survey's map of Montgomery County, Kansas, "Bates very fine sand," like many of the ever growing numbers of soil types, was of only local significance, so local, or so new, that it did not feature in a Soil Survey publication on all American soils the following year.[63]

The results of the U.S. Soil Survey's first twelve years' work were collected in the 1913 edition of the "Soils of the United States." The index of soil types alone filled twenty pages of small text in double columns. The chapter on the "Great Plains Region" occupied eighty-three pages, plus a pull-out sheet with a helpful "key" to the region's soils. The chapter contained a bewildering array of sixty-seven "soil series," and rather more "types" and "phases," as well as different textural groups, such as "fine sand" and "gravelly sandy loam" and others. Among the dozens of "soil types," and this is a random example, was "Albion loamy coarse sand," which was "derived from the more gravelly phases of unconstituted Tertiary deposits" in "undulating to rolling topography" and covered 11,968 acres in parts of Kansas. Whitney was very pleased with the volume, which ran to 791 pages plus an attractive map in a pocket at the end, and warmly recommended it to the Secretary of Agriculture: "the large amount of new information gathered, makes the bulletin considerably larger than the earlier works and a very complete handbook of the soils of the United States." The author of the chapter on the Great Plains region was Macy Lapham, who had been with the Soil Survey since 1899. The volume's editorial team was led by Marbut, shortly after his

[61] The map of Montgomery County can be found with other county soil maps in Kansas Historical Society, State Archives, Topeka, Kansas, Maps, Unit no. 218477, acc. no. 2010–065.01 US Bureau of Soils, Soil Maps of Kansas (7), 1902–1915. See also F. V. Emerson and C. S. Waldrop, "Soil Survey of Montgomery County, Kansas," *USDA Bureau of Soils Field Operations, 1913* (1915), available online at www.nrcs.usda.gov/Internet/FSE_MANUSCRIPTS/kansas/montgomery KS1915/montgomeryKS1915.pdf, accessed March 6, 2018.

[62] Glinka's map is at the back of K. D. Glinka, *Pochvovedenie* (Spb: A. F. Devrien, 1908).

[63] Three other soil types in the Bates series were mentioned in Curtis F. Marbut, Hugh H. Bennett, J. E. Lapham, and M. H. Lapham, "Soils of the United States," *USDA Bureau of Soils Bulletin*, no. 96 (1913), 383.

appointment to head the Soil Survey, but before his conversion to Russian soil science.[64]

Lapham later looked back at the volume and the early years of the Soil Survey. In his memoir published in 1945, Lapham recalled how he had started as a survey worker in the field before promotion to inspector. Most of his time was spent in the western states. In 1908, for example, he undertook a reconnaissance survey of western North Dakota with George Coffey and Thomas D. Rice. It was arduous work. They traveled on horses and in buggies as well as trains. In less populated areas, they slept out in the prairies and cooked their own food. Lapham described how he had learned the job, for example using a six-foot soil auger to bore for samples, in the field. In the survey's early years, workers identified boundaries between different "soil types" by noting variations in texture, color, structure, and mineral character through frequent borings. When "soil series" were introduced, they looked for soils with similar parent materials, but different textures. He noted that "mapping ... was admittedly incomplete, crude, and lacking in detail." From the vantage point of 1945, he wrote that the classification system they used in 1913 was later extensively revised and brought into conformity with the "dictates of modern soil science." (He specifically referred to the Russian principles of soil horizons and profiles.) When he was in Washington, DC, Lapham got to know Whitney. He remembered how, when they were preparing the 1913 edition of the "Soils of the United States," Whitney had invited him and others to his house at weekends to work on it. Lapham found his chief in his own home to be "gracious, witty, and entertaining, and withal a delightful host," in contrast to his dignified, aloof, and reserved manner in the office.[65]

In 1928, the year after Whitney's death, Marbut also looked back to the early days of the Soil Survey. By then he was very familiar with Russian soil science, which informed his critique. Marbut described Whitney's belief in the importance of soil texture, and that as a result the Survey began as "a differentiation of soils on the basis of fineness or coarseness of the particles in the surface layer." Soon, he noted, surveyors in the field became convinced of the importance of other soil characteristics. Nevertheless, no one had thought about "the desirability of differentiating the soils first on general characteristics and mapping them in broad groups and then breaking these up into smaller groups" (i.e. the Russian approach).

[64] Marbut et al. "Soils of the United States," 3, 381–464, 428–9, 771–91.
[65] Lapham, *Crisscross Trails*, pp. 3, 11–13, 16, 28–9, 70, 73, 80, 99.

It was realized quite quickly that the Survey's detailed local work was incompatible with making a soil map of the entire United States, rather than "a series of isolated soils maps of small areas." The key problem was correlating similar types of soils in different regions. Marbut concluded: "A soil map constructed on any other basis would have no value." He was critical also of the geological characteristics used to define the "soil provinces" created in 1906, but abolished in 1916.[66] Thus, Whitney's stubborn insistence on his physical concept of soils, resisting contrary ideas, impeded the development and value of the U.S. Soil Survey.

Whitney had some support from other American scientists, for example Henry Luke Bolley, a plant specialist from North Dakota, and Thomas Chrowder Chamberlin, a geologist from Illinois.[67] But, their support served only to provoke further ire from his opponents. Hilgard, whom Lapham found generally "sympathetic, courteous, and kindly,"[68] had no patience with Whitney, and criticized him and his work to anyone whenever the opportunity arose. Hilgard's criticisms of Whitney's work are significant, not just because of his own innovative ideas on soils, but because they were written in the early years of the U.S. Soil Survey, before Marbut had overhauled its scientific approach. Hilgard's frustration may have been exacerbated by the knowledge that things could have worked out so differently. In 1889, Hilgard had turned down an offer to serve as Assistant Secretary of Agriculture. Three years later, Dabney was appointed to this position, and then hired his mentee, Whitney, to head the new Division of Agricultural Soils.[69] Hilgard's criticisms centered on Whitney's faulty science and his management style that inhibited criticism by removing people who disagreed with him. Hilard's critique of Whitney indicates the difficulties faced by anyone who challenged him, as Marbut went on to do.

The first contact between the older scientist, Hilgard, and younger man, Whitney, in 1890 was amicable, but they quickly fell out over Whitney's insistence on soil physics alone as the key factor in soil fertility.[70] In April 1892, Hilgard wrote to Franklin H. King, a soil scientist at the University of Wisconsin:

[66] Marbut's "statement" was quoted in Weber, *Bureau*, pp. 91–3.

[67] For example, H. L. Bolley, "Interpretations of Results Noted in Experiments upon Cereal Cropping Methods after Soil Sterilization," *Science* 33, no. 841 (February 10, 1911), 229–232; T. C. Chamberlin, "Soil Productivity," *Science* 33, no. 841 (February 10, 1911), 225–227.

[68] Lapham, *Crisscross Trails*, pp. 78–9.

[69] Helms et al., "Founding," 15; Amundson, "Philosophical Developments," 160–1.

[70] See Amundson, "Philosophical Developments;" Hans Jenny, *E. W. Hilgard and the Birth of Modern Soil Science* (Pisa: Collana Della Rivista "Agrochimica: 1961), 89–102.

> [Whitney] ought to have lived a century ago, when people started out, as he does, with a thesis to defend through thick and thin, regardless of what has been done before in the same and in other lines. In reading it I have sometimes rubbed my head to make sure that I was living at the end of the 19th century, when induction and experimental proof are supposed to be the basis of science, instead of evolution from one's internal consciousness. . . . Whitney has run wild on Maryland soils, and from them generalizes about the world at large. . .[71]

Hilgard criticized Whitney's views to journalists in San Francisco: "No one outside the bureau of soils has been found to agree with him."[72] His annoyance spilled over into letters to his family. In December 1907, in a chatty letter to his sister Therese, he described Whitney as a "humbug" and mentioned that "all men of science laugh at his vagaries."[73]

Hilgard was incensed by Whitney's hostility to ideas that differed from his own and to people who held them. In 1911, he wrote to Dr. Paul Schwartz of Berlin to recommend King to serve on an international committee:

> I do not mention ... any one connected with the Bureau of Soils at Washington, for the reason that [they] ... are hampered in the expression of their opinions and the kind of work they do by the arbitrary rules of the Chief, Professor Whitney, who will not tolerate any difference of opinion in any one of his subordinates. Professor King not long ago was in the employ of the Bureau of Soils and, because of the results of his observations and experiments did not agree with the views of the Chief, was forced to resign.[74]

Hilgard was referring to events in 1904, when Whitney had compelled King to leave his post as Chief of the Division of Soil Management at the Bureau of Soils, as he had had the audacity to contradict his chief's views on the paramount importance of soil physics.[75] (King had confided in Hilgard about his experiences.[76]) Hilgard knew of other episodes in which

[71] Hilgard Papers, Hilgard, E. W.: Outgoing Letters, Letterpress copy books, vol. 19, pp. 55–6, Hilgard to King, April 23, 1892.

[72] Hilgard Papers, Carton 6, File: Clippings – Milton W. Whitney, "Professors Differ Concerning Soils. Hilgard of Berkeley Sharply Criticizes Unsupported Theory Advanced by Whitney," *San Francisco Chronicle*, November 27, 1903.

[73] Hilgard Papers, Hilgard, E. W.: Outgoing Letters, Box 3, Hilgard to Therese, December 7, 1907.

[74] Hilgard Papers, Hilgard, E. W.: Outgoing Letters, Box 3, Hilgard to Dr. Paul Schwartz, March 14, 1911.

[75] See Edward R. Landa, "Potash, Passion, and a President: Early Twentieth-Century Debates on Soil Fertility in the United States," in *The Soil Underfoot: Infinite Possibilities for a Finite Resource*, ed. G. J. Churchman and E. R. Landa (Boca Raton, FL: CRC Press, 2014), pp. 269–88.

[76] Hilgard Papers, Hilgard, E. W.: Incoming Letters, Box 14, King to Hilgard, November 18, 1903; King to Hilgard, January 27, 1904; King to Hilgard, December 6, 1904; King to Hilgard, February 28, 1905.

Whitney tried to silence scientists at the Bureau of Soils who held other views. When Alway wrote to Hilgard about a meeting on soil classification in Omaha, Nebraska, in December 1909 (see pp. 200–1), he noted that: "Prof. Coffee [sic] of the Bureau of Soils" was present, but he "had been instructed by Dr. Whitney, I believe, not to take any part in the discussion."[77] At this time, Coffey was having doubts about the Soil Survey's system for classifying soils and had been reading about Russian soil science, which he thought pointed towards a better way.[78]

The American Reception of Russian Soil Science to 1914

Coffey was one of the first Americans to take an interest in Russian soil science. Most American scientists remained largely unaware of it for several decades after Dokuchaev and his colleagues came up with their innovative approach in the 1870s. The reception and acceptance of Russian soil science in the United States came up against several barriers. Russia was not somewhere many agricultural scientists in the United States would look to for such innovations, since many considered Russia to be "backward." Some American soil scientists, in particular Whitney (who had not studied in Europe), paid little attention to foreign research. Marbut wrote in 1928: "At the time the [soil] mapping was begun [in the United States in 1899] . . . Whitney, . . . did not know of the work of the Germans and the Russians, but if he had known of it, it would not have furnished a basis for the work which he did afterward."[79] American knowledge of the Russian work was hindered by the language barrier. A. G. McCall, who replaced Whitney as chief of the Bureau of Soils in 1927, wrote "the work of the Russian soil scientists had very little if any influence upon the shaping of the early work of the Bureau of Soils, because . . . our scientists did not have ready access to the Russian reports because of the language difficulty."[80] Dokuchaev's seminal monograph on the Russian Chernozem of 1883 was not published in English translation until 1967.[81] Many of the Russian soil scientists' publications were in Russian, because their

[77] Hilgard Papers, Hilgard, E. W.: Incoming Letters, Alway to Hilgard, December 15, 1909.

[78] G. N. Coffey, "Value of the Field Study of Soils," *Proceedings of the American Society of Agronomy* 1 (1907–9), 168–75; G. N. Coffey, "Physical Principles of Soil Classification," *Proceedings of the American Society of Agronomy* 1 (1907–9), 175–85. See also pp. 217–18.

[79] Quoted in Weber, *Bureau*, p. 911. Simonson repeated Marbut's assertion in *Historical Highlights*, p. 5.

[80] A. G. McCall, "The Development of Soil Science," *AH* 5, no. 2 (1931), 52.

[81] V. V. Dokuchaev, *Russian Chernozem*, trans. N. Kaner (Jerusalem: Israel Programme for Scientific Translations [for USDA-NSF], 1967).

research was sponsored by Russian scientific societies and provincial councils for internal use, or they were text books written for students at Russian educational institutions. Articles in the journal *Pochvovedenie*, founded in 1899, were in Russian only, with no abstracts in other languages, for its first decade as the editor lacked the resources to provide them. In this period, few American scientists, besides Russian–Jewish émigrés, could read Russian. A further barrier was the different geographical configurations of the steppes, which run from east to west, and the Great Plains, that are orientated from north to south. For the reasons explained earlier, this not only made it easier for Russian scientists to see the connection between soils and the environments in which they formed, but harder for American scientists both to see these connections and to apply the Russian classification system to their soils.

The main barrier, however, was that once knowledge of the Russian innovation spread among American scientists, it challenged existing hierarchies and institutions, most notably Whitney, his Division/Bureau of Soils, and the U.S. Soil Survey. Some scientists ignored the Russian ideas and marginalized American scientists who took an interest in them. Whitney and his associates sought also to marginalize work by American scientists, such as Hilgard and King, that differed from prevailing orthodoxies. Under Whitney's leadership, into the 1920s, the U.S. Soil Survey continued to identify and map an ever growing number of "soil types," based mainly on texture and geology, that defied attempts to combine them in larger groups to represent on a soil map of the United States. This was long after Russian soil scientists had devised their broader conception of soils that readily allowed the mapping of overarching soil groups across large areas.

By the 1890s, the Russian genetic soil scientists were aware that their work was innovative and universal in application, and tried to disseminate it around the globe. They presented papers at international conferences and met foreign scientists while on visits abroad, including in the United States. These activities led to the publication of a small number of shorter works on Russian soil science in French, German, and English, which were then the main international languages of science. They prepared exhibits for the world's fairs, including the Expositions Universelle in Paris, France, in 1889 and 1900, and those in the United States.[82]

[82] See P. V. Ototskii, "Iz sudeb russkogo pochvovedeniia," *Pochvovedenie*, no. 3 (1909), appendix, 14; Jean Boulaine, "V. V. Dokouchaev et les débuts de la pédologie," *Revue d'histoire des sciences* 36, no. 3–4 (1983), 285–306; I. P. Vtorov, "Soil as a Museum Exhibit in Russia," in *13th International*

The first Russian to present the new soil science in the United States was Dokuchaev's former student, Krasnov, who delivered paper on the "Black Earth" of the Russian steppes to the 5th International Geological Congress in Washington, DC, in August, 1891. It was published in English in an American scientific journal. Krasnov summarized the studies by Dokuchaev and his colleagues. He explained their focus on the environmental conditions in which the soil had formed. But, he omitted the methodology of studying soil profiles and horizons. Krasnov concluded that he hoped Russian and American "geologists" would work together to investigate the "origin of the American prairies and of the Russian steppes," to "find in America relations between soil, climate, relief, and vegetation analogous to those of Russia, as well as a dependence on climate analogous to ours." He speculated that "most of the American prairies are merely at a less advanced stage of evolution than those of Russia."[83] Krasnov wrote to Dokuchaev that his paper had aroused great interest among the American geologists as the Russian work was new to them. He added that the American's data were "much worse than ours," and asked Dokuchaev for the program of Russian research and classification system to pass on to the Americans.[84]

One American scientist wrote that he was "greatly interested in Professor Krasnof's paper, as I have studied the American 'black-prairie soils'." This was Hilgard. In his comments on Krasnov's paper, he remarked: "on the whole, I agree entirely with him in his conclusions as to the conditions under which such soils may be formed." He noted that he, Hilgard, gave more emphasis to the underlying geology, in particular "calcareous" limestones, marly rocks or gravels, but agreed that the "other concurrent conditions" (i.e. Dokuchaev's soil-forming factors) were also necessary.[85] Hilgard wrote to Jean Vilbouchevitch (a Russian-born French scientist) that he had "had the pleasure of meeting ... Prof. Krasnoff, who read a very interesting paper on the Tschernosem of Central Russia." He continued that he was trying to convince him "that the formation of black earths

ERBE Symposium: Cultural Heritage in Geosciences, Mining and Metallurgy, Libraries – Archives – Museums: Proceedings, ed. Daniel Harvan (Banská Štiavnica: Slovak Mining Museum, 2016), pp. 238–40; David Moon, "The International Dissemination of Russian Genetic Soil Science (Pochvovedenie), 1870s–1914," *Historia provinciae – The Journal of Regional History* 2, no. 1 (2018), 75–91.

[83] A. N. Krassnof, "The 'Black Earth' of the Steppes of Southern Russia," trans. H. Stein, *Bulletin of the Geological Society of America* 3 (1892), 68–81, quotations from 73.

[84] V. V. Dokuchaev, *Sochineniia*, 9 vols. (Moscow–Leningrad: AN SSSR, 1949–61), vol. 8, p. 499.

[85] Hilgard's comments were reported in Krassnof, "The 'Black Earth'," 80.

is essentially dependent upon the presence of a large proportion of lime in the soil.[86] Hilgard's interest in Krasnov's paper was not shared by many American scientists. There was no mention of it in reports on the Congress in the journal *Science*.[87]

A wider opportunity to acquaint Americans with Russian genetic soil science was provided by the World's Columbian Exposition in Chicago in 1893. The Russian government promoted its achievements at the fair. It published a five-volume set on Russian industries, including agriculture, with the assistance of John Crawford, the U.S. Consul General in St. Petersburg. The chapter on soil was written by Pavel Kostychev whose views differed from Dokuchaev's. (Dokuchaev had been the first choice of author, but was not available.) Kostychev explained how "European Russia" was divided into the "region of the Chernoziom or 'black earth'" in the southeast and the "non-Chernoziom lands" in the north and northwest, and that this "difference in soils . . . corresponds almost exactly" to the division between steppe and woodland. Kostychev emphasized the underlying geology, the chemical composition, and physical properties, i.e. texture, of soils ahead of climate. This reflected a disagreement between Kostychev and Dokuchaev on the importance of climate in soil formation. Kostychev made only brief mention of soil "layers," i.e. horizons, that were central to Dokuchaev's approach.[88]

Russian genetic soil science was displayed to better effect at the Chicago fair in the soil exhibit prepared by Dokuchaev and Sibirtsev. It contained a collection of soil "monoliths": large glass tubes containing soils from around the Russian Empire revealing their profiles and the different horizons. They were accompanied by maps, charts, diagrams, and publications.[89] Dokuchaev and Sibirtsev prepared two publications in English. The first was a review of the collection of monoliths. The second was a

[86] Hilgard Papers, Hilgard, E. W.: Outgoing Letters, Letterpress copy books, vol. 18, June 1889–February 1892, Hilgard to M. I. Vilbouchevitch, November 15, 1891. Hilgard used different spellings of Krasnov's name. On Hilgard's correspondent, see: "Jean Vilbouchevitch (1866–1907)," *Revue de botanique appliquée et d'agriculture coloniale* 3, no. 21 (1923), 367–8.

[87] Persifor Frazer, "The International Geological Congress in Washington," *Science* 18, 457 (November 6, 1891), 258–9; D. S. Martin, "The International Geological Congress," *Science* 18, 459 (November 20, 1891), 290–1.

[88] John Martin Crawford (ed.), *The Industries of Russia*, vol. 3, *Agriculture and Forestry*, trans. Crawford (Spb: Department of Agriculture, Ministry of Crown Domains, 1893), pp. 18–41; RGIA, f.399, op.1, 1892, d.238, l.26. On the disagreement with Dokuchaev, see P. A. Kostychev, "Po voprosu o proiskhozhdenii chernozema," *SKhiL* 147 (1884), 259–82.

[89] *World's Columbian Exposition 1893 Chicago: Catalogue of the Russian Section* (Spb: Imperial Russian Commission, Ministry of Finance, 1893), p. 103.

slim volume: *The Russian Steppes. Study of the Soil in Russia, Its Past and Present*, containing an abridged translation of Dokuchaev's short book, *Our Steppes: Past and Present*, written after the drought and crop failure in 1891 (see p. 197 n.31), together with a survey of Russian soil research. The two publications contained concise explanations in English of the Russian theory of soil formation, the methodology, including analysis "levels," (horizons) in cross sections (profiles). They also surveyed practical applications, such as the soil maps prepared for the Ministry of State Domains and the Nizhnii Novgorod method for land evaluation (see p. 196).[90]

The soil exhibit at Chicago does not seem to have done much to facilitate the dissemination of Russian soil science in the United States. Crawford previewed the entire Russian exhibit at the fair in an American periodical. He mentioned that there would be a complete collection of Russian agricultural produce, but did not refer to the soil exhibit. He was more interested in the "large number of models of the Russian fleet" that would be on display.[91] During the fair, the *Chicago Tribune* noted that the Russian agricultural exhibit was "one of the most important and best arranged," but did not mention the soils.[92] The *Sunday Inter Ocean* described "Russia's Exhibit" as: "A Fine Showing of Arts and Industries," including the display of "agricultural produce," but had no words about the soil exhibit.[93] Hubert Bancroft's lavishly illustrated *Book of the Fair* described the Russian agricultural exhibit, but not the soils.[94] A four volume history of the fair published in 1897–8 made favorable references to many of the Russian exhibits, for example, agricultural produce, fertilizers, precious stones, coal, pamphlets on the construction of the Trans-Siberian railway, over a hundred paintings, including Ilia Repin's portrayal of cossacks in the steppe, and the model ships, but omitted the soil exhibit.

[90] W. W. Dokoutschaieff and N. M. Sibirtzeff, *Short Scientific Review of Prof. Dokoutschaeff's and His Pupil's Collection of Soils, Exposed in Chicago in the Year 1893* (Spb: Evdokimov, 1893); V. V. Dokuchaev, *The Russian Steppes. Study of the Soil in Russia, Its Past and Present*, ed. John Martin Crawford (Spb: Department of Agriculture, Ministry of Crown Domains, 1893). (The spellings of Dokuchaev's name are in the originals.)

[91] J. M. Crawford, "Europe at the World's Fair. II. Russia," *The North American Review* 155, no. 432 (1892), 626–31.

[92] "Czar Land Treasure: Russia's Display One of Features of the Fair," *Chicago Tribune*, July 24, 1893.

[93] "Russia's Exhibit: A Fine Showing of Arts and Industries," *Sunday Inter Ocean*, July 2, 1893, Part 2, 13.

[94] Hubert Howe Bancroft, *The Book of the Fair: An Historical and Descriptive Presentation of the World's Science, Art and Industry, As Viewed through the Columbian Exposition at Chicago in 1893* (New York: Bounty, 1894), pp. 209, 341, 372–3, 383.

The author was not indifferent to soils, since he mentioned the American soil exhibits.[95] Even the Russian official report on its section at the Chicago fair missed out the soil exhibit.[96]

Nor did most American soil scientists take much interest in the Russian soil exhibit at Chicago. Simonson wrote that few American scientists seem to have read Dokuchaev's report prepared for the fair. He did not find a copy of it until the mid 1960s and located only one citation to it.[97] Copies of the English-language publications the Russians prepared for the fair are held by the USDA's National Agricultural Library, but were acquired later. The library's copy of the translations of Dokuchaev's work on the Russian steppes and studies of the soil arrived in the spring of 1902. Dokuchaev and Sibirtsev's review of the soil monoliths reached the library only in October 1927.[98] Charles Kellogg, Marbut's successor as chief of the U.S. Soil Survey, observed in 1974: "Apparently few American soil scientists noted [the Russian soil exhibit]."[99] This seems surprising, since the exhibits of American soils at the Chicago fair, included soils from California, organized by Hilgard, and Maryland, prepared by Whitney.[100]

There may have been an exception in Knoxville, Tennessee. Charles F. Vanderford, a Confederate army veteran who worked at the Tennessee agricultural experiment station, reported in 1897 how he had collected samples of "virgin soils" 3.5 feet deep in "prisms" that showed the "soil, sub-soil, and under-stratum." He gathered data on the geology, flora and fauna, climate, and agricultural use of the soils. He noted the impact of the climate on the parent rock, and wrote about the "many hundreds of years during which well-ordered natural forces have been preparing these soils." He also analyzed the chemical composition, organic matter, and the

[95] Rossiter Johnson, *A History of the World's Columbian Exposition Held in Chicago in 1893*, 4 vols. (New York: D. Appleton and Co., 1897–8), vol. 3, pp. 23, 52 (American soils), 62, 143, 163, 167, 256, 257, 409.

[96] P. I. Glukhovskii, *Otchet general'nogo kommissara Russkogo otdela Vsemirnoi kolumbovoi vystavki v Chikago* (Spb: Kirshbaum, 1895).

[97] Simonson, "Early Teaching," 12.

[98] Call Numbers: Special Collections, 56.26 D68 and 56.26 D68S. I am grateful to Amy Morgan, Special Collections Librarian, National Agricultural Library, for information on accession dates, emails to author, April 11 and 12, 2017.

[99] Charles E. Kellogg, "Soil Genesis, Classification, and Cartography: 1924–1974," *Geoderma* 12, no. 4 (1974), 348. See also Perrin Selcer, "Patterns of Science: Developing Knowledge for a World Community at Unesco" (2011), available online at https://repository.upenn.edu/edissertations/323, 364, accessed March 9, 2018

[100] Ball, *Federal*, vol. 1, p. 174; Roy W. Simonson, "Soil Science at the World's Columbian Exposition, 1893," *Soil Horizons* 30, no. 2 (1989), 41–2.

"mechanical condition," or texture, of the soil samples.[101] Vanderford did not explain where he got his ideas from, yet the parallels with Russian soil science are so striking – three layers (or horizons) of soil "preparing" due to "natural forces" in the environment over time – that it is difficult to believe they were coincidental. Kellogg noted that Vanderford's publication included "photographs of men collecting and preserving samples of soil profiles in boxes by methods similar to those used by Dokuchaev."[102] He suggested elsewhere, with no direct evidence, that Vanderford got the idea for the "soil prisms" from the Russian exhibit at Chicago.[103] This was certainly possible, since the Tennessee Agricultural Experiment Station exhibited at the Chicago fair.[104] It is likely, moreover, that Whitney knew about Vanderford's work. Dabney, the president of the University of Tennessee and Whitney's former mentor, acknowledged Whitney's "advice and assistance" in making the "mechanical analyses."[105] The apparent Russian influence on Vanderford's work was an isolated case, and the Confederate army veteran's work received little attention.[106]

In time, some American agricultural scientists began to take note of the Russian innovation in soil science. In 1899–1900, the state agricultural experiment stations' central periodical, *Experiment Station Record* (*ESR*) included two short notices about Russian soil science and the first issue of the journal *Pochvovedenie*.[107] The following year, *Experiment Station Record* published a translation of abridged versions of two articles by Sibirtsev under the title "Russian Soil Investigations." They explained Dokuchaev's pioneering work, the zonal system of soil classification, the theory of soil formation, and presented analyses, including the chemical and mechanical composition, of different types of soils, together with an explanation of the different soil "horizons."[108] The editorial notes

[101] C. F. Vanderford, "The Soils of Tennessee," *Tennessee Agricultural Experiment Station Bulletin* 10, no. 3 (1897), 33, 34, 41, 49, and passim. See also F. Ray Sibley, *Confederate Artillery Organizations: An Alphabetical Listing of the Officers and Batteries of the Confederacy, 1861–1865* (El Dorado Hills, CA: Savas Beatie 2015), n.p.

[102] Kellogg, "Soil Genesis," 348.

[103] Kellogg's suggestion is quoted in Gardner, *National Cooperative Soil Survey*, p. 18.

[104] "Sixth Annual Report of the Agricultural Experiment Station of the University of Tennessee to the Governor, 1893," 15, available online at http://trace.tennessee.edu/utk_agannual/74, accessed April 10, 2017.

[105] Vanderford, "Soils of Tennessee," ii–iii (introduction by Dabney).

[106] Kellogg, "Soil and Society," *Soils and Men: Yearbook of Agriculture 1938* (Washington, DC: USDA, 1938), p. 883, n. 5.

[107] P. V. Ototskii, "Science of Soils," *ESR* 11 (1899–1900), 434; A. Sovietov and N. Adamov, "Contributions to the Study of Russian Soils," *ESR* 11 (1899–1900), 623.

[108] N. M. Sibirtsev "Russian Soil Investigations," *ESR* 12 (1900–1), 704–12, 807–18. The article included Sibirtsev's observations on similarities between the Russian work and Hilgard's.

summarized Russian soil science, how it differed from other work on soils, and pointed to its applicability to the western part of the United States:

> [Dokuchaev] has founded a new school of soil investigation, the fundamental idea of which is the conception of the soil as an independent natural body. With the collaboration of Sibirtzev, this idea has been utilized in ... a ... genetic or natural classification of soils, which, in the study of soil formations, requires a differentiation between parent rock species and the cultivated horizon. His classification differs fundamentally from the petrographic [i.e. geological] and physico-chemical classifications commonly followed by investigators who have dealt with soils which have been profoundly modified under culture, rather than with those in a largely virgin condition, as in Russia and in the western United States.[109]

The abridged translations of Sibirtsev's articles were prepared by Peter Fireman (1863–1962), who taught chemistry at George Washington University in Washington, DC. He was born into a Jewish family in Lipovets in the steppe region of today's Ukraine. He was educated at universities in Odessa, in the Russian Empire, and in Germany and Switzerland, before moving to the United States in 1882.[110] Fireman's role as translator illustrates the important part in the transmission of scientific ideas from Russia played by American scientists of Russian–Jewish origin as they were among the few people in the United States at this time who had the necessary language skills and scientific training.

Fireman's translation of Sibirtsev's essays was only the second article, after Krasnov's, on Russian soil science published in English in an American scientific journal. Simonson, writing in the 1980s, noted: "Looking at [Fireman's] translation now, I think it was good. Yet it made little or no impression on American soil scientists."[111] This was harsh. It was cited in his doctoral dissertation by Coffey, who worked for the U.S. Soil Survey at the Bureau of Soils.[112] Coffey's background was in geology and chemistry. After joining the Soil Survey in 1900, he participated in surveys in the Great Plains, from Texas to South Dakota, prairie states further east, and his native North Carolina. Between 1905 and 1909, he was in charge of soil classification at the Bureau, and from 1909 to 1911 he headed the Bureau's Great Plains Division. During his work for the Survey, Coffey

[109] "Editorial Notes," *ESR* 12 (1900–1), 701.
[110] "Fireman, Peter," *National Cyclopedia of American Biography*, vol. 52 (New York: James T. White & Co., 1970), pp. 270–2.
[111] Simonson, *Historical Highlights*, p. 47.
[112] Coffey's dissertation was published by the Bureau of Soils. George N. Coffey, "A Study of the Soils of the United States," *USDA Bureau of Soils Bulletin*, no. 85 (1912), 32–3.

came to realize the need for a better system of soil classification. This prompted him to study for higher degrees at George Washington University.[113] It also prompted him to find a way through the language barrier to learn about the Russian classification system and the science that underpinned it. In addition to Fireman's translation, he read a paper in French by Sibirtsev presented to the 7th International Geological Congress in St. Petersburg in 1897.[114]

Coffey wrote in his dissertation that "Dokouchayev" had "founded a new school of soil investigation." He presented Sibirtsev's zonal system of classification, and explained how "the Russians make their primary classification upon the origin of the soil – not so much the geological as the climatic and organic origin." Coffey contrasted the Russian classification system with those of France, Germany, and Japan, which were based on geology, and the "largely physical" classification of the U.S. Bureau of Soils. Coffey made positive references to Hilgard's work, and listed Vanderford's study in his bibliography.[115] He drew further on the Russian work in defining soil as "an independent natural body ... differing essentially from the rock which underlies it." He emphasized the character of the parent material as well as "soil forming agencies." In the latter, echoing the Russian theory, he included the parent rock, moisture, temperature, climate, and hinted at the importance of topography and "age," i.e. time. But, in contrast to the Russians, he did not include vegetation. Nor did he analyze horizons in soil profiles, but he did refer to differences between "soil" and "subsoil." He ambitiously proposed a new classification system for American soils based on both parent material and soil-forming agencies, which was thus a compromise between geological and Russian genetic principles. He presented the soils of the United States on a map in five broad divisions, including "arid soils" in the west, "dark-colored prairie soils" extending from Texas into Canada in the middle, and "light-colored timbered soils" in the east. His system contained sub-divisions based on geological origin.[116] Coffey recognized that his classification was

[113] See Eric C. Brevik, "George Nelson Coffey, Early American Pedologist," *Soil Science Society of America Journal* 63, no. 6 (1999), 1485–93; Helms, "Early Leaders," 35–9.

[114] V. N. [sic] Sibirtzew, "Étude des sols de la Russie. Avec une carte et deux planches," *Congrès géologique Internationale compte rendu de la VII session, St. Petersbourg, 1897*, vol. 2 (Spb: Imprimerie de M. Stassuléwitsch, 1899), pp. 74–125. Coffey cited it in his dissertation (Coffey, "A Study," 32–3) and summarized it in Coffey, "The Development of Soil Survey Work in the United States with a Brief Reference to Foreign Countries," *Proceedings of the American Society of Agronomists* 3 (1911), 127–9.

[115] Coffey, "A Study," cited Hilgard on 6, 14, 30, 31, 39. Vanderford's work is listed on 113.

[116] Coffey, "A Study," 7, 8, 13–23, 31, 35–7, 40–4 and passim. The map is at the end of the volume.

"preliminary" and that he needed more information to finalize it. But, he was politely dismissive of the classification system used by the U.S. Soil Survey.[117] His system, like the Russian and in contrast to Whitney's, started with broad, overarching soil groups that could readily be shown on a map of the whole country, and only then moved on to sub-groups.

Coffey's dissertation was rarely cited over the following two or three decades (although it may have been read).[118] He did not, however, apply his innovative ideas to the surveys he carried out under Whitney, who would not have permitted it. For example, a survey in western Kansas he conducted in 1910 was quite conventional.[119] His co-author, Thomas Rice, will feature later as one of the first American converts to studying soil profiles in the Russian manner. Nevertheless, by allowing himself to be influenced by the ideas of Hilgard and the Russian soil scientists, rather than those of his chief, Coffey ended his career at the Bureau of Soils. In the words of Douglas Helms, he "could not dislodge Whitney and the Bureau of Soils from their geological moorings." He resigned in 1911 to take up a post at the Ohio Agricultural Experiment Station.[120] Whitney permitted the publication of Coffey's dissertation, but disassociated himself and his Bureau from its contents. Whitney defended the Bureau's current classification system, pointing to "difficulties ... at the present time and with our limited knowledge" in "construct[ing] a general map based upon the characteristics of the soil itself." He concluded: "In publishing [Coffey's report], however, the Bureau of Soils does so for the purpose of offering it to the scientific world as a contribution to the subject, without indorsing [sic] the scheme of classification proposed and without accepting all the conclusions from the facts cited."[121]

Coffey's engagement with Russian soil science is evidence that it was possible at this time for American scientists open to different approaches to learn about it from publications in English (and French) available in the United States. Two other American scientists who worked for the USDA recognized the significance of Russian soil science at this time: the plant explorers Niels Hansen and Mark Carleton. Hansen visited the Russian soil exhibit at the fair in Chicago and met the Russian soil scientist Vasilii Vil'iams, who was a member of the Russian delegation. Four decades later

[117] Coffey, "A Study," 27–9, 34. See also Gardner, *National Cooperative Soil Survey*, pp. 49–50.
[118] Brevik found only one citation between 1912 and the 1930s, "George Nelson Coffey," 1489, 1490.
[119] E.g. George Nelson Coffey and Thomas D. Rice, "Reconnaissance Soil Survey of Western Kansas," *USDA Bureau of Soils* (1912).
[120] Helms, "Early Leaders," 39–40. [121] Coffey, "A Study," 2.

Hansen recalled meeting "Williams" in Chicago and wrote: "The science work in soils [in the Soviet Union] is considered superior to that anywhere in the world."[122] Carleton's interest in similarities between the soils and climate of the Great Plains and steppes predated his to visits to Russia to collect grain varieties at the end of the 1890s. In 1895, he had ordered samples of "wheat soils" from Russia and sent them to Whitney's Bureau of Soils for analysis.[123] In line with the Bureau's methods, these samples of "Chernozem" from the steppes, together with similar soils from the Great Plains and the Argentinean "steppe," were analyzed to reveal their physical properties.[124] In April 1898, Carleton informed Whitney that he was writing about soils in a bulletin on "wheat environments." He was concerned that Whitney "might ... think that I should scarcely presume to write on the subject of wheat soils." Carleton explained that he was focusing on "our Great Plains, the Russian Chernozem, and the Argentine Pampas," and emphasized the importance of soils for his topic, wheat. He ended by enthusing about the Russian studies: "I believe Russian is next to German in importance for many in [the USDA]."[125] I could not find a reply from Whitney, or the bulletin Carleton referred to, in the USDA archives. Carleton retained his interest in Russian soil science. In 1916, he wrote about the Russians' attention to the vegetation, "temperature and moisture" in "black soil formation." He concluded: "It would be of great benefit to American students to have at least abstracts of these publications put into English."[126]

The positive response to Russian soil science by Hansen and Carleton may reflect the fact that they were specialists in another branch of science and worked for another section of the USDA. Thus, they were not subordinate to Whitney, who did not tolerate deviation from his conception of soils among his staff. Another American scientist who was receptive to Russian soil science, and was not beholden to Whitney, was

[122] SDSU ASC, UA 53.4, N. E. Hansen Papers, Series 2. Helen Hansen Loen Collection, Box 3, Folder 93, Niels Hansen to his father, June 8, 1893 (trans from Danish); Box 4, Folder 174, p. 244 (part of unpublished manuscript: "Russia as Observed by an Agricultural Explorer: 1934–37").

[123] NARA CP, RG 54, Finding Aid A1, Entry 58, Division of Vegetable Pathology and Physiology: Correspondence of M. A. Carleton, 1891–1900, Folder: M. A. Carleton – 1895, Carleton to B. T. Galloway, November 1, 1895, Carleton to Galloway, November 9, 1895.

[124] Mark Alfred Carleton, *The Small Grains* (New York: Macmillan, 1916), pp. 243–5.

[125] NARA CP, RG 54, Finding Aid PI-66, Entry 180, Records of the Bureau of Chemistry and Soils, General Records – Letters received, 1894–1901, Box 2, General correspondence, C–E 1894–96, Carleton to Milton Whitney, April 30, 1898.

[126] Carleton, *Small Grains*, pp. 235–7. Carleton included in his bibliography Krasnov's dissertation in Russian, work by other Russian scientists, e.g. Kostychev, in German editions, but no work by Dokuchaev. Carleton, *Small Grains*, pp. 654–60.

Hilgard. He tried to find out more about the Russian work. Since the 1870s, he had been corresponding with several Russian scientists. A recurring theme in his letters was regret that his lack of knowledge of Russian meant he could not read more about Russian soil science. In January 1892, Aleksandr Voeikov suggested Hilgard write directly to Dokuchaev. Bearing in mind that, unlike many Russian scientists at this time, Dokuchaev had limited knowledge of western European languages, Voeikov wrote:

> As to Prof. Dokutschaef, . . . You may write the address in German and the letter in French. German is less familiar to him, but can certainly find people to translate your communication. Perhaps you would best do so: write a short letter in French and . . . questions in German, and send also copies of your works, especially what is published in French or German. I am sure he will be very glad to communicate with you.[127]

Sadly for historians of soil science, it seems the two pioneers – one Russian, one American – did not correspond with each other.[128]

Voeikov encouraged Hilgard to visit Russia when he was in Europe in 1892, advising him that he would not suffer from the climate if he brought the right clothing. Voeikov offered to help him with translations of Russian studies, and explained that, since many Russian specialists spoke German (Hilgard's mother tongue), he would be able to speak with them. Voeikov was disappointed when Hilgard did not visit, remarking: "You are spoiled by the climate of California."[129] Hilgard read what he could about Dokuchaev's work in western European languages. On May 24, 1894, he wrote to Vilbouchevitch, who was about to visit St. Petersburg: "I am glad to know that you will be at the headquarters of the soil work of Russia . . . I have been looking over lately all I have on Dokutshaieff's work, and find it as closely parallel to my own as I could wish," but pointed out that he disagreed on some points concerning analysis of organic matter.[130] In May 1911, Hilgard wrote to Pavel Ototskii, the editor of the Russian soil

[127] Hilgard Papers, Hilgard, E. W.: Incoming Letters, Box 23, File: Voeikov, Aleksandr Ivanovich, Voeikov to Hilgard, Jan [?] 6/18, 1892.

[128] There is no correspondence between them in Hilgard's Papers: Hilgard, E. W.: Incoming Letters, Box 8 (Coo-E Misc, incl D Misc); Hilgard, E. W.: Outgoing Letters, Box 1, 1848–83, Box 2, 1884–1904. There is none in Dokuchaev's papers, PFA RAN, fond 184, Dokuchaev, V. V., nor in his published correspondence. Dokuchaev, *Sochineniia*, vol. 8.

[129] Hilgard Papers, Hilgard, E. W.: Incoming Letters, Box 23, File: Voeikov, Aleksandr Ivanovich, Voeikov to Hilgard, Dec 18/30 1892; Voeikov to Hilgard, October 22/November 3 [1893?].

[130] Hilgard Papers, Hilgard, E.W.: Outgoing Letters, Letterpress copy books, vol. 20, February 1894–October 1896, pp. 57–8, Hilgard to Vilbouchevitch, May 24, 1894.

science journal, that the new policy of including abstracts in translation would increase appreciation of "the fine work that is being done in Russia."[131]

Hilgard's approval of Russian soil science was consistent with his disdain for Whitney's work (see p. 208–10). In his letter to Ototskii, he contrasted the quality of the Russian work and

> the frequent publications of the Bureau of Soils at Washington . . . [which] usually contain so much ill digested matter and unwarranted conclusions – owing to the wholly one sided point of view of the Chief, Professor Whitney, that it would take a great deal of time and space to controver them all.[132]

Hilgard sometimes cited Russian studies (in translation),[133] and was pleased to learn that his own work was appreciated in Russia.[134]

Tulaikov's year-long trip to the United States in 1908–9 (see pp. 93, 109, 116–17), like Krasnov's visit in 1890, provided an opportunity for American scientists to learn about Russian soil science, if they were minded to do so. While he was in Berkeley, Tulaikov wrote an article on "The Genetic Classification of Soils" for a British journal, but it attracted little attention in the United States.[135] Tulaikov visited the Bureau of Soils in Washington, DC. Whitney received his Russian guest with courtesy and acquainted him with the work of the U.S. Soil Survey, but seems not to have taken advantage of his Russian visitor to enquire about Russian soil science. In an article he wrote for his Russian colleagues, Tulaikov was very critical of Whitney's Soil Survey. He wrote that the concept of soil horizons and detailed descriptions of soil profiles that characterized Russian work were "alien" to American soil scientists. He was critical also of the U.S. Soil Survey's classification system, but speculated that soils of the Marshall series might resemble Chernozems. Tulaikov regretted that the "successes" of Russian soil science remained largely unknown to foreigners,

[131] Hilgard Papers, Hilgard, E. W.: Outgoing Letters, Box 3, Hilgard to Professor P. Ototzky [Ototskii], May 11, 1911.

[132] Hilgard Papers, Hilgard, E. W.: Outgoing Letters, Box 3, Hilgard to Professor P. Ototzky [Ototskii], May 11, 1911.

[133] See, for example, Eugene W. Hilgard, *Soils: Their Formation, Properties, Composition, and Relations to Climate and Plant Growth in the Humid and Arid Regions* (New York: Macmillan, 1906), p. 130 (reference to work by Kostychev), p. 265 (reference to work by Ototskii).

[134] Hilgard Papers, Hilgard, E. W.: Outgoing Letters, Box 3 1905–15, Hilgard to Therese, March 12, 1908; Hilgard to Rosa, March 27, 1908.

[135] N. M. Tulaïkoff, "The Genetic Classification of Soils," *The Journal of Agricultural Science* 3 (1908), 80–5. On the limited American reception, see Simonson, *Historical Highlights*, p. 48.

but attributed this in part to the language barrier.[136] On his visit to the Bureau of Soils, Tulaikov also met Marbut,[137] who had not yet taken over as head of the Soil Survey, but he seems also to have shown little interest in Russian soil science at this time.

Tulaikov found a more receptive audience when he met soil scientists in the Great Plains. In June 1909, he visited the Nebraska Agricultural Experiment Station in Lincoln. Alway, one of the station's scientists, wrote to Hilgard:

> I was very pleased to hear from you through Prof. Tulaikoff who brought your letter of introduction and spent two days with me. I learned a great many valuable things from him regarding the soil of the semi-arid Russian steppes and found that we had a number of similar investigations underway and that our results were very similar, especially in the case of the heavy soils of Saskatchewan.

Alway evidently enjoyed his discussions with Tulaikov rather more than with some of his American colleagues. In 1907, Alway complained to Hilgard about one them, Keyser, who had had fallen under the influence of the "Bureau of Soils men" who were "surveying" part of Nebraska. Alway objected that their surveys were superficial in that they did not go deep into the soil. He was relieved when Keyser left Lincoln for a post in Colorado[138] For Alway and Hilgard, the "Bureau of Soils men" were their enemies as they were working under Whitney, whose concept of soils they rejected. They found the work of Russian soil scientists, which was based on different environmental conditions in different regions, much more amenable. However, they were not in positions to challenge Whitney. Alway will feature in the next chapter as he was one of the American soil scientists who were receptive to the Russian ideas when Marbut, who was in a position to take on Whitney, promoted them among workers in the U.S. Soil Survey.

Conclusion

Before 1914, when Glinka's book on Russian soil science in German was published, most American soil scientists had been unreceptive to the Russian innovation. Since the 1870s, Russian soil scientists had created a

[136] N. M. Tulaikov, "Pochvennye issledovaniia v Soedinennykh Shtatakh," *Pochvovedenie* 10, 4 (1908), 293–322; Tulaikov, *Ocherki po sel'skomu khoziaistvu v Soedinennykh Shtatakh* (Moscow, 1912), pp. 118–22.

[137] N. M. Tulaikov, "Pamiati d-ra K.F. Marbut," *Pochvovedenie* 31, no. 4 (1936), 513–15.

[138] Hilgard Papers, Hilgard, E. W.: Incoming Letters, Box 4, Alway to Hilgard, June 19, 1909; Alway to Hilgard, October 7, 1907.

new scientific discipline that emphasized soils as independent, natural historical bodies. They conceived and classified soils in the contexts of the environments in which they had formed and emphasized five soil-forming factors: climate; organic matter (including vegetation); parent rock; relief; and time. The prevailing understanding of soils in the United States at this time, however, was based on geology and soil physics. Least receptive to the Russian innovation was the USDA's Bureau of Soils headed by Maryland-born Milton Whitney. Since 1899, Milton's Bureau had been conducting a survey of the soils of the United States that classified them on the basis of their texture, i.e. particle size, and their geological origins.

At this time, only a few American soil scientists took an interest in the Russian ideas. Charles Vanderford in Tennessee seems to have been influenced by the Russian soil exhibit at the world's fair in Chicago in 1893. But, his work had little influence. George Coffey at the Bureau of Soils, who had experience of the Great Plains, read about Russian soil science in translation and drew on it in his dissertation. Whitney, who was his chief, refused to be persuaded and Coffey left the Bureau. There was interest in the innovative Russian approach among soil scientists in the Great Plains, such as Frederick Alway of Nebraska. They found it more applicable to the soils of their region than the concepts of soils held by men, such as Whitney, whose experience was largely limited to the different soils that had formed in the contrasting conditions of the eastern United States. The German-born and educated Eugene Hilgard also took a keen interest in the Russian work and corresponded with Russian scientists. His own work, which resembled the Russians', also challenged orthodoxies among American soil scientists, in particular Whitney. But, he was geographically isolated in California and died, aged 83, in 1916. Outside his home state, recognition of the importance of Hilgard's work in the United States grew only after his death.[139]

Hilgard was not the only American soil scientist to correspond with his Russian counterparts. The English-language publications the Russians prepared for their exhibit at the World's Columbian Exposition in Chicago in 1893 attracted the attention of another American. He wrote to Dokuchaev for advice on soil classification and mapping. In the St. Petersburg branch of the Archive of the Russian Academy of Sciences, Dokuchaev's correspondence file contains a letter sent in November 1897 addressed to "Prof. V.V. Dokuchaer" (sic). Unexpectedly, it was from Milton Whitney, the head of the USDA's Division of Soils. Whitney wrote that he had seen "the

[139] See Gardner, *National Cooperative Soil Survey*, p. 13; Jenny, *E. W. Hilgard.*

admirable pamphlet you prepared to accompany the collection of the Russian soils at the Columbian exposition." He requested copies of his works on soil classification and mapping "for the official use of this Division." Whitney stated that the Division had just worked out a "system of classification of soils . . . based upon their texture and physical properties," and was about to begin detailed study and mapping of soils."[140] There is no copy of a reply from Dokuchaev to Whitney in the archives in St. Petersburg, nor was I able to locate one in the USDA collections in the National Archives in College Park. Nevertheless, Dokuchaev reported the episode in a publication in Russia in 1898. He described Whitney as the "famous American soil scientist and geologist, who heads all soil matters in his country." Quoting Whitney's letter in Russian translation, Dokuchaev wrote "even the most practical nation in the world . . . the United States . . . has now resolved to embark on systematic *soil surveys* in its country." He hinted at the contents of a reply to Whitney: "it is very possible that the great transatlantic republic will accomplish the Nizhnii Novgorod method [for land evaluation] more quickly than even greater Russia!"[141] Even if Dokuchaev did not reply to Whitney recommending his method, Whitney would have known about it from the pamphlet prepared for the Chicago exposition (see p. 225).

The letter from Whitney to Dokuchaev contradicts an assertion made by Marbut in 1928 that Whitney did not know about the Russian work (see p. 210). Marbut may have been drawing on a passage from Whitney's introduction to the annual publication of Soil Survey reports for 1918:

> When soil surveying was begun in this country there were no precedents to follow. No work of a similar kind had been done up to that time in any country. The results obtained by the Russians prior to that time, fundamental and basic as their work was, were not well known in any country outside of Russia and, had they been better known, could not have been applied in this country, as the work was of a general character and not concerned with details.[142]

Thus, we have clear evidence that Whitney did know about the Russian work before the U.S. Soil Survey started under his direction, and had written to Dokuchaev to find out more, but chose to disregard it.

[140] PFA RAN, f.184, op. 2, d.94, Uitnei, Mil'ton (Milton Whitney), pis'mo k V.V. Dokuchaevu. November 8, 1897. The file contains the original letter and a Russian translation.

[141] V. V. Dokuchaev, *K voprosu o pereotsenke zemel' evropeiskoi i aziatskoi Rossii s klassifikatsii pochv* (Moscow: Snegirovaia, 1898), pp. 113–14, 114, n. 1. On the Nizhnii Novgorod method, see pp. 196, 225.

[142] Whitney, *Field Operations of the Bureau of Soils*, p. 2.

CHAPTER 6

Soil Science II

Introduction

In 1897, while planning the U.S. Soil Survey, Milton Whitney wrote to Vasilii Dokuchaev asking about the Russian system for classifying and mapping soils. He was prompted to write by the English-language publications on Russian soil science he had seen at the Russian exhibit at the World's Columbian Exposition in Chicago in 1893 (see pp. 213–14.) Since the 1870s, Russian soil scientists had devised a new way of classifying soils into large, overarching groups, for example the Chernozem, or black earth of the steppes, that could readily be represented on maps of large areas. They understood soils as "natural historical bodies" that formed through the interaction of climate, plant and animal organisms, parent rock, relief, over time. The processes of soil formation were revealed by analyzing cross sections, or "profiles," of soils that comprised layers, or "horizons," from the parent rock in the C horizon at the bottom, through a transitional B horizon, to the top soil in the A horizon (see Figure 5.1). Their theory showed how different types of soils had formed in different environmental conditions around the vast Russian Empire. Russian soil scientists applied their theory on a global scale and identified Chernozem in the North American Great Plains, where conditions resembled the steppes. The Russian innovation became the basis of modern soil science.

Whitney ignored the Russian scientific innovation when he set up the U.S. Soil Survey that began in 1899. Surveyors made their way around the United States, classifying soils into an ever-growing number of local "soil types," based mostly on their texture, and combining similar soils with different textures into "soil series." They tried, without success, to group "soil series" into soil provinces and regions based mainly on their geology. In 1913, the U.S. Bureau of Soils published the Survey's findings in a bulletin entitled "Soils of the United States." It contained a lot of raw data and ran to almost 800 pages.[1] The Soil Survey's system of classifying soils did not allow the identification of larger groups that could be represented on a coherent, and useful, soil map of the United States.

Whitney was successful in obtaining federal and state government funding for the Soil Survey. But, he lacked the vision to overcome the shortcomings in the science that underlay it. His own experience was limited largely to the eastern United States, where soils had formed in a humid climate under forested vegetation, and where the processes of soil formation had been obscured by centuries of farming. Whitney had little experience elsewhere, where he would have found quite different soils that had formed in the semi-arid grasslands and arid deserts of the western United States. These soils, moreover, had not yet been transformed by generations of farming, making it easier to detect how they had formed. Whitney's limited geographical experience marked him out from the Russians, whose work he disregarded, and some American soil scientists. Eugene W. Hilgard had analyzed soils in humid Mississippi and arid California. George Coffey had experience in both the Great Plains and eastern states. Both were familiar with the Russian work. Frederick J. Alway in Lincoln, Nebraska, considered the U.S. Soil Survey's approach inappropriate to the Great Plains, and was open to the Russian ideas. Whitney was dead set against ideas that contradicted his own and against people who entertained them. When Coffey, whom Whitney had put in charge of soil classification, recognized the value of the Russian science, his career with the Soil Survey was over, and Whitney disassociated himself from his work. In 1913, instead of Coffey, Whitney appointed as head of the U.S. Soil Survey a geologist from Missouri called Curtis F. Marbut. This looked like an attempt to keep the survey grounded in its geological roots.[2]

By 1914, therefore, there seemed little prospect of changes in the U.S. Soil Survey to take account of the Russian innovation in soil science. The main barrier was Whitney. The Russian science contradicted his understanding of soils and his Soil Survey's methodology, and so threatened his status. There were other obstacles. The language barrier prevented most American scientists reading about the Russian work in detail as there were only a few, shorter publications in the western European languages. The lack of diplomatic relations between the United States and the new Soviet

[1] Curtis F. Marbut, Hugh H. Bennett, J. E. Lapham, and M. H. Lapham, "Soils of the United States," *USDA Bureau of Soils Bulletin*, no. 96 (1913).

[2] See Douglas Helms, "Early Leaders of the Soil Survey," in *Profiles in the History of the U.S. Soil Survey*, ed. Helms, Anne B. W. Effland, and Patricia J. Durana (Ames: Iowa State Press, 2002), p. 40. On Hilgard, Coffey, and Alway, see Chapter 5, pp. 200–1, 217–19, 223.

state between 1917 and 1933 hindered scientists who worked for the U.S. government visiting the Soviet Union. Applying a classification system based on the Russian science in the United States would be complicated, moreover, since the environmental regions differed from those in Russia as a result of the Rocky Mountains (see p. 201).

This second chapter on soil science will trace how these barriers were overcome. Unexpectedly, Marbut went on to lead a "Russian Revolution" in the U.S. Soil Survey's scientific underpinning and methodology. He read a book published in German by the Russian soil scientist Konstantin Glinka in 1914. Marbut became convinced by the Russian approach when he tested it in the Great Plains between 1917 and 1922. He recognized similarities between the soils of the North American grasslands and the Russian steppes, where Dokuchaev had devised his new soil science in the 1870s–80s. Thus, the Great Plains and the steppes played a vital, if unconscious, role in our story. Marbut disseminated his ideas among Soil Survey workers and gradually overcame barriers to their acceptance. American soil scientists gained a greater understanding of Russian soil science at the first two International Congresses of Soil Science held in Washington, DC, in 1927 and in Leningrad and Moscow in the Soviet Union in 1930. Both were followed by excursions around their respective countries to see different types of soils in the environments in which they had formed. By the mid 1930s, soil scientists at the USDA classified the soils of the Great Plains as Chernozems and used a methodology and concept of soils that drew on the Russian science.

The key to recognizing the influence of Russian soil science in this chapter are references to the three main features that were, gradually, adopted by the U.S. Soil Survey: analysis of "soil horizons" in "soil profiles"; the theory of soil formation as a result of the interaction of "soil-forming factors"; and classification of soils into "great soil groups," such as Chernozem that formed in semi-arid grasslands.

The Dissemination of Russian Soil Science in Europe and North America: Konstantin Glinka and Curtis Marbut

International Conferences of "Agrogeology" before 1914

While Whitney was resisting the new Russian soil science in the United States, soil scientists in European countries, including the Russian Empire and Germany, were working towards common systems for conceptualizing, classifying, and mapping soils. The first conference of "agrogeology" was held in Budapest in the Austro-Hungarian Empire in April 1909.[3] Glinka presented a paper on soil zones and soil types in European and Asiatic Russia, drawing on his expeditions, illustrated by a collection of soil samples and a map of their geographical distribution. He reported to his Russian colleagues that their works were cited with approval by several delegates. During post-conference excursions, participants saw "Chernozems" and other soils in the Hungarian plain,[4] which is a "bay" of the "continental ocean" of the steppes to the east.[5] There were no Americans present, but Hilgard, who disagreed sharply with Whitney, sent a paper on chemical analysis.[6]

The second agrogeological conference was held in Stockholm, Sweden, in 1910. More progress was made towards common approaches. There was one Russian delegate, but again no Americans. Hilgard and R. H. Loughridge sent a paper reviewing geological, physical, chemical, climatic, and vegetative systems for classifying soils. Hilgard, characteristically, criticized Whitney and his system:

> the physical point of view alone conspicuously fails as the sole basis of a general classification, such as has been attempted by Professor Whitney, chief of the Soil Bureau of the U.S. Department of Agriculture. In this he attempts to indicate hundreds of soil varieties by local names coupled with merely physical designations, with utter disregard of differences of climate in which these soils have been formed, and also of their conspicuous chemical differences.

Hilgard and Loughridge urged work towards uniform methods, but felt a common classification system was premature.[7] A third agrogeological conference, planned for St. Petersburg in 1914, was not held because of international tensions on the eve of the First World War.[8] Soil science

[3] See E. Michéli and M. Fuchs, "'Bridging the Centuries: 1909–2009': Centennial Meetings on the Occasion of the 100th Anniversary of the 1st International Conference of Agrogeology," *Agrokémia és Talajtan* 59, no. 1 (2010), 195–202.

[4] K. D. Glinka, "Pervaia agrogeologicheskaia konferentsiia v Budapeshte," *Pochvovedenie*, no. 2 (1909), 125–40.

[5] Lajos Rácz, *The Steppe to Europe: An Environmental History of Hungary in the Traditional Age*, trans. Alan Campbell (Cambridge, England: White Horse Press, 2013), pp. 23, 33.

[6] Bancroft Library, Berkeley, California, The Hilgard Family Papers (c. 1848–1945), BANC MSS C-B 972 (hereafter Hilgard Papers), Carton 4, File: Agrogeological convention, Budapest, 1909.

[7] E. W. Hilgard and R. H. Loughridge, "The Classification of Soils," *Verhandlungen der Zweiten Internationalen Agrogeologenkonferenz* (Stockholm: Nordiska Bokhandeln, 1911), p. 229.

[8] "History of the Organization of the International Society of Soil Science," *SS* 25, 1 (1928), 3–4; Hilgard and Loughridge, "The Classification of Soils"; A. A. Iarilov, "Po povodu vtorogo mezhdunarodnogo agrogeologicheskogo s"ezda v Stockgol'me," *Pochvovedenie*, no. 4 (1910), 367–83.

conferences resumed after the war and culminated in the Congresses of 1927 and 1930.

Glinka's Book in German

Before Russian soil science could have more influence internationally, including in the United States, scientists needed to know more about it. A key role was played a book entitled *Die Typen der Bodenbildung: Ihre Klassifikation und Geographische Verbreitung (The Types of Soil Formation: Their Classification and Geographical Distribution)* written by Glinka and published in German in Berlin in 1914. This was the first lengthy presentation of the Russian scientific innovation in a western European language. Glinka wrote it, in collaboration with the German soil scientist Hermann Stremme, to place the Russian work before a "Western European reader." In Glinka's words: "I attempt to sum up all observations which have been made by Russian investigators in the field of soil genesis and the geography of soil types as well as the conclusions they have drawn from them."[9] He covered aspects of the Russian work that were less well known outside Russia. He focussed on theory rather than practical applications. And Glinka presented his own interpretation, rather than the range of views among his compatriots.[10] He chose German as it was the most popular language, marginally ahead of English and French, for scientific publications in the early twentieth century. German was fairly well known among American scientists at this time.[11]

Since Glinka's book was so important in disseminating Russian soil science in the United States, its contents merit summarizing. Glinka explained the significance of soil horizons in soil profiles. He reviewed the evolution of Russian systems for classifying soils in comparison with those in other countries. He described the differences between the geographical distribution of moisture in the Russian Empire and North

[9] K. Glinka, *Die Typen der Bodenbildung: Ihre Klassifikation und Geographische Verbreitung* (Berlin: Verlag von Gebrüder Borntraeger Verlagsbuchhandlung, 1914), p. [iii]. The English text quoted here comes from Marbut's translation, K. D. Glinka, *The Great Soil Groups of the World and Their Development* (Ann Arbor, MI: Edwards Bros, 1927), p. iii. Marbut's translation of the title was "free." Roy W. Simonson, "Early Teaching in USA of Dokuchaiev Factors of Soil Formation," *Soil Science Society of America Journal* 61 (1997), 12.

[10] See Jan Arend, "Russian Science in Translation: How *pochvovedenie* was brought to the West, c. 1875–1945," *Kritika* 18 (2017), 683–708; Arend, *Russlands Bodenkunde in der Welt: Eine ost-westliche Transfergeschichte, 1880–1945* (Göttingen: Vandenhoeck und Ruprecht, 2017).

[11] Michael D. Gordin, *Scientific Babel: The Language of Science from the Fall of Latin to the Rise of English* (London: Profile Books, 2015), pp. 6, 98–103, 181.

America, as a result of the Rocky Mountains' rain shadow, and the significance of these differences for soil formation in the two countries. Glinka analyzed the characteristics and formation of the main groups of soils, spending most of his time on what he termed mature ("ectodynamomorphic") soils, in which the parent material had been transformed into soils by the external conditions, in particular moisture, heat, and vegetation.

Glinka divided "mature soils" according to the moisture conditions in which they had formed: "optimum moisture conditions," such as the red Laterite soils of the tropics; "average moisture conditions," for example "Podsol soils" of forested regions; "inadequate moisture conditions," including "Chestnut soils" of arid steppe regions; and "temporary excess of moisture," in particular "Solonetz soils" with high content of alkaline salts that formed in parts of arid regions where moisture had accumulated. In the middle, Glinka put "Tschernosem [Chernozem] soils" under "Soils Developed under Moderate Moisture Conditions." There was sufficient moisture for rich grassy vegetation, but insufficient for the rapid decomposition of organic substances. As a result, abundant supplies of carbonic acid accumulated in a layer of alkaline "lime carbonate" (calcium carbonate) in the B horizon. This alkaline layer was a defining characteristic of Chernozems. Glinka's classification was based on analyzing the horizons in the profiles of different soils. His types, or groups, such as Tschernosem (Chernozem), included soils with different textures. While his discussion of Tschernosem drew largely on field work in Russia, he noted from first principles that in North America it occurred in Dakota (sic), Nebraska, and Texas. (It is not clear why he omitted Kansas and Oklahoma.) The second part of Glinka's book discussed soils in "European," "Asiatic," and "mountainous" Russia.[12]

The publication of Glinka's German book in 1914 was the key to the "Russian Revolution" in the U.S. Bureau of Soils. The revolution's leader was the man Whitney had appointed to prevent such an occurrence.

Marbut's Appointment to Head the U.S. Soil Survey

Whitney met Marbut on one of his rare trips away from Washington, DC, to deliver a lecture in Columbia, Missouri, in January 1909. Marbut was professor of geology at the University of Missouri and, since 1905, director

[12] Glinka, *Die Typen*. The spellings, such as "Tschernosem," are transliterations from Cyrillic using the German system, and hence differ from "Chernozem" using the system for English.

of the state Soil Survey. Marbut's wife arranged a dinner for their visitor. Whitney was struck by Marbut and appointed him "Special Agent in the Bureau of Soils" to lead a reconnaissance soil survey of the Ozark Mountains region of Missouri and Arkansas. Whitney then appointed him "Scientist in Charge of the Soil Survey" at the Bureau in Washington, DC, with effect from September 1, 1910. Two months later, Whitney put Marbut in charge of the entire U.S. Soil Survey. The University of Missouri tried to lure Marbut back. In April 1912, Whitney wrote to Secretary of Agriculture James Wilson for support to keep him:

> He has been most diligent and efficient ..., has inspired the utmost confidence in his men, has raised the standard of the field work, ... and has inspired the confidence and esteem of the Committee on Agriculture and the Senators and Representatives ... He is in my judgment the most capable and ... efficient man in the United States for this position. He is ... an indomitable worker and is recognized as an authority in geology and in soil surveying.

(It is noteworthy that Whitney emphasized his authority in geology ahead of soil surveying.) Marbut was offered a pay rise and, after some hesitation, he accepted. His appointment as "Scientist in Charge of the Soil Survey" was formalized in June 1913.[13]

Marbut's transition from geology professor at a Midwestern university to head of the U.S. Soil Survey in Washington, DC, coincided with a family tragedy, which may partly explain his devotion to his work over the following years. In March 1909, Marbut's wife died suddenly aged only forty-four. She left him with five children, ranging from his daughter Louise, who was seventeen, to a boy of three. Louise later wrote that the death of his wife (and her mother) left him "bereft and stricken beyond words," and torn between his children and his scientific ambitions. In 1910, moved his family to Washington, where they lived for two years. Louise recalled that her father walked five miles to work and back each day and worked in the office even on Sundays and holidays. They all moved back to Columbia in the summer of 1912. When Marbut decided to

[13] Louise Marbut Moomaw, "Curtis Fletcher Marbut," in *Life and Work of C. F. Marbut, Soil Scientist: A Memorial Volume*, ed. H. H. Krusekopf ([Columbia, MO]: Soil Science Society of America, [1942]), pp. 17–20; SHSM MC, Marbut, Curtis Fletcher (1863–1935), Papers, 1852–1963 (C3720) (hereafter Marbut Papers [C3720]), Folder 2, Milton Whitney to Secretary of Agriculture (copy), January 26, 1909; Whitney to Marbut (copy) February 15, 1909; Note that the papers are missing on Marbut's appointment putting him in charge of the Soil Survey on October 28, 1910; Folder 3, (Whitney) to Secretary of Agriculture, (copy), April 15, 1912; (Secretary of Agriculture) to Marbut, June 1, 1913; Folder 152, Louise Moomaw Manuscript, c. 1940, pp. 19–21.

accept Whitney's offer and return to Washington, he left his family behind and lived a "monastic life" in a rented room.[14] Separated from his children, he wrote regularly to them, especially Louise. She drew on their letters, not all of which have survived, in two memoirs of her father. One was published and the other, longer and franker, was not. Their correspondence and Louise's memoirs are major sources for this chapter.

Marbut had wide-ranging experience before his move to Washington. He was born and grew up on a farm in Barry County in the Ozark region of southwestern Missouri. He retained links with the region, owning a farm that was run by his brother, and returning regularly. As a young man, he had considered moving to North Dakota and taking up wheat farming. But, his future wife vetoed the notion. Instead, he studied for a BS at the University of Missouri. After graduating in 1889, he worked for the Missouri Geological Survey. He broadened his experience, intellectually and geographically, in 1893, when he went to Harvard University for a master's degree in geology under Nathaniel Southgate Shaler. He was influenced also by Hilgard's work. Marbut returned to Columbia in 1895 to resume work for the Geological Survey and teach geology at the university. He added new horizons to his profile on a trip to study the geology and geography of Europe in 1899–1900. His itinerary included the Alps, the Scandinavian fjords, and less obviously for a geologist, the Hungarian plain. Louise wrote much later that her father's training and experience had made him receptive to the concept of soils as natural and historical bodies.[15]

For Marbut's first few years at the Bureau of Soils he adhered to Whitney's scientific approach and classification system. He presented a geologist's view of soils and soil formation in his introduction to the Bureau's bulletin, the "Soils of the United States," of 1913.[16] Some soil surveyors objected to the appointment of a geologist to head the survey. Thomas D. Rice recalled later:

[14] Moomaw, "Curtis Fletcher Marbut," 17–20; Marbut Papers (C3720), Folder 152, Louise Moomaw Manuscript, c. 1940, pp. 19–20; Folder 153, Louise Moomaw Manuscript, c. 1940, pp. 10–11.

[15] Moomaw, "Curtis Fletcher Marbut," 11–18; Marbut Papers (C3720), Folder 152, Louise Moomaw Manuscript, c. 1940; Folder 153, Louise Moomaw Manuscript, c. 1940, pp. 4–7; Helms, "Early Leaders," 43; David Rice Gardner, *The National Cooperative Soil Survey of the United States* (Washington, DC: USDA, Natural Resources Conservation Service, Resource Economics and Social Sciences Division, 1998) ("Thesis presented by David Rice Gardner to the Graduate School of Public Administration, May 1957, Harvard University"), p. 79.

[16] Curtis F. Marbut, "Introduction," in "Soils of the United States," *USDA Bureau of Soils Bulletin*, no. 96 (1913), 1–16.

> We did not think that the soil survey needed a geologist, at least not one of the hard rock or fossil-hunting type. We feared that this was a backward step. Even when Marbut took charge of the Soil Survey, it seemed to us that he still regarded the soil as an interesting geological formation that had been overlooked by other geologists.[17]

Macey Lapham, who worked with both Marbut and Whitney, wrote in his memoir: "[Marbut] came to us as a geologist with an interest in soils, and was strongly convinced of the close relationship between geology and soils." Marbut's cool reception from some personnel was exacerbated because he criticized aspects of the Survey's work and censured some staff for their limited training in geology. Lapham was more impressed by Marbut when he joined him in the field in California in June 1911. He noted that he was "frugally minded and had attained his manhood and accomplishments the hard way," "preferred to and sometimes did travel cheaply," and "walked a lot on field trips."[18] Marbut later recalled difficulties he had experienced in the Bureau in Washington after he joined the Soil Survey. For example, he had hesitated over whether to refer a technical matter to Whitney, but after an anxious night, resolved it himself. More generally, he noted that he had: "Floundered and groped in darkness without a well defined ray of light for 4 years."[19]

Marbut and Glinka's Book

During his first four years in Washington, DC, Marbut improved his knowledge of soil science by extensive reading. He was aware that many survey workers were looking for a better system for studying and classifying soils than Whitney's. Marbut's daughter recalled that during the Sundays and holidays he spent in his office, he read works by German soil scientists, and in their writings he picked up hints of the Russian work. Marbut noted that, in contrast to the Bureau's very detailed classification system, German scientists were organizing soils into more general groups, and the Russians were placing them into still larger groups. The "ray of light" Marbut had been seeking appeared in the library of the United States Department of

[17] Marbut Papers (C3720), Folder 162, Miscellaneous Manuscripts and Notes, Thomas D. Rice, "Doctor Marbut's Services in the Soil Survey" (Appreciation at the Annual Dinner of the American Soil Survey Association, Washington, DC, December 20, 1934), 3. See also Elmer H. Johnson, "Prof. Dr. Curtis Fletcher Marbut," *Pochvovedenie* 31, no. 4 (1936), 484.

[18] Macy H. Lapham, *Crisscross Trails: Narrative of a Soil Surveyor* (Berkeley, CA: W. E. Berg, 1949), pp. 94–5.

[19] Marbut Papers (C3720), Folder 144, typed notes taken from handwritten notes on back of envelope, dated November 7, 1932.

Agriculture (USDA) in "the latter part of 1914." In his own words: "In 1914 a book on the great soil groups of the world and characteristics and development was published in Berlin. This book described the work in soil investigations done by the Russians up to that time." Marbut studied Glinka's book and translated it from German. He devoted his nights, holidays, and Sundays to the task. By the spring of 1917, he had translated the first part of the book on the characteristics and variation of soil types. (He did not translate the second part on Russian soils.)[20]

Marbut described how Glinka's book transformed his understanding of soils:

> It showed that the soil has been so profoundly changed in becoming a soil that it differs in a great many respects, and to a very great extent from its parent geological material. It showed that the study of the soil as soil and without other than mere incidental reference to the geological formations beneath it yields results of the very greatest value.

Marbut realized that American soils were not identical to Russian soils, and that the Russian classification system could not simply be applied in the United States. But, it provided a basis for establishing relationships between soils "in widely separated areas" and working out "broader groupings" that were more satisfactory than those of the U.S. Soil Survey at that time. Marbut also learned from Glinka's book that soil was made up of distinct layers created during its development that needed to be analyzed separately. He compared analyzing soil horizons in soil profiles with a zoologist analyzing different parts of an animal's anatomy separately and not mixing them together. Glinka's book, he concluded, "gave a new direction to the work of the Soil Survey."[21]

The impact of Glinka's book on Marbut, and on the U.S. Soil Survey he headed, cannot be exaggerated. He was aware of Coffey's work, which had drawn on Russian soil science. But, Coffey had read translations of Sibirtsev's work, while Marbut's chief Russian influence was Glinka. There were differences between these two former students of Dokuchaev. Glinka moved away from Sibirtsev's concepts of zonal, intrazonal and azonal soils

[20] Marbut Papers (C3720), Folder 153, Louise Moomaw Manuscript, c. 1940, n.p., 12; Folder 144, Typed notes taken from handwritten notes on back of envelope, dated November 7, 1932; Marbut quoted in Gustavus A. Weber, *The Bureau of Chemistry and Soils: Its History Activities and Organization* (Baltimore, MD: The Johns Hopkins Press, 1928), pp. 91, 94; Glinka, *Great Soil Groups*.

[21] Marbut quoted in Weber, *Bureau of Chemistry and Soils*, pp. 94, 97. See also Moomaw, "Curtis Fletcher Marbut," 20; Carleton R. Ball, *Federal, State, and Local Administration Relationships in Agriculture*, 2 vols. (Berkeley: University of California Press, 1938), vol. 1 , p. 204.

(see pp. 197–8), because they were based on geography rather than the soils themselves. Glinka placed analysis of actual soils ahead of their geography.[22] The Russian concepts were attracting attention among some American scientists at this time. The American Society of Agronomy's committee on soil classification, which included Marbut and Coffey, discussed "soil horizons" in 1916.[23] The first American soil scientists to cite Glinka's 1914 book were Hugh H. Bennett and Rice in their survey of Alaskan soils of 1915.[24]

Marbut's Field Work in the Great Plains and the Origins of a New Classification System

The Great Plains

Marbut was keen to test the Russian soil science in the field, especially in the prairies and Great Plains. Louise described how, on an inspection and field trip in "the Middle West," he traveled from western Texas to North Dakota through "one of the most distinctive soil belts in the United States," and studied the "Chernozem" or "black earth." Marbut had "a soil sampling device he had read about in [the] Russian literature," which he used to take monoliths, or columns, of soils "encased in a frame" that preserved the various layers in their original positions relative to each other. It was "a cumbersome device and the samples [were] bulky, but Marbut was learning about horizons and the soil profile."[25] In Glinka's book, which Marbut translated, the Russian scientist had written:

> A block of soil of considerable size can be secured using the Rispoleshenski apparatus. This consists of an open iron box with sharp edges. . . . It would be advisable to take samples of the various depths, in their order from the surface downward with the same box, accumulating in this way a sample of the whole profile.

Glinka explained the designations of soil horizons, from A to C, and the need to take further samples with "every change of relief, parent rock, plant

[22] Glinka, *Great Soil Groups*, p. 29; T. R. Paton and G. S. Humphreys, "A Critical Evaluation of the Zonalistic Foundations of Soil Science in the United States. Part I: The Beginning of Soil Classification," *Geoderma* 139, nos. 3–4 (2007), 261.

[23] C. F. Marbut, "Report of the Committee on Soil Classification," *Journal of the American Society of Agronomy* 8 (1916), 387–90.

[24] Hugh H. Bennett and Thomas D. Rice, *Soil Reconnoissance [sic] in Alaska: With an Estimate of the Agricultural Possibilities* (Washington, DC: USDA Bureau of Soils, 1915), pp. 51, 198.

[25] Moomaw, "Curtis Fletcher Marbut," 20.

formation, change in color." As well as describing the soil profiles, "the investigator must describe" their geographical position, the relief of the locality, and the plant cover. This would allow the main soil types of the area to be determined.[26]

Marbut followed Glinka's approach during his trip in the Great Plains. Louise wrote that her father

> often said that in that summer ... digging pits and driving a soil auger in the drouth-baked earth ... somewhere on the plains of western Kansas, he reached the conviction that he was on the way to getting the knowledge he needed. He began to see soil as made up of parts which have evolved together as the result of the conditions surrounding them.

When Marbut saw for himself the process of soil formation revealed by the horizons in the profiles he had dug with his Russian-inspired device and analyzed in following Glinka's instructions, he appreciated the originality of the Russian approach. The Russian theory of soil formation was quite different from the processes of sedimentation, transposition, and deposition he knew from his geological training and had previously applied to soils. From Glinka, he learned also the importance of climate, in particular rainfall, and that the vast United States provided a fertile environment for distinguishing between different soils.[27] His daughter wrote:

> That summer's experience sent him back to Washington more eager than ever to study the Russians. Dohuchaev [sic] and Sibirtzev [sic] were no longer just names, they were students, field men who had been looking for the very clues to a new system of human knowledge ... now that he was beginning to see. It sent him back more eager than ever to his translation of Glinka.[28]

It is not clear when Marbut made this trip to the Great Plains. His daughter, in both memoirs of her father, stated it was in the summer of 1914.[29] This may have been a slip for 1917. In his own notes Marbut recalled his "first trip to Great Plains in 1917."[30] Curtis rather than Louise Marbut was probably correct. The summer of 1914 was too early for Marbut to have read Glinka, since it arrived in the USDA library only in

[26] Glinka, *Great Soil Groups*, pp. 11–13. (Marbut's translation.)
[27] See Moomaw, "Curtis Fletcher Marbut," 20.
[28] Marbut Papers (C3720), Folder 153, Louise Moomaw Manuscript, c. 1940, p. 20.
[29] Moomaw, "Curtis Fletcher Marbut," 20; Marbut Papers (C3720), Folder 153, Louise Moomaw Manuscript, c. 1940, p. 19.
[30] Marbut Papers (C3720), Folder 144, Marbut literary papers, 1931–5, Typed notes taken from handwritten notes on back of envelope, dated November 7, 1932.

"the latter part of 1914." A list of Marbut's field trips prepared later does not mention one to the "Middle West and Great Plains" in the summer of 1914, but does include one in the summer of 1917. Further, Marbut's papers contain a set of 6 × 4 inch cards that seem to date to 1917 with field notes on soil profiles, horizons, and locations for various places in the Midwest.[31] Over the following years, Marbut made further field trips around the Great Plains and elsewhere in North America that to some extent replicated those Dokuchaev had made in the steppes and other parts of the Russian Empire in the 1870s–80s. In the summer of 1920, Marbut studied the soils of the northern plains; in May and June 1921, he investigated soils in the deserts of southern California, Nevada, and Arizona; in October 1922, he traveled through western Kansas, eastern Colorado, Oklahoma, and west Texas in the southern plains. In the summer of 1925, he ventured north to study the soils of Saskatchewan and Alberta in the Canadian prairies. He explored other locations, such as Central America, where soils had formed in very different conditions, further broadening his experience.[32]

Marbut invested a lot of effort, physical and mental, in testing Glinka's approach to soils in the Great Plains. He wrote to Louise on August 1, 1920, from Billings, Montana, describing how he was riding in a high seat he had installed at the back of the truck, which allowed him "to take notes on the way and see much more." He added: "It shakes and jolts pretty bad some times and the small boys laugh sometimes when they see us go by." A month later, from Pierre, South Dakota, he mentioned that his overcoat was worn out and he lost his top coat. Nevertheless, they were collecting "lots of samples" of soil.[33] His daughter connected Marbut's "discoveries" on his field trip across the northern plains in 1920 and "his formulation of the principles of a new science."[34] She described how, in testing the methodology with his colleague, agricultural economist Oliver E. Baker, Marbut came to appreciate the significance of climate, especially the moisture, in soil formation:

[31] SHSM MC, Marbut, Curtis Fletcher (1863–1935), Papers, 1901–43, 1981–3 (C3721) (hereafter Marbut Papers [C3721]), Folder 38, Miscellaneous, CV of Curtis Fletcher Marbut prepared for Meeting of Missouri Association of Soil Scientists, September 9, 1983, Sabbaticals and Field trips; Folder 30, Field notes, c. 1917.

[32] Marbut Papers (C3721), Folder 38, Miscellaneous, CV of Curtis Fletcher Marbut prepared for Meeting of Missouri Association of Soil Scientists, September 9, 1983, Sabbaticals and Field trips; Moomaw, "Curtis Fletcher Marbut," 23.

[33] Marbut Papers (C3720), Folder 18, Marbut to Louise, September 1, 1920.

[34] Marbut Papers (C3720), Folder 155, Louise Moomaw Manuscript, c. 1940, [9].

> Dr. Baker was carrying with him land office maps showing long-time weather records and when the two would find a good soil profile near one of the points where weather records were recorded on the map, Baker would get from ... [Marbut] a guess as to the amount of average yearly rainfall. Dr. Marbut, judging by the soil profile, the depth of lime accumulations and the color, particularly of the surface layer, never missed by more than two inches, and usually came within one inch.[35]

After his field trip across the southern plains in 1922, accompanied again by Baker, Marbut wrote a detailed study of the "Soils of the Great Plains."[36] He published it not as an official bulletin of the Bureau of Soils, where it might have received a disclaimer from Whitney, but in the *Annals of the Association of American Geographers*. Marbut gave "predominant attention" to the character of the soil. He defined the Great Plains region by soils characterized by "(1) the presence, on some horizon of the soil ... profile, of a zone of alkaline salt accumulation, usually, not exclusively, lime carbonate [calcium carbonate] and (2) a relatively dark colored surface soil." He represented broad groups of soils on a map of the region and included diagrams of soil profiles indicating the "carbonate horizon" (i.e. the alkaline layer). He discussed soil texture, but less prominently than in official U.S. Soil Survey reports. Marbut directly compared the soils of the Great Plains that he had analyzed himself with the soils of the steppes he had read about in Glinka's book: "The Black Belt [in the eastern plains from the Dakotas, through Kansas, to Texas] seems to be the equivalent of the Black Earth or Chernozem of the Steppes in European and Asiatic Russia."[37] Louise called her father's "Great Plains Study" 'the great basic research project of his life."[38] (Emphasis in the original.)

Towards a New Classification System

Marbut's investigations of Great Plains soils were part of his efforts to reconceptualize soils in the light of the Russian science. He thought it all through during an enforced break in the fall of 1919. While on leave for the apple harvest on his farm in Barry County, Missouri, he was thrown from a wagon. He broke his elbow and a rib, wrenched his back and hip,

[35] Marbut Papers (C3720), Folder 154, Louise Moomaw Manuscript, c. 1940, [21].
[36] For a selection of letters from Marbut to his daughter on the field trips to the southern Great Plains in 1922, see Marbut Papers (C3720), Folder 21, Marbut to Louise October 16, 1922; Marbut to Louise, October 20, 1922; Marbut to Louise, October 31, 1922; Folder 22, Correspondence, 1922–3, Marbut to Louise, November 5, 1922.
[37] Curtis F. Marbut, "Soils of the Great Plains," *AAAG* 13 (1923), 41–66 (quote from 47).
[38] Marbut Papers (C3720), Folder 157, Louise Moomaw Manuscript, c. 1940, p. 50.

and was hospitalized for three weeks.[39] Characteristically, he used the time to devise a new system for classifying soils to place those of the United States in larger groups that could readily be represented on maps of the entire country. This would overcome the problem of Whitney's classification system that had led to a plethora of "soil types" and "soil series," but no coherent way of combining them into larger groups. Marbut wrote about his ideas to Robert L. Pendleton, an American scientist working in India. I have not been able to locate Marbut's letter, but have inferred its contents from Pendleton's replies. Marbut asked Pendleton for samples of Laterite soils that formed in the tropics (which Glinka had referred to), and enquired where he would place them in his proposed classification. Pendleton replied

> he was very much interested in the matters you discussed in your [long letter]. The matter of reclassifying the soils of the United States is surely a vital one, and . . . I believe that the tentative scheme that you suggested is along the right lines. I judge that you have been studying Glinka and other . . . Russian authors.[40]

Pendleton shared Marbut's dissatisfaction with Whitney's system and was familiar with Russian soil science. He had recently defended his PhD dissertation at Hilgard's former department at Berkeley. He had cited an article in English by Nikolai Tulaikov and Peter Fireman's translation of Sibirtsev's work.[41]

Marbut also discussed his ideas on classification with Baker during their field trip across the northern plains in 1920. Baker asked Marbut to outline on a map of the United States the regions covered by his broad soil groups, including: "Podsol soils"; "Black Prairie (Chernoziom) Soils"; and the "Eastern Boundary of layer of Lime accumulation" (the alkaline layer) that Marbut, following the Russians, deemed important for determining Chernozems. Baker was interested as it "would greatly help me . . . in studying the Economics of land utilization in the U.S."[42]

[39] Moomaw, "Curtis Fletcher Marbut," 22; Marbut Papers (C3720), Folder 17, Memo for Mr. Reese, Chief Clerk, (Bureau of Soils), December 10, 1919.

[40] Marbut Papers (C3720), Folder 2, Robert L. Pendleton to Marbut, May 31, 1920; Pendleton to Marbut, May 22, 1921.

[41] Robert Larrimore Pendleton, "Are Soils Mapped under a Given Type Name by the Bureau of Soils Method Closely Similar to One Another?" *University of California Publications in Agricultural Science* 3, no. 11 (1919), 369–498. See also George F. Carter, "Obituary: Robert L. Pendleton," *GR* 48, no. 1 (1958), 124–6.

[42] Marbut Papers (C3720), Folder 18, O. E. Baker to Marbut, n.d. (between September 13 and October 5, 1920); Marbut to Louise, September 13, 1920. Baker drew on Marbut's classification system and explanation of the formation of the black earth in O. E. Baker, "The Agriculture of the Great Plains Region," *AAAG* 13, no. 3 (1923), 109–67, esp. 112.

In his report on their work in the northern plains in 1920, Marbut explained how their attention had been confined "to a description of the basis on which the soils were differentiated into a few broad groups . . . on the basis of soil characteristics alone, with practically no reference to the parent material from which the soils were derived." He continued: "Two soil characteristics were utilized . . . the color of the soil and the depth to the zone of carbonate accumulation [the alkaline layer]." The color of the "second [B] horizon" varied from black to light brown and the "zone of carbonate accumulation," which was "universally present in the profile of the mature upland soil," ranged in depth in the "second [B] horizon" from 7 to about 24 inches. Marbut was, therefore, replacing the methodology of the U.S. Soil Survey with one derived from Glinka.[43]

In 1921, Marbut wrote about his proposed system for classifying soils in an article published, not as an official bulletin of the Bureau of Soils, but in the *Proceedings of the Society for the Promotion of Agricultural Sciences.* He considered the implications of adding "great soil groups" above the Bureau's existing "soil series" and "soil types." He gave an example from the Great Plains and the prairies to the east. In the original survey, a "soil series" named "Carrington" had been identified in the plains of North Dakota. It was described as "a dark-colored soil derived from glacial-drift material," i.e. its geological origins, but no other characteristics were described. Later, soils of similar color and origins in Indiana, Wisconsin, and Iowa were added to the "Carrington soil series." When more detailed surveys that analyzed more soil characteristics were carried out, however, they revealed differences among the various "Carrington" soils. In contrast to those in the prairie states, the "Carrington" soils in North Dakota were "black," rather than "dark-colored," and had a looser structure. Critically, in North Dakota, "not only were the subsoils unleached of their carbonates but abundant accumulations of lime carbonate were found to be universally present in the lower part of the profile." The alkaline layer in the B horizon of soils in semi-arid climates was considered by Glinka, and subsequently Marbut, as a defining characteristic of Chernozems. Marbut explained that all the soils in the "Carrington" series had the same geological parent material, but that "they had become entirely different soils" due to "the different conditions of development," such as the lower precipitation in North Dakota that allowed the alkaline layer to form. He redesignated the North Dakota soils as "Barnes" soils, a new "soil series,"

[43] C. F. Marbut, "Soil Reconnaissance of the Northern Great Plains," *BAASSW* 1 (1921), 38.

while those further east retained the name "Carrington."[44] In the 1930s, they were assigned to different "great soil groups": the "Barnes" soils of the Dakotas were "Chernozem" soils, while the "Carrington" soils further east were "prairie" soils that had formed in different conditions, including higher precipitation.[45] Before these changes in classification were made, however, Marbut needed to get his new system accepted by the Bureau of Soils and soil survey workers in the field.

Marbut Leads a "Russian Revolution" in U.S. Government Soil Science

While he was translating Glinka's book, studying soils in the Great Plains and elsewhere, and thinking through a new classification system, Marbut disseminated his evolving ideas to workers in the U.S. Soil Survey. He circulated parts of his translation to Soil Survey workers before he completed it in 1917, and in March 1920 offered the entire translation to the Department of Agriculture at the University of Minnesota in St. Paul.[46] Marbut used the annual meetings of the American Association of Soil Survey Workers, which was formed in 1920, to spread his new ideas away from Whitney in the Bureau of Soils in Washington, DC.

The Reaction Inside the Bureau of Soils

Meanwhile, inside the Bureau in Washington, Marbut's colleagues' reactions to his reception of Glinka's book and conversion to Russian soil science were largely hostile. The author of an official history later described the impact of the "new concept of soil" from Russia: "In a period of rapid change not all minds act and react alike. Some move rapidly to new standpoints, while others are conservative or even reactionary." The ensuring diversity of opinion on "proper methods of classification," moreover,

[44] C. F. Marbut, "The Contribution of Soil Surveys to Soil Science," *Proceedings of Society for the Promotion of Agricultural Science* 40–41 (1921), 128–9.

[45] NARA CP, RG 54, Finding Aid A1, Entry 82, Soil Survey Division, General Records 1922–42, Box 1 General to Dr. Marbut, File: Dr. Marbut, 1931–2, Byers, Chief, Division of Soil Chemistry and Physics, to Marbut, March 5, 1931; C. F. Marbut, "Soils of the United States," in *Atlas of American Agriculture: Physical Basis Including Land Relief, Climate, Soils, and Natural Vegetation of the United States*, ed. O. E. Baker, part III (Washington, DC: USDA, 1935), pp. 62–3, 72–3.

[46] See Charles E. Kellogg, "Russian Contributions to Soil Science," *Land Policy Review* 9, no. 2 (1946), 9–14; Marbut Papers (C3720), Folder 144, Notes on back of envelope, dated November 7, 1932; Folder 18, (Illeg.) Division of Soils, Dept of Agriculture, The University of Minnesota, St. Paul, to Marbut, March 10, 1920; Folder 154, Louise Moomaw Manuscript, c. 1940, [3].

created problems.[47] Marbut found himself in disagreement with Whitney, the chief of the Bureau and his boss, who persisted in his quite different concept of soils and classification system. We can find a hint of Marbut's position in a letter from his daughter in July 1915: "Your letter bothers me, of course, because it seems the situation there is worrying you and I don't know whether it is worth while living away years of one's life in such or not."[48]

Marbut and Whitney were courteous in their dealings with each other, but this outward display of correct behavior concealed a growing tension we can detect in Marbut's private writings and letters. He made notes about a meeting in April 1915:

> About 10 o'clock this morning in a long conference with Chief Whitney I put the following questions to him:
>
> Do you intend to place any limitations whatever on my opportunity to study soils in the field – to limit or restrict in anyway the kind of work I have been doing? He replied, "No."
>
> Do you intend to prevent my passing on the correlation of the soils in every area [i.e. identifying similar soils across wider areas] before it is finally approved for publication? His reply was No.
>
> The conference was very agreeable and pleasand [sic]. He stated that he intended that he and I should follow up the work of the inspectors and check up its accuracy in every case and hold them rigidly responsible for the work.[49]

Marbut's notes suggest that underneath the pleasantries were a stiffness and tension.

There were less veiled tensions between Marbut and others at the Bureau. In his notes on meetings held between May and July 1915, Marbut recorded: "Attempt to interfere with Committees (Bowie)"; "Attempt to distort facts by preconceived definitions (Everglades)"; "Attempt to have men tell falsehoods about work that does not agree with his peculiar view (Kalendar)." In July, he learned that John R. Bowie planned to sue him for "Criminal Libel and also have me dismissed because I am alleged to have stated that Tumulty was making appointments mainly because of their Catholic faith."[50] His daughter recalled in

[47] Ball, *Federal*, pp. 204–5.
[48] Marbut Papers (C3720), Folder 5, Louise to Marbut, July 25, 1915. I could not find Marbut's letter to his daughter.
[49] Marbut Papers (C3720), Folder 141, from the notebook 1915.
[50] Marbut Papers (C3720), Folder 141, from the notebook 1915.

her unpublished, franker, memoir that in the summer of 1915 "certain malcontents . . . were attempting to influence Chief Whitney to divide the responsibility for inspection of survey work . . . between three men, one of whom would have been Marbut." This would have defeated Marbut's aim of working out the relationships between soil types and series across the entire United States that was essential for devising a new system for classifying soils into larger groups. She continued: "Marbut believed in his new criteria and wanted to develop them in his own practice, though he anticipated that it might be a long while before they would be generally accepted. He would need to proceed with *tact and caution* [my emphasis DM]." When Whitney called a meeting, the "malcontents" backed down and Marbut convinced Whitney to leave him in charge of "the work for the country as a whole."[51]

Looking back in 1932, Marbut made notes on the back of an envelope about the atmosphere in the Bureau. Recording "very unpleasant struggles" in 1917, he wrote:

> Opposition in the Bureau and by my own men. – A long struggle. – Was called theoretical, a high brow. – A Russian worshipper etc. I saw I could not press the matter too hard – else it would go to Chief of Bureau [Whitney] and I would have been overruled. I had to work in secret. Chief had reputation of getting rid of men who did not agree with him. I told him I would follow his policy but think for myself.'[52]

The fact that he made these notes many years later, after he had overcome the opposition, suggests that he was still dwelling on the tensions. Louise wrote in her unpublished memoir that he had not been

> sure of firm support and cooperation in the organization... His chief [Whitney] was an admirable gentleman and a generous executive, – generous to the degree that he had a faculty [sic] of getting into his staff men who disagreed with him. He was given to seeing view points other than his own to the extent that he was influenced to make first one decision and then an opposite one. Yet at the same time, he was given to summarily dismissing men who disagreed with him.[53]

Hanging over Marbut were the fates of Franklin King and Coffey, whose careers at the Bureau of Soils had ended when they disagreed with Whitney. Marbut had added worry in October 1918, when he heard that

[51] Marbut Papers (C3720), Folder 153, Louise Moomaw Manuscript, c. 1940, 20–[21].
[52] Marbut Papers (C3720), Folder 144, typed notes taken from handwritten notes on back of envelope, dated November 7, 1932.
[53] Marbut Papers (C3720), Folder 153, Louise Moomaw Manuscript, c. 1940, [17–18].

his son, Fiske, who was serving with the American Expeditionary Force in France, had been wounded when a shell hit his billet, leaving him hospitalized and blind for a few days.[54] Nevertheless, Marbut persisted, with tact and caution, in pursuing his objective. In October 1921, he wrote to his friend, Dr. H. J. Waters, editor of the *Weekly Kansas City Star*: "I am doing my best to bring about the inauguration of some reforms, not only in the work of the Soil Survey but also in the form of its publications. I do not know whether I shall succeed or not. I certainly shall not fail from lack of effort."[55]

Interludes Away from the Bureau of Soils

Periods of time away from Washington, DC, gave Marbut the chance to discuss his evolving ideas more freely. Starting in 1922, he gave an annual series of lectures on the geography of soils of the United States at the Graduate School of Geography at Clark University in Worster, Massachusetts. W. Elmer Ekblaw, the Professor of Geography, recalled how over the years Marbut's lectures developed from a "traditional treatment of the origin of soils, their relation with the parent material" to a presentation of newer ideas reflecting both his own research and work by other scientists in United States and abroad. In his lectures, Marbut cited the work of Hilgard as well as Russian scientists, including "Dokutschaeff" and Glinka. He presented his growing interest in the Russian theory of soil genesis and system of classifying soils into large groups.[56] Marbut took leave from the Bureau, with Whitney's permission, to deliver the lectures.[57]

Scientific conferences in Europe gave Marbut space to think through his ideas and the opportunity to meet the Russian soil scientists whose work had had such a profound influence on him. In April 1922, he attended the 3rd International Conference of Pedology in Prague, Czechoslovakia (which followed on from the agrogeological conferences before 1914.) Marbut met Glinka for the first time and they became friends.[58] Marbut

[54] Marbut Papers (C3720), Folder 14, Fiske Marbut to Marbut, October 8, 1918. Fiske recovered.

[55] Marbut Papers (C3720), Folder 20, Marbut to Dr. H. J. Waters, October 26, 1921.

[56] W. Elmer Ekblaw, "Dr. Curtis Fletcher Marbut at Clark," *BASSA* 17 (1936), 3–5.

[57] Marbut Papers (C3720), Folder 20, Marbut to Dr. W. W. Atwood, President, Clark University, January 18, 1922; Folder 22, Administrative Assistant, USDA to Whitney, January 19, 1923. The lectures Marbut presented to the USDA Graduate School in 1928 (see pp. 189–90) were the latest version of his Clark lectures. Moomaw, *Life and Work of C. F. Marbut*, p. 257.

[58] "The History of the International Society of Soil Science 1924–1974," *Bulletin of the International Society of Soil Science* 45 (1974), 3–4; Marbut Papers (C3720), Folder 154, Louise Moomaw Manuscript, c. 1940, [3].

wanted to use his trip to Europe to see the Russian Chernozems that had inspired the Russian soil scientists. This was complicated due to the lack of diplomatic relations between the United States and the Soviet Union. He sought help from Raphael Zon, a Russian–Jewish émigré at the USDA's Forest Service, who had Russian contacts. Zon wrote to Dmitrii Borodin, at the Russian Bureau of Applied Botany in New York City, to introduce Marbut. He explained that Marbut was keen to visit Russia as he was "a great admirer of the work of Russian soil scientists," had translated Glinka's book, and wanted to collect soil samples. Borodin forwarded the letter with a request for help to Leonid Prasolov, a soil scientist in Petrograd. Borodin gave Marbut a letter addressed to the Soviet representative in Berlin. Marbut did not manage to visit the Soviet Union, however, as he was taken ill and experienced serious delays in securing a Soviet visa.[59] He was able to visit Hungary, Romania, and Serbia after the conference.[60]

While he was in Hungary, Marbut wrote a long letter, at intervals over a fortnight, to Whitney back in Washington, DC. The tone of the letter is more relaxed than their usual businesslike communications. Marbut described the trip from Cherbourg to Prague, noting the types of soil and geological formations he saw. He explained to his chief how he had saved money by traveling on trains without sleeping cars and staying in modest hotels. His account of the conference was brief, made no mention of meeting Glinka, and suggested that he had found informal discussions, in German, more interesting than the formal sessions, which he had struggled to understand. He described the post-conference excursion across Czechoslovakia and Hungary to see different soils and local agriculture. He described the soil in the Danube valley near Bratislava as "very dark in color," "almost black," and with a high lime content. It reminded Marbut of the river-bottom land in Missouri, Iowa, and Nebraska. Near Budapest, he saw alkali soils similar to those "we have in the west." They traveled east towards the border with Romania, where we "visit[ed] several

[59] See MHS, MN, P1237, Raphael Zon Papers, 1887–1957 (hereafter Zon Papers), Box 4, Folder 5, Zon to Borodin, April 6, 1922 (copy); ARAN, f.687, op.4, d.77, l.3, L. Prasolov to Marbut, April 1, 1922; l.4, Marbut to Prasolov, July 3, 1922; d.476, l.38, Tulaikov to Prasolov, June 3, 1922; d.541, l.1, Zon to Borodin, April 6, 1922 (copy); TsGANTD Spb, f.318, op.1-1, d.23, l.160, Borodin to Prasoloff, May 15, 1922; D. N. Borodin, "Introduktsiia novykh kul'tur Russkim S.Kh. Agenstvom (Biuro) v Amerike iz Novogo Sveta v SSSR i dr. raboty," *Obozrenie Amerikanskogo Sel'skogo Khoziaistva: Digest of American Agriculture* 2, nos. 4–5 (November 15, 1925), 40; Marbut Papers (C3720), Folder 20, Marbut to Louise, April 8, 1922.

[60] Marbut Papers (C3720), Folder 20, Marbut, Budapest, Hungary, to Louise, May 5, 1922; Marbut, Bucharest, Romania, to Louise, May 19, 1922; Marbut, Belgrade, Serbia, to Louise, May 28, 1922.

different soils . . . and collect[ed] a sample of a rich dark soil that to me is very interesting. The rainfall is 32 inches . . . and the region a beautiful one of black prairie soils." He thought the soil in this part of the Hungarian plain was "almost identical with that of central Nebraska." He remarked also on the soils' texture and geological parent material, perhaps for the benefit of Whitney who still considered them paramount in classifying soils. However, much of Marbut's letter is imbued with the ideas of Russian soil science and the concept of great soil groups that could be found in regions with similar environmental conditions around the globe, such as the Great Hungarian Plain (which resembled the Russian steppes) and the Great Plains of North America.[61]

It is not clear whether Marbut sent the letter, but if he did, Whitney does not seem to have held against him his attempt, from afar, to persuade him of the significance of Russian soil science and the existence of larger soil groups. Indeed, he recommended Marbut for his first pay rise since 1913:

> Prof. Marbut is recognized as the foremost authority in this country on the classification of soils and has an international reputation in this branch of science. He is at present studying the soils of Europe for the purpose of determining their relationship to the soils of the United States.[62]

Marbut returned to Europe in the spring of 1924 for the 4th International Conference of Pedology in Rome. He presented a paper on soil classification. Drawing on Glinka's book and his own field work, Marbut argued that soils be classified according to their intrinsic characteristics and the moisture conditions in which they had developed, rather than with reference to climatic terms, such as "arid" or "humid." Together other American scientists present, he was involved in organizing the new International Society of Soil Science and planning future Congresses.[63]

Marbut's Tactful and Cautious "Russian Revolution"

Marbut's growing expertise in soil classification did not convince Whitney of the need for change. In October 1924, in a polite letter to Marbut about U.S. Soil Survey publications, he wrote:

[61] Marbut Papers (C3720), Folder 21, Marbut to Whitney, April 28, May 4, May 11, 1922.
[62] Marbut Papers (C3720), Folder 21, Whitney to the Secretary of Agriculture (copy), June 9, 1922.
[63] C. I. [sic] Marbut, "The Classification of Arid Soils," *Actes de la IV-Conference Internationale de Pédologie, Rome, 12–19 Mai 1924*, 3 vols. (Rome: Imprimerie de Institute International d'Agriculture, 1926), vol. 2, pp. 3, 362–75, available online at www.iuss.org/publications/congress-reports-since-1909/, accessed October 25, 2019; Moomaw, "Curtis Fletcher Marbut," 23.

> My own opinion is that the Russian system of classification is only of broad application, as they have not attempted to apply it to detailed work such as we have been carrying on and their climatic belts are quite different from the belts occurring in the United States. I think the basis of soil classification as worked out in the detailed surveys of this Bureau is very much broader, is very much more complete, and is very much better adapted to American conditions than anything that has been tried in foreign countries.

He concluded: "I am at all times glad to discuss these things with you and with your associates," but his letter suggests a closed mind. Whitney urged Marbut to write a publication about the Russian classification system separately, i.e. not as part of the official Survey.[64] Whitney's letter demonstrated awareness of the geographical differences between Russia and North America on account of the Rocky Mountains. It also showed the contrast between his approach to classification, which started at the bottom with local "soil types," and the Russians', which began at the top with larger soil groups.

Although Whitney had recommended Marbut for a pay rise and was polite to his face, he was franker in his official assessment of Marbut's work in December 1924. Whitney was generally complimentary, but, added:

> He is below average in cooperativeness for he is so imbued with his own work as to find it difficult to bring himself into a frame of mind as to permit of his taking up subjects that others have dealt with... He has not in himself developed employees of the highest possible caliber as he has been too absorbed in his investigational work to give attention to this.[65]

The "investigational work" he was absorbed in was trying to bring together the two different systems for classifying soils. The Bureau of Soil's original system defined local "soil types" largely by texture and then grouped them into "soil series" based mainly on similar underlying geologies. Marbut wanted to add a higher order of classification, great soil groups, such as Chernozem, that drew on the Russian classification. These great soil groups were identified from analysis of a range of characteristics in the horizons of soil profiles. The great soil groups would replace the soil provinces and regions that had been tried, but abolished in 1916. However, Marbut's solution entailed repeating the surveys of "soil series" to investigate a wider range of characteristics.

[64] Marbut Papers (C3720), Folder.25, Whitney to Marbut, October 18, 1924.

[65] Marbut Papers (C3720), Folder 25, Chief of Bureau [Whitney] to Dr. W. W. Stockberger, Dept Personnel Classification officer, December 15, 1924.

Marbut Wins over Soil Survey Workers

Thus, by the early 1920s, Marbut had thought through a new classification system that drew heavily on the Russian theory. But, to introduce his new system to the U.S. Soil Survey, Marbut needed not only to overcome Whitney's opposition, but to win over the surveyors who carried out the work in the field. He had to convince them of the need to repeat the surveys they had carried out since 1899. Over the 1920s and into the 1930s, Marbut worked hard away from the Bureau and Whitney to do precisely this. He attended the annual meetings of the American Association of Soil Survey Workers (renamed the American Soil Survey Association in 1923) that were held in different cities each fall. Although the association had been founded in part "to counteract" his "growing influence,"[66] Marbut turned its meetings into forums to discuss the Russian concepts and their application in the field. (Whitney must have been aware of what Marbut was doing, as the Association's proceedings were published, but did not try to stop him.)

At the Association's first meeting in Chicago in November 1920, Marbut reported on his field trip across the northern plains and explained his new approach to soil classification based on analyzing soil profiles.[67] He intended to use the next meeting, in Lansing, Michigan, in late 1921, to present his new classification system. Before he did so, however, he took the precaution of going over Whitney's head and discussing his intention with E. W. Ball, the Director of the USDA's Office of Scientific Work. Ball urged Marbut to go ahead:

> It seems to me, in view of the fact that a soil survey is supposed to be [a] complete and finished product in itself and that its future value will depend so largely on the confidence the workers may have in the accuracy of its conclusions, that it is highly important that a full discussion at this time should be had of just what data is essential to inspire this confidence.[68]

Marbut followed Ball's advice. At the meeting in Lansing he began by criticizing the Soil Survey's approach. He asked how many soil surveyors and agronomists

[66] Roy W. Simonson, *Historical Highlights of Soil Survey and Soil Classification with Emphasis on the United States, 1899–1970* (Wageningen, The Netherlands: International Soil Reference and Information Centre, 1989) p. 35. See also Gardner, *National Cooperative Soil Survey*, p. 82.

[67] Marbut, "Soil Reconnaissance"; "American Association of Soil Survey Workers," *ESR* 44, no. 3 (March 1921), 300.

[68] Marbut Papers (C3720), Folder 20, E. W. Ball, Director, Office of the Director of Scientific Work, USDA, Washington, DC, to Marbut, November 15, 1921.

> know how many fundamentally different soil profiles have been developed within the area of the United States? How many of them really know that there are characteristic soil profiles or soil sections in this country, each prevailing over a wide area of country, and each just as characteristic of the region over which it prevails as the character of the native vegetation, the character of the agriculture, or the character of the climate?

He referred to the assumption in which "we have all found [satisfaction]" that the fundamental characteristics of soils were based on their geological origins, for example, "a granite soil." He continued: "Scientific history will probably record no greater mistake, none with more profound effect in delaying the advent of the period of real investigation, than this one." Although he did not refer to Russian soil science directly, his ideas showed their influence: "The soil is a natural body, developed by natural forces acting through natural processes on natural materials." Further, every soil "develops a series of horizons or what might be called a soil . . . profile." He explained how soils could be classified into broader groups by analyzing features in the horizons. Marbut described ten different soil profiles in the United States east of the Rocky Mountains. He combined them into two much larger groups: those with and those without "carbonate accumulations," i.e. the alkaline layer. He represented the ten profiles and two groups on a map that placed the profiles with carbonate accumulations in the Great Plains region.[69] While Marbut postponed decisions on naming the two larger groups, some of the soils in the Great Plains resembled those the Russians classified as Chernozems, and some further east as Podzols. Marbut was thus well on the way from Whitney's physical classification to one based on Russian soil science.[70]

At the following meeting in Urbana, Illinois, in November 1922, the Association had a Russian guest. Tulaikov presented a paper summarizing the principles of Russian soil science and soil survey work in his country. He concluded by noting similarities between the soils of the United States and Russia, and appealing for collaboration between "American and Russian soil investigators" in "developing a scientific soil classification" for "the entire surface of the earth."[71]

[69] C. F. Marbut, "Soil Classification," *BAASSW* 3 (1922), 24–32. For an earlier draft, see Marbut Papers (C3720), Folder 141, Marbut, Soil Classification. Paper read at the Second Annual Meeting of the American Association of Soil Survey Workers, East Lansing, Michigan, 1921.

[70] See Roy W. Simonson, "Lessons from the First Half Century of Soil Survey. I. Classification of Soils," *SS* 74, no. 3 (1952), 251; Simonson, *Historical Highlights*, pp. 52–3; Gardner, *National Cooperative Soil Survey*, pp. 95–6.

[71] N. M. Tulaikov, "The Condition of Soil Survey Work in Russia toward the Beginning of the Year of 1923," *BAASSW* 4, no. 2 (1923), 166–9. He was on an official visit to study American

Marbut was succeeding in convincing workers in the field of the need for change in the U.S. Soil Survey. At the Association's meetings over the following years, soil surveyors reported their progress in adopting the new methodology for analyzing horizons in soil profiles, and using Marbut's system for classifying soils. A few examples from different parts of the United States can illustrate this. At the 1923 meeting, M. M. McCool explained that in Michigan they were considering analyzing soil profiles, "as promulgated by Dr. C. F. Marbut."[72] The following year, Richard Bradfield of Marbut's home state of Missouri reported:

> A group of soil profiles, typical of some of the more widely distributed types found in Missouri, have been sampled by Professor H. H. Krusekopf according to methods recommended by the Bureau of Soils [i.e. Marbut]. The profile was exposed by digging a trench large enough to permit easy working and accurate observation of the exposed horizons. Samples were taken of the surface, subsurface, subsoil and substratum.[73]

Krusekopf explained how he analyzed horizons designated A, B, and C. He referred to climate, vegetation, parent material, topography, and time in explaining the distribution of different soils across states from New York to Missouri. He pointed out how lower rainfall and higher evaporation in the "black prairies" had retarded the decay of the organic matter and leaching of mineral salts in the soil (i.e. the alkaline layer characteristic of Chernozems). Krusekopf's report showed the influence of the Russian soil science that Marbut was propagating.[74]

The influence of the Russian science was most marked in the prairies and Great Plains. J. C. Russell and L. G. Engle of the University of Nebraska presented a paper to the 1924 meeting on "Soil Horizons in the Central Prairies," based on observations in Kansas and Nebraska. They reported that due to the "uniform variation of vegetation with climate" across the prairies from east to west, "it seems possible that there is a greater uniformity in soil character over the prairie area than has hitherto been realized." They analyzed the distribution of organic matter, nitrogen, lime, and clay in horizons, designated with letters from A to C, in the profiles. They found layers of calcium and magnesium carbonates (alkalines) in horizons below

agriculture. N. M. Tulaikov, *Sovremennoe polozhenie sel'skogo khoziaistva v Soedinennykh Shtatakh Severnoi Ameriki* (Moscow: Novaia Derevnia, 1923), p. 6.

[72] M. M. McCool, "The Value of Soil Survey as a Basis for Soil Studies," *BASSA* 5, no. 2 (1924), 108–11. See also P. E. Brown, "Problems in Mapping and Classifying Iowa soils," *BASSA* 5, no. 1 (1924), 20–3.

[73] Richard Bradfield, "Variations in Chemical and Mechanical Composition in some Typical Missouri Soil Profiles," *BASSA* 6, no. 2 (1925), 127

[74] H. H. Krusekopf, "The Brown Soils of the North Central States," *BASSA* 6, no. 2 (1925), 146–8.

the "surface soil." The nature and depth of the "calcareous horizons" varied with the precipitation, and were higher in soil profiles in the west where rainfall was lower. Overall, however, soil profiles were similar across the prairies, despite "widely varying rock materials." Drawing on data in Glinka's book, they compared the soils of Kansas and Nebraska with Russian Chernozems, but found more differences than similarities. They attributed these to climatic differences, and argued that "the soils of the Red River Valley in Minnesota and North Dakota ... come much nearer paralleling the Russian Chernozems than do the prairies of eastern Kansas and Nebraska." Russell and Engle suggested that their findings corresponded to those of Marbut in his study of Great Plains soils published in 1923. They concluded: "Out of the chaos of multitudinous soil series and types soil scientists are beginning to construct order, founded on fundamentals."[75] The fundamentals were those of Russian soil science. Russell's and Engle's conclusions from their survey of soils in the prairies and Great Plains echoed those of Dokuchaev and his colleagues during their field work in the steppes nearly half a century earlier.

Soil surveyors discussed the implications of the new approach for the old system of classification. At the 1925 meeting, Rice (one of the first Americans to cite Glinka's book, see p. 236) described his efforts to bring the old "soil series" into line with the new methodology. He analyzed soil characteristics in the horizons of "typical profiles" of the main "soil series," for example the Carrington and Barnes series, of "a few of the more important dark colored prairie soils occurring in Minnesota, Iowa, North Dakota, South Dakota and Nebraska."[76] One of his colleagues in Nebraska, F. A. Hayes, analyzed the profiles of the old Grundy "soil series" and divided it into two new series.[77] Soil surveyors reported similar work throughout the United States at Association meetings in the mid to late 1920s.[78]

The Association addressed specific issues in adopting the new approach to soils. A committee on "research projects" discussed methods for studying soil horizons and profiles and gathering data on the soil-forming factors of climate, vegetation, parent material, and topography. A committee on

[75] J. C. Russell and L. G. Engle, "Soil Horizons in the Central Prairies," *BASSA* 6, no. 1 (1925), 1–18.

[76] T. D. Rice, "Profile Studies in Representative Soils in the Northern Prairies States," *BASSA* 7 (1926), 1–14.

[77] F. A. Hayes, "The Grundy Soils of Nebraska," *BASSA* 8 (1927), 112–23; Hayes, "Revision of the Grundy Series of Nebraska," *BASSA* 9 (1928), 86–99.

[78] See, for example, J. O. Veatch, "Northern Podsol Soils in the United States," *BASSA* 6, 1 (1925), 24–8; Macy H. Lapham, "Some Western Problems in Soil Classification and Mapping," *BASSA* 6, no. 1 (1925), 48–54.

terminology considered terms, such as "horizons" and "profiles," and names for the new soil groups. In some cases, they adopted the Russian terms, for example Podsol and Tschernozem/Chernozem.[79] In 1928, a soil surveyor from Indiana remarked that the Association's meetings had lost the balance between discussing the mapping of soils, which was his interest, and the study of the soil itself, which was taking up more and more time. He noted, however, that Marbut had pointed out "we can't map until we have created the something which we map – and our mapping units are the products of our soil science studies."[80] At the same meeting, Rice presented a paper entitled: "What Is a Soil Series?" in which he explained how "series" were defined by analyzing soil profiles and horizons.[81] This was all a striking departure from Whitney's understanding of soils and showed how far they had come.

The gradual change in approach can be detected in the reports on individual soil surveys that the Bureau of Soils published each year. The reports from 1900 to 1928 show how, for the first two decades, surveyors took samples of soils at predetermined depths without regard to the relationship between the layers in terms of soil formation. Then, from the early 1920s, some began to use the Russian understanding of soil "horizons" as parent material, transition layer, and top soil that occurred at different depths in different soils in different environmental conditions. Analysis of the frequency of the words "profile" and "horizon" in the annual volumes of reports reveals that "profile" appeared from the start, but for a long time was used to mean "soil sections" taken at fixed depths preset by the Soil Survey, not for the Russian conception of a body of soil with genetically related horizons. The word "horizon" came into use around 1921, but for several years designated layers defined by predetermined depths, rather than their place in the soil profile.[82]

The "Russian Revolution" Accomplished

The "Russian Revolution" in U.S. government soil science was accomplished in 1926. In that year, the introduction to the Bureau of Soil's

[79] "Report of the Committee on Research Projects," *BASSA* 8 (1927), 35–52; "Report of the Committee on Terminology. Soil Terminology," *BASSA* 8 (1927), 66–93; Chas. F. Shaw, "A Glossary of Soil Terms (Report of the Committee on Nomenklatura)," *BASSA* 9 (1928), 23–58; Austin L. Patrick, "Report of Progress of the Committee on Nomenklatura," *BASSA* 14 (1933), 53–4.

[80] T. M. Bushnell "Aerial Photography and Soil Survey," *BASSA* 10 (1929), 23–28.

[81] T. D. Rice, "What Is a Soil Series?," *BASSA* 9 (1929), 125–30.

[82] See Milton Whitney et al., *Field Operations of the Division/Bureau of Soils 1899–1922*, multiple vols. (Washington, DC: USDA Bureau of Soils, 1900–28), available online at https://catalog.hathitrust.org/Record/007417874, accessed June 11, 2017.

annual publication of survey reports summarized the changes that Marbut had been working to implement. On methodology, the volume's introduction emphasized analysis of soil "horizons," designated A, B, C, in soil "profiles" to identify the characteristics of soils and determine their classification. Concerning soil formation, it subordinated "geological materials from which soils developed" to "real soil features . . . produced . . . by processes recognized as soil-making processes." And regarding classification, it recognized "great soil groups" covering large geographical areas as a higher order above "soil types" and "soil series." The text noted laconically: "This report marks a definite step forward in soil study." The reports published in the 1926 volume were for surveys carried out in 1921.[83] The delay was normal, since it took time to write up the reports and prepare them for publication, but it also indicates that some surveyors had been employing the new methodology since Marbut had proposed it to the Association of Soil Survey Workers at the start of the decade. Thus, his work among the surveyors away from the Bureau in Washington, DC, had paid off. The contrast between the introduction to the 1926 volume and that for the previous year could hardly be more striking. The introduction to the 1925 volume was little more than a list of surveys, rather than an exposition of their scientific and methodological approach.[84]

The named author of the introduction to the 1926 volume, which presented the "definite step forward," was Whitney, as Chief of the Bureau of Soils. In reality, its author was probably Marbut as the changes summarized were those he had proposed and Whitney had resisted. It is unlikely that in 1926 Whitney had finally acquiesced in a fundamental change to the Soil Survey. He had rejected Russian soil science when he was planning the survey in the 1890s and had restated his objection in his polite letter to Marbut as recently as October 1924. In 1926, moreover, Whitney was working on a manuscript on the "composition and origin" of American soils. It showed that his views had not changed. He treated soils as weathered rocks and focused on their "chemical composition" and "mechanical constitution," i.e. texture. In some places the word "profile" had been crossed out, perhaps in case it was confused with the Russian concept. The manuscript contains many pages summarizing analyses of 845 "soil types," while admitting they covered only ten percent of the

[83] Milton Whitney, "General Review of the Work," *Field Operations of the Bureau of Soils, 1921* (Washington, DC: USDA Bureau of Soils, 1926), pp. xiii–xxvii. "General Review of the Work" was the title given to the introduction.

[84] Milton Whitney, "General Review of the Work," *Field Operations of the Bureau of Soils, 1920* (Washington, DC: USDA Bureau of Soils, 1925), pp. xiii–xvi.

territory of the United States east of the Rockies. It is testament to the impossibility of the task that Whitney had assigned the Soil Survey in 1899, that his manuscript ended abruptly, unfinished, on page 261. It was found among Whitney's papers after his death in November 1927. His former colleagues considered publishing it with a preface praising Whitney for his long service and concluding (and obliquely condemning it): "This publication should be of great interest to all interested in the analytical data of Soil Chemistry and Physics." Instead, they left it in the file where I came across it unexpectedly in the National Archives.[85]

Since Whitney had not relented, it is likely that Marbut was able finally to introduce the changes he wanted, because by 1926 Whitney had been seriously ill with heart disease for over a year.[86] The "letter of transmittal" to the Secretary of Agriculture at the start of the volume that summarized the changes was dated June 17, 1926.[87] A few weeks later, Marbut wrote to his daughter: "Prof. Whitney is getting worse and worse. He has not been to his office for three weeks and when I left Tuesday he had not been to bed for 3 or 4 days nights – sits in his chair all the time." Marbut also wrote to Louise about the plans to merge the Bureau of Soils into a larger Bureau incorporating chemistry under A. G. McCall. Marbut confided: "McCall will make a very good chief I think. He will not interfere much I think. He is somewhat fat and will take things easy. He has never developed a particular line of work of his own and . . . will have no special predilections for any branch of work."[88] Marbut was looking forward to running the Soil Survey without having to exercise caution lest he alienate his chief. He may have anticipated this greater freedom by taking advantage of Whitney's illness to include his changes in the introduction to the volume of Soil Survey reports published in 1926.

Thus, under Marbut's leadership, the U.S. Soil Survey was adopting an understanding of "great soil groups," such as Chernozem, inspired by the Russian classification system and science of soils. The 1926 volume

[85] NARA CP, RG 54, Finding Aid A1, Entry 82, Soil Survey Division, General Records 1922–42, Box 2 Mrs. Peltcher to Whitney Manuscript, File: Bureau of Chemistry and Soils, Soil Investigations, Whitney, Manuscript, "The Soils of the United States, Central America, and South America: Their Composition and Origin."

[86] "Milton Whitney," in *Proceedings and Papers of the First International Congress of Soil Science, June 13–22, 1927, Washington, DC, USA* ed. P. R. Dawson, R. B. Deemer and Albert R. Merz (Washington, DC: American Organizing Committee, ICSS, 1928), vol. 2, p. xxix. (The *Proceedings and Papers* of the First Congress are available at, www.iuss.org/meetings-events/world-soil-congress/, accessed November 1, 2019. The volume numbers in this and other references are those on this website.)

[87] Whitney, "Letter of Transmittal," *Field Operations of the Bureau of Soils 1921*, p. v.

[88] Marbut Papers (C3720), Folder 29, Marbut to Louise, September 1926.

contained what Jacob Joffe (see p. 258) considered the first full descriptions of soil profiles in the U.S. Soil Survey.[89] An example was Hayes' survey of Antelope County, Nebraska. Hayes was on his way towards the new system. He analyzed the "three distinct horizons" in the "typical profile" of the "black soils" of well-drained uplands. He noted the "high content of carbonate, principally lime carbonate [i.e. the alkaline layer] in the lower subsoil" (i.e. the B horizon). He explained:

> The most important characteristics of the soils of the region are a result directly or indirectly of climatic influences, particularly those which determine the available supply of moisture. The parent materials vary widely ... but the soil-forming processes, acting through long periods of time, have produced soils having certain common characteristics.

He paid attention to vegetation, noting how the "black soils" had formed under grasses. Hayes then explained how he had differentiated the "black soils" and other "groups of soils" into "soil series," based partly on the soil profiles, and "soil types" according to the texture of the surface soil.[90] Thus, Whitney's original criterion, texture, came right at the end.

Marbut presented the latest version of his classification system for soils of the United States at the 1st International Congress of Soil Science in Washington, DC, in 1927. He divided American soils into "two great groups": "pedalfers" to the east and "pedocals," with higher content of lime carbonate" in the profiles (i.e. the alkaline layer), to the west. The dividing line ran along the eastern boundary of the Great Plains and reflected the available moisture. Marbut recognized that several "sub-groups" of soils inside these "two great groups" had been "defined by Russian investigators nearly half a century ago," including "Podsolic soils" to the east and "Tschernosems" in the Great Plains. He further argued that, because of the geographical differences in climatic belts between North America and Eurasia due to the Rocky Mountains, there were sub-groups of American soils that did not exist in Eurasia, for example prairie soils. These were dark-colored humid soils that developed under grassland, but lacked the alkaline layer of "Tschernosems" of more arid regions to the west.[91]

[89] J. S. Joffe, "Soil Profile Studies I: Soil as an Independent Body and Soil Morphology," *SS* 28 (1929), 49.

[90] F. A. Hayes et al., "Soil Survey of Antelope County, Nebraska," in *Field Operations of the Bureau of Soils 1921* (1926), pp. 770–1.

[91] C. F. Marbut, "A Scheme for Soil Classification," in *Proceedings and Papers of the First International Congress of Soil Science, June 13–22, 1927, Washington, DC, USA*, ed. P. R. Dawson, R. B. Deemer and Albert R. Merz (Washington, DC: American Organizing Committee, ICSS, 1928), vol. 5, pp. 1–31. On the geographical differences, see also C. F. Marbut, "Geography at the First International Congress of Soil Science," *GR* 17, no. 4 (1927), 663–4 and p. 201.

Thus, Marbut's soil classification system for the United States drew heavily on Russian soil science, but he adapted it to the environmental conditions in North America. In Marbut's system, soils were assigned to the "great soil groups" not according to the climate of the region, for example, humid or semi-arid, but by analyzing the characteristics of soil profiles, hence the importance of the alkaline layer in defining Chernozems. In 1929, he explained this to Horace G. Byers, Chief of the Division of Soil Chemistry and Physics. Byers was analyzing some soil samples from the Gulf coast of Texas, and enquired about the local climate to assist in classifying them. Marbut explained that, although the soils were near

> the boundary between semi-arid and humid soils ... we are defining semiarid in terms of soil characteristics only, and not giving it a climatic definition. ... We ... have long since abandoned any attempt to classify soils on the basis of climate. We may use climate to explain the development of the characteristics on which our classification is based, but that is a quite different matter from basing the classification on climate.[92]

Four years later, Byers replied to an enquiry from a state agricultural experiment station about an investigation into soil moisture in different soil textures published in 1920. Byers explained that the investigation had been made "before the importance of the soil classification into great soil groups of the world was recognized."[93]

Intervention by Russian Émigré Soil Scientists

Over the 1920s and into the 1930s, American soil surveyors were trying to assimilate the Russian-inspired methodology for analyzing horizons in soil profiles. For example, in his survey of Antelope County, Nebraska, published in 1926, Hayes identified the horizons in the profiles by their depths in inches,[94] but did not use the letters A, B, and C that characterized Russian surveys. The letters indicated that what mattered was not just how deep the horizons were, but how they were related to each other in the process of soil formation from parent material in the C horizon at the bottom, through the transitional B horizon to the soil in the A horizon at

[92] NARA CP, RG 54, Finding Aid A1, Entry 82, Soil Survey Division, General Records 1922–42, Box 1 General to Dr. Marbut, File: Dr. Marbut, 1929–30, Horace G. Byers, Chief, Division of Soil Chemistry and Physics to Marbut, January 31, 1929; Marbut to Byers, February 2, 1929.

[93] NARA CP, RG 54, Finding Aid A1, Entry 82, Soil Survey Division, General Records 1922–42, Box 1 General to Dr. Marbut, File: Dr. Marbut, 1933, Byers to Mr. G. W. Patteson, January 28, 1933.

[94] Hayes et al., "Soil Survey of Antelope County," 770–1.

the top (see Figure 5.1). American soil surveyors were assisted in understanding the Russian concepts, especially the A–B–C horizons, by two American scientists of Russian origin: Jacob Joffe and Constantine Nikiforoff. Both presented papers at the American Soil Survey Association meeting in 1930.

Joffe was a professor at Rutgers University and a researcher at the New Jersey Agricultural Experiment Station. He had a sound knowledge of Russian soil science, because he could read Russian and was in contact with Russian scientists. He was born in 1886 into a Jewish family in Lithuania, then in the Russian Empire, and immigrated to the United States in 1907.[95] He was, thus, another of the Russian–Jewish émigrés, who facilitated the transfer of scientific knowledge from Russia. In his paper to the Association in 1930 on "Russian Studies on Soil Profiles," he explained:

> The soil profile as it reveals itself in a vertical cut of a virgin soil with the characteristic build and constitution of the layers, known as horizons, is the alpha and omega of the Russian system of soil studies. The specificity of the soil profile, its morphological features and characters, its uniqueness of any particular climatic zone, and hence its uniform geographic distribution served as the basis upon which the founders of the Russian school of soil science – Dokuchaev, Sibirtzev, and their followers – have developed and built the concept of the soil as an independent natural body.

He explained the Russian understanding of profiles, made up of horizons A, B, and C, and the connections between them. He gave examples of soil profiles he had seen on the excursion following the Soil Science Congress in the Soviet Union in 1930 (see pp. 268–70). Joffe described how Russian soil scientists had redirected attention from simply analyzing "the mass of the soil" to the "geographicity" of soils and "factors involved in soil formation" identified by Dokuchaev. Joffe stressed the importance of climate, in particular moisture, in soil formation, noting also Hilgard's views on the role of precipitation. He explained the systems of classifying soils in soil zones (or "great soil groups") developed by Russian soil scientists, and the importance of profiles in identifying different soils, for example, the alkaline layer that developed in the B horizon of Chernozem soils as a consequence of

[95] See Jacques Cattell (ed.), *American Men of Science: A Biographical Directory*, 10th edition, *The Physical and Biological Sciences, F–K* (Tempe, AZ: The Jacques Cattell Press, 1960), p. 2006; Zaynvl Diamant, "Yankev-Shmuel Yofe (Jacob Samuel Joffe)," Yiddish Leksikon, January 3, 2017, available online at https://bit.ly/2p2OSJo, accessed March 7, 2017.

limited rainfall. On the other hand, Podzols, with their characteristic profiles, had formed in more humid, forested regions.[96]

The other American scientist of Russian origin who helped in the assimilation of the Russian understanding of soil profiles was Nikiforoff. He fled from Russia to the United States in 1921, after the White army, in which he served, was defeated in the Russian Civil War. He had trained in the Russian school of soil science, receiving his doctorate from Dokuchaev's former university in St. Petersburg in 1912. Before the First World War, and service as an officer in the tsarist army, he worked for the Russian Department of Agriculture and Land Improvement. He surveyed soils in Siberia and the steppes of the Don region. He analyzed soil profiles as combinations of interconnected horizons to determine genetic types of soils, including Chernozems, and their geographical distribution.[97] After he moved to the United States, he spent his first few years working as a taxi driver and farm laborer in New York and Minnesota, because he had no documentation on his academic qualifications. In 1927, he met one of the Soviet soil scientists who were in the United States for the Soil Science Congress, who verified his background. Nikiforoff was promptly hired by the University of Minnesota as an instructor in soil science and soil surveyor.[98]

Nikiforoff was ideally qualified and experienced to explain the Russian concept of soil horizons and profiles to his new American colleagues. In September 1930, P. E. Brown of Iowa State College wrote to Alway of the University of Minnesota asking for a paper on soil profiles for the annual meeting of the Soil Survey Association. Alway recommended "our Dr. Constantine Nikiforoff," who was surveying soils in the valley of the Red River of the north, and could give "a first class paper on his studies of these American soil profiles from the Russian view point."[99] Nikiforoff's presentation to the Association

[96] J. S. Joffe, "Russian Studies on Soil Profiles," *BASSA* 12 (1931), 12–39 (quotation from 12). See also Joffe, "Soil Profile Studies I," 39–54; Joffe and L. L. Lee, "Soil Profile Studies II: Methods used in the Profile Survey of New Jersey," *SS* 28 (1929), 469–79.

[97] K. K. Nikiforov, "Morfologicheskoe opisanie chernozema severnoi chasti Donskoi oblast'," in *Trudy Dokuchaevskogo Pochvennogo Komiteta 4*, ed. L. I. Prasolov (Petrograd, 1916).

[98] J. McKeen Cattell and Jaques Cattell (eds.), *American Men of Science*, 6th edition (New York: The Science Press, 1938), p. 1047; Obituary: "C. C. Nikiforoff, 92, Retired Soil Scientist with USDA," *Washington Post*, April 14, 1979; Jan Arend, *Russlands Bodenkunde in der Welt: Eine ost-westliche Transfergeschichte 1880–1945* (Göttingen: Vandenhoeck und Ruprecht, 2017), pp. 157–8.

[99] UMA, Collection Number uarc 173, Department of Soils Records, 1911–90s (hereafter UMA, Dept. of Soils Records), Correspondence, Box 2, Folder 3, Nikiforoff, C. C., 1928–30, P. E. Brown to F. J. Alway, September 16, 1930; Alway to Brown, October 1, 1930.

explained the interconnectedness of the horizons in soil profiles, and how Russian scientists had developed the system of A, B, and C horizons to add "subhorizons," designated A1, A2, etc.[100]

Nikiforoff greatly assisted the U.S. Soil Survey in the transition to the new approach. He maintained contacts with Soviet scientists. They used him to exchange scientific literature with their American counterparts and, importantly, he translated Russian works into English. For example, in March 1928, Sergei Neustruev of the Dokuchaev Soil Institute in Leningrad proposed an exchange of publications with the Department of Agriculture of the University of Minnesota. Alway replied that they would like the Russian publications, as "we could make excellent use of . . . [them]," especially as we "have a Russian member on the staff."[101] Nikiforoff advised American surveyors in the field and gained a reputation for his commitment. Lapham noted that, while they did not always agree, "we admired his tireless energy and diligence, and respected his opinions."[102] Nikiforoff was in great demand. In the spring of 1930, he visited Hayes in Nebraska, who had written to the University of Minnesota requesting that he come to Lincoln "[i]n view of the fact that [he]. . . is well acquainted with the prairie soil profiles, both in this country and in Russia . . . in the hope that he may be able to help us in our study of soil profiles."[103] Nikiforoff visited the University of Manitoba in Winnipeg, in the Canadian prairies, following an invitation from J. H. Ellis. In 1928, Ellis had written to Professor P. R. McMiller of the University of Minnesota that he had "at last found a real Tschernosem," which "would have made Glinka homesick." (Glinka had been in North America in 1927 for the Soil Science Congress and excursion, see pp. 264–8) Ellis continued: "After your association with Dr. Nikiforoff, I presume you will be talking in a weird and strange terminology that may be somewhat mystifying to one like myself who has to keep company with his own thoughts."[104]

[100] C. C. Nikiforoff, "History of A, B, C," *BASSA* 12 (1931), 67–70.

[101] UMA, Dept. of Soils Records, Correspondence, Box 2, Folder 3, Nikiforoff, C. C., 1928–30, Neustreiev to Dept of Agricuture, University of Minnosota, March 23, 1928; Alway to Miss Hariet Sewall, July 27, 1928.

[102] Lapham, *Crisscross Trails*, p. 214.

[103] UMA, Dept. of Soils Records, Correspondence, Box 2, Folder 4, Nikiforoff, C.C. 1928–31, Hayes to Coffee, April 10, 1930; Folder 3, Nikiforoff, C. C., 1928–30, Hayes to Nikiforoff, May 5, 1930.

[104] UMA, Dept of Soils Records, Correspondence, Box 2, Nikiforoff, C. C., 1928–30, Folder 3, Ellis to McMiller, October 11, 1928; Ellis to McMiller, November 14, 1928; Ellis to Nikiforoff, April 21, 1930; (Anon., Alway, or McMiller?) to Nikiforoff, June 4, 1930.

From 1929, Nikiforoff was in charge of a soil survey of the Red River Valley in northwestern Minnesota.[105] He was a hard task master, perhaps drawing on his experience as an officer in the tsarist and White armies. Eric Kneen of the University of Nebraska complained: "I'll never forgive Nicki for being mad as the devil because it once took me from 8 am to 6 pm (steady hiking, no lunch) to go 6½ miles in some real tough stuff."[106] Nikiforoff worked with several scientists, including Rice, who wrote that he was looking forward "to a profitable study of the soils there under your guidance."[107] This prompted Marbut to hold "a sort of school in the Red River Valley for some of the Bureau men."[108] Soil surveyors from several states visited the valley of the Red River in Minnesota to work with the Russian émigré soil scientist. It took the American soil scientists several years to reclassify soils from the old "soil series" to fit the new system. Nikiforoff advised on which soils to include in which new series, a task that some American surveyors were finding difficult.[109] According to Kneen: "Nikiforoff of course worked all over the county," as he did not trust one of the surveyors, adding that it was difficult to find the boundaries between the different soil types.[110] In November 1936, Charles E. Kellogg (acting Chief of the U.S. Soil Survey since Marbut's retirement in 1934) convened a meeting with Nikiforoff and other surveyors, including Alway, Hayes, and Rice, to discuss progress in the soil survey of the valley of the Red River of the North. Nikiforoff was charged with writing up one report for all eight counties surveyed.[111] Nikiforoff's report was published in 1939. Following detailed analysis of "soil types" and "soil series," including the Barnes series, he wrote at length on the "Morphology and Genesis of Soils," analyzing horizons, profiles, and soil-forming factors, and explaining the

[105] UMA, Dept. of Soils Records, Correspondence, Box 2, Folder 4, Nikiforoff, C. C. 1928–31, Alway to Wolfganger, February 11, 1931.

[106] UMA, Dept. of Soils Records, Box 5, Folder: Red River Valley soils, 1936–9, Kneen to "Mac" (McMiller), September 14, 1936.

[107] UMA, Dept. of Soils Records, Correspondence, Box 2, Folder 4, Nikiforoff, C. C. 1928–31, Rice to Nikiforoff, June 15, 1929.

[108] UMA, Dept. of Soils Records, Correspondence, Box 2, Folder 4: Nikiforoff, C. C. 1928–31, Alway to Marbut, May 9, 1930.

[109] UMA, Dept. of Soils Records, Box 5, Folder: Red River Valley soils, 1936–9, Alway to Hide, August 24, 1936; Hide to McMiller, August 26, 1936.

[110] UMA, Dept. of Soils Records, Box 5, Folder: Red River Valley soils, 1936–9, Kneen to "Mac" (McMiller), September 14, 1936.

[111] UMA, Dept. of Soils Records, Box 5, Folder: Red River Valley soils, 1936–9, Kellogg, Memo for file, November 16, 1936.

distribution of "soil series" into two "great zonal groups": Chernozems and Podzols.[112]

Thus, Nikiforoff had two careers as a soil scientist. In his first, in Russia before 1914, he completed his training and gained experience of surveying the Chernozem of the Russian steppes. In his second, in the United States, he oversaw the surveying, analysis, and classification of soils in the American Great Plains. In the process, he greatly assisted in the reclassification of the U.S. Bureau's "soil types" and "series" to fit into larger soil groups based on Marbut's adaptation of the Russian classification system. In 1931, Marbut hired Nikiforoff to work with him in the Bureau of Chemistry and Soils in Washington, DC. Marbut, who had led the "Russian Revolution" in U.S. government soil science, valued having a genuine Russian soil scientist on his staff.[113]

Russian Soil Science in the United States Outside the USDA Soils Bureau

In time, a growing number of American soil scientists outside the USDA had come to appreciate the importance of the Russian innovation in soil science. Whitney, who rejected it, was increasingly out of step. Rutgers University in New Jersey became a center for disseminating the new Russian science in the United States and of contacts between American and Russian, later Soviet, soil scientists. Jacob Lipman, who held senior posts at Rutgers and the New Jersey Agricultural Experiment Station, was a Russian-Jewish émigré.[114] He attracted more Jewish émigrés from the Russian Empire to Rutgers, including Joffe and Selman Waksman.[115] Lipman was founding editor of the first American journal in the field, *Soil Science*, which began publication in 1916.[116] One of his assistant editors

[112] C. C. Nikiforoff, A. H. Hasty, G. A. Swenson, et al., *Soil Survey. (Reconnaissance). The Red River Valley Area. Minnesota* series 1933, no. 25 (Washington, DC: USDA in cooperation with the University of Minnesota Agricultural Experient Station, April 1939).

[113] UMA, Dept. of Soils Records, Correspondence, Box 2, Folder 4: Nikiforoff, C. C. 1928–31, Alway to Marbut, May 9, 1930. Nikiforoff had sent Marbut preliminary findings of the Red River Valley survey. Folder 3, Nikiforoff, C. C., 1928–30, Marbut to Nikiforoff, October, 7, 1930. For more on Nikiforoff and Marbut, see Arend, *Russlands Bodenkunde*, pp. 157–62.

[114] Selman Abraham Waksman, *Jacob G. Lipman: Agricultural Scientist, Humanitarian* (New Brunswick, NJ: Rutgers University Press, 1966), pp. 3–56. See also, p. 106.

[115] On Waksman, see Paul Israel, Dawn Faint, Margaret Swift Cunningham et al., "Guide to the Selman A. Waksman Papers, 1916–1977: Biographical Sketch of Selman A. Waksman," Special Collections and University Archives, Rutgers University Libraries, available online at www2.scc.rutgers.edu/ead/uarchives/waksmanf.html, accessed April 26, 2016. On Joffe, see p. 106.

[116] David Moon and Edward R. Landa, "The Centenary of the Journal Soil Science: Reflections on the Discipline in the United States and Russia around a Hundred Years Ago," *SS* 182, no. 6 (2017), 203–15.

was Nicholas Kopeloff, a Russian scientist who had joined the New Jersey experiment station in 1914.[117] The journal's international consultants included Tulaikov, director of an agricultural experiment station in the steppe region of Russia,[118] and a periodic visitor to the United States. In the journal's early years, Lipman published articles by American and Russian scientists that showed awareness of each other's work. The first articles to cite Russian work, from German translations, were studies of the loess soils of Nebraska by Alway and colleagues. They compared the Nebraska soils with the Chernozem of Russia.[119] Lipman authored the first article in the journal to cite literature in the Russian language.[120] Tulaikov published articles, which cited American work, in 1923.[121] The journal continued to publish articles by American scientists that drew on Russian work, for example, a study of soil profiles in the northern Great Plains that cited Marbut's translation of Glinka's book.[122]

The International Congresses of Soil Science in the United States and Soviet Union in 1927 and 1930

While American soil scientists assimilating the Russian innovation, they had the opportunity to meet their Soviet counterparts at the first two International Congresses of Soil Science in Washington, DC, in 1927 and in Leningrad and Moscow in 1930. Both were followed by excursions around their respective land masses to see the different soils that had formed in the varied environmental conditions.[123] The first Congress was organized by Lipman, who was elected to the role at the Pedological Conference in Rome in 1924.[124] He traveled to Moscow in May 1926 to

[117] C. R. Woodward and I. N. Waller, *New Jersey's Agricultural Experiment Station, 1880–1930* (New Brunswick: New Jersey Agricultural Experiment Station, 1932), pp. 115, 562.

[118] *SS* 1, no. 1 (1916), n.p.

[119] F. J. Alway and Guy R. McDole, "The Loess Soils of the Nebraska Portion of the Transition Region: I. Hygroscopicity, Nitrogen and Organic Carbon," *SS* 1, no. 3 (1916), 197–238; F. J. Alwaym and Morris J. Blish, "The Loess Soils of the Nebraska Portion of the Transition Region: II. Humus, Humus–Nitrogen and Color," *SS* 1, no. 3 (1916), 239–58.

[120] J. G. Lipman et al., "Sulfur Oxidation in Soils and Its Effect on the Availability of Mineral Phosphates," *SS* 2, no. 6 (1916), 499–538.

[121] N. M. Tulaikov, "The Soil Solution and Its Importance in the Growth of Plants," *SS* 15 (1923), 229–34; Tulaikov and M. S. Kuzmin, "On the Question of Obtaining the Soil Solution," *SS* 15 (1923), 235–40.

[122] James Thorp, "The Effects of Vegetation and Climate upon Soil Profiles in Northern and Northwestern Wyoming," *SS* 32, no. 4 (1931), 283–301.

[123] For fuller accounts of the congresses and excursions, see Arend, *Russlands Bodenkunde*, pp. 125–75; I. A. Krupenikov, *Istoriia pochvovedeniia: ot vremeni ego zarozhdeniia do nashikh dnei* (Moscow: Nauka, 1981), pp. 224–8.

[124] "History of the Organization of the International Society of Soil Science," *SS* 25, no. 1 (1928), 4.

discuss the organization of the Congress and arrangements, including some funding, for the Soviet delegation to travel to the United States. Both Lipman's visit to the Soviet Union and the Soviet delegation's trip to America were complicated by the absence of diplomatic relations between the two countries.[125]

Lipman's motives can be found in a letter he wrote to A. A. Iarilov, the chair of the Soviet organizing committee:

> It is my belief that in the field of theoretical soil science the scientific workers of the USSR will play for a long time a leading role in international science. However, what we do know of what they have accomplished is very fragmentary and altogether incomplete. It is, therefore, our desire to obtain at the International Congress in Washington a complete, clear and well defined picture of what they have done during the last 50 years not only through books, articles and reports, but by direct conversations with the authors and investigators themselves.

He asked them to bring a display of monoliths of soil profiles, charts, and photographs for display at the Congress.[126] The American preparations stressed the Russian influence on American soil science. Marbut arranged for his English translation of Glinka's book to be published in time for the Congress.[127] Glinka wrote to Marbut: "I shall be delighted, if that work will help us to find the same language for the study of soils of our countries."[128] Delegates were issued with a diary containing an essay on "The new concept of soils" that highlighted the Russian contribution and its influence on the U.S. Soil Survey.[129] Also on display was George Coffey's soil map, recognizing his pioneering interest in Russian soil science.[130]

The Soviet delegation of twenty-one was the largest foreign group. It was headed by Glinka and included Tulaikov and other leading Soviet soil scientists.[131] Following Lipman's request, their "chief object" was "to give

125 ARAN, f.487, op.1, 1927, d.8, l.2 and passim, see also dd.5, 10.

126 "Ot Mezhdunarodnogo Obshchestva Pochvovedov i Organizatsionnogo Komiteta I Mezhdunarodnogo Kongressa v Vashingtone," *Biulletini pochvoveda*, nos. 5–7 (1926), 59; ARAN, f.582, op.4, d.175, l.3.

127 Glinka, *Great Soil Groups*.

128 Marbut Papers (C3720), Folder 28 Correspondence, 1926, K. Glinka, Leningrad, to Marbut, March 22, 1926. (Original handwritten in English.)

129 Leonid Prasolov's diary is preserved in his archive. ARAN, f.687, op.2, d.4, Dnevnikovye zapiski o prebyvanii na 1-om mezhdunarodnom kongrese pochvovedov, ll.33ff.

130 L. Prasolov, "Kartografiia pochv na I mezhdunarodnom kongresse pochvovedov v Vashingtone," *Pochvovedenie*, no. 1–2 (1928), 146.

131 "Foreign Official Delegates," in Dawson et al. (eds.) *Proceedings and Papers*, vol. 2, pp. xiii–xiv; "21 Soviet Scientists at Soil Conference: Delegation at Washington Sessions Was the Largest Ever to Leave Russia," *NYT*, June 19, 1927.

the . . . Congress an idea of a half century's work of Russian investigators in the sphere of theoretical soil science," but they also presented papers on applied topics.[132] They had been advised by a Soviet scientist who visited the United States in 1926 that American soil scientists knew about the Dokuchaev school of soil science through Marbut's translation of Glinka's book. And, since few American scientists, except émigrés, knew Russian, they should present their papers in English.[133] Following this advice they brought copies of a set of pamphlets in English, entitled "Russian pedological investigations," on such topics as Dokuchaev's ideas, soil classification, soil genesis, cartography, soil surveys in Russia since the 1870s, and agronomy.[134] On the final day of the Congress, the Soviet scientists led a session on the genesis, morphology, classification, and mapping of soils.[135] Joffe, who spoke Russian, looked after them and translated their papers.[136] Glinka's paper on the history of Russian soil science was read out by Lipman.[137]

Marbut's main role at Congress was as president of the Commission on Classification, Nomenclature, and Mapping of Soils. He delivered a paper on soil classification that showed the enduring influence of Glinka's work, but also his evolving differences with the Russian approach, which in turn reflected the geographical differences between North America and Russia. Marbut noted how:

> In 1914 the results of Russian soil investigations [were] made known to workers of Western Europe and America through a summary published in German [i.e. Glinka's book]. This caused in Europe and to a less extent in

[132] A. A. Yarilov [sic], *Brief Review of the Progress of Applied Soil Science in USSR* (Leningrad: Publishing office of the Academy, 1927), 1, footnote.

[133] A. F. Lebedev, "V Biuro Upolnomochennykh Vsesoiuznykh S"ezdov po pochvovedeniiu (pis'mo iz Ameriki)," *Biulletini pochvoveda*, nos. 5–7 (1926), 54–5.

[134] J. N. Afanasiev, *The Classification Problem in Russian Soil Science* (Leningrad: Publishing Office of the Academy, 1927); K. D. Glinka, *Dokuchaiev's Ideas in the Development of Pedology and Cognate Sciences* (Leningrad: Publishing Office of the Academy, 1927); S. S. Neustruev, *Genesis of Soils* (Leningrad: Publishing Office of the Academy, 1927); L. I. Prasolov, *Cartography of Soils* (Leningrad: Publishing Office of the Academy, 1927); N. M. Tulaikov, *Russian Pedology in Agricultural Experimental Work* (Leningrad: Publishing office of the Academy, 1927). See also, p. 190 n.14.

[135] Dawson et al. (eds.) *Proceedings and Papers*, vol. 2, pp. xl, 116–36; ARAN, f.687, op.2, d.4, Congress handbook with programme.

[136] J. S. Joffe, "Russian Contributions to Soil Science," in *Soviet Science: Symposium at the 1951 Philadelphia Meeting of the American Association for the Advancement of Science*, ed. R. G. Christman (Washington, DC: AAAS Publications, 1951), p. 61; R. B. Deemer, "Preface," in Dawson et al. (eds.) *Proceedings and Papers*, vol. 5 [vi].

[137] Charles A. Shull and Frank Thone, "The First International Congress of Soil Science," *Plant Physiology* 2, no. 4 (1927), 376.

> this country a stampede to the Russian point of view in which the soil type is defined in terms of the climatic forces which are supposed to have brought about the development of its general characteristics.

He explained how this had replaced the previous emphasis on geology, but that more recently soil scientists were moving towards classifying soils on the basis of the characteristics of soils themselves, which were determined by analysis of the horizons in the soil profiles.[138] The delegates debated whether soil classification should be based on the external soil-forming factors, which was the original Russian position, or soil characteristics, which was a development of it. Marbut was supported by Rice, one of his American colleagues. One of the Russians, Iakov Afanas'ev, asserted "the classification problem [has] . . . formally and historically . . . been solved by the Russian school of soil science from the days of Dokuchaiev by three genetic methods: geographic, morphological and chemical." Showing perhaps similar stubbornness to Whitney, he would not allow one of these to be selected as the most important in any system of classification.[139]

Marbut's other major contribution was organizing and leading the post-Congress excursion around North America. They travelled from Washington, DC, through the South and Marbut's home state of Missouri, across the prairies and Great Plains in Kansas, and over the Rocky Mountains to California. From there they headed north to British Columbia, crossed the Canadian prairies to Winnipeg, then traveled along the Red River Valley through Minnesota, on to Iowa, and through the Midwest back to Washington. Marbut explained: "By selecting [this] route . . . some of the great north–south belts of soils were seen in two places . . . especially . . . the belts of prairie soils and . . . the Black Earths [Chernozems]. The latter were seen in Kansas and also in . . . Canada." Marbut referred to "the Black Earth or Tschernosems" as "the most important of [the] great soil groups," "as defined by the Russians," and compared their origins in the United States and Russia. Pits had been dug along the route to allow delegates to view soil profiles. The scientists debated the types of soils they saw. The Russians considered the "prairie

[138] C. F. Marbut, "A Scheme for Soil Classification," in Dawson et al. (eds.) *Proceedings and Papers*, vol. 5, pp. 1–11.

[139] T. D. Rice, "Should the Various Categories in a Scheme of Soil Classification be Based on Soil Characteristics or on the Forces and Conditions Which Have Produced Them?" in Dawson et al. (eds.) *Proceedings and Papers*, vol. 5, pp. 108–12; J. Afanasieff, "Soil Classification Problems in Russia," in Dawson et al. (eds.) *Proceedings and Papers*, vol. 5, pp. 498–501. See also C. F. Marbut, "Fifth Commission. Classification, Nomenclature, and Mapping of Soils," First International Congress of Soil Science: A Summary of the Scientific Proceedings, *SS* 25, no. 1 (1928), 51–60.

soils" near Kansas City to be "Chernozems" that had been "degraded" by a change to a wetter climate since they had formed. Marbut could not accept this, as the soils lacked the alkaline "zone of lime carbonate accumulation" in the B horizon that he considered defined Chernozems. There was no disagreement further west around Hays, Kansas, where, in Marbut's words: "Soils considered to be the equivalent of the southern Tschernosems in Russia were examined . . . This conclusion, previously arrived at by the American workers, was confirmed by those delegates familiar with Russian soils."[140] While in Kansas the itinerary included the state's "Hard Winter Wheat Belt." The delegates were told it was the "largest individual cropping area in Kansas," and that the previous year over nine million acres had been sown with the hard red winter wheat that had been introduced from the Russian steppes in the 1870s.[141]

The Soviet scientists' participation at the Congress gave their American counterparts the chance to learn more about their work. Two Americans praised the exhibits the Russians brought with them:

> Russia made a very favorable impression by its large and varied display. There were maps showing their soil surveys, periodicals, publications, text books, soil sections of typical Russian soils, portraits of the great Russian scientists who, more than any others have been responsible for the development of Pedology. It was a delight to see this galaxy of Russian stars – DOKUCHAIEV, GLINKA, . . . – a group of whom any nation would be proud. The exhibit of books by these leaders of thought and research was a revelation to those who examined them.[142]

V. C. Finch of the University of Wisconsin thanked Marbut for his "excellent translation of Professor Glinka's monumental work," and asked how his university could acquire a copy of the set of pamphlets the Soviet delegation had brought to the Congress.[143] These pamphlets were used in soil science courses from the early 1930s at two institutions in the Great

[140] A. G. McCall, "The Organization of the Trans-Continental Excursion," in *Proceedings and Papers First International Congress of Soil Science, June 13–22, 1927, Washington, DC, USA*, 5th Souvenir Volume, *Transcontinental Excursion and Impressions of the Congress and of America*, ed. Selman A. Waksman and R. B. Deemer (Washington, DC: American Organizing Committee, ICSS, 1928), pp. 17–21 (map on p. 20); C. F. Marbut, "The Excursion," in ibid., pp. 43, 46–7, 61–70; A. A. Iarilov, "Na kongresse i o kongresse," *Biulleteni pochvoveda*, nos. 5–8 (1927), 134.

[141] Marbut Papers (C3720), Folder 177, Periodicals and Pamphets, V.6, C. F. Marbut, "The Transcontinental Excursion under the Auspices of the American Soil Survey Assocation: Descriptions, Discussions and Interpretations of Soils and Soil Relationships along the Route of the Excursion," (c. 1927), Appendix III, R. I. Throckmorton, "The Hard Winter Wheat Belt of Kansas."

[142] Shull and Thone, "The First International Congress," 380–1.

[143] Marbut Papers (C3720), Folder 34, V. C. Finch to Marbut, October 27, 1927.

Plains: the University of Nebraska in Lincoln by J. C. Russell and North Dakota Agricultural College in Fargo by Kellogg.[144] Nikolai Vavilov, who visited the United States again in 1930–1, noted that Glinka's book (presumably Marbut's translation) was considered a main text on soil science.[145] The Congress sparked interest among American scientists in working with their Soviet counterparts. In 1928, Oliver E. Baker (who had accompanied Marbut on his field trips across the Great Plains) wrote to Vavilov in Leningrad to suggest collaborating in a broad survey of soil resources of the world. Baker wrote: "Russia is the best prepared country to do so. . . . No other country, except our own, has so extensive and well matured soil surveys, and no other country, . . ., is so deeply interested in the subject of natural resources from the standpoint of scientific research."[146] American scientists were saddened by news of Glinka's death a few months after the Congress. "GLINKA was not well at the time of the congress," two Americans wrote, "and his recent death leaves a place among the Russian workers that cannot be filled."[147] Waksman wrote a laudatory obituary for the journal *Soil Science*.[148] Marbut sent sincere condolences to Glinka's colleagues.[149]

Soviet soil scientists returned the hospitality at the 2nd International Congress of Soil Science in Leningrad and Moscow in 1930. They organized an epic excursion to see different soils, visit scientific institutes, and meet scientists around their vast country. The American delegation, comprising seventy-three scientists, was by far the largest international group and testament to the American interest in Russian soil science.[150] Marbut was eager to go and at last to see the Chernozems of the steppes. He had to travel in a private capacity with funding from other sources, as the U.S. government would not permit its employees to visit the Soviet Union in their official capacities (see p. 77). At the Congress, Marbut introduced the discussion of the Commission on Classification, Nomenklatura and Mapping with a paper on the "Relation of Soil Type to the Environment."

[144] Roy W. Simonson, "Early Teaching," 11–16. [145] TsGANTD Spb, f.318, op.1-1, d.364, l.84.
[146] TsGANTD Spb, f.318, op.1-1, d.278, ll.60–1, Baker to Vavilov, August 25, 1928.
[147] Shull and Thone, "The First International Congress," 383.
[148] See S. A. Waksman, "Professor K. D. Glinka," *SS* 25, no. 1 (1928), 1–2.
[149] Marbut Papers (C3720), Folder 35, Marbut to the Dokuchaiev Pedological Institute, Leningrad, USSR, November 4, 1927.
[150] "Introduction," in *Trudy II Mezhdunarodnogo Kongressa Pochvovedov, Leningrad–Moscow, July 20–31, 1930, Protokoly, Ekskursii*, ed. D. G. Vilenskii, V. A. Kovda, and A. A. Iarilov (Moscow: Gos. izdatel'stvo Sel'sko-Khoziaistvennaia literatura, 1935). (The *Trudy* [proceedings and papers] of the second Congress are available online at www.iuss.org/meetings-events/world-soil-congress/, accessed November 1, 2019.)

Continuing the discussion from the previous Congress, he raised the relative importance of climate and vegetation as soil forming factors, and the role of soil characteristics in defining different soils. This prompted a lively exchange with the Soviet scientists. While there were disagreements between Marbut and Soviet scientists, their conceptual framework was that of Russian soil science, which Marbut was adapting to American conditions.[151] The Commission discussed international collaboration in investigating, collecting sample profiles, and mapping soils according to a common system based on Russian soil science.[152]

The Soviet scientists had prepared another exhibit of soil monoliths in boxes six-ten inches wide and three-twenty feet deep showing different soil profiles from around the Soviet Union. The variety of soils was demonstrated even better during the post-Congress excursion. The scientists traveled south and southeast from Moscow, enabling them to study – in large pits excavated specially – the profiles of "Podsolic soils" in the forested region, and "darker-colored soils" in grassland areas of the forest–steppe that lacked the alkaline accumulation characteristic of the "Chernozems of the true steppe." Further southeast, where the climate was semi-arid, they saw genuine Chernozems. While traveling southwest from Saratov on the Volga to Rostov-on-Don (covering similar ground to Dokuchaev's expeditions in the 1870s), they saw the transition from Chernozems to the Chestnut (*kashtanovye*) soils of the more arid steppe. Later, returning to Moscow through Ukraine, they studied more Chernozem profiles.[153] Marbut commented that participants in the Congress and excursion had "obtained a full picture of ... the entire system of soil science."[154] Charles Shaw of the University of California wrote that the Congress and excursion had been "highly successful." "The foundation upon which the Russian school of soil science is based," he observed, "is better understood, and the usefulness to

[151] C. F. Marbut, "Relation of Soil Type to the Environment," *Trudy II Mezhdunarodnogo Kongressa Pochvovedov: V Komissiia Klassifikatsiia, Geografiia i Kartografiia Pochv*, ed. L. I. Prasolov, D. G. Vilenskii (Moscow: 1932), pp. 1–6. On these disagreements, see A. G. McCall, "The Development of Soil Science," *AH* 5, no. 2 (1931), 53–5; NAL, Special Collections, Manuscript Collection 91, Charles Edwin Kellogg Papers (hereafter Kellogg Papers), Box 29, Folder 592, Kellogg correspondence, 1933–41, Kellogg, Chief, Division of Soil Survey to Dr. C. H. Spurway, Soils Section, Michigan State College of Agricuture, March 4, 1936; Krupenikov, *Istoriia pochvovedeniia*, p. 235.

[152] Prasolov and Vilenskii (eds.), *Trudy II Mezhdunarodnogo Kongressa Pochvovedov: V Komissiia*, "Minutes of the Vth Commission," xix–xxiii.

[153] Charles F. Shaw, "Geography at the Second International Congress of Soil Science," *GR* 21, no. 1 (1931), 146–8; "Excursion dans la Partie Européenne de l'URSS," in Vilenskii et al. (eds.) *Trudy II Mezhdunarodnogo Kongressa Pochvovedov*, pp. 99–146 (including map on p. 100).

[154] ARAN, f.487, op.1, d.27, l.159.

the rest of the world of the schemes of soil classification and their supporting theories can be better evaluated."[155] Several American delegates thanked the Soviet organizers. Joffe wrote that, not forgetting "some difficulties," every American delegate was enraptured with the Congress and excursion, and "believe me, this is not only my opinion."[156]

Marbut was very satisfied with what he learned from both Congresses and excursions. In 1927, after he returned home from the excursion around North America, he wrote: "I have gotten many suggestions (scientific) from the excursionists and I am convinced they have had some from the trip. It was fully worth while from the scientific standpoint and from the personal standpoints, pleasant for everyone."[157] Three years later, on his way home from the Soviet Union, Marbut wrote that the excursion had been arduous, but: "Scientifically it was worth it all. Every man with training sufficient to profit by it gained a great deal. The Russian scientists had planned the trip well and had made arrangements for showing the things that are significant."[158] He wrote to Iarilov: "I became acquainted with a part of the world which is very interesting for us at the present time, and widened my circle of acquaintances."[159] A few months later Marbut started to study Russian so that he could read more of the Russian soil scientists' work (see p. 125). The enthusiasm for Russian soil science among Marbut and other American scientists inspired by the Congresses in 1927 and 1930 stands in sharp contrast to the near indifference, except ironically from Whitney, prompted by the Russian soil exhibit at the World's Columbian Exposition in Chicago less than four decades earlier.

Conclusion

In an authoritative essay on the "Soils of the United States" published by the USDA in 1935, Curtis Marbut presented a revised version of his system for classifying American soils. It showed how he had adapted the Russian system, which he had learned about from Konstantin Glinka's book, to the different environmental conditions of North America. Marbut classified the soils of most of the Great Plains as "Chernozems," and

[155] Shaw, "Geography," 148.
[156] ARAN, f.487, op.1, d.29, l.19, Iakov Izrailovich Ioffe to Iarilov, November 12, 1930.
[157] Marbut Papers (C3720), Folder 32, Marbut to Louise and Leroy (Moomaw), July 24, 1927.
[158] Marbut Papers (C3720), Folder 50, Marbut, Berlin, to Louise, September, 1930.
[159] ARAN, f.487, op.1, d.16, ll.3–4; f.487, op.1, d.29, l.35, Marbut, Washington, DC, to Iarilov, n. d. (1930).

presented the "great soil groups," including Chernozem, on a map of the United States (See Map 6: Map of the Distribution of the Great Soil Groups, Curtis Marbut, 1935). His understanding of American soils was based also on his field work, during which he had "familiarized himself with at least general features of the soils of the whole country." He thanked his chief, Milton Whitney, for allowing him to undertake the "great amount travel" required.[160] In contrast to Whitney, who had limited experience outside the eastern United States, Marbut's extensive field work allowed him to appreciate how quite different soils had formed in quite different environmental conditions. He divided the soils of the United States into two broad groups: "pedalfers" east of the Great Plains, and "pedocals," including Chernozems, in the plains and to the west.[161] This resembled Glinka's division of soils into groups that formed in different moisture conditions. The bibliography of the 1935 edition of the "Soils of the United States" was a who's who of international soil science. It listed works by some older American soil scientists, such as Eugene W. Hilgard, Americans influenced by the Russian work, including George Coffey, Russian émigrés working in the United States, for example, Jacob Joffe, Constantine Nikiforoff, and Selman Waksman, and a number of Russian soil scientists, including Vasilii Dokuchaev, the founder of the innovative Russian school of soil science in the 1870s, as well as Glinka, and others who attended the Congress in Washington in 1927. The bibliography contained only one item by Whitney.[162]

Marbut's retirement was postponed twice, on orders from President Franklin Roosevelt, so he could complete the "Soils of the United States." In 1935, he crossed the Atlantic once more to attend the 3rd International Congress of Soil Science in Oxford, England. Afterwards he traveled on, across Europe, via Moscow and the Trans-Siberian railroad to China, where he was to advise the government on a soil survey. He was taken ill with pneumonia on the way, however, and died in Harbin, Manchuria, in August 1935 aged 72.[163]

Marbut's successor as head of the U.S. Soil Survey was Charles E. Kellogg (1902–80). He was born, brought up, and educated in Michigan. He moved to Fargo, North Dakota, in the Great Plains in 1930 to work as assistant professor of soils at the State Agricultural College. Four years later, he moved to Washington, DC, to work for the Soil Survey. In July 1934, he became acting head and formally replaced

[160] Marbut, "Soils of the United States," 2.
[161] Marbut, "Soils of the United States," 14.
[162] Marbut, "Soils of the United States," 3–4.
[163] Moomaw, "Curtis Fletcher Marbut," 26–7.

Marbut on his death the following year. Kellogg shared many of his predecessor's ideas about soils. His doctoral research had been on soil profiles.[164] In North Dakota, as well as teaching Russian soil science (see pp. 267–8), he cooperated with the Soil Survey in analyzing horizons in the profiles of Solonetz soils in the west of the state. His correspondence showed a detailed knowledge of the Russian-inspired methodology.[165]

In 1937, Kellogg revised Marbut's classification system of 1935 in a new edition of the Soil Survey's manual. He explained the methodology for collecting and analyzing soil horizons and profiles in the context of soil formation under the influence of natural vegetation, climate, relief, parent material and age. He explained the hierarchy in the classification system, from "soil types" and "phases" at the bottom, "soil series," defined as soils with similar genetic horizons, in the middle, and larger soil groups, such as Chernozems, at the top. "Texture," the main criterion for classifying soils in Whitney's original system, took up less than two of the manual's 136 pages.[166] The new edition of the manual replaced that of 1913, which Kellogg described as "hopelessly out of date."[167] Roy Simonson, who worked as a soil surveyor at this time, wrote that he found the new manual "easy to follow" as most it was "already part of my working knowledge from undergraduate courses taught by Kellogg and from working in soil surveys supervised by Thomas D. Rice."[168] Kellogg's revised classification system, which he connected with the pioneering Russian work under Dokuchaev, was also published in the USDA's Yearbook for 1938.[169] Thus, by the 1930s, the Russian-inspired approach had become central to the U.S. Soil Survey's methodology and classification system, and was being passed on by American scientists to a new generation of soil survey workers. While the U.S. Soil Survey continued to revise its methodology and classification system, by the late 1930s the "Russian Revolution" in

[164] William M. Johnson, "Charles E. Kellogg," *Soil Horizons* 21 (1980), 19–21; Helms, "Early Leaders," 43–6.

[165] NARA CP, RG 54, Finding Aid A1, Entry 82, Soil Survey Division, General Records 1922–42, Box 1. General to Dr. Marbut, File: ND Experiment Station, Kellogg, 1933–34, Charles E. Kellogg, Proforma "Soil Profile," August 3, 1931; Kellogg to Horace G. Byers, Bureau of Soils, USDA, October 20, 1932; Kellogg to Byers, October 28, 1932.

[166] Charles E. Kellogg, "Soil Survey Manual," *USDA Miscellaneous Publication*, no. 274 (1937).

[167] Kellogg Papers, Box 29, Folder 592, Kellogg to Mr George Kent, May 13, 1936.

[168] Simonson, *Historical Highlights*, p. 37.

[169] Mark Baldwin, Charles E. Kellogg, and James Thorp, "Soil Classification," *Soils and Men, Yearbook of Agriculture 1938* (Washington, DC: USDA, 1938), pp. 979–1001; The Soil Survey Division, "Soils of the United States," in ibid., pp. 1019–161. On Kellogg's revisions to Marbut's classification, see Paton and Humphreys, "A Critical Evaluation: Part 1," 257–67; Paton and Humphreys, "A Critical Evaluation: Part II: The Pragmatism of Charles Kellogg," 268–76.

U.S. government soil science, initiated by Marbut, was being consolidated by Marbut's successor.

Towards the end of his life, Marbut wrote the introduction to a book on soil science by the Russian–Jewish émigré Jacob Joffe that drew heavily on "the wealth of material . . . in the contributions of the Russian pedologists [soil scientists]."[170] Marbut asserted:

> Every American pedologist should familiarize himself with at least the broad lines of the history of the development of pedological science in Russia. Russian work . . . determined a definite relationship between the soil and the environment in which the soil is found, thus showing the soil to be related, on the plane of development, to biological bodies and not wholly physical.
>
> At the time when western Europe was still engaged in the futile assertion that the soil . . . is dominated in its general features by the materials out of which it has been built, the Russian workers had already shown that the soil is the product of process rather than of material and is, therefore, a developing body rather than a static body. They allied the soil to life rather than to death.

Marbut explained how, as a result of geographical differences between Eurasia and North America that were caused by the Rocky Mountains (see p. 201), the Russians' conclusions on the soils of their grasslands could not be applied directly to those of the United States. Nevertheless, he agreed with the "fundamental principles of the science of the soil," for which "the world is indebted to Russian scientists."[171]

Others were equally fulsome in their praise of Marbut. Jacob Lipman (another Russian–Jewish émigré) wrote:

> The work of Marbut stands out as one of the distinct contributions made by American workers. He was among the first of the Western workers to appreciate the new ideas about soils. Equipped with the sciences of geology, mineralogy, and geography, Marbut quickly orientated himself to the complexities of the Russian system of soil science and adopted the methods in a comprehensive study of the soils of the United States.[172]

Lipman described Marbut's new soil classification as a "genetic system" patterned after Glinka's idea, and "a most important milestone in soil

[170] Jacob S. Joffe, *Pedology*, 2nd edition (New Brunswick, NJ: Pedology Publications, 1949)[1st published 1936], p. viii.

[171] Curtis F. Marbut, "Introduction," in Joffe, *Pedology*, pp. ix, x. There is a handwritten Russian translation of Marbut's introduction in the papers of Arsenii Iarilov. ARAN, f.582, op.1, d.130.

[172] J. G. Lipman, "A Quarter Century of Soil Science," *Journal of the American Society of Agronomy* 25 (1933), 15.

classification in the United States, ... [that] led to the extensive study of soil profiles in this country."[173]

Macy Lapham, whose career in the Soil Survey spanned Marbut's revolution, gave eloquent testimony to his achievement in his memoir. When Marbut was appointed to head the Soil Survey, some workers had been cool at the appointment of a geologist. But, Lapham wrote:

> As the years went by he became less of a geologist and more of a soil scientist. He proved himself to be a man who was not afraid to change his mind, even to acknowledging error if convinced he had been in the wrong. As he came under the influence of the Russian school of soil science, in which climate and vegetation are stressed in soil development, he completely reversed his attitude as to the dominating influence of parent geological materials on the character of soils.[174]

Relating a visit by Marbut to surveyors in the field:

> [he] brought us something of the new school of soil science ... that he had but recently gleaned from the works of Glinka and other Russian soil scientists. It was the beginning of modernization in soil science and of soil classification of this country; our eyes were opened to a new field in the study of soil development and morphology. It placed the soil survey upon a stable scientific basis and quickened the interest and perception of the men in the field.

He described how, under Marbut's guidance, they came to understand the sequence of horizons in soil profiles: "We were a little slow to adopt the strange Russian terms..., but discovery and identification of some of the morphological features that had been pointed out to us was like playing with a new toy."[175]

Soviet soil scientists were as saddened by Marbut's death as Americans had been by Glinka's. The journal *Pochvovedenie* devoted a special issue to Marbut's memory.[176] Iarilov described him as a "Columbus for our time," who had "acclimatized" the soil science worked out by Dokuchaev and his collaborators to the United States.[177] Kellogg sent copies of the USDA's *Atlas of American Agriculture*, containing Marbut's essay on the "Soils of the United States," to several Soviet scientists. Leonid Prasolov thanked Kellogg: "We are deeply grateful to you for this sending, it is of a great value for our work." He continued: "Owing to a certain similarity between the soils of the territory of the USSR and those of the USA – the American

[173] Lipman, "A Quarter Century of Soil Science," 16. [174] Lapham, *Crisscross Trails*, pp. 94–5.
[175] Lapham, *Crisscross Trails*, pp. 132, 133. [176] *Pochvovedenie* 31, no. 4 (1936).
[177] A. Iarilov, "K. F. Marbut (nekrolog)," *Pochvovedenie* 31, no. 1 (1936), 142–5; ARAN, f.582, op.1, d.126, ll.6–21, Stat'ia A.A. Iarilova "Kolumb sovremennosti."

detailed soil maps are also very valuable for our work."[178] Prasolov's words conveyed the parallels between soils of the two countries, including the Chernozem of the steppes and Great Plains. They also showed how, as result of Marbut's "Russian Revolution," American and Soviet scientists could share their work on similar terms.

On the other hand, not all Americans scholars were enamored with Marbut's contribution or Russian soil science. Kansas historian James Malin noted that Dokuchaev's student Sibirtsev had cited Hilgard's work. He argued, accurately, that this was evidence that relations with Russian soil science were not just a "one-way proposition."[179] Hans Jenny, who held Hilgard's former chair at the University of California, Berkeley, championed Hilgard's cause in a book he wrote while engaging in a lengthy correspondence with Malin over the 1950s. Jenny emphasized the importance and innovative nature of Hilgard's work. He argued that Hilgard had independently and in advance of Dokuchaev identified many of the same issues in soil formation. He admitted that Hilgard "did not display the relations of soil and the genetic factors as neatly and as philosophically as the Russian scholars." And further that Hilgard "was not a system-making thinker" and was "suspicious of blueprints for nature." Nevertheless, Jenny argued that Hilgard and Dokuchaev were "the twin fountainheads of modern pedology [soil science]."[180] Hilgard was aware of the originality of his work and why he had come to realize the importance of climate. In November 1913 he wrote: "no one before me had systematically investigated the influence of climate upon soil formation . . . Having worked for 18 years in the extremely humid climates of the Cotton States [including Mississippi], I was forcibly struck with the contrast of conditions here [in California]."[181] However, Hilgard was in correspondence with some Russian scientists (but not Dokuchaev) and there was no rivalry between them (see p. 221). While Marbut had come across Hilgard's work when he was a master's student at Harvard, it was Glinka's book published in German in 1914 that converted him to

[178] ARAN, f.687, op.4, d.55, ll.2, 3, Prasolov to Kellogg, February 12, 1936. See also Iu. A. Liverovskii, "Atlas Amerikanskikh pochv K. F. Marbuta i ego znachenie dlia pochvovedeniia," *Pochvovedenie*, no. 9 (1944), 410–18.

[179] James C. Malin, *The Grasslands of North America: Prolegoma to Its History* (Lawrence, KS: privately published, 1947), p. 219.

[180] Hans Jenny, *E. W. Hilgard and the Birth of Modern Soil Science* (Pisa: Collana Della Rivista "Agrochimica: 1961), p. 77 and passim.

[181] Hilgard Papers, Hilgard, E. W.: Outgoing Letters, Box 3, Hilgard to nephew, November 26, 1913. See also Eugene W. Hilgard, "A Report on the Relations of Soil to Climate," *US Department of Agriculture. Weather Bureau Bulletin*, no. 3 (1892); Hilgard, *Soils: Their Formation, Properties, Composition, and Relations to Climate and Plant Growth in the Humid and Arid Regions* (New York and London: Macmillan, 1906).

new ways of conceptualizing soils. Thus, it was the Russian work that informed the "Revolution" Marbut led in U.S. government soil science.

The Soil Survey was still in transition in the 1930s when it, and the new Soil Conservation Service under Hugh H. Bennett, tried to deal with the Dust Bowl in the Great Plains. Revised soil surveys of the region were still in progress.[182] When the Federal Writers' Project gathered information about soils in Kansas in the mid 1930s, all they had available were surveys conducted according to the old methods that emphasized texture and geological origins.[183] One of the measures to address the Dust Bowl was planting shelterbelts of trees in the Great Plains to reduce the force of the wind. Cultivating trees in semi-arid grasslands, as Russian forestry scientists well knew from a century's experience, is difficult as many tree species cannot cope with the alkaline layer in the B horizon of the soils that form in such grasslands. American foresters knew they needed to identify areas with suitable soils for planting trees, but some were also aware that the surveys they needed did not yet exist. In May 1935, H. D. Cochrane, the Shelterbelt Project's public relations chief, wrote to the administrative director, Paul Roberts, of the necessity for: "More thorough soil classification" of the region designated for tree planting.[184]

Although revising the soil surveys would take many years, in 1930 Raphael Zon congratulated Marbut on the award of the Cullum Medal by the American Geographical Society:

> I cannot begin to tell you how pleased I was ... I know of no one who deserves this honor as much as you do. ... it is only the exceptional man who can change the outlook in attitude of a whole science. You certainly have done for American soil science what Dokuchaev and Glinka have done for Russian soil science.[185]

Zon played a similar role in the transmission of Russian forestry science to the United States.

[182] See Arend, *Russlands Bodenkunde*, pp. 179–206; Douglas Helms, "Hugh Hammond Bennett and the Creation of the Soil Erosion Service," *Journal of Soil and Water Conservation* 64 (2009), 68–74; Kenneth M. Sylvester and Eric S. A. Ripley, "Revising the Dust Bowl: High above the Kansas Grasslands," *EH* 17 (2012), 603–33, esp. 615.

[183] WSUL SCUA, Federal Writers' Project, Kansas, Box 4, File 13 Harper County, Arthur Foster, Geology. Soils and Sub-soils; Box 5, File 8 Harvey County, Rees E. Thomas, Soils and Sub-soils; Box 10, File 1, Reno County, Katherine Sturgeon, General Description of Reno County. Rock Formation, 1936.

[184] NARA KC, RG114, Entry 187 Dept of Agriculture. Forest Service. Region 2, General Records 1934–41, Box 5, Folder Inspection, South Dakota, Over three years, H. D. Cochran, Chief of Public Relations, Memorandum for Mr Roberts, May 20, 1935.

[185] Zon Papers, Box 6, Folder 8, Zon to Marbut, March 3, 1930.

CHAPTER 7

Shelterbelts I

Introduction

On July 11, 1934, as drought and dust storms enveloped the Great Plains, President Franklin Roosevelt signed an executive order designating $15,000,000 from emergency relief funds "to the Secretary of Agriculture for the planting of forest protection strips in the Plains region as a means of ameliorating drought conditions." The order was distributed on July 21.[1] The press release, under the names of Secretary Henry A. Wallace and Chief Forester Ferdinand A. Silcox, explained: "This will be the largest project ever undertaken in this country to modify climate and other agricultural conditions in an area that is now constantly harassed by winds and drought." The aims of the Shelterbelt Project were not to "change all the forces of the weather," which "man" could not do, but to "ameliorate the effects of the weather on a large scale" by decreasing the "surface velocity of the wind" so that the soil would be "held in place" and soil moisture conserved. The project was "based upon the long-time experience of several European countries," including "Russia," and on expertise in the U.S. Forest Service, in particular at the Lakes States Forest Experiment Station in St. Paul, Minnesota and its branch station in the plains at Denbigh, North Dakota.[2]

On July 22, 1934, *The New York Times* reported: "President Roosevelt gave the signal today for the beginning in the Great Plains area . . . of the

[1] Copies of the order (later revoked) are in NARA KC, RG 114, Entry 187 Dept of Agriculture. Forest Service. Region 2, General Records 1934–41, Box 6, Folder D-Legislation, Federal, Over Three Years, Executive Order (no. 6793) Allocating Funds from the Appropriation to meet the emergency and necessity for relief in stricken agricultural areas. Franklin D. Roosevelt, The White House, July 11, 1934; and NSHS SAMD, RG 3941.AM: Paul Henley Roberts Papers, 1891–1971 (hereafter Roberts Papers), Box 2, Folder 10, Raymond Marsh to Paul Roberts, February 2, 1967 (documents attached).

[2] E. L. Perry, *History of the Prairie States Forestry Project* (Washington, DC: U.S. Forest Service, 1942), pp. 13–16.

largest reforestation project ever undertaken outside Soviet Russia – an experiment in climate control to combat the ravages of drought."[3] A week later, a lengthy article in the same paper drew parallels with the Russian steppes: "Such conditions ... which have prevailed this year remind observers of the 'black storms' which at times ravage the Ukraine black-earth belt of Russia [sic], when dust laden clouds darken the sky." The article continued: "Prairie planting [of trees] in Russia during the past seventy years has given evidence of tangible effects on atmospheric humidity and precipitation. The Russian plantings and ... existing windbreaks throughout Kansas and Nebraska show markedly improved conditions for crop growth."[4] The article's byline named Chief Forester Silcox, but it was written by Raphael Zon, the director of the Lakes States Forest Experiment Station. Zon had drawn on Russian experience in the steppes to convince the president that planting shelterbelts in the Great Plains would mitigate the drought.[5] Zon was a Russian–Jewish immigrant and was familiar with Russian forestry research.[6]

The reference to "black storms" in Ukraine almost certainly came from an article with that title published in a Soviet scientific journal in 1930. After describing dust storms in the steppes, the author wrote about experience in Ukraine of "planting trees as protection for crops" to combat the "injurious effects of the winds."[7] The article was one of several Russian and Soviet publications on steppe forestry and shelterbelts that were located in the Library of Congress by a team of translators working for the U.S. Forest Service in late 1933 and early 1934. The publications were used, under Zon's direction, in preparing a technical bulletin for the American Shelterbelt Project.[8]

The executive order, Zon's role, and the significance of Russian experience in the steppes were known in American forestry circles before the order was published. On July 14, 1934, the editor of *American Forests* wrote to Zon:

[3] "Tree Belt in West to Fight Droughts," *NYT*, July 22, 1934.

[4] F. A. Silcox, "To Insure against Drought, a Vast Plan Takes Shape," *NYT*, July 29, 1934.

[5] On July 3, 1934, Edward Munns, Chief of the Division of Silvics in the Forest Service, wrote to Zon that Silcox wanted to release: "Your article on the Shelterbelt ... to the press." MHS, P1237, Raphael Zon Papers, 1887–1957 (hereafter Zon Papers), Box 9, Folder 1, Ed [E. N. Munns] to Raphael Zon, July 3, 1934.

[6] On Zon, see pp. 305–10.

[7] S. O. Vorob'ev, "Chernye buri na Ukraine," *Trudy po sel'sko-khoziaistvennoi meteorologii* 21, no. 7 (1930), 268–80.

[8] The periodical containing Vorob'ev's article is held by the Library of Congress at QC989.R49 T7. On the translation project and the technical bulletin, see pp. 321, 329–32.

> In view of Roosevelt's shelter belt project, which you were so instrumental in putting in concrete form, I would like to run a shelter belt article in our September number, presenting the merits of the project, the conditions to be met, results shown by shelter belt plantings in this country and abroad, notably Russia, which you used in your report, etc.[9]

According to Edgar L. Perry, who in 1942 wrote an official history of the Shelterbelt Project, the press initially greeted it with a mixture of "enthusiasm," especially in the Great Plains, some "pretty wild stories" based on imagination and the meager data in the press release, but also some skepticism and concerns over the cost.[10] As we shall see, there was considerable skepticism among some foresters who doubted the feasibility of planting trees in a region where few grew naturally and the supposed benefits in modifying the climate. The executive order of July 11 was revoked as a result of legal objections to the president's proposed use of emergency relief funds for a longer-term project. The project never received congressional support. The Forest Service kept it going, with presidential support, but it subsisted on smaller annual allocations from emergency funds and New Deal job-creation programs, in particular the Works Progress Administration (WPA). In the summer of 1934, a budget of $1,000,000 was set for the first year. The total sum spent never quite reached the $15,000,000 originally proposed. The project was ended in 1942, when planting shelterbelts was transferred to the Soil Conservation Service.[11] In attempt to move beyond the controversy, in 1935, the "Shelterbelt Project" was renamed the "Prairie States Forestry Project." The original name continued in informal use and will be used here.[12]

The Shelterbelt Project is another example of the "Russian roots" of Great Plains agriculture. In the summer of 1934, as the Dust Bowl worsened, President Roosevelt had allocated substantial funds to the risky undertaking of planting trees in the Great Plains. He had been advised by the Chief Forester and Forest Service scientists, including Zon, that

[9] Zon Papers, Box 9, Folder 1, Ovid Butler to Zon, July 14, 1934. Zon declined the invitation. Zon to Butler, July 18, 1934.

[10] Perry, *History*, pp. 16–19.

[11] Perry, *History*, pp. 12–13, 54–7. There had been tensions between the Soil Conservation Service and Prairies States Forestry Project, both of which were planting shelterbelts. In 1940 Perry commented on the failure of compulsion to prevent tensions: "By threat of punishment you can compel two small boys to quit tearing down each other's snow man, but only by making them believe that they can have a bigger and better snow man can you compel them to pitch in and build one together." NARA KC, RG 114, Entry 187, Box 5, Folder Inspection, South Dakota, 1940, Perry Memorandum for Director, July 26, 1940.

[12] See Wilmon H. Droze, *Trees, Prairies, and People: Tree Planting in the Plains States* (Denton, TX: Texas Woman's University, 1977), p. xxii (Introduction by Paul Roberts).

Russian experience in the steppes had shown that such belts of trees could moderate the region's climate. Later in the year, however, critics noted that the project had been launched "without any apparent effort to ascertain the views either of the profession of forestry or of the scientists in ... the national and state governments who had been experimenting for years with shelterbelts in this region."[13] Nor did the president seek the advice of Soviet forestry scientists.

Evaluating the importance of Russian and Soviet experience in the American Shelterbelt Project is complicated for two main reasons. First, accounts of the project were distorted by the competing egos of Zon, the project's technical director, and Paul Roberts, its administrative director. The second reason is the difficulty in locating the primary sources. Zon and Roberts presented their own, contradictory, versions of the project's origins and history. This is important here as Zon was a major, but not the only, conduit for Russian forestry expertise to the United States, while Roberts was more interested in American experience. The two men fell out shortly after their appointments as co-directors in the summer of 1934. Thereafter, Zon was keen to promote his role and the significance of Russian expertise. As well as championing both in the article he wrote for *The New York Times*, he strongly advocated the project at the time of its inception, when it came under attack later in 1934, and over subsequent years.[14] A passage in the memoir of Gifford Pinchot, who was Zon's mentor, reads: "The great Shelter Belt Plan begun under President Franklin D. Roosevelt [was] so brilliantly suggested and successfully directed by Raphael Zon."[15] However, Zon had helped Pinchot complete his memoir when he was seriously ill and edited it after Pinchot's death. Forest historian Char Miller noted: "Some in the Forest Service who watched the then-retired Zon toil over his editorial work ... wondered just whose perspectives the completed manuscript reflected." This mattered since Pinchot was an influential figure. A prominent conservationist, he had served as head of U.S. government forestry (1898–1910) and for two terms

[13] H. H. Chapman, "Digest of Opinions Received on the Shelterbelt Project," *JoF* 32 (1934), 956.

[14] Droze, *Trees*, pp. 93–4; Raphael Zon, "The Plains Shelterbelt Project," *Minnesota Conservationist*, no. 18 (November 1934), 4–5, 17; Raphael Zon, "Shelterbelts: Futile Dream or Workable Plan," *Science*, NS, 81, no. 2104 (April 26, 1935), 391–4; Zon Papers, Box 9, Folder 1, Raphael Zon to Anna Zon, November 9, 1934; Box 10, Folder 7, "How Shelterbelts Temper the Hot Winds of Kansas," talk by Zon before Kansas club of Minneapolis, January 30, 1939; "'Last Laugh' goes to Zon's trees," *St. Paul Sunday Pioneer Press*, Magazine Section, October 16, 1938; Roberts Papers, Box 2, Folder 9, Roberts to Ed Munns, June 8, 1965.

[15] Gifford Pinchot, *Breaking New Ground* (Island Press: Washington, DC, 1998) [1st edition 1947], 238.

as governor of Pennsylvania (1923–7, 1931–5).[16] Zon's self-promotion was quite successful and informed perceptions that he played the key role in the Shelterbelt Project.

Zon's significance in the Shelterbelt Project, and that of Russian expertise, were questioned by Roberts, who directed the project from its administrative headquarters in Lincoln, Nebraska from 1934 to 1942. In his retirement, Roberts was frustrated to discover that Zon was being given credit that he believed belonged to him and other foresters. He was dismayed when, in 1957, he read Pinchot's memoir, and more dismayed when he learned that Zon had edited it.[17] Roberts' pride was wounded again in 1965. A cousin asked for his recollections for a regional history project, but he was disappointed to learn that she was really interested in Zon's contribution, "since he is such a primary figure in the history of forestry," and in "what influence the earlier Russian shelter belts had on ours."[18] Roberts resolved to write his own history of the Shelterbelt Project. He consulted former colleagues for their recollections. One of his main informants was Edward N. Munns, who, as Chief of the Division of Silvics in the Forest Service in Washington, DC, had been deeply involved in the project's origins.[19] Roberts worked on his history until his death in 1971. During his final illness he handed copies of his papers to Wilmon H. Droze, a historian at the University of Texas at Arlington, who completed the task with the support of the Forest Service. Droze's book, which reflects Roberts' views, appeared in 1977.[20]

[16] Char Miller. *Gifford Pinchot and the Making of Modern Environmentalism* (Washington, DC: Island Press, 2001), pp. 363–5 and passim; Harold K. Steen (ed.), *The Conservation Diaries of Gifford Pinchot* (Durham, NC: Forest History Society, 2001), pp. 208–21; Zon Papers, Box 13, Folder 1, Zon to Elmer McClain, May 17, 1945; Pinchot to Zon, May 24, 1945; Pinchot to Zon, May 27, 1945; Folder 3, more correspondence between Pinchot and Zon; Folder 4, correspondence between Zon and Pinchot's widow regarding the book.

[17] Roberts Papers, Box 2, Folder 6, Roberts to Lee [Kneipp], [1957?]; Folder 8, Roberts to Edward Munns, May 20, 1963. Roberts wrote a memorandum correcting the account in Pinchot's memoir and giving due credit to himself and Carlos Bates. Folder 8, Roberts to Arthur Carhart, Conservation Library Center of N. America, Denver, July 2, 1963; Memorandum on Genesis of Shelterbelt Project, July 2, 1963.

[18] Roberts Papers, Box 2, Folder 9, A.[melia] R.[oberts] Fry to Roberts, January 29, 1965; Roberts to Fry, April 13, 1965; Fry to Roberts, April 20, 1965.

[19] Roberts Papers, Box 2, Folder 9, Roberts to John [Emerson?] and Hank [Lobenstein?], June 7, 1965. Correspondence between Roberts and Munns is cited in this and the following chapters. On Munns' post, see T. W. Norcross, *Handbook of Erosion Control Engineering on the National Forests* (Washington, DC: USDA Forest Service, 1936), p. iii.

[20] Roberts initially considered Droze as a competitor and was unhelpful, but relented during his final illness. Roberts Papers, Box 2, Folder 10, Droze to Paul Roberts, August 15, 1967; Paul Roberts to Droze, August 22, 1967; Box 3, Folder 12, Paul Roberts to Droze, April 17, 1970; Mrs. Roberts to Droze, March 25, 1970, Droze to Mrs. Roberts, April 8, 1970, USDA Forest Service memo on Shelterbelt History to be written by Droze; Droze, *Trees*, pp. xi–xii.

A second complication in assessing the role of Russian expertise is difficulty locating all the primary sources on the Shelterbelt Project. They are scattered around several archives, not all have been catalogued, and some were removed. Roberts explained the problem back in 1937: "Prior to ... the Spring of 1936, the Project was operated without specific divisional responsibility in the Washington office. As a result of this, much valuable information and correspondence regarding the Project ... has never been assembled in the Washington office, but is scattered under various designations in the files."[21] He came up against the problem again in the 1960s, when he was researching his history. Working from his retirement home in Arizona, he had help from former colleagues in Washington, DC, and contacts in the federal government, who located some of the documents.[22] Roberts' enquiries of the National Archives in Washington, DC, revealed the existence of his headquarters office files from 1934 to 1941. He learned that some documents on the project's origins had been sent to the Franklin D. Roosevelt Library in Hyde Park, New York, which had published a selection.[23] These were not the only documents taken from the National Archives. According to Munns, Melville Cohee of the Soil Conservation Service removed the early correspondence "as he wanted to get the autograph of F.D.R."[24] Copies of the documents Roberts was able to get hold of and his correspondence with former colleagues about their recollections are preserved at the Nebraska State Historical Society in Lincoln, and provide important sources for these two chapters.

Roberts did not visit the National Archives in Washington, DC, and Droze, who did, could not find the key documents on the project's origins.[25] Joel Orth, a historian who made an exhaustive attempt to find sources, noted: "A search of ... National Archives II, College Park, Maryland [where the USDA files had been relocated] revealed little

[21] NARA KC, RG 114, Entry 187, Box 6, Folder Inspection, General, over three years, Roberts to Chief, Forest Service, February 3, 1937.

[22] For example, Roberts Papers, Box 2, Folder 10, Raymond Marsh to Roberts, February 2, 1967.

[23] Roberts Papers, Box 3, Folder 11, Richard S. Maxwell, National Archives and Records Service, to Roberts, April 4, 1968; Elizabeth B. Drewry, Franklin D. Roosevelt Library, Hyde Park, NY, to Roberts, April 24, 1968; Edgar B. Nixon (ed.), *Franklin D. Roosevelt and Conservation, 1911–1945*, 2 vols. (Hyde Park, NY: NARA, 1957).

[24] Roberts Papers, Box 2, Folder 8, Munns to Roberts, 28 May 1963. See also Peggy James, "Conservation Pioneer Melville H. Cohee," *Journal of Soil and Water Conservation* 63, no. 4 (2008),107A–8A.

[25] For the documents Droze did find in the National Archives, see Droze, *Trees*, pp. 262, 268, 271, 281, 284, 288.

additional information."[26] My attempts to locate materials on the origins of the Shelterbelt in the National Archives in College Park, even with the help of archivists who took me into the stacks, also revealed little. Orth discovered that the records of Roberts' headquarters office had been temporarily misplaced, but some were preserved in the National Archives in Kansas City, Missouri,[27] which I have consulted. To add to the complications, some of the records of Roberts' headquarters are kept, uncataloged, in a shed at the National Agroforestry Center in Lincoln, Nebraska.[28] To counterbalance Roberts' version of the story, I used Zon's papers (although not all of those concerning the project have survived) in the Minnesota Historical Society in St. Paul and Pinchot's papers in the Library of Congress.

There are several accounts of the project's history, starting with Perry's official report of 1942, and a few academic studies.[29] The aim of this pair of chapters is to assess the extent to which the Shelterbelt Project drew on Russian experience of steppe forestry. They will consider the role of Zon and others in transmitting Russian expertise to the Great Plains. In doing these things, they will mediate between Roberts' and Zon's representations of their roles. This first chapter on shelterbelts will consider: Russian experience of forestry and shelterbelts in the steppes; American forestry in the Great Plains before the Shelterbelt Project; and the transfer of relevant Russian experience, and trees, to the United States. The following chapter will analyze the immediate origins of the project and its launch in 1934, planting the shelterbelt, and debates about the project.

Russian Forestry and Shelterbelts in the Steppes to the 1930s

Planting shelterbelts in the steppes began in the early nineteenth century, while Russian experience of cultivating trees in the region dated back to

[26] Joel Jason Orth, "The Conservation Landscape: Trees and Nature on the Great Plains," (PhD dissertation, Iowa State University, 2004), 183, n. 6.

[27] NARA KC, RG 114, Entry 187, 114.9.2 Records of the Prairie States Forestry Project.

[28] Joel Orth, "The Shelterbelt Project: Cooperative Conservation in 1930s America," *AH* 81 (2007), 354, n. 6. Orth made use of them. I visited the National Agroforestry Center in the spring of 2007, but had insufficient time to more than glance at them.

[29] Perry, *History*; Droze, *Trees*; Orth, "Conservation Landscape"; Thomas R. Wessel, "Roosevelt and the Great Plains Shelterbelt," *Great Plains Journal* 8, no. 2 (1969), 57–74; Allan J. Soffar, "The Forest Shelterbelt Project, 1934–1944," *Journal of the West* 14 (1975), 95–107; Robert Gardner, "Trees as Technology: Planting Shelterbelts on the Great Plains." *History and Technology* 25 (December 2009), 325–41; Sarah Thomas Karle and David Karle, *Conserving the Dust Bowl: The New Deal's Prairie States Forestry Project* (Baton Rouge: Louisiana State University Press, 2017).

the late seventeenth century.[30] When migrants moved beyond the humid, forested heartland of Russia around Moscow to the semi-arid steppes, they encountered a vast grassland that was largely, but not totally, devoid of trees. They made isolated attempts to plant trees in the steppe over the eighteenth century. There were more systematic efforts, by the Russian authorities, estate owners, and agricultural settlers, from the beginning of the nineteenth century. Among the settlers were Slavs and foreigners from central and southeastern Europe. The Russian government required foreign settlers, many of whom were Germanic and included Mennonites, to plant trees on the land it granted them.

There were several motives behind planting trees in the steppes. The first was practical: Many of the settlers came from forested environments and most were accustomed to using wood for many purposes, such as construction, making farm implements and items of everyday use, and as fuel. A second motive was aesthetic: the authorities and settler population desired to create a wooded landscape that resembled their homelands. Over the eighteenth and nineteenth centuries, moreover, naturalists around the world, including the Russian Empire, argued that trees and forests bestowed benefits on regions and the people who lived in them. Forests, it was widely believed, moderated the climate, attracted rain clouds and increased precipitation, provided shelter from the wind, conserved moisture in the soil, stopped snow blowing away, and so retained melt water. Forests were also believed to regulate water levels in rivers and reduce flooding. Europeans considered forested environments to be "healthy" and deforested lands liable to desiccation and desertification. In Russia, beliefs that planting trees "improved" the environment and drawing up plans for afforestation were most common during and after the droughts that periodically afflicted the steppes – leading to dust storms, crop failures, and famines – such as those of 1832–3 and 1891–2.

It was difficult to grow trees in conditions in which grasses were the predominant natural vegetation. In the steppes, trees grow naturally in

[30] The first part of this discussion is based, except where other references have been provided, on David Moon, *The Plough that Broke the Steppes: Agriculture and Environment on Russia's Grasslands, 1700–1914* (Oxford: Oxford University Press, 2013), pp. 173–205; Moon, "Planting Trees in Unsuitable Places: Steppe Forestry in the Russian Empire, 1696–1850," in *Eurasian Environments: Nature and Ecology in Imperial Russian and Soviet History*, ed. Nicholas Breyfogle (Pittsburgh, PA: University of Pittsburgh Press, 2018), pp. 23–42. See also M. V. Loskutova and A. A. Fedotova, "Forests, Climate, and the Rise of Scientific Forestry in Russia: From Local Knowledge and Natural History to Modern Experiments (1840s–early 1890s)," in *New Perspectives on the History of Life Sciences and Agriculture*, ed. D. Phillips and S. Kingsland (Cham, Switzerland: Springer Verlag, 2015), pp. 113–37.

river valleys and ravines, where there is more moisture, shelter from the wind, and different types of soil. Throughout the steppes, trees also grow naturally on sandy soils, which are porous and allow the roots access to water. Over the decades, settlers, naturalists, agricultural specialists, and scientific foresters worked out why it was so hard to cultivate trees in the rest of the steppe region. They identified the continental climate, with sharp swings between hot and cold temperatures, periodic droughts, high winds, wild fires, and damage by grazing animals and pests, including rodents and insects. As the incoming agricultural settlers displaced the indigenous pastoralists, conditions became more favorable for trees. This was because the nomadic pastoralists regularly burned the steppe, killing off young trees, to encourage the growth of fresh grasses for their livestock, which then grazed on and damaged saplings. But, the difficulties in growing trees in the steppes persisted. Some foresters realized there was another inhibiting factor. They found that, after a number of years, some trees that had survived other hazards started to die off. In the first half of the nineteenth century, several specialists argued that the trees' roots had entered a layer of salt (i.e. alkaline) below the surface of the soils (but not sandy soils) of much of the steppes. Later in the nineteenth century, Russian soil scientists analyzing cross sections, or profiles, of soils identified an alkaline layer in the B horizon of the soils of much of the steppe region. The Russian innovation in understanding how soils formed enabled scientists to work out that, due to the low rainfall and high evaporation, there was insufficient moisture to leach out the salts that created the alkaline layer. This contrasted with the more humid and cooler regions to the north, where higher rainfall and lower evaporation meant there was sufficient water to wash out the salts, allowing the formation of acidic soils in which trees thrived (See Chapters 5 and 6).

Some people, including agricultural specialists, however, refused to accept that there were natural barriers to tree growth in the steppes, and persisted in trying to increase tree cover in the region. There were some serious failures. For example, from the 1830s, there were sustained attempts by the authorities to plant trees in the arid steppes in the south of the Don Cossack Territory and nearby lands of nomadic Kalmyks. The attempts failed and were abandoned by the 1850s. Meanwhile, the basis for scientific steppe forestry was being laid by some estate owners and Mennonite farmers. They used labor-intensive methods to prepare the soil for planting, including plowing or digging deeply, which partly removed the alkaline layer, and to protect saplings from pests, weeds, drought, and winds. The techniques were refined at state forestry plantations (the equivalent of American forest experiment stations) at Velikii Anadol' and Berdiansk that were established in the 1840s. Steppe foresters devised techniques for planting and cultivating trees to increase their survival rates, identified species of trees that could grow in the conditions, in particular the droughts and alkaline subsoils, and worked out combinations of trees and shrubs that worked well together.

Over the second half of the nineteenth century, hopes for large-scale afforestation of the steppes receded in the face of the difficulties and expense. More extravagant hopes to moderate the climate and increase rainfall by planting trees also receded, but did not disappear entirely, after studies questioned their likelihood. Instead, forestry specialists and informed government officials advocated planting trees for more particular purposes, such as inhibiting erosion by water and wind. Trees and shrubs were planted along the sides of ravines. Areas of loose sand were planted with trees and shrubs to bind them and prevent the wind blowing the sand over surrounding land. A third particular purpose for planting trees and shrubs in the steppes was to shelter fields from the wind. To this end, trees and shrubs were planted in shelterbelts on the windward side of arable fields and around settlements. Pioneers of shelterbelt planting in the steppes in the early nineteenth century included estate owners in different parts of the region and Mennonites in the Molotschna colony, who used the technique as part of a range of methods to conserve moisture following the drought and famine of the early 1830s.[31]

A few decades later, the drought, dust storms, crop failure, and famine of 1891–2 prompted studies into ways of making steppe agriculture more resilient. The techniques considered included shelterbelts. In 1892, the Southern Russian Agricultural Society commissioned an agronomist, A. A. Bychikhin of the University of New Russia in Odessa, to investigate the influence of such belts planted by estate owner A. A. de Karrier at Kamenovka in Kherson province in the 1870s. Bychikhin reported that they had reduced the force of the wind, increased moisture in the soil, which in turn had led to higher crop yields that had repaid the cost of planting the belts. He reported also that de Karrier's estate had been spared the impact of the drought of 1891.[32] In the 1890s, more systematic and controlled experiments into the influence of shelterbelts were carried out as part of a "Special Expedition" to the steppes led by soil scientist Vasilii

[31] See David Moon, "Cultivating the Steppe: The Origins of Mennonite Farming Practices in the Russian Empire," *Journal of Mennonite Studies* 35 (2017), 251–78.

[32] A. A. Bychikhin, *Znachenie zashchitnykh nasazhdenii dlia stepnoi polosy* (Odessa, 1893).

Dokuchaev and funded by the Forestry Department of the Ministry of State Domains. The expedition's scientists were based on three field, or experiment, stations, including Velikii Anadol' and Kamennaia Step' ("Stone Steppe" in English). They investigated the steppe environment, generating data for use as controls for their experiments into drought-resilient farming techniques.[33]

The scientist assigned to test shelterbelts was Georgii Vysotskii (1865–1940). He was from Chernigov province in the forest–steppe region in the north of today's Ukraine, and was educated at the Petrovskaia Agricultural Academy in Moscow. While primarily a forestry scientist, he had expertise in soil science and meteorology. Vysotskii was one of the major specialists on steppe forestry and shelterbelts. From the 1890s to the 1930s, he researched and wrote extensively on the practicalities of planting such belts and their influence on adjacent areas.[34] His work is worth summarizing, especially since he became involved in the controversy over the American Shelterbelt Project in 1935. Vysotskii explained the aims of shelterbelts as: protecting fields from the harmful effects of strong, dry winds; retaining snow so that the melt water would augment moisture in the soils for crops; and destroying the "dead zone" in the soil where water did not penetrate and alkaline salts harmful to trees accumulated. Over the years, he gathered data on the extent to which shelterbelts achieved these aims. His conclusions became firmer as the shelterbelts he was testing grew in height. He was well aware that conditions in the steppe region were harmful to trees. Following the drought of 1891, he observed widespread "drying out" of trees, especially those planted artificially, in the Velikii Anadol' plantation. Concerning the ability of shelterbelts to protect fields from the wind, he noted that on level land they had an impact on an area to a distance of ten times the height of the belt. But, even in areas where the soil and climatic conditions were not too harmful for trees, belts could reach a maximum height of only 35–42 feet, and thus protect land up to 420 feet away. He calculated that belts large enough to be effective would take up a significant area – fourteen percent – of the fields they sheltered. He noted also that, rather than moderating the climate, forests tended to intensify daily variations in temperature. From further tests he concluded that, contrary to expectations, shelterbelts dried out the soil underneath

[33] See David Moon, "The Environmental History of the Russian Steppes: Vasilii Dokuchaev and the Harvest Failure of 1891," *Transactions of the Royal Historical Society*, 6th series, 15 (2005), 149–74; Loskutova and Fedotova, "Forests, Climate, and Rise of Scientific Forestry," 131–2.

[34] See G. N. Vysotskii, "Georgii Nikolaevich Vysotskii i ego trudy (avtobiografiia)," in *G. N. Vysotskii, Izbrannye sochineniia*, 2 vols. (Moscow: AN SSSR, 1962), vol. 1, 11–32.

them and up to fifty feet away in the summer, with negative consequences on grain yields. Rather destroying the "dead zone" in the soil, moreover, shelterbelts increased concentrations of alkaline salts that were harmful to trees in the groundwater.

In 1924, Vysotskii concluded that shelterbelts had negative as well as positive consequences for crops, and that his findings "strongly shake faith in the all round benefits which supporters of planting shelterbelts attribute to them." He questioned the conclusions of individual specialists. He doubted the positive influences on crop yields reported by Bychikhin in 1892, noting that the higher yields were recorded near the middle of the estate, where land was normally cultivated more intensively than in outlying fields. He criticized a report by Soviet scientist G. M. Tumin on his findings at the Kamennaia Step' experiment station. Vysotskii argued that the effects Tumin attributed to shelterbelts were a result of the topography. More dismissively, he argued that the higher rainfall Tumin reported in sheltered fields was due to more effective rain gauges.[35]

Vysotskii's conclusions were more circumspect than those of many of his Russian and Soviet contemporaries. In the first years of the twentieth century, some specialists had been cautious about reaching premature conclusions from young, and low, shelterbelts.[36] Others were more positive. Nikolai Adamov, a meteorologist in Dokuchaev's "Special Expedition," concluded that rainfall was higher over forests than steppe. Several scientists who observed the effects of shelterbelts in the expedition's experiment stations, including Kamennaia Step', noted that even young belts retained large quantities of snow, which then irrigated adjacent fields with meltwater in the spring.[37] Vysotskii's research and

[35] G. N. Vysotskii, "Biologicheskie, pochvennye i fenologicheskie nabliudeniia v Veliko-Anadole (1892–93 gg.)," *Trudy opytnykh lesnichestv* 3 (1901), 1st pagn., 1–306; Vysotskii, "Lesnye kul'tury Mariupol'skogo Opytnogo Lesnichestva," *Trudy opytnykh lesnichestv* 3 (1901), 2nd pagn., 1–244; Vysotskii, "*O vzaimnykh sootnosheniiakh mezhdu lesnoiu rastitel'nost'iu i vlagoiu preimushchestvenno v iuzhno-russkikh stepei*, part 1 (Spb: Slovo, 1904); Vysotskii, "Lesnye kul'tury stepnykh opytnykh lesnichestv s 1893 po 1907 g.," *Trudy po lesnomu opytnomu delu v Rossii* 41 (1912), 1–557; Vysotskii, "Leso-vodnye ocherki," *Zapiski Belorusskogo Gosudarstvennogo Instituta Sel'skogo Khoziaistva* 3 (1924), 11–44. Vysotskii published a series of articles on soils in the steppe region, including the "dead zone" and tree growth, starting with G. N. Vysotskii, "Pochvennye zony Evropeiskoi Rossii v sviazi s solenosnost'iu gruntov kharakterom lesnoi rastitel'nosti," *Pochvevedenie* 1, no. 1 (1899),19–26.

[36] For example, A. Batsiev, "Vliianie lesnykh polos na silu i vlazhnost' vetra," *Trudy opytnykh lesnichestv 1905 god*, no. 3 (1905), 203–54.

[37] N. P. Adamov, "Meteorologicheskie nabliudeniia v otkrytnykh lesnichestvakh 1896–1899 godov," *Trudy opytnykh lesnichestv, nauchnyi otdel* 1 (1901), 106–7, 342–3; N. A. Mikhailov, "Nabliudeniia nad snezhnym pokrovom v sviazi s vlazhnost'iu pochvy zimy 1901/2 i 1902/3 gg. v raione Kamenno-Stepnogo opytnogo lesnichestva," *Trudy opytnykh lesnichestv 1905 god*, no. 3 (1905),

reluctance to give in to a "natural temptation" to proclaim the unqualified benefits of forests in the steppe earned him the respect of some foresters, for example Mikhail Tkachenko, who had visited the United States (see pp. 304–5). Vysotskii's conclusions were taken seriously by the Soviet government, which commissioned a critical assessment of data on shelterbelts and further research.[38]

The drought, crop failure, and famine in the Volga region in 1921–2 prompted renewed interest in shelterbelts and other measures to combat droughts. Over the 1920s and early 1930s, there were several studies on shelterbelts. Many drew on existing data rather than new research. Occasional reference was made to American experience. One author referred to the work of "Bets" (American forester Carlos Bates, see pp. 293–4), and there was discussion of North American tree species. The authors of the Soviet studies repeated existing advice that, when planting shelterbelts, attention needed to be paid to the soil, climate, species of trees and mixtures of species in the belts, the orientation of belts, and the spacing of trees inside them. Most of these Soviet studies were less cautious than Vysotskii's, paid less attention to or denied the negative consequences he had identified, and did not always present their data in full. The authors reached positive conclusions that shelterbelts could: assist in the struggle against droughts that were holding back Soviet agriculture; moderate microclimates in adjacent fields; weaken the force of the wind; retain snow; and reduce evaporation and thereby increase the humidity of the air and moisture content of the soil near the belts. Some argued that, over time, the additional moisture made the soil more fertile. Several writers, such as Tumin (who challenged Vysotskii's criticisms of his earlier work) asserted that shelterbelts increased precipitation, except in the summer. G. M. Leont'evskii stated in the conclusion to an article that the benefits of shelterbelts fourteen meters (forty-six feet) high could be felt over distances up to 500 meters (1,640 feet). This was three and a half times more than Vysotskii had calculated. Leont'evskii's conclusion also contradicted data in studies he had cited in the body of his article referring to belts only up to nine meters (thirty feet) high that had been shown to have an influence over a distance of 200 meters (656 feet).[39]

63–201; G. F. Morozov, "Vliianie zashchitnykh lesnykh polos na vlazhnost' pochvy okruzhaiushchego prostranstva," *Trudy opytnykh lesnichestv 1902 god*, no. 1 (1902), 244–6.

[38] M. E. Tkachenko, "Uspekhi lesovodstva za poslednee desiatiletie (prodolzhenie)," *Lesovod*, no. 3 (1928), 32, 33.

[39] See, for example, N. I. Sus, *Lesomelioratsiia i bor'ba s zasukhoi na Iugo-Vostoke* (Leningrad: Izdanie gos. nauchno-melioratsionnogo in-ta, 1926); Sus, "Polosnye lesnye posadki v dele bor'by s zasukhoi

An outcome of these positive, and not always well-founded, studies of the benefits of shelterbelts was that the second Soviet Five-Year Plan for economic development of 1933–7 included ambitious targets for planting trees, including shelterbelts, in the drought-prone Volga region.[40] The plan was promoted by publications on forestry that were propagandistic rather than scientific. They were full of assertions, but low on citations of the scientific literature, and lacked data. In a collection of essays of 1933 entitled *The Forest in the Service of the Construction of Socialism*, E. Rikman stated, without evidence, that shelterbelts could increase yields by 200–400 percent. In an article on "Forest in the Service of Agriculture," Z. Goviadin concluded: "The mission of the whole of society of our country and especially regions affected by drought is to mobilize to fulfill the tasks set by the government."[41]

Some of the American publicity surrounding the Shelterbelt Project in 1934 (see pp. 277–8) drew on the more positive Russian and Soviet assessments, including those of a propagandistic nature of the 1920s and early 1930s. Vysotskii's more cautious conclusions were not taken much into account by American foresters advocating shelterbelts.

American Forestry and Shelterbelts in the Great Plains to the 1930s

There was American experience of planting trees, including shelterbelts, in their semi-arid grasslands, long before the Shelterbelt Project.[42] Settling

i neurozhaiami," *Puti sel'skogo khoziaistva*, no. 2 (1926), 165–82; N. N. Stepanov, *Stepnoe lesorazvedenie, izdanie vtoroe, pererabotannoe i dopolnennoe* (Moscow: Novaia derevnia, 1927); A. A. Shapovalov, *Vliianie sostava nasazhdenii na razvitie drevesnykh porod v lesnykh polosakh kamennoi stepi* (Voronezh: Izd-vo "Kommuna," 1930); Vorob'ev, "Chernye buri na Ukraine"; E. A. Danilov, *Zashchitnye lesnye polosy* (Moscow–Leningrad: Gos izdat-vo sel-khoz. i kolkhozno-kooperativnoi literatury, 1931); N. P. Leont'evskii, "Rol' drevesnykh zashchitnykh polos v povyshenii urozhainosti," *Zhurnal geofiziki* 4, no. 1 (1934), 128–40 (for data and assertions on the distance at which belts were effective, see 129, 139). For references to American experience, see Leont'evskii, "Rol' drevesnykh zashchitnykh polos," 133, 135; Danilov, *Zashchitnye*, pp. 73, 84–6.

[40] N. M. Gorshein, Ia. D. Panfilov, Ia. T. Godunov, A. S. Barabanshchikov, and P. I. Kolomeitsev, *Agrolesomelioratsiia v oroshaemom zavol'zhe*, part 4 (Moscow: Lesnoe tekhnicheskoe izd-vo, 1934).

[41] E. Rikman, "Lesokul'turnaia rabota," in *Les – na sluzhbu sotsialisticheskomu stroitel'stvu: sbornik rabot kollektiva nauchnykh rabotnikov VHILAMI* [Vsesoiuznyi nauchno-issledovatel'skii lesokul'turnyi agrolesomeliorativnyi institut] (Moscow–Leningrad: Gos. Izdat-vo Kolkhoznoi i Sovkhoznoi literatury, 1933), p. 18; Z. Goviadin, "Les – na sluzhbu sel'skomu khoziaistvu," in *Les*, 40.

[42] Except where otherwise indicated, the next four paragraphs are based on Droze, *Trees*, pp. 3–47; W. Kollmorgen, "The Woodsman's Assaults on the Domain of the Cattleman," *AAAG* 59 (1969), 215–39; Willis Conner Sorensen, "The Kansas National Forest, 1905–1915," *KHQ* 35 (Winter 1969), 386–7; Brian Allen Drake, "Waving 'A Bough of Challenge' Forestry on the Kansas Grasslands, 1868–1915," *GPQ* 23 (2003), 19–34; Gardner, "Trees as Technology," 325–30; Joseph Giacomelli, "The Meaning of Uncertainty: Debating Climate Change in the Gilded-Age

and planting trees in the American plains in the second half of the nineteenth century resembled similar developments in the steppes of the Russian Empire that began several decades earlier. In the United States, when Euro-Americans moved from the humid, forested lands of the eastern states and parts of Europe to the Great Plains, they encountered a vast grassland with a semi-arid climate that was largely bereft of trees. For many settlers this was an unfamiliar environment from which they felt something – trees – was missing. Most were accustomed to using timber for buildings, fences, farm implements, furniture, firewood, and other purposes. Many missed the familiarity and comfort of woodlands. David Fairchild described his feelings as a boy after his family moved from Michigan to Kansas: "The occasional groves of cottonwoods, box elders, and maples were a poor substitute for the great forests of Michigan. But their scarcity increased our interest in each individual tree." He was more explicit when he described the northern plains as "treeless wastes."[43] There was much discussion in the United States of the ideas that were widespread around the globe at this time that planting trees would moderate the climate and increase rainfall. For a while these alleged benefits were pushed by the federal and state governments, railroad companies, and local boosters, which all had an interest in encouraging the settlement of the plains. Men with scientific knowledge, such as Bernhard E. Fernow, chief of the U.S. Division of Forestry from 1886 to 1898, and Charles E. Bessey, a botanist at the University of Nebraska, gave credibility to such beliefs.

Just as in the steppes, and for similar reasons, cultivating trees in the Great Plains proved difficult in an environment where few grew naturally outside river valleys, ravines, and on sandy soil. River valleys lined with cottonwood trees (a type of poplar) contrasted with the treeless grasslands on the high plains. Throughout the region, the low and unreliable rainfall, extremes of heat and cold, high winds, some types of soil, and livestock and pests all inhibited tree growth. The Plains Indians had hindered the growth of trees by burning the plains to promote fresh grasses for the bison they hunted, and the bison also grazed on and destroyed young saplings.

United States," *E and H* 24 (2018), 237–64; Charles A. Gillett, "Forests and Forestry in North Dakota," *JoF* 25, no. 1 (1927), 38–43; Julie Courtwright, "'When We First Come Here It All Looked Like Prairie Land Almost': Prairie Fire and Plains Settlement," *WHQ* 38 (2007), 157–79. On American knowledge of the alleged environmental benefits of trees, see George Perkins Marsh, *Man and Nature: Or Physical Geography as Modified by Human Action*, ed. David Lowenthal (Cambridge, MA: Belknap Press of Harvard University Press, 1965) [reprint of 1864 edition].

[43] David Fairchild, *The World Was My Garden: Travels of Plant Explorer* (New York: Scribner's Sons, 1938), pp. 9–10, 439, 446.

But, the difficulties continued after the forced removal of most of the Plains Indians and the near demise of the bison. In the 1920s, Charles A. Gillett, a forester in North Dakota, wrote that in the west of the state some trees lived only fifteen-twenty years. He attributed this to the lack of soil moisture. Despite the difficulties, the early settlers planted trees to provide for their needs. The federal and state governments encouraged them. The Timber Culture Act of 1873 granted settlers titles to 160 acres if they planted trees on forty acres. The act failed, even after the acreage was cut, and was repealed in 1891. Many settlers lacked the knowledge and experience to cultivate trees in the conditions.[44] Drawing on her childhood experiences in Webster County in central Nebraska in the 1880s, the novelist Willa Cather had one of her characters, an immigrant from the eastern United States, worry about the few trees that did grow in the plains: "Trees were so rare in that country, and they had to make such a hard fight to grow, that we used to feel anxious about them, and visit them as if they were persons."[45] In the mid 1930s, Webster County was in the zone designated for planting shelterbelts.

Successive chiefs of U.S. forestry, Fernow and Pinchot, brought scientific expertise and bureaucratic power to plains forestry and, in the first decade of the twentieth century, forestry received strong support from President Theodore Roosevelt. Back in 1891, the Fernow's Division of Forestry planned a network of forest reserves in the Great Plains. Bessey led research into the viability of planting trees in the Sandhills of northwestern Nebraska. The aims were to provide timber for the population, stabilize the Sandhills, and ameliorate the climate by protecting the land from the wind. In the early twentieth century, Pinchot, as chief of the renamed Bureau of Forestry, expanded the Sandhills plantings. In 1903, a tree nursery was established to supply seedlings, mostly conifers, for the plantations and the local population. The Sandhills plantations and nursery survived losses to fire in 1910. They survive to this day as the Nebraska National Forest and Charles E. Bessey Nursery.[46] The results in the Nebraska Sandhills were not repeated when the Forest Service (renamed

[44] Droze, *Trees*, pp. 23–7; Gardner, "Trees as Technology," 329–30.

[45] Willa Cather, *My Antonia* (Oxford: Oxford University Press, 2006) [1st published 1918], p. 22.

[46] See Joel Orth, "Directing Nature's Creative Forces: Climate Change, Afforestation, and the Nebraska National Forest," *WHQ* 42 (2011), 197–217; Robert Gardner, "Constructing a Technological Forest: Nature, Culture, and Tree-Planting in the Nebraska Sand Hills," *EH* 14 (2009), 275–97; Gardner, "Trees as Technology," 331–2. See also, USDA Forest Service, Charles E. Bessey Tree Nursery, available online at www.fs.usda. gov/detail/nebraska/about-forest/districts/?cid=stelprdb5343059, accessed September 14, 2017.

again in 1905) tried to expand the network of forest reserves in the plains. In July 1905, Theodore Roosevelt established a reserve in sand hills in southwestern Kansas. Around 80,000 seedlings were planted the following spring. Two years later, the reserve was increased in size and renamed the Kansas National Forest. Hundreds of thousands more seedlings were planted. But, scientific expertise and government backing were not sufficient to overcome the environmental challenges. Successes alternated with failures as many seedlings were killed by wild fires, droughts, and hot summer winds. In 1915, the forest was abolished by President Woodrow Wilson, who was less enamored with plains forestry than Theodore Roosevelt.[47] The failure of the Kansas National Forest mirrored the unsuccessful attempts over half a century earlier to plant trees in the arid steppes of Don Cossack and Kalmyk lands in the Russian Empire.

Planting trees in western Nebraska and Kansas at the end of the nineteenth and early twentieth centuries gave the Forest Service valuable experience, both positive and negative. One of the foresters was Royal Shaw Kellogg (1874–1965). Born in New York State, he moved to western Kansas as a child, where his father was among the few settlers who succeeded in planting sufficient trees to gain the title to 160 acres under the Timber Culture Act. Kellogg was educated at the Kansas Agricultural College in Manhattan and joined the Bureau of Forestry in 1901. He wrote two bulletins on forest planting in Nebraska and Kansas, published in 1904 and 1905, prior to the failure of the Kansas forest. He argued that with intelligent selection of trees and careful cultivation to conserve soil moisture it was possible to overcome the adverse conditions in the west of the states. He doubted forests would increase precipitation – pointing out there was no proof – but did believe that forests could moderate extremes of temperature, check the wind, and reduce evaporation.[48]

Another forester with experience of Nebraska and Kansas was Carlos G. Bates (1885–1949). He was born in Topeka, Kansas, attended the University of Nebraska in Lincoln, and spent his career in the Rocky

[47] Sorensen, "The Kansas National Forest."

[48] Royal Shaw Kellogg, "Forest Planting in Western Kansas," *U.S. Bureau of Forestry Bulletin*, no. 52 (1904); Kellogg, "Forest Belts of Western Kansas and Nebraska," *U.S. Forest Service Bulletin*, no. 66 (1905). On the author, see Elwood R. Maunder, "Oral History Interview I With Royal S. Kellogg, Palmetto, Florida, April 16, 1955, Forest Industry Foundation, Inc., St. Paul, Minnesota, available online at https://foresthistory.org/wp-content/uploads/2016/12/KelloggRS-1955.pdf, accessed October 30, 2019; LoC MD, Gifford Pinchot Papers (hereafter Pinchot Papers), Box 1008, File: Breaking New Ground, Personal Narratives, Kellogg, R. S. As we shall see, by 1934 Kellogg had changed his mind.

Mountains and Midwest.[49] Bates co-authored a study, published in 1913, of the Nebraska and Kansas plantations, which concluded that such plantings would fail only in exceptional conditions.[50] Another advocate of planting trees in the plains, and one who persisted in the belief that trees could ameliorate the climate, was Zon.[51] Zon and Bates worked together for much of their careers, and played key roles in launching the Shelterbelt Project in 1934. In his memoir, Pinchot (or possibly Zon as editor) noted Bates' contribution to the Nebraska National Forest, and that the Shelterbelt Project was "its direct lineal descendant."[52]

By the late nineteenth and early twentieth centuries, hopes for wider-scale afforestation of the Great Plains were receding in the face of the difficulties. Many, but not all, foresters were ceasing to believe planting trees would significantly affect the climate and increase rainfall. Instead, foresters paid more attention to limited plantings for specific purposes, such as shelterbelts. In 1894, Isaac Cline, a Texas meteorologist, who noted the harmful influence of hot summer winds on vegetation and crops, suggested that "a generous growth of hardy timber" could "lessen their injurious effects."[53] In 1904, Royal Kellogg advocated shelterbelts. He felt it a "safe practical assumption" that they protected ground up to ten to fifteen times their height.[54] This change in emphasis towards smaller-scale plantings for particular purposes such as shelterbelts resembled a similar change, slightly earlier, in the Russian steppes.

Planting trees and shrubs to shelter land from the wind began soon after Euro-Americans began to settle the plains in the mid nineteenth century. Settlers soon learned the value of the practice. In 1880, William McCracken of Sedgewick County reported to the Kansas State Horticultural Society: "Timber-belts, properly arranged, afford a beneficial protection to field and garden crops." J. W. Robson of Dickinson County went further "those settlers who planted shelter-belts and groves are fixtures on their farms, while those who never planted a tree have

[49] Richard S. Sartz, "Carlos G. Bates: Maverick Forest Service Scientist," *Journal of Forest History* 21 (1977), 31–9.

[50] C. G. Bates, and R. G. Pierce, "Forestation of the Sand Hills of Nebraska and Kansas," *USDA Forest Service Bulletin*, no. 121 (1913).

[51] See Orth, "Directing Nature's Creative Forces," 216.

[52] Pinchot, *Breaking New Ground*, pp. 238, 309. On Zon's role in editing the memoir, see p. 280.

[53] I. M. Cline, "Summer Hot Winds on the Great Plains," *ESR* 5 (1893–4), 1035–7.

[54] Kellogg, "Forest Planting," 14–16. The growing emphasis on shelterbelts can be seen by comparing two bulletins published twenty-five years apart: Charles A. Keffer, "Experimental Tree Planting in the Plains," *USDA Forestry Division Bulletin*, no. 18 (1898); Fred R. Johnson and F. E Cobb, "Tree Planting in the Great Plains Region," *USDA Farmers' Bulletin*, no. 1312 (1923).

pulled up stakes and gone elsewhere."[55] E. J. Fautch, of Ransom County, North Dakota, recalled much later how in 1890 the first settlers had planted groves of trees that had been "serving the settlers for many years as shelter from wind and snow."[56] Railroad companies, such as the Atchison, Topeka, and Santa Fe, planted trees along their lines across the plains to protect them from snowdrifts. State governments offered bounties to settlers who planted shelterbelts. Such laws were enacted in Kansas in 1874, Nebraska in 1879, South Dakota in 1890, and North Dakota in 1905. The Canadian government started planting shelterbelts in its prairie provinces in 1901.[57] A decade or so later, when Fairchild of the USDA visited Saskatchewan, he was impressed: "I convinced myself that trees really would grow and make low forests and shelterbelts on these treeless wastes."[58]

The USDA began to promote shelterbelt planting. In 1913, its Bureau of Plant Industry (not the Forest Service) established the Northern Great Plains Field Station at Mandan, North Dakota, to support agriculture in the more arid, or dryland, conditions of the northern plains west of the 100th meridian. The station had strong support from Senator P. J. Macumber of North Dakota. His knowledge of plains forestry was out of date, however, as he told Fairchild that he would like to see his "almost treeless" state "covered with trees."[59] From 1916, the Mandan station planted shelterbelts in the western Dakotas and eastern Montana and Wyoming. It provided farmers with trees and instructions on planting and caring for them.[60] In 1925, the chief of the Bureau of Plant Industry reported:

> In previous years, soil blowing presented a very serious problem, which has now been practically solved through the protection afforded crops by shelterbelts supplemented by proper cultural treatments of the soil. The shelterbelts [also] ... afford the shade and comfort associated with the country home. Years of experiment in developing trees adaptable to that region under the cooperative shelterbelt distribution conducted at the ...

[55] *Kansas Horticultural Report for the Year 1880* (Topeka, KS: G. Martin, 1881), pp. 256, 192, and passim. The society's reports for other years contain much discussion of shelterbelts and windbreaks.

[56] E. J. Fautch, City View Farm, quoted in NARA KC, RG 114, Entry 187, Box 1, File D-Cooperation, Joint Congressional Committee, 1940, Charles G. Bangert, Enderlin, ND, on behalf of Enderlin Community Forest and citizens of North Dakota, "Prairie States Forestry Project," to Joint Committee on Forestry, U.S. Congress. n.d. [1940].

[57] Thomas R. Wessel, "Prologue to the Shelterbelt," *Journal of the West* 6(1967) 120–1, 126, 127.

[58] Fairchild, *World*, p. 439. [59] Fairchild, *World*, p. 370.

[60] See Wessel, "Prologue," 127–9; Gillett, "Forests and Forestry," 39–42.

> Northern Great Plains Field Station, Mandan, N. Dak., have produced trees that can be grown successfully in the region. Farmers are realizing the importance of shelter belts, and there is now a demand for trees that exceeds the supply.[61]

By 1935, 1,700 shelterbelts had been planted, seventy percent of which had survived, demonstrating what was possible in this unpromising environment. Further support from the federal government came from the Clarke–McNary Act of 1924, which funded state nurseries to distribute trees to farmers at cost price for shelterbelts and other purposes. A state forest nursery was established at Bottineau, North Dakota. These last two programs, however, had limited success.[62]

Bates wrote a report, published in 1911, on Forest Service studies of the effects on crops in fields adjoining existing shelterbelts, rows, and groves of trees in Kansas, Nebraska, Minnesota, and Iowa. He challenged the "now common argument that they do more harm than good." Bates noted that tree roots reduced soil moisture and crop yields in land right next to them. But, further away, belts of trees calmed the wind, thus protecting against soil blowing and snow drifting, had beneficial effects on air and soil temperatures, reduced evaporation, and helped retain soil moisture. He calculated that trees protected an area from evaporation to a distance of 15–20 times their height, depending on the width of the belts and the wind velocity. Bates concluded that "great benefit" may be expected, in particular in dry seasons, and that increased crop yields compensated for the cost of planting the trees and the land they took up. He summarized his conclusions in a publication for farmers published in 1917 and revised in several editions between 1917 and 1944.[63] Bates was more positive about shelterbelts than his Russian counterpart Vysotskii, but did acknowledge some drawbacks.

Thus, planting shelterbelts in the plains was not new when the Shelterbelt Project was launched in 1935. Perry remarked in his 1942 history of the Project: "It is difficult to understand why all this accumulated

[61] NARA CP, RG 54, Finding Aid A1, Entry 51, Division of Plant Exploration and Introduction: Reference Files, 1914–51, "Report of the Chief of the Bureau of Plant Industry 1925," 28. See also from the same file: "Report of the Chief of the Bureau of Plant Industry 1920," 28; "Report of the Chief of the Bureau of Plant Industry 1924," 39–40.

[62] E. J. George, "Growth and Survival of Deciduous Trees in Shelter-Belt Experiments at Mandan, N. Dak., 1915–34," *USDA Technical Bulletin*, no. 496 (1936); Wessell, "Prologue," 127–9; Gillett, "Forests and Forestry," 39–42.

[63] Carlos B. Bates, "Windbreaks: Their Influence and Value," *USDA Forest Service Bulletin*, no. 86 (1911), revised under the same title as *USDA Farmers' Bulletin*, no. 788 (1917) and *USDA Farmers' Bulletin*, no. 1405 (1944).

knowledge ... was not sooner translated into an aggressive tree-planting campaign."[64] Also not new in the 1930s was American familiarity with trees and shrubs from the steppes and knowledge of Russian experience in planting shelterbelts.

The Introduction of Trees and Shrubs from the Steppes to the Great Plains

From at least the 1870s, trees and shrubs suitable for shelterbelts were being introduced to the Great Plains from the steppes of Eurasia. They arrived by various routes. The Mennonites who moved to the Great Plains in the 1870s brought over trees and shrubs as well as their wheat. Kansas journalist Noble L. Prentis, who visited a Mennonite community in 1882, wrote: "A very common shrub was imported from Russia and called the wild olive ...; but the all-prevailing growth was of the mulberry, another Russian idea, which is used as a hedge, a fruit tree, for fuel, and as food for the silk-worms."[65] Russian olive and Russian mulberry proved useful throughout the plains in hedges, shelterbelts, and for other purposes.[66] Russian trees attracted interest in agricultural colleges and experiment stations in the prairies and plains in the 1880s–90s. The Iowa Agricultural College imported Russian poplar (*Populus certinensis*) and willows. Following tests in Iowa and South Dakota, they proved useful in windbreaks. By the 1890s, private nurseries in South Dakota, Minnesota, and Iowa were growing and selling species imported from Russia.[67]

The success of Russian trees and shrubs in the plains attracted interest in the USDA, including Fairchild, who was in charge of plant introductions from abroad from the late 1890s. In his memoir, he linked his official responsibilities with recollections of his boyhood in Kansas (see p. 291): "The hours under those Kansas cottonwoods later gave me a real understanding of the shelter-belt experiments ... and added zest to the quest which we made in China and Siberia for the hardiest trees and

[64] Perry, *History*, p. 4.

[65] Reprinted in Noble L. Prentis, *Kansas Miscellanies*, 2nd edition (Topeka, KS: Kansas Publishing Company, 1889), p. 171.

[66] "Russian Mulberry (*Morus alba tatarica*)," *U.S. Forest Service Circular* 83 (1907); WSUL SCUA, Federal Writers' Project, Kansas, Box 10, File 2. Reno County, Sturgeon, Katherine, 1936, Racial Groups: The Mennonites, p. 4.

[67] C. A. Keffer, "Forest Trees, Fruits and Vegetables," *South Dakota Agricultural College and Experiment Station Bulletin*, no. 23 (1891), 132–3, 138; Keffer, "Experimental Tree Planting," 23.

shrubs for the treeless wastes of the Dakotas."[68] Plant explorers Niels Hansen and Frank Meyer shipped back many trees and shrubs from Eurasia for the Great Plains. In 1897–8, Hansen collected species of poplar (related to cottonwoods), caragana, known as Siberian pea tree (*Caragana arboresueus*), willow, maple, and mulberry. He took particular interest in Siberian larch (*Larix siberica*) as Russian foresters told him it was hardier than European larch. Several of the species Hansen collected were planted in the plains, including his home state of South Dakota, to test their suitability.[69]

During his Eurasian expeditions for the USDA, Meyer paid attention to local climates and soils to "give us a clue where to obtain material . . . for the Western . . . United States." In December 1911, he visited experimental forester A. D. Voeikov in Simbirsk province. (They had been put in touch by Zon, who was from Simbirsk.) Meyer noted that the climate was "not congenial to trees," but listed those that grew there, including types of poplar, oak, willow, and Siberian pea tree. Like Hansen, Meyer was keen to collect samples of Siberian larch on account of its hardiness. He learned from Voeikov that trees from the Central Asian uplands did well in southern Russia, as they were very resistant to drought and cold, whereas species from Manchuria and the eastern United States suffered badly from spring frosts.[70] In December 1912, Meyer sent back twelve species of poplar he had bought from Voeikov. He explained that they "may show quite some variation as regards hardyness [sic] and adaptability to such an uncongenial climate as around Mandan and some . . . may prove . . . very desirable shade trees, new to the North West."[71] The USDA forwarded them to Ellery Chilcott at the Mandan field station in North Dakota, where he "expect[ed] to get a couple of thousand feet of wind breaks planted and a few acres . . . for nursery work."[72]

[68] Fairchild, *World*, pp. 9–10.

[69] Niels Ebbesen Hansen, *The Banebryder (The Trail Breaker), The Travel Records of Niels Ebbesen Hansen 1897–1934*, ed. Helen Hansen Loen (n.p: privately printed, 2002), pp. 9, 18; N. E. Hansen, "The Shade, Windbreak and Timber Trees of South Dakota," *South Dakota Agriculture Experiment Station Bulletin*, no. 246 (March 1930), 20–1; Keffer, "Experimental Tree Planting," 24, 37.

[70] NARA CP, RG 54, Finding Aid (FA) NC-135, Entry 135I, Records of Frank N. Meyer, Plant Explorer, 1902–1908, Box 7, Folder [large, orange, no number], Meyer to David Fairchild, USDA, December 6, 1911; Meyer to Fairchild, December 16, 1922 [mistake for 1911]. A. D. Voeikov is not to be confused with the climatologist Aleksandr Voeikov mentioned elsewhere.

[71] NARA CP, RG 54, FA NC-135, Entry 135I, Box 8, Folder [no name], Meyer to P. H. Dorsett, December 22, 1912; Bill from A.D. Voeikov, November 23, 1912.

[72] NARA CP, RG 54, FA NC-135, Entry 135I, Box 9, Folder 16 January–21 May 1913, Plant Introducer in charge of plant introduction field stations, USDA, to Meyer, January 20, 1913.

Meyer was especially interested in elms, for example *Ulmus pumila* and *Ulmus turkestanica*, and sent back types that could withstand alkaline soils.[73] He continued his explorations in Siberia in January and February 1913, "while a biting cold wind blew across the scantily wooded, snow covered steppe." He was interested in Siberia because he considered it similar to the "North Western Plains" of the United States. After a "well-earned bath and change of clothes" in Tomsk, he traveled east to Chita, from where he sent back a list of seeds available from a local supplier, including a hardy elm (*Ulmus pumila var.*), which was "of value especially in practically treeless regions like the N.W. Plains."[74] The search for hardy elms to plant in shelterbelts in the northern plains was complicated by confusion between what turned out to be two different types. Meyer traveled from eastern Siberia to Manchuria and northern China. Near Peking (Beijing) he collected a species he named Chinese elm. He used the same scientific name, *Ulmus pumila*, that he had used for similar elms in the harsher climate of eastern Siberia. The USDA planted Meyer's Chinese elms at the Mandan field station and the state forest nursery at Bottineau in North Dakota. Chinese elms were planted extensively in the Great Plains with "considerable success" and were "very satisfactory" in shelter-belts for a few years. However, the trees Meyer introduced from China suffered from winter killing in the northern plains.[75] In 1923, the USDA alerted plains farmers that two different strains of "Chinese elm" had been introduced from different locations, and that one was hardier than the other.[76] In 1930, Hansen warned against planting the Chinese elms Meyer had introduced from the Peking region as they lacked hardiness. He assigned these elms the name *Ulmus parvifolia* to avoid confusion with the "North Chinese elm," for which he used the scientific name *Ulmus pumila* that Meyer had given to elms in eastern Siberia. This latter species, which was hardier, proved popular in the Dakotas.[77] The confusion

[73] NARA CP, RG 54, Finding Aid (FA) NC-135, Entry 135I, Records of Frank N. Meyer, Plant Explorer, 1902–1908, Box 7, Folder [large, orange, no number], Meyer to Fairchild, USDA, December 6, 1911.

[74] NARA CP, RG 54, FA NC-135, Entry 135I, Box 9, Folder 16 Jan–21 May 1913, Meyer to Fairchild, February 1, 1913; Meyer to Fairchild, February 15, 1913.

[75] S. S. Burton, "Chinese Elm (*Ulmus pumila*)," *JoF* 27 (1929), 303–5.

[76] Johnson and Cobb, "Tree Planting," 13.

[77] Hansen, "The Shade, Windbreak and Timber Trees," 46–8. When I visited in 2018, I checked which varieties had been planted in the Saskatoon Forestry Farm, which dates back to 1913, in Saskatchewan and a few weeks later in the near identical conditions at the Botanical Garden in Astana/Nur-Sultan, Kazakhstan, founded in 2018. In the former, I came across some mature examples of the hardier *Ulmus pumila*, but in the latter I found some young *Ulmus parvifolia*.

created problems when some of the less hardy types of elm were planted in shelterbelts in the 1930s.

Trees from the steppes also made their way to the Great Plains through contacts between government scientists. The hardy Siberian larch was of particular interest. In January 1930, in a routine enquiry, Harry Harlan of the Bureau of Plant Industry of the USDA in Washington, DC, wrote to Nikolai Vavilov in Leningrad about "source[s] in Russia of seed of the Siberian larch (*Larix Siberica*) and of any pines, spruces or firs . . . native to Russia," as the Dry Land Field Station at Cheyenne, Wyoming, wanted Russian trees to test in shelterbelts. Vavilov sent two kilograms of the Siberian larch seed.[78]

By the 1920s the USDA was recommending several species of trees and shrubs from Eurasia, including the steppes, to plains farmers for hedges and shelterbelts, including: Siberian pea tree/caragana; Russian olive; Russian mulberry; Russian golden white willow; Tamarix; and, with caution, Chinese/Siberian elms.[79] A decade earlier, a Russian agricultural specialist visiting the United States noted that some trees from Russia had proved suitable for arid parts of the plains.[80]

The Transfer of Russian Experience and Expertise to the United States to the 1930s

United States government specialists were familiar with Russian experience in steppe forestry and acknowledged its relevance to the Great Plains as far back as 1868. Joseph Wilson, the Commissioner of the General Land Office, reflected on "forest culture" in the light of the extension of settlement "across the treeless prairies of Kansas and Nebraska." He was familiar with the perceived environmental benefits of forests, the consequences of deforestation, and tree planting practices in Europe. He was aware also, from general literature, that "a large portion of the [Russian] Steppes are as destitute of trees as our own western plains." He noted recent attempts at "wooding of the steppes," in places to "fix" the sands, and that the "Emperor of Russia has undertaken the reclamation of the steppes." If trees could be "successfully cultivated" in the steppes, Wilson

[78] TsGANTD, f. 318, op.1-1, d. 369, ll.44, 46; NARA CP, RG 54, FA PI-66, Entry 30 Records of the Division of Cereal Crops and Diseases, Foreign Correspondence, 1900–34, Box 8, File Russia, L–Z, Robert Wilson, Great Plains Field Station, Cheyenne, Wyo. to Harlan, January 6, 1930; Harlan to Vavilov, April 25, 1930.

[79] Johnson and Cobb, "Tree Planting," 5, 10, 21.

[80] V. Benzin, "Sukhoe zemledelie," *SKhiL* 241 (April 1913), 705.

saw "not the remotest reason why forest culture should fail on any part of the great American plains."[81]

Practical knowledge of steppe forestry, including shelterbelts, was brought to the Great Plains by immigrants, including the Mennonites. They had been among the pioneers of tree and shelterbelt planting in the steppes in the early nineteenth century, and followed the same practices after they moved to the United States. The journalist Prentis compared what he saw on a visit to some Mennonites in Kansas in 1882 with his previous visit shortly after their arrival in the mid 1870s: "A great change had taken place":

> The most surprising thing ... is the growth of trees. I left bare prairie; I returned to find a score of miniature forests... several acres around every house were set in hedges, orchards, lanes and alleys of trees – trees in lines, trees in groups, and trees all alone. In many cases, the houses were hardly visible from the road, and in a few years will be entirely hidden in the cool shade.

Heinrich Richert, one of the Mennonites, attributed their success in growing trees to constantly plowing the land. Prentis and Richert saw "the most promising orchards" and "immense fields of the greenest wheat."[82] The Mennonites' practices spread around the region. A few decades later, the USDA cereal specialist Mark Carleton – who was familiar with Mennonite practices in both the Great Plains and the steppes – recommended planting "long, narrow belts of trees running east and west," including Russian mulberry, to prevent "soil blowing."[83]

The transfer of Russian knowledge to the United States was facilitated by visits by forestry specialists to each other's countries. In 1896, Vasilii Tikhonov, a Russian government forester with experience in the steppes, visited North America. He met Fernow, the chief of the U.S. Division of Forestry, in Washington, DC, and accompanied him on a tour around the United States. This gave him the opportunities to learn about American forestry and share his knowledge of Russian practices. In the Great Plains,

[81] Joseph Wilson, *Report of the Commissioner of General Land Office for the year 1868* (Washington, DC: Government Printing Office, 1868), pp. 173–98, quotations from pp. 181, 195–7.

[82] Quoted in Prentis, *Kansas Miscellanies*, pp. 171–82. For another contemporary account of Mennonites planting trees on their arrival in the United States, see Delbert Plett (ed.), *Pioneers and Pilgrims: The Mennonite Kleine Gemeinde in Manitoba, Nebraska and Kansas, 1874 to 1882* (Steinbach, Manitoba: D.F.P. Publications, 1990), p. 82.

[83] Mark A. Carleton, *The Small Grains* (New York: Macmillan, 1916), p. 356. See also Larry Rutter, "Mennonites and Shelterbelts," *Kansas Canopy: Newsletter of the Kansas Forest Service*, no. 42 (Spring 2012), 8–9.

he visited agricultural colleges and experiment stations in Brookings, South Dakota, and Lincoln, Nebraska, and forest reserves in western Nebraska and southwestern South Dakota. While in what he called the "steppe" region of the United States, Tikhonov observed that efforts to grow trees to shelter the land and houses had been generally unsuccessful, because the trees died after fifteen-twenty years. He noted the limited experience of American foresters in "steppe forestry." Americans he met expressed great interest in Russian experience of growing trees in their plains, and asked if the Russian Ministry of Agriculture could prepare English translations of Russian studies. He concluded that with their fifty years' experience of "steppe forestry" the Russians were far ahead of the Americans, but both could learn from each other. Fernow expressed interest, and Tikhonov suggested the USDA send people to Russia to study steppe forestry.[84]

American specialists visited Russia and brought back information about forestry in the steppes, including erosion-protection practices and shelter-belts. While collecting plants and trees in the Russian Empire in 1897, Hansen visited the Repetking experiment station, in the Central Asian desert, to see test plantings to bind drifting sands.[85] Charles Bessey visited the Russian Empire in 1903 with his son Ernst, who was a USDA plant explorer. The elder Bessey noted that some farmers in the steppes planted trees as "windbrakes [sic]."[86] More important was the visit to Russia in 1902 by Pinchot, then chief of the U.S. Bureau of Forestry (See p. 85). Pinchot's itinerary was arranged by Nikolai Nesterov, a Russian forester he had met when both were students in France a decade or so earlier. In St. Petersburg, with a letter of introduction from President Theodore Roosevelt, Pinchot visited the Forestry Department of the Russian Ministry of Agriculture and the Forestry Institute. He went to Moscow, visiting several forestry estates en route, before heading south, via the forest–steppe to the southern, treeless steppe. He made notes on steppe forestry practices, including their history going back to the early nineteenth century, and the importance Russians attached to planting trees and shrubs to control erosion. He noted the types of soil (including "tchernoziom"), the amount of rainfall (400–450 mm in the forest–steppe,

[84] V. Tikhonov, "Ocherk lesnogo khoziaistva severnoi Ameriki (iz otcheta po komandirovke)," *SKhiL* 189 (April 1898), 106–7, 116–17; (May 1898), 247–51. See also V. Tepiakov, "Vasilii Andreevich Tikhonov (1849–1913)," *Ustoichivoe lesopol'zovanie* 49, no. 1 (2017), 44–6.

[85] Hansen, *The Banebryder*, p. 20.

[86] Charles E. Bessey, "Some European Forest Notes," *JoF* 8 (1910), 205; "Biography of Ernst Athearn Bessey," Bessey Family Papers, Michigan State University Archives and Historical Collections, available online at http://archives.msu.edu/findaid/116.html, accessed September 29, 2017.

350 mm further south), and species of trees (for example Siberian larch, but also trees introduced from North America). The southern steppes, with their fertile soil, reminded him of "our own prairies." Nesterov explained that few trees grew naturally in the steppe region because of the alkali in the soil. He cited a case of trees dying after twenty years when their roots reached "an alkali stratum" in the soil. Pinchot noted that the tree plantations he saw were "doing superbly," except "those on alkali soils," but thought fire the main cause of the treelessness of the steppes.

Pinchot visited all three field stations of Dokuchaev's special scientific expedition of the 1890s, including Velikii Anadol' and Kamennaia Step', where Vysotskii had made many of his studies. Pinchot paid particular attention to the Kamennaia Step' plantation and described the "shelter belts." He recorded the methods the Russian foresters used, especially regular cultivation of the soil to control weeds. He noted the different combinations of trees and shrubs planted to assist the growth of the shelterbelts. He was not convinced that some of the Russians' techniques were cost effective and thought some belts too wide, but did record that planting trees and shrubs together in a ratio of two shrubs for every tree could be of value in the "Western United States." In his notes he made at the time of his visit, Pinchot wrote the following about the shelterbelts:

> The purpose of all this plantation is double, to supply timber, and to afford protection to agricultural crops. Careful measurements are in progress, of strips ten feet wide parallel to the plantations, to ascertain the effect of protection from wind. There are no final results as yet, but the indications are strong that the crops vary in quality with their nearness to the shelter belt.[87]

In Pinchot's memoir, published in 1947, the following sentence was added: "The Russians had ... begun the planting of shelter belts by the government, whereby they beat us ... by some forty years."[88] This sentence does not appear in the notes Pinchot made at the time or in his diary. It may have been added by Zon during his "substantial copy-editing" of Pinchot's manuscript after his death (see p. 280).

Pinchot's visit to the Russian Empire was part of the growing connection between American and Russian forestry specialists that assisted

[87] Pinchot Papers, Box 715, File: "Notes taken by Mr. Pinchot on his trip through Russia and Siberia. Prepared by Mr. Pinchot 1902." The account of his visit to the steppes is in ff.45–65. There is a bound copy, with the same pagination, in Box 563, File: "Forests, Foreign: Russia, Siberia, China." See also Steen (ed.), *Conservation Diaries of Gifford Pinchot*, pp. 119–21; Pinchot, *Breaking New Ground*, pp. 11, 215–16.

[88] Pinchot, *Breaking New Ground*, p. 216.

exchanges of information and expertise on steppe forestry, shelterbelts, and the relationship between forests and climate. In early 1902, shortly before he left for Russia, Pinchot had a meeting with Zon,[89] and Zon likely gave him advice on the country of his birth. Over the following few years, Zon was in touch with Nesterov,[90] and, most likely via Zon, Bates learned about and cited a study by Nesterov on the influence of forests on wind speed and direction.[91] Thus, by the early twentieth century, Pinchot, Zon, and Bates were familiar with Russian experience of planting trees and shelterbelts in the steppes and parallels with American experience in the Great Plains.

We can trace the dissemination of information about Russian forestry, and steppe forestry including shelterbelts, among American foresters by examining the American *Journal of Forestry* (*JoF*). There was little attention to Russian forestry until the 1920s. In 1907, the journal published reports on studies by Pavel Ototskii and Vysotskii showing that forests lowered ground water levels: results that cast doubt on the value of shelterbelts.[92] The journal did not cite the original Russian studies, but summaries from German periodicals, suggesting that the language barrier impeded further knowledge of Russian forestry. Most attention in the journal to forestry outside the United States was paid to western Europe, British territories, including India and Canada, American territories such as the Philippines, and China and Japan.[93]

Prior to the 1920s, the references in *Journal of Forestry* to Russian forestry experience seem to have been a result of direct contacts between foresters. For example, the journal's surveys of current periodical literature included recent issues of the Russian forestry journal (*Lesnoi zhurnal*) following Pinchot's visit to Russia in 1902.[94] A summary of the work of Russian forestry specialist Tkachenko appeared in *Journal of Forestry* in 1913, after his visit to the United States. The journal also published a general article on forestry in Russia based on a conversation with Tkachenko. Concerning steppe forestry, Tkachenko explained that planting

[89] Zon Papers, Box 2, Folder 4, Raphael Zon to Anna [Zon], January 30, 1902.

[90] In 1910, Zon sent Nesterov a copy of one of his publications. Zon Papers, Box 3, Folder 1, List of people to whom copies of Zon's "Forest Resources of the World" was sent to, December 15, 1910.

[91] Bates, *Windbreaks*, p. 40; N. S. Nesterov, "O vliianii lesa na silu i napravlenie vetra," *Lesopromyshlennyi Vestnik: Zhurnal lesnogo khoziaistva, lesnoi promyshlennosti i torgovli lesom* (February 21, 1908), no. 8: 73–8 and no. 9: 85–90.

[92] "Periodical Literature," *JoF* 5 (1907), 325–7; 417–19.

[93] See A. B. Recknagel and Theodore S. Woolsey, "European Study for Foresters," *JoF* 10 (1912), 417–439.

[94] "Periodical Literature," *JoF* 1 (1902), 33–4; "Periodical Literature," *JoF* 2 (1903), 74.

trees was successful for thirty or forty years, but then the trees began to die due to alkaline salts in the soil. He explained that species were selected that could cope with alkaline subsoils and planting methods were devised to assist them.[95] For the following decade, there was little mention in *Journal of Forestry* of Russian forestry, possibly as a result of communication problems between Russia and much of the world between the outbreak of the First World War in 1914 and the end of the Russian Civil War in 1921. The USDA, however, took the initiative in reestablishing contacts by writing to Vysotskii in 1923.[96]

American knowledge of Russian steppe forestry and shelterbelts increased from the early 1920s and had become common currency by the early 1930s, when Russian experience played a key role in the decision to launch the Shelterbelt Project. An American forester who played an important part in this growing awareness of Russian forestry was Zon. At this point it is worth introducing Raphael Zon (1874–1956) in more detail.[97] Like several American scientists who acted as conduits for Russian knowledge, Zon was a Jewish émigré from the Russian Empire. He was born in 1874 in the Volga city of Simbirsk, on the boundary of the steppe region. He attended the city's classical gymnasium (where one of his near contemporaries was Lenin[98]) and entered Kazan' University to study medical and natural sciences in 1892. He fled Russia in 1896, after being sentenced to ten years' imprisonment for radical activities, and after two years in Belgium and England, immigrated to the United States. In 1901, he graduated in forestry from Cornell University, where he came to the attention of Pinchot and Fernow. They assisted him find work at the Bureau of Forestry in Washington, DC. Zon and Pinchot became lifelong friends, corresponded regularly and worked together to promote

[95] "Forest Influence on the Soil," *JoF* 11 (1913), 423–4; "Notes on Forestry in Russia," *JoF* 12 (1914), 567–77. On his visit, see M. Tkachenko, *Lesa, lesnoe khoziaistvo i drevoobrabatyvaiushchaia promyshlennost' Severo-Amerikanskikh Soedinennykh Shtatakh: Otchet zagranichnoi komandirovki* (Petrograd: 1914).

[96] NARA CP, RG 54, Finding Aid PI-66, Entry 30, Records of the Division of Cereal Crops and Diseases. Foreign Correspondence, 1900–34, Box 8, File Russia, C. W. Warburton, Acting Cerealist in charge, to Mr. George Visozki [Vysotskii], Manager, Steppe Scientific Station, State Protection Estate Askania Nova, Russia, August 15, 1923.

[97] Except where otherwise indicated, this account is based on P. O. Rudolf, "R. Zon, Pioneer in Forest Research," *Science*, NS, 125, no. 3261 (June 28, 1957), 1283–4; Norman J. Schmaltz, "Raphael Zon: Forest Researcher," *Journal of Forest History* 24 (1980), 24–39, 86–97; Jeremy Young, "Warrior of Science: Raphael Zon and the Origins of Forest Experiment Stations," *Forest History Today* 16, nos. 1–2 (2010), 4–12.

[98] Zon later recalled that they knew Lenin's family well. Zon Papers, Box 10, Folder 9, Zon to Louis Adamic, August 2, 1939.

conservation and forestry research.[99] Zon and his wife, Anna (a fellow Russian–Jewish émigré) worked hard at their English and to make their way in their adopted homeland.

Zon worked in the Bureau of Forestry (from 1905 the Forest Service) in Washington, DC until 1923. Pinchot appointed him chief of the Office of Silvics in 1907. Zon championed scientific research and the establishment of the first Forest Experiment Station (in Arizona) in 1908. Scientific forestry grew apace following Zon's appointment as chief of the Office of Forest Investigations in 1914, a post he held until 1920. In 1907–9, Zon worked closely with Bates, and they remained in contact after Bates moved to a post in Denver.[100] Zon engaged in his own research and published extensively. In 1923, partly as a result of a personality clash with William Greeley, the chief of the Forest Service, Zon was appointed director of the new Lakes States Forest Experiment Station in St. Paul, Minnesota, where he remained until his retirement in 1944. To Zon's elation, Bates moved to St. Paul as senior silviculturalist in 1927. Zon's station was responsible for research in the lakes states of Minnesota, Wisconsin, and Michigan, to regenerate the forests that had been felled indiscriminately over the previous decades. Nevertheless, Zon and Bates maintained an interest in shelterbelts, the influence of forests on climate and water, and planting trees in the Great Plains. In 1931, Zon established a branch station at Denbigh, North Dakota, where Bates led experiments in planting trees, including windbreaks, to control erosion.[101]

Zon was regarded in the Forest Service, the USDA as a whole, and more widely among American scientists, as a key contact on matters concerning Russia. He knew Russian, was acquainted with some Russian scholars, and read Russian publications. At the start of his career, he had translated works on Russian forestry for American foresters, including Fernow.[102] In 1905, he was listed in *Journal of Forestry* as one of three authors of reviews of periodical literature, which contained occasional references to Russian

[99] Zon recalled attending lecture by Pinchot at the New York College of Forests, Axton, Adironacks, in April 1900. Pinchot Papers, Box 1009, File: Breaking New Ground, Personal Narratives, Zon, Raphael, Zon to Pinchot, November 12, 1912; Zon Papers, Box 2, Folder 2, Diary of time spent in field work in forests in up-state New York [April 1900]. Their papers contain many letters between them.

[100] For example, they worked on a joint publication. Carlos G. Bates, and Raphael Zon, "Research Methods in the Study of Forest Environment," *USDA Bulletin*, no. 1059 (1922).

[101] Paul O. Rudolf, *History of the Lake States Forest Experiment Station* (St. Paul, MN: North Central Forest Experiment Station, Forest Service, USDA, 1985), pp. 10, 32–4; Sartz, "Carlos G. Bates," 38; Droze, *Trees*, pp. 60–1.

[102] Zon Papers, Box 2, Folder 5, Exchange of letters between B. E. Fernow (dean of the New York State College of Forestry at Cornell) and Zon, February 14, 1905, February 27, 1905.

work.[103] In April 1934, when he was involved in proposing the Shelterbelt Project, he ordered twenty-eight books by Soviet specialists on agriculture and renewed his subscriptions to two Soviet forestry journals, *Forest Economy and Forest Exploitation* (*Lesnoe khoziastvo i lesoeksploatatsiia*) and *Forest Industry* (*Lesnaia promyshlennost*').[104] Zon drew on work by Russian scientists in his own writings. In 1935, he came into conflict with Vysotskii, who accused him of plagiarizing his work two decades earlier (see pp. 217–18). In the offending publications, Zon argued that forests increased rainfall, deforestation reduced it, and trees protected against erosion.[105] Zon shared his knowledge of Russian steppe forestry and shelterbelts with Bates. In 1912, Zon wrote a short paper for him on based on Russian studies. He explained how hopes in Russia that planting trees in "solid bodies" in the steppes would increase the humidity of the climate had not been realized. Moreover, trees planted in large blocks began to die off when their roots reached a "dead horizon." This was due to "mineral [alkaline] salts" in the soil that were "unfavorable to tree growth." Zon wrote that the "tendency now is to plant [trees] in . . . windbreaks" to "protect against the drying effect of the wind," and assist in retaining snow in the winter. He concluded with a list of tree species planted in the steppes.[106]

Zon promoted Russian research among American foresters during his tenure as editor-in-chief of *Journal of Forestry* from 1923 to 1928. He published several articles on Russian forestry, including one of his own that drew on Russian research into the relationship between forests and climate.[107] Under Zon's editorship, the journal carried reviews of Soviet works

[103] *JoF* 3 (1905), 290.

[104] Zon Papers, Box 8, Folder 7, H. E. Gordon, Amkniga Corporation, 258 Fifth Avenue, NY to Zon, and reply, both on April 21, 1934.

[105] Raphael Zon, "Forests and Water in the Light of Scientific Investigation," Appendix V to Final Report of the National Waterways Commission,1912, Senate Document No.469, 62d Congress, 2d Session: 205–332. It was reprinted by the Forest Service, with a revised bibliography, in 1927, and also translated into Russian. Zon Papers, Box 7, Folder 6, I. I. Rostchin, Transcaucasian Institute of Water Regulation and Forestry, Tiflis, to Zon, February 20, 1931, Zon to Rostchin, May 27, 1931; Raphael Zon, "The Relation of Forests in the Atlantic Plain to the Humidity of the Central States and Prairie Region," *Science* NS 38, no. 968 (July 18, 1913), 63–75; Zon, "How the Forests Feed the Clouds," in *Science Remaking the World*, ed. Otis W. Caldwell and Edwin E. Slosson (Garden City, NY: Doubleday, 1924), pp. 212–22.

[106] NARA CP, RG 95 Records of the Forest Service, Forest Research Divisions, Finding Aid PI-18, Entry 115, Research Compilation File, 1897–1935, Box 89, File Description and Resources of Russia, Report "Forestation in the Semi-Humid Regions of Russia," by Zon for C. G. Bates, October 8, 1912.

[107] John D. Guthrie, "Some Notes on the Forests of Northern Russia, *JoF* 22 (1924), 197–204; "Paper Production and Consumption in Russia," *JoF* 23 (1925), 78–9; V. B. Shostakovitch, "Forest Conflagrations in Siberia," *JoF* 23 (1925), 365–71; Raphael Zon, "Forestry Versus Climate," *JoF* 26 (1928), 711–13.

written by a staff member at his St. Paul forest experiment station. Suren Rubenian Gevorkiantz could read Russian as he was born in the Russian Empire (he was of Armenian origin).[108] Zon ensured his colleagues in the Forest Service were aware that Russian and Soviet forestry experience was relevant in the United States. In April 1930, he wrote to Earle H. Clapp, the assistant chief forester in charge of research, proposing to travel to the Soviet Union to attend the Soil Science Congress. Zon wrote: "The results of prairie planting in the southern steppes of Russia could be almost duplicated in the prairie region of our own country." If he were to go, he could "dig... up the facts which should be of benefit to our own work."[109] Zon did not go, because at this time the federal government prohibited its employees from traveling to the Soviet Union in their official capacities.[110]

Such was Zon's interest in the country of his birth that in March 1933 he asked Pinchot to recommend him to President Franklin Roosevelt to serve on a commission on reestablishing diplomatic relations with "Russia." Pinchot declined, as he was not recommending anyone to the new president, but complimented Zon on his "unique knowledge of Russia and USA."[111] After relations were established in November 1933, Zon and his wife were invited to a reception at the new Soviet embassy in Washington, DC. He turned it down, because of the distance from St. Paul and the demands of work, but concluded: "May we extend our sincerest wishes to you and Mrs. Troyanovsky for a happy, enjoyable, and long stay in Washington for the mutual benefit of our countries. With friendliest greetings..."[112] Zon was kept up to date with developments in the Soviet Union by letters from his and his wife's families who lived there and from Soviet scholars. These letters conveyed, sometimes in coded language, the difficulties Soviet citizens were experiencing and also, possibly for the benefit of the censors, propagandistic reports on the successes of state intervention to promote industrial development under the Five-Year

[108] *JoF* 26 (1928), 1037–9, 1043–6. "Gevorkiantz, S.[uren], R.[ubenian]," in *American Men of Science: A Biographical Dictionary*, 7th edition, ed. Jacques Cattell (Lancaster, PA: The Science Press, 1944), 642; Zon Papers, Box 5, Folder 5, Zon to Theodore S. Woolsey, October 17, 1925.

[109] Zon Papers, Box 7, Folder 1, Zon to Mr. E. H. Clapp, April 21, 1930. On Clapp, see Harold K. Steen, *The U.S. Forest Service: A Centennial History* (Durham, NC: The Forest History Society/ University of Washington Press, 2004), pp. 137, 140–1.

[110] Zon Papers, Box 7, Folder 1, R. J. Stuart to Zon, May 8, 1930; Zon to Stuart, May 15, 1930. See pp. 77–8.

[111] Pinchot Papers, Box 708, File Roosevelt, Franklin, D. (Pres.) 1933: Russia, Zon to Pinchot, March 17, 1933, Pinchot to Zon, March 20, 1933 (copy); Zon Papers, Box 8, Folder 4, Pinchot to Zon March 20, 1933 (original).

[112] Zon Papers, Box 8, Folder 7, Zon to Ambassador and Mrs. Troyanovsky, Washington, DC, April 5, 1934.

Plans, the modernization of the villages under the collectivization of agriculture, and the construction of major works of infrastructure, such as the Moscow–Volga canal. Indeed, Zon's sister's son was none other than Genrikh Iagoda, the head of the NKVD (Stalin's secret police) from 1934 to 1936. In early 1934, Zon wrote to Iagoda to intercede on behalf of his wife's brother, Monia, who seems to have been arrested, and died in unclear circumstance the following year.[113] Despite his relatives' troubles, Zon was generally positive about developments in the Soviet Union and admitted to "sentiment" in his attitude to Russia.[114] It is not clear from his writings the extent to which he thought the U.S. government should emulate the Soviet government in intervening in the economy. Zon was clearer in his defense of Soviet forestry against what he saw as "unfair," "prejudiced," and "false" portrayals in the west. In 1939, he offered to write an article, "for the sake of scientific truth and fair play," for *Soviet Russia Today*, a magazine published by the American "Friends of the Soviet Union."[115]

Zon was a fascinating and strong-willed character who did not shy away from saying what he thought and fighting for what he believed in. He relished argument and maintained friendships with men such as Royal Kellogg and Herman H. Chapman, with whom he fiercely disagreed. He was a mercurial character who expected his staff to work hard, show initiative, and be prepared to defend their ideas. It is a sign of his ego that, while he was editor of *Journal of Forestry*, it published an article

[113] Zon's Papers contain many letters from his and his wife's relatives in the Soviet Union. A copy of Zon's letter to Iagoda in early 1934 is not preserved in his papers, but is referred to in other letters. Zon's sister-in-law, Sonia, asked him to write to Iagoda about her husband's "illness," which may have been code for arrest. Following Zon's letter, Monia was assigned to work as a supervisor on the Moscow–Volga canal (built by forced labor). In late 1935, Sonia wrote to Anna Zon that he died suddenly in "hospital" after the return of his "illness," which may also have been coded language. Zon Papers, Box 8, Folder 5, Sonia Puziriskaia to Anna Zon, December 10, 1933; Folder 6, Monia Puziriskii to Anna Zon, n.d. [1934]; Monia Puziriskii to Anna Zon, January 5, 1934; Monia Puziriskii to Anna Zon, February 28, 1934; Box 9, Folder 1, Monia Puziriskii and Sonia Puziriskaia to Anna Zon, July 18, 1934; Folder 4, Sonia Puziriskaia to Anna Zon, December 11, 1935. In 1936, chief forester Silcox wrote to Zon asking for contacts in Moscow and Leningrad as he was planning to visit. Among the people Zon suggested was his nephew "Henry Yagoda," who was "Secretary for Domestic Affairs." Zon Papers, Box 9, Folder 6, Silcox to Zon, June 25, 1936, Zon to Silcox, June 29, 1936.

[114] See, for example, his letters to assistant forester Clapp, Pinchot, and chief forester Silcox in April–June, 1936. Zon Papers, Box 7, Folder 1, Zon to E. H. Clapp, April 21, 1930, Box 9, Folder 6, Zon to Pinchot, May 15, 1936, Zon to Silcox, June 29, 1936.

[115] Zon Papers, Box 10, Folder 7, Jessica Smith, *Soviet Russia Today*, to Zon, March 28, 1939; Folder 8, Zon to Jessica Smith, April 20, 1939; Zon to J. W. Pincus, May 25, 1939. Zon may, or may not, have known that the "Friends of the Soviet Union" was a Communist Front organization. Louis Nemzer, "The Soviet Friendship Societies," *Public Opinion Quarterly* 13, no. 2 (1949), 265–84.

praising his personality, using such words as "brilliant," "original," "genius," "vision," "imagination," "honest," "fearless," "kindliness," "unselfishness," and reporting Bates saying "of course we all love Zon."[116] People either loved or hated Zon. Bates and others at his station in St. Paul, for the most part, seem to have fallen into the former category. Roberts, his co-director of the Shelterbelt Project, likely fell into the latter, but was too professional to say so outright. Zon's superiors at the Forest Service in Washington, DC, were also divided, but were probably glad that he was in St. Paul, Minnesota. A recent historian may have gone too far when he summed Zon up as "arrogant, cantankerous, always in trouble, and hard to like."[117] However, he did get things done. And one of those things was promoting Russian and Soviet research on forestry science and shelterbelts in the steppes to the extent that they were well known among American foresters, including senior figures in the Forest Service, by the early 1930s. His connections, notably with Pinchot, and persistence helped turn this knowledge into a major factor behind the implementation of the American Shelterbelt Project.

Conclusion

Leaving Mr. Zon aside for a short while, we can review attitudes among American foresters to Russian forestry, steppe forestry, and shelterbelts from the pages of *Journal of Forestry* in the early 1930s (after Zon's term as editor). The journal was committed to balance and presented diverse opinions. A major concern among American foresters at this time was the threat of competition from unrestricted exports of cheap lumber by the Soviet government to pay for industrialization.[118] In 1930, the journal published a review of a book by Vysotskii. The reviewer, Sergei Aleksandrovich Wilde, was Russian (of Dutch descent) who had recently arrived in the United States. He emphasized Vysotskii's negative views on steppe forestry: "the artificial planting of the prairie [i.e. steppe] is connected with almost insuperable difficulties, and as a rule has no positive results ... [and] may bring worse moisture conditions for the whole region."[119] The

[116] Edward Richards, "Raphael Zon: The Man," *JoF* 24 (1926), 850–57.
[117] Young, "Warrior of Science," 4–12 (quotation from 10). See also Char Miller, "Militant Forester: Raphael Zon," *Forest Magazine* (Winter 2005), 22–3; Schmaltz, "Raphael Zon."
[118] W. H. Brown, "A Menace to Forestry: Russia," *JoF* 28 (1930), 1170–3.
[119] S. A. Wilde, Review of G. N. Vysotskii, Uchenie o lesnoi pertinentsii (Forest Influences), *JoF* 28 (1930), 762. See also C. B. Davey, "S. A. Wilde: The Man and His Message," (1997), http://depts.washington.edu/s7/SAWilde/, accessed September 29, 2017.

following year the journal contained a review offering a more balanced assessment of steppe forestry, including Vysotskii's views.[120] An even-handed assessment of steppe forestry and shelterbelts had been presented in a concise historical overview of Russian forestry, by Tkachenko, in 1930.[121]

Earlier in 1930, Earle Clapp published a general article on the future needs of American forestry, in which he stated unreservedly:

> Russian investigation of many years seems to show that the gridironing of dry, wind-swept plains with forest belts has been responsible, amongst other benefits, for average annual increases in precipitation of 3 inches. This suggests that we may after checking Russian results under our own conditions wish to extend forest belts into the middle-western plains region.[122]

Clapp was a long-standing colleague of Zon, who may have been his source of information.

Thus, by the early 1930s, Russian and Soviet experience and expertise in steppe forestry in general and shelterbelt planting in particular were well known among American foresters, inside and outside the Forest Service. This knowledge was used by Zon and other members of the Forest Service to support the case for planting shelterbelts in the Great Plains when, for various reasons, it seemed expedient to do so in 1934.

[120] J. Roeser, Review of E. Buchholz, *Der Russische Steppen-Waldbau* (Russian Steppe Silviculture), *JoF* 29 (1931), 1099–102.

[121] M. E. Tkachenko, "Origin and Propagation of Forestry Ideas," *JoF* 28 (1930), 595–617.

[122] Earle H. Clapp, "Our Future Forest Needs," *JoF* 28 (1930), 152–3.

CHAPTER 8

Shelterbelts II

Introduction

By the early 1930s, many American foresters, including leading members of the U.S. Forest Service, knew about Russian and Soviet experience in planting shelterbelts of trees in the steppes. They were familiar with arguments that they moderated the climate. And some were aware that one of the leading Soviet specialists, Grigorii Vysotskii, had reservations about the value of shelterbelts. In the 1920s and 1930s, however, most Soviet specialists advocated the support of their government for shelterbelt planting in the steppes. There was also American experience in planting shelterbelts in the Great Plains. But, the Russian experience went back further and seemed more extensive. One of the leading American advocates of Russian and Soviet forestry science was Raphael Zon, a Jewish émigré from the Russian Empire, who was the director of the Forest Service's Lakes States Forest Experiment Station in St. Paul, Minnesota. Among his colleagues was Carlos Bates, a leading American specialist on plains forestry.

In the early 1930s, Zon and Edward N. Munns, Chief of the Division of Silvics in Forest Service in Washington, DC, argued for a U.S. government program to plant shelterbelts in the Great Plains. They used the Russian and Soviet experience to support their argument. They were backed by Bates. Their actions were expedient for several reasons. First, in November 1932, the United States elected a president, Franklin D. Roosevelt, with a long-standing interest in forestry. Second, both main candidates for the presidency proposed radical cuts in federal government spending, including the Forest Service budget, to address the financial crisis. Third, in a United States already reeling from the Depression and financial crisis, drought returned to the Great Plains in

1932, reducing vegetation cover, allowing winds to blow the top soil, creating the Dust Bowl.[1]

These unfolding economic, social, and ecological crises were the context for Roosevelt's strategy to win and maintain the presidency and Democratic control of Congress. He sought advantage over his Republican opponents, who were traditionally strong in the Great Plains and Western United States. Roosevelt launched his presidential campaign in North Dakota in January 1932, and made key speeches in Topeka, Kansas, in September of the election year. He returned to North Dakota in the summer of 1934, shortly after the Shelterbelt Project was launched, in the run up to the mid-term elections later that year. His "New Deal" benefitted the Great Plains and the West, where Republican success would threaten the Democrats' strategy. The Shelterbelt Project was part of the New Deal program of major infrastructure and conservation projects, most famously the big dams, in the West during the 1930s. Roosevelt's and the Democratic Party's electoral success in the region throughout the 1930s and into the 1940s are testament to their strategy's success.[2]

Launching the Shelterbelt

The Great Plains Shelterbelt Project was launched twenty months after Roosevelt's election in November 1932. This had created an opportunity for American foresters to promote their cause as the new president was keenly involved in forestry. He had overseen planting hundreds of thousands of trees on his estate at Hyde Park in New York State. He promoted reforestation when he was a New York state senator in 1911–13 and governor in 1929–32. He recognized the importance of forestry science by seeking advice from the State College of Forestry at Syracuse. He also consulted Gifford Pinchot, former chief of U.S. forestry, despite Pinchot's allegiance to the Republican Party. Even while president, Franklin Roosevelt referred to himself as a "tree farmer" and sought refuge among his trees at difficult times.[3] When he

[1] See Geoff Cunfer, *On the Great Plains: Agriculture and Environment* (College Station: Texas A&M University Press, 2005), pp. 44–5; Hurt, R. Douglas. *The Big Empty: The Great Plains in the Twentieth Century* (Tucson: University of Arizona Press, 2011), pp. 91–4.

[2] See David M. Wrobel, *America's West: A History, 1890–1950* (Cambridge, England: Cambridge University Press, 2017), pp. 130–51 158–61, 192; Sarah T. Phillips, *This Land, This Nation: Conservation, Rural America, and the New Deal* (Cambridge, England: Cambridge University Press, 2007). See also, p. 313.

[3] See Brian Black, "The Complex Environmentalist: Franklin D. Roosevelt and the Ethos of New Deal Conservation," in *FDR and the Environment*, ed. Henry L. Henderson and David B. Woolner (New York: Palgrave Macmillan, 2005), pp. 19–47.

accepted the nomination at the Democratic National Convention in Chicago in July 1932, Roosevelt spoke about unemployment, abandoned farms, and the prospects of soil erosion and timber shortage. He proposed to employ "a million men" in "reforestation." He was, however, speaking about the territory east of the Mississippi.[4] Following discussion with foresters, Roosevelt indicated that "reforestation" covered forestry work more broadly.[5] He also directed his attention west of the Mississippi. In a story that has often been told, but is hard to pin down to an original source, when Roosevelt was campaigning in Montana in the summer of 1932 his train was held up near Butte. While the candidate and accompanying journalists were surveying the empty plains from the heat of their railroad car, Roosevelt allegedly remarked: "Boys, when I'm President I'm going to plant up all this country to trees." When the Shelterbelt Project was underway, he considered it "his baby."[6]

Plains forestry and shelterbelts were not, however, the main concerns of Zon, Pinchot, and their allies on the eve of Roosevelt's inauguration. Rather, they looked back to the cause of conservation they had championed during Theodore Roosevelt's presidency, and hoped his namesake would pursue a similar policy. On December 2, 1932, Zon asked Pinchot to lead a new fight for conservation. Pinchot replied: "You bet I will."[7] On January 20, 1933, Pinchot, who was on first name terms with the incoming president, sent him a proposal for conservation, reforestation, and extending public ownership of forest lands.[8] Thus, Pinchot, and indirectly Zon, were in contact with Franklin Roosevelt about forestry from the very start of his presidency.[9]

A more urgent cause for the Forest Service was protecting its funding. The entire research budget of the United States Department of Agriculture

[4] Edgar B. Nixon (ed.), *Franklin D. Roosevelt and Conservation, 1911–1945*, 2 vols. (Hyde Park, NY: NARA, 1957), vol. 1, pp. 112–13.

[5] Nixon, *Roosevelt and Conservation*, vol. 1, pp. 113–27; Wilmon H. Droze, *Trees, Prairies, and People: Tree Planting in the Plains States* (Denton, TX: Texas Woman's University, 1977), pp. 48–77.

[6] See NSHS SAMD, RG 3941.AM: Paul Henley Roberts Papers, 1891–1971 (hereafter Roberts Papers), Box 2, Folder 8, Munns to Roberts, May 28, 1963; Droze, *Trees*, pp. 50, 62.

[7] MHS, P1237, Raphael Zon Papers, 1887–1957 (hereafter Zon Papers), Box 8, Folder 2, Zon to Pinchot, December 2, 1932 (copy), Pinchot to Zon, December 5, 1932 (original); LoC MD, Gifford Pinchot Papers (hereafter Pinchot Papers), Box 587, File Program proposal to Franklin D. Roosevelt, Zon to Pinchot, December 2, 1932 (original); Pinchot to Zon, December 5, 1932 (copy).

[8] Nixon, *Roosevelt and Conservation*, vol. 1, p. 132; Zon Papers, Box 8, Folder 3, Pinchot to Zon, January 20, 1933.

[9] Although Pinchot was a Republican, he had been "weak" in his support for his party's candidate, Hoover, in the election. E. Jeffrey Ludwig, "Pennsylvania: The National Election of 1932," *Pennsylvania History: A Journal of Mid-Atlantic Studies* 31, no. 3 (July, 1964), 349.

(USDA), including the Forest Service, was under threat as part of planned cuts in federal government spending. In Roosevelt's first 100 days, Congress enacted the Economy Act of March 20, 1933, which made massive cuts in government spending, and the National Industrial Recovery Act of June 16, 1933, which amongst other provisions appropriated large funds for relief in public works programs.[10] If the Forest Service could not protect its research budget, it could try to get emergency relief funds to support its activities. The USDA research budget was vital to Zon, since it paid for his experiment station in St. Paul. Zon had been warned of the threat to his station by Earle H. Clapp, the assistant chief forester who was in charge of research, in July 1932. Clapp urged Zon to "fight" for his station.[11] (Zon was no stranger to such battles, and had recently campaigned for more funds for "forest investigations in the northern plains."[12]) Pinchot took up the cause. On April 17, 1933, he wrote to Roosevelt, urging him not to cut the USDA's entire research budget. Around the same time Clapp informed Zon that the budget director "has met with such protest from all sections of the country it may be necessary for him to alter his program [to eliminate research funds for agriculture and forestry] considerably." Zon immediately wrote to Pinchot: "The situation with regard to the elimination of research, . . ., is still very serious. If you should happen to be in Washington soon and see the President, the hastily conceived plan of the Director of the Budget could probably be staved off or at least modified. The Forest Service places its almost entire reliance on your intercession." Pinchot wrote to Secretary of Agriculture Henry A. Wallace urging him to support continued funding for research. Wallace forwarded Pinchot's letter to the president. Pinchot sent copies of the correspondence to Zon.[13]

[10] For the impact on the USDA, see David Rice Gardner, *The National Cooperative Soil Survey of the United States* (Washington, DC: USDA Natural Resources Conservation Service, 1998) ("Thesis presented by David Rice Gardner to the Graduate School of Public Administration, May 1957, Harvard University"), pp. 199–200. See also Anthony J. Badger, *FDR: The First Hundred Days* (New York: Hill and Wang, 2008), pp. 48–105; Julian E. Zelizer, "The Forgotten Legacy of the New Deal: Fiscal Conservatism and the Roosevelt Administration, 1933–1938," *Presidential Studies Quarterly* 30, no. 2 (2000), 334–41.

[11] Zon Papers, Box 8, Folder 1, EHC [Earle H. Clapp], to Zon, July 29, 1932.

[12] Zon Papers, Box 7, Folder 1, Zon to A. D. Stedman (journalist for St. Paul Dispatch and St. Paul Pioneer Press), April 16, 1930; Zon to Prof. F. E. Cobb, North Dakota School of Forestry, Bottineau, ND, May 28, 1930. He was not successful on this occasion.

[13] Zon Papers, Box 8, Folder 4, Pinchot to FDR, April 17, 1933; Augustine Lonergan, Senate Committee on Finance, to Mr. E. B. Field, April 25, 1933 (Copy forwarded to Zon by E. H. Clapp); Zon to Pinchot, April 29, 1933; Pinchot to Henry Wallace, May 8, 1933; Pinchot to Zon, May 22, 1933.

Zon also had support at the top of the Forest Service. On May 27, 1933, Chief Forester Robert Y. Stuart wrote to him:

> There is still great uncertainty as to how and when the axe will be wielded on the Department ... The matter is held very close to the confines of the Bureau of the Budget. I feel reasonably sure, however, that the emphasis which has been given to the need for caution in slashing research will have its effect, if that already has not been the case. Your guess is as good as mine as to what the final result will be.

On June 29, 1933, Clapp warned Zon of a possible twenty-five percent cut in research funding. He also sent him information about the fight against the cuts by agricultural experiment stations and colleges. Clapp suggested Zon try to interest "the President personally with forest research as an activity, of getting him to realize its value ... If this could be done it would undoubtedly be the key to the other difficulties we are facing." The danger of cuts grew. On June 30, 1933, Clapp sent a typed letter to Zon warning that the cuts could be worse than twenty-five percent, continuing:

> As in the past, decision as to whether any action at all be taken is entirely yours, as is also the form of action. In the last analysis it boils down to accepting without protest action proposed by the Director of the Budget, who apparently has no appreciation of what research means in forestry ..., or trying to prevent such action by all legitimate means. Obviously the odds are against prevention. It is an uphill struggle at best.

Clapp wrote in long hand at the bottom: "After reading please destroy these letters." Zon (who evidently ignored this), replied to Clapp on July 7, 1933 that he did not have to urge him to take action as he was "pretty alive to any real threat to our research activities."[14] Clapp was well aware of Zon's combative nature. A few months earlier he had reprimanded him for outspoken conduct in a spat with Herman H. Chapman of the Yale School of Forestry.[15] It is likely that this time Clapp was counting on Zon's outspokenness. Zon's situation was very serious. In July 1933, the annual budget for his Lakes States Forest Experiment Station was cut by twenty-eight percent from $98,000. Temporary staff were fired and its work cut back.[16]

[14] Zon Papers, Box 8, Folder 4, R. Y. Stuart to Zon, May 27, 1933; EHC (E. H. Clapp), to Zon, June 29, 1933; "Slash in Federal Activity Held Up," *Washington Star*, June 23, 1933 (typed copy); Clapp to Zon, June 30, 1933; Folder 5, Zon to Clapp, July 7, 1933.

[15] Zon Papers, Box 8, Folder 3, Clapp to Zon, January 10, 1933, Zon to Clapp, January 14, 1933. On the spat, see Zon Papers, Box 8, Folder 2, H. H. Chapman to Zon, December 13, 1932, Zon to Chapman, December, 26, 1932.

[16] Paul O. Rudolf, *History of the Lake States Forest Experiment Station* (St. Paul, MN: North Central Forest Experiment Station, Forest Service, USDA, 1985), pp. 13–14.

Over the following year, the Forest Service's plan to combat cuts to its research budget, Clapp's challenges to Zon to fight for his experiment station and interest the President in forestry research, Zon's and Munns' knowledge of Russian forestry expertise, including shelterbelts, and Bates' experience of tree planting in the Great Plains all came together. Crucially, they coincided with the president's interest in forestry and wish to address the crises in the Great Plains. These culminated in July 1934 with Roosevelt's endorsement of the Shelterbelt Project and the allocation of funds to Zon's station to carry out the research for it. When Paul Roberts – the project's administrative director from the summer of 1934 – started work on his history of the project in the 1960s, he was aware of, and resented, the way Zon had emphasized his role in the project (See p. 281). But, Roberts was unaware of the extent to which Zon had been involved in the maneuvering that led to the project's launch and the full importance of Russian experience.[17]

President Roosevelt was thinking about forestry in the spring and summer of 1933. He worked with Secretary of Agriculture Wallace and Chief Forester Stuart to expand national forests in the South, where unemployed labor would be mobilized by the new Civilian Conservation Corps. On July 26, 1933, Roosevelt allocated $20 million for this purpose.[18] Perhaps recalling his visit to Montana the previous summer, Roosevelt was also contemplating planting trees in the Great Plains. The precise events are blurred as not all the documentation has survived. Roosevelt bypassed normal chains of authority, consulting informally with Chief Forester Stuart, rather than through Secretary Wallace, and may not have kept full records. When Roberts asked Munns for his memories of the events of that summer in the 1960s, Munns recalled that Roosevelt "sent a note to Bob Stuart on a piece of bluish paper ... printed White House – Dear Major: What can be done to plant the great American desert, F. D. R." The president was thinking of a great wall of trees several miles deep, west of the 100th meridian, from the Canadian border to Texas to protect the plains from encroachment by the "desert" to the west.[19] The president's original note has not survived. Edgar L. Perry could not find it when he

[17] See Roberts Papers, Box 3, Folder 11, Roberts to Munns, March 14, 1968; Roberts to Munns, March 26, 1968.

[18] Nixon, *Roosevelt and Conservation*, vol. 1, pp. 163–98; Droze, *Trees*, pp. 58–9.

[19] Roberts Papers, Box 2, Folder 8, Munns to Roberts, May 28, 1963. See also E. L. Perry, *History of the Prairie States Forestry Project* (Washington, DC: U.S. Forest Service, 1942), p. 9a; Droze, *Trees*, p. 62; Joel Jason Orth, "The Conservation Landscape: Trees and Nature on the Great Plains," (PhD dissertation, Iowa State University, 2004), 145–6.

wrote his official history of the project in 1942. Munns thought Melville Cohee had taken it for his autograph collection (see p. 282).[20]

Following the president's note to Stuart, the Forest Service tried to work out what he wanted, and then tried to persuade him it would be more appropriate to plant strips of trees, or shelterbelts, in a north–south zone further east in the plains. On August 15, 1933, Stuart sent a memorandum, entitled "Forest Planting Possibilities in the Prairie Region," with an unenthusiastic covering letter, to Wallace, who forwarded it to Roosevelt. The memorandum began: "Forest plantations in the true prairie or plains region have proved effective only in the form of shelterbelts." It continued that shelterbelt plantings had been successful in many parts of the region, and "considerable numbers of tree species" had been found that were "satisfactory for this purpose." Existing "forest strips" had a marked influence on the wind, reducing velocities over distances several times the heights of the strips. This reduced evaporation, saving "considerable quantities of moisture for crop production." The strips of trees also prevented snow blowing away, conserving meltwater which was absorbed into the ground. The memorandum concluded: "Forest strips are a form of insurance against the evil effects of drouth."

In advising the president to plant trees further east, the Forest Service was drawing on its experience in the western plains, for example, the failed Kansas National Forest (see p. 293). In advising the president to plant shelterbelts, it was drawing on Russian experience:

> This conclusion had been reached in Russia 50 years ago; at that time a special commission of soil experts recommended forest-strip planting in the Ukraine as one means of overcoming drouth and promoting land settlement. This system in which the plains are gridironed with forest strips is now an accepted part of the program of land settlement in southern Russia. Reports on Russian studies in the Ukraine state that the annual precipitation on plains areas protected by a widespread system of forest strips was 3 inches (16 per cent) greater than that on similar areas not so protected. The increase in available moisture was reflected in an increase in crop yields per unit of area.

The "commission of soil experts" was Dokuchaev's "Special Expedition" of the 1890s (see p. 286). The memorandum mentioned similar results in Poland, Nebraska, and Canada, but the Russian experience was presented in most detail. Near the end, the author of the memorandum stated that shelterbelts' "beneficial effects as to climate would spread far beyond their

[20] Roberts Papers, Box 2, Folder 10, Munns to Roberts, November 2, 1966; Perry, *History*, p. 9a

immediate vicinity."[21] This controversial claim would come back to haunt the Forest Service.

The memorandum of August 15, 1933 is very important for this analysis of the "Russian roots" of Great Plains agriculture, because it made explicit reference to Russian experience of shelterbelts in the steppes, and because it was a major step towards the American Shelterbelt Project. But, who wrote it? No author is named. Edgar Nixon, who included it in a volume he edited for the Roosevelt Library, thought it was "probably written" by Zon, as he planned the project.[22] Historian Joel Orth wrote that the contents certainly suggested Zon, possibly in collaboration with Munns or others.[23] Yet, Zon's papers contain no reference to it, and he was not shy in recording his achievements. Since Russian experience of steppe forestry was well known among American foresters, partly because of Zon's efforts, the memorandum could have been written by other foresters. Perry wrote that the author was Munns.[24] This was confirmed by Munns in a letter to Roberts in the 1960s: "So, anyway, the memo was mine." Munns explained that after Stuart had received Roosevelt's note, he had asked his advice. Munns did some research in the Library of Congress, and had some exchanges with the White House, before coming up with the shelterbelt idea. Stuart involved other agencies, including the Bureau of Plant Industry, experiment stations, and the Bureau of Chemistry and Soils. No other agency wanted the job, so Stuart instructed Munns to write up the memorandum with advice from Clapp and others in the Forest Service. After Munns' memorandum had been sent to the White House, copies were circulated to foresters in the Great Plains and to Zon's experiment station in St. Paul, Minnesota. Munns wrote that he discussed it with Zon when he was in Washington only "some time after."[25]

Another version of the story of the memorandum, ascribing a major role to Zon and Bates, appeared in a history of the Lakes States Forest Experiment Station by Paul Rudolf published in 1985. Rudolf was an assistant forester at the station in the 1930s. His account was based on a letter dated 1976 from Hardy Shirley, who also worked at the station in the 1930s. According to Shirley and Rudolf, in August 1933, Roosevelt sent a note to Secretary Wallace asking whether the Forest Service could

[21] The memorandum is published in Nixon, *Roosevelt and Conservation*, vol. 1, pp. 199–203.
[22] Nixon, *Roosevelt and Conservation*, vol. 1, p. 203.
[23] Orth, "Conservation Landscape," 146–7, 183, n. 9. [24] Perry, *History*, p. 9a.
[25] Roberts Papers, Box 2, Folder 8, Munns to Roberts, May 28,1963; Folder 10, Munns to Roberts, November 2, 1966; Box 3, Folder 11, Roberts to Munns, May 11, 1968, Munns to Roberts, May 22, 1968.

plant a band of trees across the Great Plains to stop the dust storms. Wallace sent the note to Chief Forester Stuart, but his staff were "at a loss to reply." Zon, who happened to be in Washington, saw the note and apparently said that the president's idea was a good one, but needed elaboration. Zon returned to St. Paul and discussed the matter at a staff meeting. Bates provided the elaboration, laying the basis for a zone planted with shelterbelts.[26] Shirley had repeated to Rudolf a story he had included in a tribute to Zon in *The New York Times* in 1956.[27] Roberts was greatly annoyed when he read Shirley's tribute, and doubted its veracity.[28] Munns, in a letter to Roberts, emphatically refuted Shirley's story. He recalled that Zon visited Washington many times on his own initiative and "once in a while" was called in by Clapp, but not by the Chiefs of the Forest Service. Munns wrote that he sometimes sent Zon information on forestry matters, and when Zon was in Washington they had long, informal conversations. But, Munns was not aware of "real formal consultations" with Zon, and Zon had never mentioned this one to him. He confirmed that Zon knew nothing of the shelterbelt until after the initial notes between him and Roosevelt.[29] There are other problems with Shirley's story. Other sources are clear that Roosevelt sent the note Stuart, not Wallace, and there is no record that Zon visited Washington in August 1933. Elsewhere in Rudolf's book, moreover, he wrote that the staff meeting in St. Paul he referred to had taken place in 1934.[30] This later date is more likely and gives credence to part of the story, but not Zon's authorship of the memorandum (See p. 323).

The memorandum of August 15, 1933 outlining the idea for shelterbelts, which we know was written by Munns, who drew heavily on Russian experience, had almost immediate effect. On August 19, Roosevelt asked Chief Forester Stuart for costings for a series of shelterbelts from Texas to Nebraska.[31] Stuart estimated the costs at $1,982,995 for saplings and planting them, and further $20 an acre to purchase and fence the land, making total of $5,040,275. The president asked Secretary Wallace whether the cost could be cut to $1 million if the land was donated by farmers.[32] At this point, however, the plan was shelved. Foresters inside and outside the

[26] Rudolf, *History*, pp. 43–4.
[27] Hardy Shirley, "Tribute to Raphael Zon," *NYT*, November 12, 1956.
[28] Roberts Papers, Box 3, Folder 11, Roberts to Munns, March 14, 1968; Roberts to Munns, March 26, 1968.
[29] Roberts Papers, Box 3, Folder 11, Munns to Roberts, March 22, 19[68].
[30] Rudolf, *History*, pp. 20–1.
[31] Nixon, *Roosevelt and Conservation*, vol. 1, pp. 198–9.
[32] Perry, *History*, p. 10.

Forest Service were divided over its feasibility. There were discussions whether it would be the responsibility of the Forest Service or the new Soil Erosion Service under Hugh H. Bennett, who was appointed on September 19, 1933. Munns persisted and sent a new memorandum to Stuart on October 17, 1933.[33] This had no result as, a few days later, Stuart died in a fall from the window of his seventh-floor office. His replacement as Chief Forester, Ferdinand A. Silcox, was appointed in November.[34]

Not to be deterred, Munns – who is emerging as a key progenitor of the Shelterbelt Project and conduit of Russian experience – gathered information on shelterbelt plantings in the United States and, especially, abroad. According to historian Wilmon Droze, in the fall of 1933, Munns sought support from Zon and Bates as they knew about shelterbelt planting in Russia, North America, and elsewhere.[35] Munns assembled a team of five translators to carry out a "world-wide study" of shelterbelts in the Library of Congress. He obtained funding from the Civil Works Administration (CWA; another New Deal job-creation scheme). The CWA was short-lived, lasting from November 8, 1933 to March 31, 1934, which allows us to date the translation project quite precisely. The translation team's head was a Russian émigré lawyer who had sought refuge in the United States after the Bolshevik Revolution. Munns recalled that he was a "real intellectual" with "excellent" English. He was also very hard working and returned from the library "staggering in with an armload" of material. At least one more of the translators was Russian, and most of the material they located concerned Russian steppe forestry and shelterbelts. Munns remembered that the Russian material was "worthwhile," and that when the Shelterbelt Project "became a reality" in 1934, they "already had a lot of material available."[36]

[33] Droze, *Trees*, pp. 66–8; Douglas Helms, "Hugh Hammond Bennett and the Creation of the Soil Erosion Service," *Journal of Soil and Water Conservation* 64 (2009), 68–74.

[34] It is not clear whether Stuart's death was an accident or suicide brought on by work-related stress. Harold K. Steen, *The U.S. Forest Service: A Centennial History* (Durham, NC: The Forest History Society/University of Washington Press, 2004), pp. 196–8.

[35] Droze, *Trees*, pp. 71–2.

[36] Roberts Papers, Box 2, Folder 10, Munns to Roberts, February 21, 1966, Munns to Roberts, November 2, 1966; Box 3, Folder 11, Munns to Roberts, May 22, 1968. See also Droze, *Trees*, p. 69. On the CWA, see Bonnie Fox Schwartz, *The Civil Works Administration, 1933–1934: The Business of Emergency Employment in the New Deal* (Princeton, NJ: Princeton University Press, 2014). Some sources state that the translation team was funded by the Works Progress Administration (WPA), which is mistaken since it was not created until May 6, 1935, after the translations had been completed. The American Presidency Project, Franklin D. Roosevelt, Executive Order 7034 – Creating Machinery for the Works Progress Administration, May 6, 1935, available online at www.presidency.ucsb.edu/documents/executive-order-7034-creating-machinery-for-the-works-progress-administration, accessed October 31, 2019.

While Munns was collecting literature on shelterbelts around the world, in particular Russia, and talking to Zon to build his case for the plan to be revived, renewed dust storms in the Great Plains in the spring of 1934 made a more immediate case for action. The dust storms were witnessed by two senior members of the U.S. Soil Survey who had both been heavily influenced by Russian soil science. Curtis F. Marbut, who headed the Soil Survey Division, and Charles E. Kellogg, who was soon to succeed him, advocated planting shelterbelts in the Great Plains in the light of Russian experience. In May 1934, Marbut toured parts of Kansas affected by dust storms. He wrote to A. G. McCall, at the Bureau of Chemistry and Soils, to recommend measures to combat the storms. One that he felt: "has great value is the establishment of wind breaks. This has been done in Russia with apparent favorable results."[37] Kellogg, who was familiar with the northern plains as he had worked in Fargo, North Dakota, was also touring the Great Plains in the spring of 1934. On May 14, he recommended to the chief of the Bureau of Chemistry and Soils:

> Shelterbelts have been found quite practical for protection of soil and plants in parts of the USSR, having conditions of climate and soil quite similar to those in the eastern part of the Great Plains in America. Such belts of trees, planted in strips through the fields at right angles to the direction of the prevailing winds, serve to protect the soil from blowing when drought conditions prevail. Their chief value lies in the protection of plants against hot winds. Evaporation is materially reduced. The use of similar plantings of trees have [sic] not been given sufficient trial in the United States and there exists a considerable possibility of success in the more moist portions of the middle west.

He mentioned difficulties in planting shelterbelts in the drier, western parts of the northern plains, but continued: "Possibly further research may disclose varieties of trees and cultural methods such that shelterbelts can also be grown extensively in this section. But up to the present the experience of the farmers has not been encouraging."[38] Kellogg had been aware of Russian experience of planting of shelterbelts for some time. In 1930 and 1931, he wrote to Soviet scientists A. F. Lebedev and

[37] NAL, Special Collections, Manuscript Collection 91, The Charles Edwin Kellogg Papers (hereafter Kellogg Papers), Box 28, Folder 588, C. F. Marbut, Manhatten, KS, to Dr. [A. G.] McCall, [Chief of Soil Investigations, Bureau of Chemistry and Soils, USDA, Washington, DC], received May 23, 1934. (It is not clear why Marbut's letter to McCall is in Kellogg's papers.)

[38] Kellogg Papers, Box 28, Folder 588, [Kellogg], "The Mid-Western and Northwestern Drought', headed: "Radio Chief of Bureau." May 14, 1934, p. 9; MS dated June 1, 1934. See also Charles E. Kellogg, "Soil Blowing and Dust Storms," *USDA Miscellaneous Publication*, vol. 221 (March 1935), 8. On Kellogg, see pp. 271–2.

Arsenii Iarilov for information "regarding the use of rows of trees to conserve soil moisture."[39]

Kellogg's reservations contrasted with Munns' and Zon's enthusiasm for a project based in part on Russian forestry science. Munns considered the shelterbelt plan a possible solution to the worsening drought and dust storms in the Great Plains. According to Munns' recollections over three decades later, after speaking to Clapp, he wrote to Zon for help in renewing the Forest Service's interest in the plan and in jogging the president's memory. He asked Zon to write directly to Chief Forester Silcox. Zon, pressed by Bates, proposed instead to come to Washington, DC.[40] Munns' later recollections are largely corroborated by a letter from Zon to Munns at the time. On May 24, 1934, Zon wrote that he would see him in Washington on June 3 or 4 as he wanted to discuss "our investigative program ... and other matters on which I awaited in vain for your comments."[41] Zon's letter suggests that he was still agitated about his station's budget, which was due for renewal or further cuts in July, and that the two men had been in recent contact. Returning to Munns' later recollections, in March 1968, he wrote to Roberts, that Zon had not been involved in the origins of the Shelterbelt: "Later Zon told me that when he got my letter, he discussed the plan with Bates who was enthusiastic, and then had a staff meeting ... Bates had insisted that immediate action was needed ..."[42] It is likely that this is the meeting, which was actually held in May 1934, that Rudolf mistakenly dated to 1933 (see p. 320), which Shirley had told him about and had written about in his tribute to Zon in *The New York Times*. Thus, Zon was involved in the elaboration of the proposal, but not its initiation.

What we do know from contemporary evidence is that in May 1934, under Zon's direction in St. Paul, Bates, Joseph Stoeckeler, and other researchers produced a position paper entitled: "A Plan for Immediate 'Drought Relief' and for the Permanent Benefit and Protection of the Great Plains through Extensive Windbreak Planting." The plan called for

[39] ARAN, f.487, op.1, 1929, d.18, l.306, Charles E. Kellogg to A. Iarilov, February 9, 1931. There is no record of a reply in the Russian archive. Both Soviet scientists had visited the United States, which may explain how Kellogg knew them. Lebedev had attended a botanical congress in 1926. A. F. Lebedev, "V Biuro Upolnomochennykh Vsesoiuznykh S"ezdov po pochvovedeniiu (pis'mo iz Ameriki)," *Biulletini pochvoveda*, nos. 5–7 (1926), 54–7. Iarilov had attended the Soil Science Congress in 1927. See p. 94.

[40] Roberts Papers, Box 2, Folder 10, November 2, 1966; Box 3, Folder 11, Munns to Roberts, March 22, 1968.

[41] Zon Papers, Box 8, Folder 7, Zon to E. N. Munns, May 24, 1934.

[42] Roberts Papers, Box 3, Folder 11, Munns to Roberts, March 22, 1968.

planting windbreaks in a belt, or zone, 100 miles wide from Canada to the Gulf of Mexico. It was presented as "merely a modification of one proposed by the President last year, which is further demonstration of his remarkable vision." The rest of the document was a clear and concise presentation of a plan for "the amelioration of unfavorable climatic elements." The authors explained: "If the surface velocity of the wind over a wide territory can be broken and decreased even slightly, soil will be held in place, the moisture of the soil will be conserved, and havens of shelter will be created for man, beast, and bird." Bates wrote a separate paper for Zon, explaining that they were not arguing that the trees would "change the climate" or prevent drought recurring. Rather, the windbreaks would create "calm havens over considerable areas," which would have a significant effect on "all forms of life."[43] Bates' claims were more modest than those in Munns' memorandum the previous summer.

In early June 1934, Zon took the plan to Washington, DC, to try to convince Chief Forester Silcox and Clapp. Munns recalled Zon's visit in the 1960s:

> I got a call to visit Clapp and there found Zon. He was agitated and walking up and down. I gathered that Zon was pressing that the Plains Plan be dragged out and that Clapp was opposing. Apparently he [Clapp] wanted me to side with him, but I said it was a good plan ... So Clapp agreed. Then the Chief [Silcox] approved and made representations to FDR, and Zon went home. No word from the White House until about noon on a Saturday when I got a call to ... go to see the publicity officer.[44]

On Zon's role, Munns concluded: "[Zon] helped seal the F[orest] S[ervice] and get things rolling."[45]

We have shorter, contemporary, accounts of what happened in Washington, DC, from Zon himself. On June 5, 1934, he wrote to his wife Anna at home in St. Paul: "Presented my plan for drouth relief – only $72 million dollars for planting shelterbelts in the prairie region. It will not go through of course." He added that he wanted to see Pinchot.[46] He did

[43] A copy of the Plan, dated June 6, 1934, is in Zon Papers, Box 8, Folder 7. The Plan was completed in late May. Droze found a copy dated May 28, 1934 in the files of the Shelterbelt Laboratory at Bottineau, North Dakota, where he also found the second paper by Bates. Droze, *Trees*, pp. 73–6, 263, n. 86, 264, n. 87. The terms "windbreak" and "shelterbelt" were used interchangeably.

[44] Roberts Papers, Box 2, Folder 8, Munns to Roberts, May 28,1963. Munns later gave slightly different versions, but the key points were the same. Box 2, Folder 9, Munns to Roberts, June 21, 1965; Box 3, Folder 11, March 22, 1968.

[45] Roberts Papers, Box 2, Folder 9, Munns to Roberts, June 21, 1965.

[46] Zon Papers, Box 8, Folder 7, Raphael Zon to Anna Zon, from Washington, DC, June 5, 1934.

not have time to travel to Pinchot's home in Milford, Pennsylvania, but wrote to him on June 19, 1934:

> The president has been interested, for some time, in starting a continuous windbreak of several miles wide through the Plains region from North Dakota to the Gulf of Mexico. I was given the job to work out the plan in detail and it has been approved by the president in general terms but it still remains to get a slice of the drought relief money to carry out the project.

The last sentence points to Zon's urgent need to replace the funds cut from his station's budget with money from emergency programs. Pinchot replied that he "quite underst[oo]d that the great project you have in hand is the thing to be stuck to. More power to your elbow and every success."[47]

The program, which was approved by Secretary Wallace for the president on June 18, 1934, was based on the plan drawn up by the Lakes States Forest Experiment Station under Zon's direction in May, with the addition of charts presenting data on the climatic influences of shelterbelts from Russia and other countries.[48] The president did not sign the executive order launching the Shelterbelt Project until July 11, as he was on vacation, and it was not published until July 21. The delay prompted some anxious correspondence between Munns and Zon.[49] Munns and Zon were two of the men, together with Bates, who had done most to turn the president's note about planting trees in the "Great American Desert" in August 1933 into a project funded by the federal government. And, arguments based on Russian experience of planting shelterbelts in the steppes played a key role in winning support for the project in the Great Plains.

There is further evidence for the key role played by Zon. On June 26, 1934, Silcox wrote: "Zon has taken a very active and enthusiastic part in promoting the shelterbelt planting project. . . . Zon will have a very active part in the project, – so much so that in all probability he is going to outline the plans."[50] In September 1934, Silcox offered him a post in

[47] Zon Papers, Box 8, Folder 7, Zon to Pinchot, June 19, 1934, Pincot to Zon, June 26, 1934.

[48] Perry, *History*, pp. 10–11 summarized the program of June 18, 1934, which had exactly the same title as the document Zon brought from St. Paul. Perry referred to the additional data. Roberts confirmed Wallace's approval of the program on June 18 and the influence data from abroad. Roberts Papers, Box 2, Folder 9, Munns to Roberts, June 21, 1965; Box 3, Folder 11, Roberts to Ray [Marsh] August 19, 1968.

[49] Zon Papers, Box 9, Folder 1, Ed [E. N. Munns] to Zon, July 3, 1934, Ed [Munns] to Zon, July 14, 1934, Ed [Munns] to Zon, July 9, 1934.

[50] Zon Papers, Box 8, Folder 7, March–June 1934, Silcox to Ward Shepard, June 26, 1934. Shepard sent a copy to Zon. Shepard was a forester who worked for the Bureau of Indian Affairs. Archives at Yale, Ward Shepard Papers, Biographical Sketch, available online at https://bit.ly/2N3Uxbl, accessed April 19, 2018.

Washington, DC. Zon was keen to return to the center of activity from what he called his "exile among the Swedes and Finns" in Minnesota. Nevertheless, his reply reveals some hesitation:

> The only project [in St. Paul] that gives me some concern is the shelterbelt. I feel a certain moral responsibility for it, particularly to you. Being fully convinced of the feasibility and worthwhileness of the project, I urged it upon you. There is a great deal at stake for the prestige of the Forest Service on you personally. I would like to see it launched right with the assurance that we will get the most out of it.[51]

Zon thus thought his own role was key. So did Munns. In 1963, Munns wrote to Roberts, who had hoped to minimize Zon's role (see pp. 281, 317): "The question has been raised time and again as to Zon's part: he, prompted by Bates, shoved the F[orest] S[ervice] into active initiation of the project." He added that Zon disagreed with him over aspects of the plan, including planting trees along roads, and as a result he cut Munns out of further planning. "[T]hereby [Zon] took whatever credit there was for the project." Munns seems to have been more indulgent of Zon's attention-seeking than Roberts and did not bear a grudge. Munns also corrected Roberts' belief that Zon was responsible for the publicity, which was to cause problems, on the alleged climatic effects of shelterbelts. Munns stated that, from the very beginning, he had stressed their value "as windbreaks, as modifiers of local climates, as saviors of winter moisture through snow accumulation and evaporation reduction. This I believed most earnestly and still do."[52]

Russian expertise played a crucial role in the preparation and approval of the Great Plains Shelterbelt Project. Both Munns and Zon had drawn on the Russian studies as arguments for the shelterbelt. Russian expertise was cited in most of the key documents promoting the plan between August 1933 and June 1934. It was also cited prominently in the official publicity: the press release of July 21, 1934 and article in *The New York Times* of July 29, 1934. The Russian angle attracted attention in the American press in the summer of 1934. News of the project was leaked by The American Tree Association in June, before the official press release. On June 28, the *St. Louis Post-Dispatch* reported that the president was considering planting

[51] Pinchot Papers, Box 99, General Correspondence, File Zon, Raphael, Forest Service, Copy of letter from Raphael Zon to F. A. Silcox, September 25, 1934, send by Zon to Pinchot, September 26, 34. For Zon's remark about his "exile," see Zon Papers, Box 8, Folder 5, Zon to Edward Richards, August 24, 1933.

[52] Roberts Papers, Box 2, Folder 8, Munns to Roberts, May 28,1963; Postscript, May 29, 1963.

a shelterbelt of trees from Canada to the Texas Panhandle. It continued that in experiments dating back to 1836 the Russian government had found that such belts increased rainfall, tempered heat waves, checked wind velocity, and prevented wind erosion. Silcox confirmed that such a plan was under consideration.[53] On August 19, 1934, *The Kansas City Star* ran an article on the Shelterbelt. The paper asked Munns "what if any accurate information is obtainable from Russian plantings where similar experiments and efforts made?" Munns replied: "Between 25,000 and 50,000 acres have been protected in Russia by shelterbelts. Their program, [which] started as far back as 1860, is ultimately planned to protect 100,000,000 acres. In Russia as well as in the great plains of Hungary and Rumania, the shelterbelt plan has been worked to success in conserving moisture in the land."[54]

Why did Munns, Zon, Bates, Clapp, Silcox, and others in the Forest Service push for federal government funding to plant vast numbers of trees in the Great Plains, which they knew to be a difficult and risky undertaking, based on controversial Russian and Soviet research? Part of the motivation was the unfolding ecological disaster in the region. Silcox wrote that alleviating "the economic condition of the people in the Plains region" had become one of the USDA's "major concerns."[55] For Zon in particular, and the Forest Service as a whole, the motivation was partly financial. They needed money from New Deal emergency funding to make up for the cuts in their regular research budget. Zon's statement in his letter to Pinchot of June 19 (see p. 325) about getting access to "drought relief money" is evidence that funding was one of his aims in pushing for the Shelterbelt Project. Munns wrote to Zon on July 3, 1934, that, once the president had signed the executive order, he would ask for $100,000 for research.[56] When funds were made available in September, after the president's original allocation of $15 million had been cut to $1 million for the first year, $100,000 were still assigned for research, of which $95,000 was for research in the field. Part went to Zon's station, which made up for part of

[53] *St. Louis Post-Dispatch*, June 28, 1934, 21.

[54] Quoted in Roberts Papers, Box 3, Folder 2, The Prairie States Forestry Project: The Shelterbelt, chapters 4–10, chapter V "The Voice of the People," 36–7.

[55] F. A. Silcox, "The Problem," in *Possibilities of Shelterbelt Planting In the Plains Region: A Study of Tree Planting for Protective and Ameliorative Purposes As Recently Begun in the Shelterbelt Zone of North And South Dakota, Nebraska, Kansas, Oklahoma, and Texas by the Forest Service; Together With Information As to Climate, Soils, and Other Conditions Affecting Land Use and Tree Growth in the Region*, ed. Lake States Forest Experiment Station (Washington, DC: U.S. Forest Service, 1935), p. 1.

[56] Zon Papers, Box 9, Folder 1, Ed [E. N. Munns] to Zon, July 3, 1934.

the budget cut in 1933. Further funding from various New Deal agencies for temporary staff over the following years also compensated for the cut in the station's budget.[57] Thus, Zon had used his knowledge of Russian expertise and experience in planting shelterbelts in the steppes to gain support and funding for a project that would help save his Lakes States Forest Experiment Station from the threat of closure.

Planning the Shelterbelt

Planning the Shelterbelt Project was discussed at a conference in Washington, DC, on August 2, 1934. The participants included Secretary Wallace, Under Secretary Rexford Tugwell, Fred Morrell, and Zon. Zon had been appointed director of technical work, with responsibility for research, selecting the sites for shelterbelts, and the techniques for planting them. Amongst many other matters discussed, Zon was adamant that the trees be planted in the middle of sections, and not along the edges, where snow accumulating along the leeward sides of belts could obstruct roads (Zon probably learned about this from Vysotskii's work, see p. 343). A few days later, the project's administrative staff began to arrive at the headquarters in Lincoln, Nebraska. Morrell was to have been the director of administration, and Roberts the associate director, but Roberts soon replaced Morrell as he disagreed with the allocation of work between the technical and administrative directors. As administrative director, Roberts was obliged to follow the recommendations of Zon as technical director.[58] Munns later recalled that Zon had wanted his experiment station, "meaning Zon," to be "the final authority for all Belt work." He continued: "Zon liked to be in the limelight, as do all of us, but his ego went to extremes at times, and his try for taking over the Belt was merely an indication of it."[59]

Zon's appointment as technical director of the Shelterbelt Project was reported in *The New York Times,* likely reflecting his talent for self-promotion, while the same newspaper remained silent on the more mundane, if vital, administrative role to be played by Morrell and later Roberts. The *Times* reported "with more than thirty years' experience in forestry . . ., Dr. Zon will be familiar with the many technical problems

[57] See Perry, *History*, p. 31; Rudolf, *History*, pp. 13–15, 43–6, 167.

[58] Perry, *History*, pp. 28–32; Zon Papers, Box 9, Folder 1, Zon to Arthur N. Pack, July 10, 1934. For Zon's objection to planting shelterbelts along roads, see Zon, "Prospective Effects of the Tree-Planting Program," in *Possibilities*, 38.

[59] Roberts Papers, Box 3, Folder 11, Munns to Roberts, May 22, 1968.

to be faced in carrying out the vast program of tree planting for the amelioration of drought conditions in the plains area."[60] Zon began to prepare a technical bulletin to explain the project's aims, methodology, and the likely influences of the shelterbelts. Munns had already started the work and, as we have seen, overseen the translation of studies of shelterbelts in other countries, especially Russia, in the Library of Congress. Clapp transferred the task of preparing the bulletin to Zon in St. Paul, using money for field research in the project budget. Munns recalled how Zon had come to Washington, gone over all the material he had collected, and taken a lot of it away.[61]

I was not able to find the translations of the Russian material overseen by Munns in the archives I visited. The originals are listed in a bibliography of the technical bulletin,[62] however, which enabled me to locate them in Russian libraries. I used this Russian material as the sources for the second part of the section on Russian forestry and shelterbelts in the steppes in the previous chapter, starting with Bychikhin's positive assessment of the impact of shelterbelts in 1892 (see pp. 286–90). This section, therefore, summarized what the American foresters preparing the Shelterbelt Project would have learned from these Russian and Soviet studies. What they would have found out, therefore, was: There was no consensus on the influences of shelterbelts, and some of their effects were negative for soil moisture and crops; planting and cultivating shelterbelts was a difficult task; and one major specialist, Vysotskii, had serious reservations about the whole undertaking of steppe forestry. A close reading of the Soviet publications from the 1920s and early 1930s, moreover, would have revealed that some were superficial political, and propagandistic, rather than serious, scientific studies, and lacked hard data to support some of their claims.

In the fall of 1934 and early 1935, Zon, Bates, and their colleagues in St. Paul prepared the technical bulletin, entitled *Possibilities of Shelterbelt Planting In the Plains Region*. They relied on the translated material on shelterbelts in other countries, in particular in Russia, and investigations into conditions in the zone designated for planting shelterbelts in the Great Plains. Stoeckeler from Zon's experiment station and F. A. Hayes from the University of Nebraska spent two months examining the soils.

[60] "To Aid Shelter Belt Plan. Dr. Raphael Zon is Named Head of Technical Phases," *NYT*, July 31, 1934.

[61] Roberts Papers, Box 2, Folder 9, Munns to Roberts, June 21, 1965; Folder 10, Munns to Roberts, February 21, 1966; Box3, Folder 11, Munns to Roberts, May 22, 1968.

[62] P. O. Rudolf and S. R. Gevorkiantz, "Shelterbelt Experience in Other Lands," in *Possibilities*, 75–6.

Bates studied the climate. They also studied the economy and history of the plains, with particular attention to tree planting.[63] Preparing the bulletin indicated that planting the shelterbelts would be rather more complicated than Zon and Bates, and Munns, had imagined. In early January 1935, Bates wrote to Zon:

> I am frank to say that the entire subject is not simple, as the soil and climatic surveys have brought to light certain facts which make necessary ... somewhat different principles from those we have had in mind. I have not yet quite formulated my own ideas from these facts... There is one fact apparent, however, namely, that within the ... [Shelterbelt] Zone, there will be so many omissions for various reasons that the solid "belt" idea ... cannot be carried out. I therefore propose to establish an eastern boundary which is somewhat arbitrary, and western boundary which is a natural limit for any tree planting, with the thought that towards this western limit only the most favorable soils and situations can be used. ... Hope to have some ideas for you on Kansas in just a few days.[64]

The typescript, which ran to 600 pages, was ready by early May.[65] Munns saw it through to publication.[66]

The content of the translated Russian studies were summarized by Rudolf and Gevorkiantz. They surveyed the history of shelterbelt planting in Russia, from the "German farmer immigrants ... in the steppe region north of Crimea" (the Mennonites) through studies at the Kamennaia Step' field station and other stations by Dokuchaev's colleagues, including Vysotskii, in the 1890s, and more recent Soviet studies since the "severe famine" of 1921. They described the Russian methods in detail, including the combinations of trees and shrubs that Pinchot had remarked on when he visited the steppe plantations in 1902. They summarized the effects of the Russian shelterbelts in: "breaking the force of winds" and "reducing soil-blowing" possibly increasing precipitation; retaining moisture, especially from snow, in the soil; and improving soil moisture in areas protected by shelterbelts. They noted that "the data point to large and significant increases in [crop] yields from shelterbelt protected areas as compared to yields from lands exposed to the full rigors of the steppe." Rudolf and Gevorkiantz's summary was balanced and noted also the negative effects reported by Vysotskii and others who were skeptical about

[63] Droze, *Trees*, pp. 110–15; F. A. Hayes and J. H. Stoeckeler, "Soil and Forest Relationships of the Shelterbelt Zone," in *Possibilities*, 111–53; C. G. Bates, "Climatic Characteristics of the Plains Region," in *Possibilities*, 83–110.

[64] Zon Papers, Box 9, Folder 2, Carlos Bates to Zon, January 9, 1935.

[65] Zon Papers, Box 9, Folder 2, Zon to F. A. Silcox, May 8, 1935.

[66] Droze, *Trees*, pp. 117–18.

shelterbelts, such as reducing soil moisture in zones adjacent to protected areas, and the uncertainty over the impact on precipitation.

Rudolf and Gevorkiantz explained the relevance of the Russian studies to "our shelterbelt area":

> The outstanding interest which attaches to Russian experience with shelterbelts arises from the similarities of soil and climate to our own . . ., from the extensive character of a number of the plantings undertaken, and especially from the fact that planting programs and methods and the choice of species are guided by the scientific work of experiment stations. Most of all, the Russian experience is significant in that past undertakings and accomplishments, carefully evaluated, have been made the basis of a publicly sponsored program for the planting, . . ., of a system of shelterbelts that compares in magnitude with our own undertaking and far exceeds any other project of similar character in history.

In explaining the similarity between the soils of the steppes and Great Plains, they drew attention to the problem Vysotskii and Russian soil scientists had noted a few decades earlier: "As a result of the low precipitation, water does not penetrate them very deeply, and consequently concentrations of alkaline compounds and salts are formed at shallow depths, as in our Great Plains soils." These conditions, which were excellent for grasses and wheat, were not conducive to tree growth. At the time of the Shelterbelt Project, however, American soil scientists were still assimilating the approach of Russian soil scientists and so not all the available surveys of the soils of the Shelterbelt zone were up to date (See p. 337).

Gevortkiantz and Rudolf concluded that American foresters would do well to pay attention to the Russian experience and practices of "the layout of shelterbelt areas, dimensions and spacing of units, density of planting, and species composition of shelterbelts." They highlighted the Russian practices for preparing sites for planting. On species selection, they noted: "The Russians have used some of our native species successfully in shelterbelt planting, and we have found certain Russian species, such as caragana [Siberian pea tree] and Russian-olive, to be very useful components of our shelterbelt."[67]

Zon presented his conclusions from the Russian studies and investigations of the Great Plains in the technical bulletin. He acknowledged that, although "trees do not naturally grow in the grassland except on water

[67] P. O. Rudolf and S. R. Gevorkiantz, "Shelterbelt Experience in Other Lands," in *Possibilities*, 59–76 (quotations from 63, 64, 70, 72–3, 74–5).

courses," experience showed that "tree planting was feasible over much of the Plains area if proper precautions [were] taken." Shelterbelt planting was "advisable" in areas with sufficient precipitation (16 inches a year in the north, 22 inches in the south). He made recommendations on the orientation of shelterbelts, their spacing, the species of trees, the design of shelterbelts, and the preparation of the ground, planting, and care of trees.[68] On the possible effects of shelterbelts, Zon was more positive than Gevortkiantz and Rudolf. He cited studies by "foreign experiment stations" showing that grain-crop yields "may be increased by as much as 25 percent." He took care, however, to report: "The general effect of shelterbelts is not the creation of more rainfall over the area covered by tree growth, but the more economic use and conservation of the available rainfall." Thus, shelterbelt planting should be considered part of a broader program of water conservation and erosion control.[69]

The technical bulletin was well received by the Shelterbelt Project's supporters, some of whom commented on the Russian studies. President Roosevelt thanked Silcox for "the Shelter Belt volume" in February 1936. Silcox sent copies of the president's letter to Zon and Roberts.[70] The bulletin was warmly endorsed by Fred R. Johnson: "This publication is a monumental work . . . It effectively answers a lot of unwarranted criticism and definitely establishes this as a well planned, scientific project." His enthusiastic review addressed some of the barriers to the reception of Russian scientific work discussed in Chapter 2 of this book. Johnson was impressed by the material on Russian experience, which he found "most illuminating to the forester who had not had access to the translations." He continued: "The extent and scientific manner in which the Russian tests have been carried out since 1880 is amazing to the Yankee who thinks of Russia as a backward nation."[71] Johnson, who worked for the Forest Service in Colorado, was hardly objective, since he was quoted in the bulletin as a strong believer in the practicability and value of tree planting in western Kansas.[72] Roberts later described the bulletin as a "monumental accomplishment" and an "important contribution to the orientation and

[68] Zon, "What the Study Discloses," in *Possibilities*, 3–10 (quotations from 5).

[69] Zon, "What the Study Discloses," 7, 9; Zon, "Prospective Effects of the Tree-Planting Program," in *Possibilities*, 33–8.

[70] Zon Papers, Box 9, Folder 5, FDR to Silcox, February 13, 1936 (copy), Silcox to Zon, February 19, 1936; Roberts Papers, Box 1, Folder 1, Silcox to Roberts, February 19, 1936; FDR to Silcox, February 13, 1936 (copy).

[71] Fred R. Johnson, review, *JoF* 34 (1936), 541–3.

[72] John H. Hatton, "A Review of Early Tree-Planting Activities in the Plains Region," in *Possibilities*, 56.

character of the Shelterbelt." His praise suggests that he considered it the work of Bates, whom he admired, rather than Zon, whom he did not. In his official history, Perry noted there was so much demand for the bulletin from foresters and other specialists that the print run was soon exhausted. He wrote that it became the project's "bible," even though its recommendations were greatly changed over time.[73] The historian Orth, more dispassionately, considered the bulletin highly political and designed to answer the project's critics and secure congressional support. He commented that the Russian material, which he considered inconclusive on the benefits of shelterbelts, was a "red flag for critics."[74]

Once the technical bulletin was completed, however, the role of Russian expertise and experience in planting shelterbelts became less important in the American Shelterbelt Project. Russian experience from the steppes was of limited importance in the flow of technical advice from the Lakes States Forest Experiment Station in St. Paul, Minnesota, under Zon, to the project's headquarters in Lincoln, Nebraska, under Roberts, which had operational responsibility for planting the shelterbelts.

One reason for the declining importance of Russian experience was the awkward relationship between the two directors, Zon and Roberts, which quickly deteriorated. Roberts recalled the breakdown in their relations in correspondence with former colleagues in the 1950s and 1960s. Towards the end of 1934, Zon took leave of absence from his station in St. Paul and moved to Washington, DC, to act as a "general adviser" to Silcox for a few months.[75] While he was in Washington, DC, Zon continued to advise on the Shelterbelt, and annoyed Roberts by trying to "tell them what to do." As a result, Roberts asked Silcox to reduce Zon's role in the project. Roberts later admitted that he had "made an everlasting enemy of [Zon]."[76]

Zon's diminished role is borne out by the contemporary sources. Neither Zon's papers nor the files of the project headquarters in the National Archives in Kansas City contain much evidence for his active involvement from 1935. In March 1935, Zon sent a business-like letter, addressed "Dear Mr. Roberts," forwarding him documentation concerning project finances.[77] Orth, who used the uncataloged files of the project

[73] See Droze, *Trees*, p. 116; Perry, *History*, p. 42. [74] Orth, "Conservation Landscape," 162–7.

[75] Zon Papers, Box 9, Folder 1, Edward Richards to Silcox, September 18, 1934; Zon to Richards, October 25, 1934; Zon to Bob Marshall, November 14, 1934. See also Rudolf, *History*, p. 16.

[76] Roberts Papers, Box 2, Folder 9, Roberts to Ed Munns, June 8, 1965.

[77] NARA KC, RG 114, Entry 187, Box 6, Folder D-Legislation, Federal, Over Three Years, Zon, Director, LSFES, St. Paul, to Roberts, Acting Director, Plains Shelterbelt Project, Lincoln, March 8, 1935.

headquarters, noted that in 1936, Zon agreed to Roberts' proposal to plant narrower belts of trees, which went against his earlier advice. Orth found that by late 1936, Zon was letting Roberts and field personnel sort out technical difficulties themselves.[78] This followed an unsuccessful attempt by Zon in July 1936 to have Roberts removed from his post. Zon wrote to Clapp criticizing Roberts' management of key personnel and the project itself. The tone of the letter, however, lacked Zon's customary fighting spirit. He opened by saying he realized he was probably rushing in "where angels fear to tread," and signed off: "It is too hot, however, to get excited."[79] Zon did get excited in February 1937, when he sent a six-page letter, addressed abruptly "Dear Roberts," with lots of underscorings, containing seven rules for planting trees along roads to prevent snow drifting on to them. Zon signed off: "I trust you will agree with me that we do not want to be guilty of making road problems more serious than haphazard individual planting and building have already made them." Roberts, confident in his authority, simply forwarded the letter to Dr. George E. Condra at the University of Nebraska, saying he considered the advice to be "provisional" and: "Naturally this whole question will be subject to a great deal of further detailed scientific work."[80]

Zon's role, and the importance of the Russian experience, declined further. Technical support from the Lakes States Station was provided mainly by Bates, Stoeckeler, and David Olson, and by Ernest George at the Mandan Dry Land Station in North Dakota. Zon played little part.[81] Reports by Forest Service inspectors on the project in late 1936 and late 1937 made little mention of Zon or Russian expertise, indicated that Bates was in charge of technical advice, and that the research office in St. Paul and administrative headquarters in Lincoln were working together well.[82]

[78] Orth, "Conservation Landscape," 242, 272, n. 7.

[79] Zon Papers, Box 9, Folder 7, Zon to Earle Clapp, July 18, 1936. There was a heat wave in southern Minnesota, including St. Paul, in July 1936. Temperatures exceeded 100°F for a fortnight. See Nancy Yang, "A Look Back At the Relentless Heat Wave of July 1936," *MPR News*, July 6, 2016, available online at www.mprnews.org/story/2016/07/06/heat-wave-1936, accessed April 16, 2018.

[80] NARA KC, RG 114, Entry 187, Box 2, Folder D-Cooperation, over three years, Zon to Roberts, February 16, 1937; Roberts to Condra, February 27, 1937.

[81] Roberts Papers, Box 2, Folder 6, Roberts to Lee [Kneipp], 1957? (sic); Box 2, Folder 8, Roberts to Munns, May 20, 1963; Box 2, Folder 9, Roberts to Ed Munns, June 8, 1965.

[82] NARA KC, RG 114, Entry 187, Box 6, Folder Inspection, General, over three years, L. S. Gross, Forest Supervisor, Inspection Report, Shelterbelt Project, November 17–December 19, 1936, submitted December 28, 1936, 12, 25, 34; Roberts to Chief, Forest Service, February 3, 1937; R. E. Marsh, Acting Assistant Chief, and J. H. Price, Associate Regional Forester, R-5, Comments on Prairie States Forestry Project in connection with general R-9 Inspection, December 10, 1937, 9; Chief, Forest Service, Silcox to Roberts, April 20, 1938. The positive official evaluations may explain characterizations of the relationship between Zon and Roberts as good (for example, Sarah

However, this successful cooperation could have been because Zon declined actively to be involved with the Lincoln office. On several occasions, Roberts wrote, correctly, to Zon, only for the reply to come from Bates. In spring of 1939, for example, Roberts wrote to Zon to invite him or Bates to attend a meeting in Laramie, Wyoming. Bates replied on Zon's behalf, at first declining, then agreeing to go. However, after speaking to Zon, he finally declined.[83] The Zon volcano was not quite extinct. He continued to defend the project and support attempts to get more secure funding.

With Bates in charge of technical advice, American expertise and new research came to the fore, and the technical bulletin, which drew heavily on Russian studies, receded into the background. The inspectors' report in December 1937 noted that the bulletin had focused on knowledge available at the time it was prepared, but: "An active program of research, largely financed by emergency ... funds has since been prosecuted under the guidance of the [Lakes States Forest Experiment] Station" in cooperation with the Lincoln office. Tests into the "influence" of shelterbelts were being carried out in Huron and Miller, in South Dakota, Towner, North Dakota, and Mangum, Oklahoma. Hundreds of feet of portable and adjustable artificial barriers, between twelve and twenty feet high, were being used to study the effect of wind breaks on wind movement and to determine the best orientation for belts. Representatives from both Lincoln and St. Paul agreed on the need for a "substantial continuing program of research" on nursery practices, methods and timing for planting trees, the space between rows, economic studies of benefits and costs, as well as continuing the work to determine the influences of shelterbelts.[84] A few weeks later, Roberts agreed that more studies of the economics and influences of shelterbelts were needed. He stressed that the project was "fundamentally a huge experiment" and not "just an administrative planting job."[85] In April 1939, Bates reported to Roberts on their research

Thomas Karle and David Karle, *Conserving the Dust Bowl: The New Deal's Prairie States Forestry Project* [Baton Rouge: Louisiana State University Press, 2017], pp. 99–100) that are belied by private sources on tensions between them.

[83] NARA KC, RG 114, Entry 187, Box 6, Folder Inspection, General, over three years, Roberts to Zon, March 16, 1939; Bates to Roberts, March 30, 1939; Bates to Roberts, April 10, 1939; Bates to Roberts, April 21, 1939.

[84] NARA KC, RG 114, Entry 187, Box 6, Folder Inspection, General, over three years, R. E. Marsh and J. H. Price, Comments on Prairie States Forestry Project in connection with general R-9 Inspection, December 10, 1937, 10–11.

[85] NARA KC, RG 114, Entry 187, Box 6, Folder Inspection, General, over three years, Roberts to Marsh, February 20, 1938.

concerning planting techniques and measuring the influence of shelterbelts on crop yields to determine whether they justified the cost of planting them. They were paying attention to moisture conservation and experimenting with different crops and different designs and orientations of belts. They were still analyzing the data, however, and had only provisional recommendations.[86] Roberts was aware of the need for more research in October 1939,[87] over five years after the project had been approved on the basis largely of Russian expertise.

Thus, once the Shelterbelt Project was underway, it relied far less on Russian experience in the steppes, but instead on continuing research being conducted, largely under Bates' direction, in the Great Plains.[88] The Forest Service continued to monitor and translate relevant Russian publications. A further eight that had appeared since Munns' translation project in 1933–4 were included in a list produced by the Lincoln office in 1938. Although they concerned topics the project's researchers were investigating, for example, the influences of shelterbelts on crop yields, wind, drought, and microclimate, and the widths of belts, Bates did not cite them in his publications.[89]

One transfer from the steppes of Russia and Eurasia to the Great Plains that remained important in the Shelterbelt Project was species of trees and shrubs that had been introduced to the United States over the preceding decades (see pp. 297–300). The press release of July 21, 1934, had stated that "Trees of native origin will be used," and gave examples of American broad-leaved and coniferous species.[90] The technical bulletin contained a longer list that also included trees and shrubs from the steppes, such as: Russian olive (*Elaeagnus angustifolia*); Siberian pea tree (*Caragana arborescens*); Tatarian honeysuckle (*Lonicera tatarica*);

[86] NARA KC, RG 114, Entry 187, Box 10, Folder D-Programs, Northern Great Plains Research Committee, 1939, C. G. Bates, "Research of the Prairie States Forestry Project," April 21, 1939. Bates published the results of the studies in 1945: Carlos G. Bates, "Shelterbelt Influences. I. General Description of Studies Made," *Jo F* 43, 2 (1945), 88–92; Bates, "Shelterbelt Influences. II. The Value of Shelter-Belts in House-Heating," *JoF* 43 (1945), 176–96.

[87] NARA KC, RG 114, Entry 187, Box 10, Folder D-Programs, Northern Great Plains Research Committee, 1939, Roberts to State Directors, October 18, 1939.

[88] On the research conducted over the course of the project, see also Perry, *History*, pp. 44–53; Droze, *Trees*, pp. 119–41; Orth, "Conservation Landscape," 188–271.

[89] NARA KC, RG 114, Entry 187, Box 10, Folder D-Programs, Northern Great Plains, Research Sub-Committee, 1941, "A Partial List of Forest and Range Research and Allied Subjects Applicable to the Economic and Social Problems of the Great Plains Region." PSFP, July 16, 1938; Folder D-Programs, Northern Great Plains Research Committee, over three years, Roberts to Zon, July 20, 1938; Bates, "Shelterbelt Influences. I"; Bates, "Shelterbelt Influences. II."

[90] Perry, *History*, p. 15.

Russian mulberry (*Morus alba tatarica*); Tamarisk (*Tamarix*); and also Chinese elm (*Ulmus pumila*).[91] In October 1935, nurserymen in South Dakota were sent instructions on cultivating Russian olive, Russian mulberry, and Chinese elm for shelterbelts.[92] Chinese elm proved easy to cultivate, grew quickly, and was resistant to drought. But, many were killed during a cold snap in the winter of 1940.[93] There seems to have been confusion between the Chinese elms and hardier elms from Siberia (see pp. 298, 300). Most Chinese elms planted in the shelterbelts seem to have been derived from samples collected by Frank Meyer near present-day Beijing in 1908, rather than the hardier elms from eastern Siberia that were better suited to the northern plains.[94] In 1934, Niels Hansen (who assigned the elms from China and Siberia different scientific names) collected more of the hardier sort near Shilka in eastern Siberia, which he noted was 830 miles north of the location of the imports from China. He shipped some back to Brookings, South Dakota, but they were too late for widespread planting in shelterbelts in the 1930s. [95] Overall, while the shelterbelts contained some trees and shrubs from Eurasia, many were indeed native species. Cottonwoods, which grew naturally in river valleys, were preferred. In part this was because they grew quickly and made young belts look successful, even though they were not the species best suited for the longer term.[96]

Another area in which the Shelterbelt Project benefitted from Russian influences, but not as much as it might, was the underlying soil science. At the time the project was launched, the U.S. Soil Survey was still adapting its methodology to the Russian science and revising its surveys of the soils of the United States (See Chapter 6). We can gauge how far the Shelterbelt Project took advantage of Russian soil science by analyzing the chapter in the project's technical bulletin on "Soil and Forest Relationships of the Shelterbelt Zone" by Hayes and Stoeckeler. Hayes, a soil scientist from the University of Nebraska in Lincoln, had been "converted" to Russian soil

[91] D. S. Olson and J. H. Stoeckeler, "The Proposed Tree Plantations, Their Establishment and Management," in *Possibilities*, 18–22.

[92] NARA KC, RG 114, Entry 187, Box 5, Folder Inspection, South Dakota, Over three years, Max Pfender, Associate Forester, Brookings, SD, Memorandum to Nurserymen, and Fall Planting Schedule, October 23, 1935.

[93] Orth, "Conservation Landscape," 261–2, 274, n. 34.

[94] Isabel Shipley Cunningham, *Frank N. Meyer: Plant Hunter in Asia* (Ames: Iowa State University Press, 1984), pp. 261, 280–1.

[95] Niels Ebbesen Hansen, *The Banebryder (The Trail Breaker), The Travel Records of Niels Ebbesen Hansen 1897–1934*, ed. Helen Hansen Loen (n.p.: privately printed, 2002), pp. 102–3.

[96] Orth, "Conservation Landscape," 251–3; Prairie States Forestry Project, *Trees That Temper the Western Winds* (Washington, DC: United States Department of Agriculture. Forest service, 1938).

science in the 1920s (see pp. 256–7). His and Stoeckeler's chapter was based on a two-month survey of soils in the shelterbelt zone in late 1934, a survey by the new Soil Erosion Service earlier that year, and older reports by the U.S. Soil Survey. Their bibliography included works by Hayes, Charles Kellogg, and Marbut, who had all been heavily influenced by Russian soil science, but not the Forest Service's translations of the work of Russian forestry specialists, such as Vysotskii, who had much experience of tree planting in the steppes and detailed knowledge of soils.

Hayes' and Stoeckler's descriptions of the soils of the Shelterbelt zone showed the influence of Russian soil science with its emphasis multiple soil-forming factors and soil profiles. They described the soils of much of the zone as Chernozems, using the Russian term. They identified the "zone of calcification or lime enrichment at a comparatively shallow depth," i.e. the alkaline layer, in the profiles of many of the zone's soils, which Russian soil scientists had discovered in the steppe soils decades earlier. Hayes and Stoeckeler started with the assumption, based on observations of existing plantings in the region, that this alkaline layer might inhibit the growth of some tree species. The harmful consequences of the alkaline layer for trees were also mentioned in other chapters in the technical bulletin. After a detailed study of tree root systems in the Great Plains, however, Hayes and Stoeckeler concluded that the alkaline layer did not seem seriously to affect the species of native and introduced trees that were best adapted to the plains. Experience in the steppes, however, had shown that it took a couple of decades or more for trees to be affected (see p. 285).

Hayes' and Stoeckeler's descriptions of the soils of the zone paid great attention to soil texture, which the older methodology of the U.S. Soil Survey had stressed. Texture was important as it was a major determinant of soils' ability to absorb and retain the moisture trees needed. They concluded that trees grew better in coarser-textured, sandy soils, than finer-textured silts and clays. Thus, they recommended that shelterbelt plantings be concentrated in sandy loam soils. They divided the soils of the zone into three categories: "favorable" for tree growth, which they calculated covered fifty-nine percent of the area; "difficult," which covered thirty-nine percent; and "unfavorable," covering only four percent. They recommended particular species for certain types of soils, including trees from the steppes, such as Russian olive and mulberry. They advised that shelterbelts needed to be arranged according the conditions, especially soils and moisture, and that "successful tree growing" could be expected over large areas. "Successful" they defined modestly as an average life of thirty to sixty years and an average ultimate height of twenty-five to forty feet.[97] Their recommendations on suitable and unsuitable soils were included in information circulated to the population on the conditions needed for planting shelterbelts on their land.[98]

Hayes and Stoeckeler did a thorough job in the limited time available in late 1934, and the necessity of relying on a survey carried out for its own purposes by the Soil Erosion Service, as well as older surveys. Their findings were more sanguine than those of Russian specialists, such as Vysotskii, who were skeptical about planting trees in semi-arid grasslands. Orth concluded that their recommendations were "educated guesses."[99] The limitations of their study became apparent when planting the shelterbelts got underway. As early as May 1935 (the first planting season), two of the project's officers raised with Roberts the need for a "more thorough soil classification" to identify areas suitable for planting. But, carrying out such a survey was postponed.[100] Farmers drew on local knowledge of the soils when requesting shelterbelts. Eugene W. Osler, of Kenesaw, Adams County, Nebraska wrote: "This soil is well adapted to trees as there is enough sand to preserve moisture."[101] At times, local knowledge came into conflict with shelterbelt officers' understandings of soil suitability. In early 1936, for example, Clayton Watkins, the state director of the shelterbelt for Nebraska, wrote to the Chamber of Commerce in Trenton, Hitchcock County, that they were no longer planting shelterbelts on the "heavy soils" in their area because they had experienced problems in adjoining counties. Frank V. Uridil from the Chamber replied that they did not understand the term "heavy," and that their soils were "Keith Silt Loam," which was "well adapted to trees." The study by Hayes and Stoeckeler, however, had advised against planting trees on such finer-textured soils as silt loams, and categorized "Keith" soils in general as "fair to difficult."[102]

[97] Hayes and Stoeckeler, "Soil and Forest Relationships"; Rudolf and Gevorkiantz, "Shelterbelt Experience in Other Lands," 63; J. M. Aikman, "Native Vegetation of the Region," in *Possibilities*, 173.

[98] NARA KC, RG 114, Entry 187, Box 40, Folder L-Supervision, Nebraska, Shelterbelt Information, South Dakota, Application for shelterbelt planting; Shelterbelt Information for South Dakota landowners [n.d.].

[99] Orth, "Conservation Landscape," 221.

[100] NARA KC, RG 114, Entry 187, Box 5, Folder Inspection, South Dakota, Over three years, John H. Hatton to Roberts, May 6, 1935; H. D. Cochran to Roberts, May 20, 1935.

[101] NARA KC, RG 114, Entry 187, Box 41, Folder Adams County, [Nebraska], Requests for Shelterbelts, Eugene W. Osler, Kenesaw, Adams Co, NE, to Clayton S. Watkins, State shelterbelt director [n.d.].

[102] NARA KC, RG 114, Entry 187, Box 41, Folder Hitchcock County, [Nebraska], Requests for Shelterbelts, Watkins to W. C. Gass, Trenton, January 30, 1936; Frank V. Uridil, Trenton, to Watkins, February 7, 1936; Hayes and Stoeckeler, "Soil and Forest Relationships," 137.

Hayes' and Stoeckeler's categorization of soils into three groups may have been too general. In December 1937, Roberts identified the need for more experimental work on "planting on doubtful sites, especially fine-textured 'hard' soils," so that they could concentrate planting in areas known to be favorable for trees.[103] The following May, A. L. Ford, the project's director for South Dakota, wrote to Roberts "our most pressing need right now is more detailed knowledge of our South Dakota soil, that is, from its desirability for successful tree establishment," but that the state shelterbelt office was not set up to analyze soils.[104] Stoeckeler and Bates were carrying out further studies into cultivating trees in soils of different textures. While reiterating that coarser, porous, soils were more suitable, they recommended measures to conserve moisture in finer-textured soils.[105]

The moisture-holding capacities of soils of different textures were important, and the emphasis on texture was in line with the older methodology of the U.S. Soil Survey, but the Shelterbelt Project's scientists may have been better advised to pay more attention to soil profiles that took account of the range of factors in soil formation identified by Russian soil scientists. This was recognized by some American specialists. In 1935, M. F. Morgan, of the Connecticut Agricultural Experiment Station, explained the presence of a "zone of accumulation of calcium carbonate," i.e. the alkaline layer, in soil profiles in the Great Plains, and noted that below this zone the material was "practically dust dry." Morgan concluded: "The selection of favorable sites for tree planting in the shelterbelt would require close cooperation between the forester and the soil survey specialist if any measure of success is to be obtained."[106] A study by Myron T. Bunger and Hugh J. Thomson of the development of the roots of trees planted in shelterbelts in Oklahoma over the 1930s paid attention to the different horizons in the soil, including the "white chalky zone of calcification [alkaline] ... 5 to 11 feet below the surface." The roots of some species, for example the hardier Chinese elm (*Ulmus pumila*), could penetrate this zone to access water below it. Other species, including

[103] NARA KC, RG 114, Entry 187, Box 6, Folder Inspection, General, over three years, Roberts to Marsh, February 20, 1938. Roberts made a similar recommendation when he reviewed issues leftover from the project after its transfer to the Soil Conservation Service in 1942. Roberts Papers, Box 1, S1.F2 Correspondence 1941–50, Roberts to Chief, Forest Service, March 22, 1944.

[104] NARA KC, RG 114, Entry 187, Box 7, Folder D-Plans, General, 1938, Ford to Roberts, May 9, 1938.

[105] J. H. Stoeckeler and C. G. Bates, "Shelterbelts: The Advantages of Porous Soils for Trees," *JoF* 37 (1939), 205–21.

[106] M. F. Morgan, "Soil Factors in Relation to Proposed Plains Shelter Belt Plantings," *JoF* 33 (1935), 137–8.

Russian mulberry, were less able to do so. Some farmers used dynamite to break the alkaline layer. Bunger and Thomson, who emphasized the importance of supplies of moisture for the trees, recommended combinations of species and spacing between them to increase the chance of success.[107] While these studies did not cite Russian publications directly, they did cite works, including chapters in the technical bulletin, which had drawn on Russian studies. In March 1939, demonstrating knowledge of Russian soil science, the Land Use Committee of the Nebraska College of Agriculture reported: "Soils vary widely in their characteristics from east to west primarily because of climatic differences, which also have a striking influence on crop production practices," and that the "profiles of normal mature soils are good indicators of inherent productivity and use suitabilities."[108] Detailed surveys of all the soils of the shelterbelt zone to inform planting practices that drew on Russian soil science were, however, not available during the project.

Debating the Shelterbelt

The Shelterbelt Project was controversial right from the start. The November 1934 issue of *Journal of Forestry* was devoted to a debate about the project. Critics attacked the feasibility of planting trees in the plains, because of the alkalinity of the soil and aridity of the climate and the likely cost of such an endeavor. They also questioned the claims that shelterbelts would influence the climate and rainfall. Herman Chapman, the journal's editor, argued that, in any case, the effects of the shelterbelts would be felt up to a distance of only ten to fifteen times the height of the trees. He also objected to the lack of wider consultation among the forestry profession.[109] Royal Kellogg, who had been involved in tree planting in Kansas and Nebraska at the start of the century and had once been positive about the practice (see pp. 293–4), drew on his experience in the northeastern United States as well as the plains when he concluded: "We might conceivably cover the High Plains with trees and we might carpet

[107] Myron T. Bunger and Hugh J. Thompson, "Root Development as a Factor in the Success or Failure of Windbreak Trees in the Southern High Plains," *JoF* 36 (1938), 790–803.

[108] NARA KC, RG 114, Entry 187, Box 10, Folder D-Programs, Northern Great Plains Research Committee, 1939, Preliminary Report on Established Principles and Research Needed For More Effective Utilization of Land in Nebraska, prepared by the Nebraska College of Agriculture, Land Use Committee, Lincoln, Nebraska, March 1939, p. 2.

[109] H. H. Chapman, "Editorial: The Shelterbelt Tree Planting Project," *JoF* 32 (1934), 801–3; Chapman, "Digest of Opinions Received on the Shelterbelt Project," *JoF* 32 (1934), 952–7.

the state of Maine with buffalo grass – but we are sensible if we shall try to do neither."[110] Bates defended the project by drawing largely on American experience. He outlined the techniques they would use and the likely benefits, including reducing the surface wind velocity, conserving moisture, and retaining snow.[111] The historian Droze argued that the debate reflected wider issues dividing the forestry profession. Critics, for example Chapman, opposed government regulation of forestry, while supporters, such as Zon, favored a larger role for the government. There was also a split between opponents, such as Chapman and Royal Kellogg, who were based in the northeastern United States, and adherents, including Bates, Zon, and others, who lived in the Midwest and Great Plains.[112]

Russian experience of shelterbelts played a larger role as the debate went on. Russian-born American foresters and, interestingly, foresters from the Soviet Union joined in. Zon defended the project in the journal *Science* in April 1935, drawing in part on the Russian studies used in the technical bulletin.[113] At the end of 1935, *Journal of Forestry* published an article by Nicholas T. Mirov, who had trained as a forester in Russia before the revolution, but moved to the United States after the Russian Civil War. He outlined two centuries of Russian experience in planting trees and shelterbelts in the steppes. He referred to inconclusive debates over the climatic effects of shelterbelts, but concluded: "Russian foresters appear to have full confidence in the feasibility of tree planting in the steppe zone," as long as attention was paid to soil conditions, species selection, and cultivation techniques.[114] A measured case for influence on microclimates was made by a Soviet shelterbelt specialist in an article published in *Journal of Forestry* in 1936.[115] On the other hand, some American specialists

[110] Royal S. Kellogg, "The Shelterbelt Scheme," *JoF* 32 (1934), 974–7. See also Royal S. Kellogg, letter, *NYT*, September 16, 1934. Kellogg commented that he and Zon were "good friends," but "generally disagreed" about most things. Elwood R. Maunder, Oral History Interview With Royal S. Kellogg, Palmetto, Florida, April 16, 1955, Forest Industry Foundation, Inc., St. Paul, Minnesota, available online at https://foresthistory.org/wp-content/uploads/2016/12/KelloggRS-1955.pdf, p. 9, accessed October 30, 2019.

[111] C.G. Bates, "The Plains Shelterbelt Project," *JoF* 32 (1934), 978–91.

[112] Droze, *Trees*, pp. 84–95.

[113] Raphael Zon, "Shelterbelts: Futile Dream or Workable Plan," *Science*, NS 81, no. 2104 (April 26, 1935), 391–4.

[114] N. T. Mirov, "Two Centuries of Afforestation and Shelterbelt Planting on the Russian Steppes," *JoF* 33 (1935), 971–3. His main source was N. N. Stepanov, *Stepnoe lesorazvedenie, izdanie vtoroe, pererabotannoe i dopolnennoe* (Moscow: Novaia derevnia, 1927), which Munns' translation team had located in the Library of Congress. On the author, see N. T. Mirov, *The Road I Came: The Memoirs of a Russian–American Forester* (Kingston, Ontario: The Limestone Press, 1978).

[115] V. Bodrov, "The Influence of Shelterbelts Over the Microclimate of Adjacent Territories," *JoF* 34 (1936), 696–7.

challenged the American claims for the influence of shelterbelts that were based on Russian studies. An American geographer asserted that "no climatologist of standing has presented evidence in support of ... claims" that had been made in some American newspapers that Russian shelterbelts "increased the rainfall." He continued that "from a scientific point of view the Russian 'evidence' is woefully inadequate."[116] A forester at an experiment station in Indiana noted that the Russian studies "contain[ed] no discussion of the methods used to obtain [the] results."[117]

The most detailed critical assessment of shelterbelts published in *Journal of Forestry* around this time was by the leading Russian specialist Vysotskii:

> I have been interested in extensive experiments in establishing forests in the Russian Steppes under the unfavorable conditions of insufficient moisture and alkaline soils. I came to the conclusion that a forest growing under such handicaps is not healthy; it degenerates ... at a certain critical age – 15, 20, 30, or 40 years.

He explained how, in some conditions, including permeable, sandy soils (i.e. those recommended for shelterbelts in the United States), planting trees could reduce the total available moisture. He accepted that there was "no doubt as to the beneficial effects of shelter belts," in particular in protecting neighboring fields, but stressed the need for "good techniques." He pointed to the danger of snow drifts blocking roads on the leeward side of belts (which Zon had been so concerned about, see pp. 328, 334). He concluded that planting shelterbelts in prairie or steppe conditions was "an art which requires good knowledge of ecology and economical aspects of agriculture as well as a broad minded approach free from routine methods and ... preconceived ideas."[118] Vysotskii was more explicit about his reservations in an American journal than he would have been in his home country, where such belts, and the Russian science behind them, enjoyed official support. He wrote to his compatriot, the soil scientist Iarilov, that his article would never see the light of day in the Soviet Union.[119]

[116] Stephen S. Visher, "Climatic Effects of the Proposed Wooded Shelter Belt in the Great Plains," *AAAG* 25, no. 2 (1935), 70.
[117] Daniel Denuyl, "The Zone of Effective Windbreak Influence," *JoF* 34 (1936), 689–95.
[118] G. N. Vyssotsky [Vysotskii], "Shelterbelts in the Steppes of Russia," *JoF* 33 (1935), 781–8. It was translated from Russian by two Russian émigrés in Berkeley. Two foresters from the Lakes States Forest Experiment Station had reviewed a publication by Vysotskii in the same journal the previous year. S. R. Gevorkiantz and H. F. Scholz, Review of "How to Propagate Forests in our Steppes and How to Take Care of Them," by G. M. Vysotsky, *JoF* 32 (1934), 358–62.
[119] ARAN, f.582, op.4, d.27a, l.42, Vysotskii to Iarilov, October 27, 1935.

The extent of opposition to the Shelterbelt Project in the United States prompted Hansen to send a telegram to President Roosevelt in February 1936, urging the continuation of tree planting as part of the program to combat soil erosion. He drew on his observations of forestry in the Russian steppes.[120] In doing so, however, he was disregarding the degree to which it was the claims in Russian studies for the beneficial effects of shelterbelts that were part of the cause of the controversy in the United States. Gradually, the debate subsided. Foresters who opposed the project realized that some of their criticisms concerned misconceptions in the press, rather than in the Forest Service. Over time, some came to accept the Shelterbelt.[121] The debates were noted inside the Soviet Union by Mikhail Tkachenko (who had visited the United States, see pp. 304–5). He reported that the American discussions revealed the practical difficulties in planting shelterbelts, but that their value outweighed arguments against them.[122]

There were also mixed reactions to the Shelterbelt Project among politicians, journalists, agricultural specialists, and the populations in the plains states. To some extent, hostile reactions were part of negative feelings in the region towards the federal government, the Democratic Party, and Roosevelt's New Deal. Kansas historian Craig Miner quoted from a letter sent by Republican Congressman Clifford Hope to one of his constituents in 1934:

> most of them [federal bureaucrats] are theorists who for the first time in their lives are having an opportunity to put some of their crazy theories into effect, and believe me they like it. They are having the time of their lives, and a lot of them haven't any more concern as to what is going to happen if their experiments fail than if they were trying their ideas out on a bunch of guinea pigs.

Similar attitudes underlay reports in some Kansas newspapers. Miner wrote: "There was laughter about the proposal for a tree belt across Western Kansas," where people remembered the failed projects of earlier decades. Ironic references were made to "scenic, shaded highways in Kansas" proving more attractive vacation destinations than Colorado or California. In June 1936, when the controversy had died down, the *Great*

[120] SDSU ASC, UA 53.4, N. E. Hansen Papers, Series 2. Helen Hansen Loen Collection, Box 5, Folder 47, "Tree Planting and Soil Erosion."

[121] See Perry, *History*, pp. 19–21.

[122] M. Tkachenko, "Ocherednye zadachi i metody izucheniia vodookhranno–zashchitnikh lesov," *Lesnoe khoziaistvo i lesoeksploatatsiia*, no. 6 (1935), 34–6. (This was one of the Soviet forestry journals Zon subscribed to. See p. 307.)

Bend Daily Tribune described the shelterbelt as "an alluring and expensive undertaking that under scrutiny proved impracticable," and an example of a federal government that "is always on the way but so often doesn't know where it is going."[123]

The population of the plains showed great interest in the project, but not its Russian roots. A. C. Davidson of Holdrege, Phelps County, in central Nebraska, was in favor of Congress helping "people on small farms," rather than in the cities or their friends, but had reservations: "I think it would be throwing money away to plant trees in some places, like in Dakota, trees will not grow there in that gumbo soil." He was interested in the government planting trees on his land, ending his letter: "Will a man be here to see me or not?"[124] More forthright was Dr. M. Hoops of Springview, Keya Paha County, in north-central Nebraska:

> How does the government expect to make trees grow, when even older tree lots have died? ... It is my opinion that this is another move to ruin the farmer living in this tree belt, provided he is forced to keep his land and have government men running over his fields with trucks, making new roads over his hay meadows, leaving gates open to let his stock out, and interfering in many other ways ...[125]

While Dr. Hoops may have been beyond persuasion, others in the plains states flooded the project's state directors with requests for shelterbelts on their land.[126] The project headquarters analyzed 335 letters to the president, the secretary of agriculture, the Forest Service, and other government agencies. Nearly half, 149, were in favor, 50 were against, and 136 expressed interest, but were neutral.[127]

Roberts and other project leaders endeavored to promote the Shelterbelt Project among the plains population and to obtain more permanent funding from Congress.[128] They found that arguments based on Russian experience did not convince the population. People who attended a

[123] Craig Miner, *Next Year Country: Dust to Dust in Western Kansas, 1890–1940* (Lawrence: University Press of Kansas, 2006), pp. 262, 281.

[124] NARA KC, RG 114, Entry 187, Box 40, Folder Phelps County [NE], Requests for Shelterbelt, A. C. Davidson to Clayton W. Watkins, December 7, 1935.

[125] NARA KC, RG 114, Entry 187, Box 40, Folder L-Supervision, Nebraska, Shelterbelt Information, Nebraska, Dr. M. Hoops to Extension Service, Agricultural College, Lincoln, August 21, 1934.

[126] See, for example, NARA KC, RG 114, Entry 187, Box 40, Folder Antelope County [NE], Requests for Shelterbelt, and similar files for other counties.

[127] Perry, *History*, pp. 23–5.

[128] See, for example, NARA KC, RG 114, Entry 187, Box 6, Folder Inspection, General, over three years, Fred Schoder, Report to Assistant Chief, U.S. Forest Service on visit to Lincoln Office of Plains Shelterbelt Project, April 28–May 5, 1936.

meeting in Manhattan, Kansas, in November 1934, were "particularly anxious to know whether an effort would be made to adapt the planting to local conditions."[129] As time went on, Roberts and his colleagues gave greater emphasis to American expertise and the hard work of the Forest Service. For example, on August 7, 1935, Roberts sent a memorandum to his staff on how to deal with potential "points of attack." He stressed that information on the benefits of shelterbelts should

> be presented to reasonable people in a way to carry conviction. Such studies . . . should assemble . . . adequately supported views and experiences as we have in this country, although comparatively limited, and pertinent foreign experience showing the effects of tree belts on the different things it is alleged they will accomplish; qualifying the foreign experience as deemed advisable.

Roberts noted a problem in using the Russian studies Zon had provided, since the data they were based on were not known in the United States. Roberts hoped the data could be obtained by corresponding with the Russian scientists. Together with Bates, moreover, he wanted more experiments to gather information on the influence of shelterbelts in the United States, in part to test the results in the Russian studies.[130]

Over time, reference to the Russian experience was dropped from publicity for the general public and the tone became more down to earth. In 1937, Roberts wrote to Chief Forester Silcox that the initial publicity had led to the "circulation of erroneous conceptions of the purposes and mechanics of the program, which caused much skepticism, and even ridicule."[131] To avoid a repetition of such problems, the Forest Service prepared a "standard broadcast script" and text of a lecture about the project. In October 1938, Silcox made a broadcast on network radio entitled: "What's happened to the shelterbelt?" In contrast to the promotional material in 1934, there was no input from Zon. The scripts made no reference to Russian experience or to climatic benefits of shelterbelts. Instead, the script made clear that the aims of the shelterbelts were to prevent soil blowing and to protect fields and crops from the wind. The

[129] NARA KC, RG 114, Entry 187, Box 3, Folder D-Inspection, Kansas, over three years, H. D. Cochran, Memorandum, Inspection: Kansas, November 23, 1934.

[130] NARA KC, RG 114, Entry 187, Box 6, Folder Inspection, General, over three years, Roberts, Memorandum for Chiefs of Office and State Directors, August 7, 1935; Box 5, Folder Inspection, South Dakota, Over three years, Roberts to Mr. Ford, State Director, Plains Shelterbelt Project, Brookings, SD, June 1, 1936.

[131] NARA KC, RG 114, Entry 187, Box 6, Folder Inspection, General, over three years, Roberts to Chief, Forest Service, February 3, 1937, pp. 3–4.

scripts emphasized past American experience and the practical results of the current plantings.[132]

A similar tone was adopted in political statements by the project's leaders to secure congressional funding. In September 1938, in a business-like report for a congressional committee, Roberts highlighted the importance of the "Prairie-Plains" for the nation's agriculture, and the "peculiar" natural conditions and hazards, including the semi- or sub-humid climate and high winds. The hazards had been exacerbated by "unwise land use" that had culminated in the "recent prolonged drought." He argued that planting shelterbelts was essential to stabilize the region's agriculture by protecting fields and crops against the erosive and drying effects of the wind. Roberts cited experimental work by American specialists and the experience of American farmers. He concluded that the Forest Service had been quite successful in establishing trees in the region, even during present drought, by paying great attention to the quality of planting stock, correlating species with soil types, designing plantations, and prescribing cultural methods. Roberts made no reference to studies in Russia or general claims about moderating the climate that had provoked much skepticism.[133]

The Forest Service made a similar case when arguing for stable funding for the project at hearings by the Joint Congressional Committee on Forestry in late 1939 and early 1940. All the data on the effects of shelterbelts were provided and cleared by Bates. They were based on American experience since 1890 and plantings since the start of the project. The Forest Service also emphasized the support it had received from farmers in the Great Plains.[134] There was no mention of Russian experience, and Zon does not appear to have been involved in any of the

[132] NARA KC, RG 114, Entry 187, Box 31, Folder I-Information, Broadcasting, Nebraska, 1938, F. A. Silcox, Chief, U.S. Forest Service, "What's happened to the shelterbelt?," Radio address delivered over National Broadcasting Company network, October 12, 1938; E. L. Perry, Chief of Information and Education, PSFP, Lincoln, NE, to State Directors, October 24, 1938, "Standard broadcast script prepared by Washington office"; Folder I-Information, Lantern slides, Nebraska, 1938.

[133] NARA KC, RG 114, Entry 187, Box 1, File D-Cooperation, Joint Congressional Committee, Finished Report, 1938, ff.[1]–22 USDA, Forest Service, Prairie States Forestry Project, Plains Region Section of Report for Congressional Committee. [Roberts, September 15, 1938], quotation from f.10.

[134] NARA KC, RG 114, Entry 187, Box 1, File D-Cooperation, Joint Congressional Committee, 1940, February 12, 1940, Henry L. Lobestein, Associate Forester, to Roberts; "Data considered desirable for presentation at hearings of congressional Joint Committee on Forestry;" File D-Cooperation J.C.C. 1939, "A Plea for Great Plains Forestry," by Great Plains Forestry Committee, G. E. Condra, W. H. Brokaw, M. B. Jenkins, all of the University of Nebraska, given before the Congressional Joint Committee on Forestry, December 19, 1939, Madison, WI.

meetings or the preparations. The arguments based on the Russian studies that had played such a large role in gaining approval for the project in 1933–4 seem by the late 1930s to have been perceived as weaknesses in attempting to win public support and congressional funding.[135]

Conclusion

The Shelterbelt, or Prairie States Forestry Project, was a large-scale undertaking approved by President Franklin Roosevelt in July 1934 to combat the consequences of the extreme drought, dust storms, and human suffering in the Great Plains by planting belts of trees and, in the process, providing employment. The project has been described as an example of the "grand high modernism" identified by James C. Scott. Together with other schemes of the New-Deal period, such as the Grand Coolee Dam and the Tennessee Valley Authority, the Shelterbelt Project bears comparison with the infrastructure projects of the Soviet Five-Year Plans of the late 1920s and 1930s.[136] By the time the Shelterbelt Project ended in 1942, a total of 222,825,220 trees had been planted in belts covering 238,212 acres that, if placed end-to-end, would measure 18,599 miles, on 30,223 farms, in six states (North Dakota, South Dakota, Nebraska, Kansas, Oklahoma, and Texas). The total cost (after the comptroller general and Congress had reined in the president's initial allocation) was $13,262,250. The difficulty of the task is indicated by the survival rates of the trees planted. They varied from 51.2 percent in the dry year of 1936, to 82.3 percent, after the rains had returned, in 1941. The survival rate overall was a respectable two-thirds. Most of the trees that died were replaced and less than five per cent of the shelterbelts were complete failures.[137] Once it was underway, the project's achievements were largely due to the hard work and managerial skills of Paul Roberts, as administrative director at the headquarters in Lincoln, Nebraska, with technical support from Carlos Bates and his colleagues under their Russian-born

[135] See Allan J. Soffar, "The Forest Shelterbelt Project, 1934–1944," *Journal of the West* 14 (1975), 95–107.

[136] See Robert Gardner, "Trees as Technology: Planting Shelterbelts on the Great Plains." *History and Technology* 25 (December 2009), 325–41; Antony J. Badger, *New Deal: The Depression Years, 1933–1940* (New York: Noonday Press, 1989), pp. 171–9; Norman E. Saul, *Friends or Foes?: The United States and Soviet Russia, 1921–1941* (Lawrence: University Press of Kansas, 2006), p. 218; James C. Scott, *Seeing Like a State: How Certain Schemes to Improve the Human Condition Have Failed* (New Haven, CT: Yale University Press, 1998).

[137] Perry, *History*, appendix, pp. 1, 2, 9, 15.

director, Raphael Zon, at the Lakes States Forest Experiment Station in St. Paul, Minnesota.

When the project was launched it attracted controversy and suspicion. Over the following years, it gained much support in the plains region. In 1940, for example, a statement on behalf of the citizens of southeastern North Dakota to the Joint Congressional Committee on Forestry noted that "shelterbelt planting is receiving the wholehearted support of the country as a whole and is generally accepted as a necessary and judicious expenditure of public funds." Other local residents wrote that the shelterbelts had transformed barren and treeless localities into desirable places to live, had broken the wind, piled up snow, conserved moisture, prevented soil drifting, provided shelter for stock, and in time would provide timber.[138] Scientific studies during the project bore out some of the advantages claimed for shelterbelts. Bates' research showed that belts reduced wind velocity. On the leeward side, the reduction was greatest at a distance five times the height of the belt, and had fallen to zero only at a distance of forty times the height. Although there was much variability, crop yields were higher over a distance of eight-ten times the height of the belt. Heating bills for homes protected by shelterbelts were reduced by twenty-five percent.[139] Zon and Munns could hardly have written a more enthusiastic assessment as it endorsed most of the claims they had made in 1933–4.

The project's roots lay in Russian and Soviet experience and expertise in the steppes that were used to great effect by Munns and Zon in 1933–4 when they made the case for shelterbelts to be planted in the Great Plains. They seem to have disregarded the mixed results reported in the Russian studies of shelterbelts and reservations expressed by leading Russian specialist Georgii Vysotskii. Munns was familiar with both from the studies his team of translators had located in the Library of Congress. Zon knew about them from his reading of the Russian forestry literature. The reliance on Russian, rather than American, experience offers an interesting parallel to the story of soil science in which American soil scientists, such as Marbut, paid far more attention to the work of Russian scientists, in particular Konstantin Glinka, than the American Eugene W. Hilgard, who had been thinking along similar lines, but had not developed as complete a theory and methodology. In the case of shelterbelts, the

[138] NARA KC, RG 114, Entry 187, Box 1, File D-Cooperation, Joint Congressional Committee, 1940, Charles G. Bangert, Enderlin, ND, on behalf of Enderlin Community Forest and citizens of North Dakota, "Prairie States Forestry Project," to Joint Committee on Forestry, U.S. Congress. n. d. [1940], 1–4.

[139] Bates, "Shelterbelt Influences. I," 89, 92; Bates, "Shelterbelt Influences. II."

Russians had also acquired more experience and expertise in the steppes than their American counterparts in the Great Plains.

Untangling the precise importance of Russian and Soviet studies of shelterbelts in the American project has been complicated by gaps in the sources, which I have tried to overcome, and by the competing egos and claims by Zon, the project's technical director, and Roberts, its administrative director, which I have mediated between. After all his hard work, Roberts was understandably stung when he found that Zon had managed to take the credit for the Shelterbelt Project. When Roberts prepared his own history of the project, to put his side of the story, he tried to overcome the lacunae in the sources by asking his former colleagues for their recollections. Even Munns, who answered many of his letters, seems to have been irritated by Roberts' efforts to undermine Zon's claims. In response to one of Roberts' letters about Zon's role, Munns addressed his reply "To the Sheriff & Vice Pres of Narcissus" and repeated that it was he, Munns, who had been denied the credit after he had "worked out the whole scheme."[140] Indeed, the surviving sources and later recollections seem to confirm that when, in the summer of 1933, President Roosevelt asked the Forest Service what could be done to plant trees in the Great Plains, it was Munns, as head of silvics, who had devised a scheme to plant belts of trees. And, he had based his arguments on Russian studies, and made more available through his translation project.

Zon's importance was twofold. First, he assiduously promoted Russian forestry expertise in the United States inside the Forest Service over his career and more widely as editor of *Journal of Forestry* in the 1920s. This may explain why Munns and other leading foresters, such as Clapp, were familiar with the Russian work by the 1930s. In 1935, Zon wrote to the editors of *Journal of Forestry*: "There is no denying that the idea that forests have an effect upon the climate over wide continents is largely a Russian idea, just as forest planting in the prairies and plains is a Russian idea – a fact which I assiduously preached on many occasions."[141] Second, Zon with backing from Bates, played a crucial role in getting the Shelterbelt Project approved in 1934. In May 1934, Munns asked Zon to go to Washington, DC to try to convince the Forest Service and the president to approve his plan. Zon brought a concise document prepared by Bates and other members of his team in St. Paul, and succeeded in persuading Clapp and Silcox, who then gained the approval of Secretary Wallace and

[140] Roberts Papers, Box 2, Folder 9, Munns to Roberts, June 21, 1965.
[141] Zon Papers, Box 9, Folder 3, Zon to Herbert Smith, 9 July 1935.

President Roosevelt. In 1968, Munns wrote to Roberts: "So, without Bates and Zon, probably the Belt would not have taken shape."[142]

Munns offered an explanation for the inaccurate story told by one of Zon's staff, Hardy Shirley, that Zon had been consulted about the president's initial request to the Forest Service in August 1933:

> Knowing Zon, I can well imagine that ... [in] St. Paul he would have a station conference and tell the gang. Liking the notoriety, he probably made his trip [in June 1934] sound like he was [being] called in for this consultation. He liked to play the great man, as you know. If Hardy believed this, then that is where he got his information. ... It may be that Hardy got his idea ... all from Zons [sic] parading of himself. I loved old Zon, he was a real mentor to me. But like many another great [men], he was egotistical and liked to parade Zon and his greatness to all who would listen. ... Well, anyway, Shirley is a bit off base, but I can understand how he got there.[143]

This sort of behavior by Zon irked Roberts because they had different personalities. Clapp put his finger on it in a letter to Roberts in 1952. Clapp noted that "mankind" was divided into "two classes": those who wanted to "be somebody" and those who wanted to "do something." Clapp wrote that the latter were the ones who really counted, and put Roberts into this class.[144] Perhaps he would have put Zon in the other class. Over a decade and a half later, Roberts sought Clapp's recollections on Zon's role in launching the Shelterbelt Project. Clapp recalled: "Zon, with his knowledge of Russian experience, helped get the President's original concept of a wide, long ... north–south belt changed to one of many narrow belts. This was perhaps in 1933 or early 1934. We commented that this may have been Zon's greatest personal contribution to the Shelterbelt."[145] Clapp, who was ninety years old when he tried to remember the events over thirty years earlier, may have blurred Munns' and Zon's roles and the precise dates, however, he did recall the importance of the Russian experience at this crucial point.

On the basis of the surviving evidence from the 1930s and Roberts' persistence in badgering former colleagues for their memories in the 1960s, it can be concluded that American knowledge of the Russian studies of planting shelterbelts in the steppes played a crucial role in the origins of the Shelterbelt Project at two points: in the summer of 1933 in turning the

[142] Roberts Papers, Box 3, Folder 11, Munns to Roberts March 22, 1968.
[143] Roberts Papers, Box 3, Folder 11, Munns to Roberts March 22, 1968.
[144] Roberts Papers, Box 1, Folder 3, Clapp to Roberts, January 12, 1952.
[145] Roberts papers, Box 3, Folder 11, Marsh to Roberts, August 25, 1968.

president's vague idea to plant trees in the Great Plains into a realistic project to plant shelterbelts; and in May–June the following year in persuading the Forest Service and the president to approve the project. This conclusion is supported by repeated references to the Russian experience in the documentation from the time, including the initial publicity. The Russian experience was also very important in preparing the technical bulletin, overseen by Zon, where it was combined with American studies to produce a blueprint for the project. The Russian and Soviet roots of the American Shelterbelt were noted by Mikhail Tkachenko, who wrote in 1935 that shelterbelts were receiving international recognition even, or perhaps because of, the "current crisis" in the capitalist countries.[146]

As time went on, however, the project relied more on American experience, backed by studies led by Bates. The change from Russian to American expertise coincided with the sidelining of Zon in 1935, after he had antagonized Roberts, but may have happened anyway. As early as January 1935, when Bates was engaged in the preliminary studies, he realized that the whole enterprise was more complicated than he and Zon had thought. It seems likely that, when they had been drawing on the Russian studies of shelterbelts to get the project approved, they had emphasized the positive conclusions, including possible climatic effects, and paid less attention to the reservations expressed by some Russian scientists. After the cuts in the Forest Service's research budget, they had the special motivation of securing funding from New Deal emergency programs to maintain their experiment station, and their jobs, in St. Paul. A year later, in 1935, Vysotskii's measured article published in the American *Journal of Forestry* brought the complications home to American foresters. While some of the more extravagant claims that shelterbelts could change the climate had been made by the press, albeit drawing on the language of the initial publicity, the project leaders had repeatedly emphasized their more limited aims of reducing wind erosion, evaporation, and harm to crops, and not increasing rainfall.[147] Nevertheless, without the Russian experience of planting shelterbelts in the steppes, and the transfer of their expertise to the United States, it is unlikely that the Shelterbelt Project would have seen the light of day.

Among the challenges to cultivating shelterbelts was the risk that the young trees would be inundated by weeds known as Russian thistle and then covered in snow drifts in the winter. When Munns was asked how the Forest Service would prevent this by the *Kansas City Star* in August

[146] Tkachenko, "Ocherednye zadachi," 34. [147] See Perry, *History*, p. 5.

1934, he replied that it was proposed to build high fences to hold the drifts and catch the thistle before they reached the plantings.[148] Few such fences were built and Russian thistles "as big as balloons" tumbled across the plains, over and under ordinary wire fences, and stacked up against shelterbelts.[149] The Russian thistle, also known as tumbleweed, was another contribution to the Great Plains from the steppes, and is the subject of the next chapter.

[148] Roberts Papers, Box 3, Folder 2 The Prairie States Forestry Project: The Shelterbelt, chapter V "The Voice of the People," 27.

[149] Mark Fiege, "The Weedy West: Mobile Nature, Boundaries, and Common Space in the Montana Landscape," *WHQ* 36 (2005), 22.

CHAPTER 9

Tumbleweed

Introduction

In 1934, Roy Rogers and the Sons of the Pioneers recorded the classic cowboy ballad "Tumbling Tumbleweeds." It presented a benign, romantic, image of a lonesome cowboy "drifting along," surrounded by tumbleweeds, under the "prairie moon."[1] The song was written by Bob Noble, who was familiar with the prairies and the West as he was born in Winnipeg, Manitoba, in 1908, spent his childhood in various parts of Canada, and later moved to Tucson, Arizona, and Los Angeles, California.[2] In 1935, the song was performed by Gene Autry in the movie *Tumbling Tumbleweeds*, and has since been recorded by artists as varied as Bing Crosby, the Supremes, and Bob Wills and his Texas Playboys.[3] In 2010, the Western Writers of America voted it number eight in their "all-time best Western songs."[4]

Tumbleweeds have featured in other movies about the Great Plains and American West. William S. Hart starred in a 1925 movie, called simply "Tumbleweeds," set on the Kansas–Oklahoma border during the land rush of 1893.[5] Gary Cooper and Jean Arthur co-starred with some tumbleweeds in the 1936 Cecil B. DeMille film "The Plainsman," in

[1] The lyrics of the song can be found at: "Tumbling Tumbleweeds," Library of Congress, Notated Music, available online at www.loc.gov/item/ihas.100008234/, accessed July 2, 2019.

[2] "Bob Nolan: Early Life and Career (1908–1931)," available online at https://bit.ly/20xBD2W, accessed December 13, 2017.

[3] See Joseph Kane (1935), *Tumbling Tumbleweeds*. USA: Republic Pictures (information available online at www.imdb.com/title/tt0027139/, accessed January 14, 2018); Richard Aquila, *The Sagebrush Trail: Western Movies and Twentieth-Century America* (Tucson: University of Arizona Press, 2015), pp. 92–3, 105–8, 113; and the list of artists, available online at http://bobnolan-sop.net/Weeds%27n%27Water/Weeds.htm, accessed January 14, 2018.

[4] Skip Skipson (2018), The Top 100 Western Songs, All the Songs, available online at https://western100.com/songsbyrank.htm, accessed November 1, 2019.

[5] King Baggott (1925), *Tumbleweeds*. USA: William S. Hart Productions (information available online at www.imdb.com/title/tt0016461/, accessed January 14, 2017); Aquila, *Sagebrush Trail*, p. 52.

which they traveled by stagecoach across Kansas.[6] Tumbleweeds have appeared in literature. In *My Ántonia*, first published in 1918, Willa Cather described one of her leading human characters returning to Nebraska, two decades after the novel's main action in the 1890s. North of the town of "Black Hawk" (Red Cloud), he saw the weeds "blowing across the uplands and piling against the wire fences like barricades."[7] For some writers tumbleweeds signified desolation and even danger.[8] Scientists also associated tumbleweed with desolation. In the western Dakotas and Montana, Oliver E. Baker described how weeds, including tumbleweed, had taken over fields that had been abandoned after the crop failures in the dry years of 1917–19.[9] This image of desolation was far removed from the romanticism in the cowboy ballad, but closer to the reality. Nevertheless, benign tumbleweeds have tumbled through the Great Plains and the West of America's, and the movie-going and cowboy-song-loving world's imagination.

And yet, this tumbleweed that became an icon of the American West had arrived in the northern plains, by chance, from the steppes of the Russian Empire in the 1870s.[10] After it turned up, tumbleweed spread and became a harmful, invasive species. Contemporaries considered it a "weed": a plant that appeared in environments disturbed by human activity, such as farmland, that people considered undesirable as it competed with cultivated or native plants they deemed useful.[11] The tumbleweed was part of the far larger transfer of plants from the Eurasian steppes to the North American plains that included the wheat varieties discussed in Chapter 4. In contrast to the other transfers and influences from the

[6] Cecille B. DeMille (1936), *The Plainsman*. USA: Paramount Productions (information available online at www.imdb.com/title/tt0028108/, accessed January 14, 2017).

[7] Willa Cather, *My Ántonia* (Oxford: Oxford University Press, 2006), p. 195.

[8] See, for example, KSHS SA, MS Coll. 13 Vivien Aten Long [1885–1968] Papers, Box 6 Unpublished and Published Prose and Poetry, Folder Unpublished Prose, "Tumbleweed Town."

[9] O. E. Baker, "The Agriculture of the Great Plains Region," *AAAG* 13, 3 (1923), 133–4.

[10] The paradox is often remarked upon, see, for example: George Johnson, "The Weed That Won the West: How Did an Invader from the Russian Steppes Become a Symbol of the American West?" *National Geographic*, (December 2013), available online at https://on.natgeo.com/2NGvSZ, accessed December 15, 2017.

[11] On "weeds," see Frieda Knobloch, *The Culture of Wilderness: Agriculture As Colonization in the American West* (Chapel Hill: University of North Carolina Press, 1996), pp. 113–45; Clinton L. Evans, *The War on Weeds in the Prairie West: An Environmental History* (Calgary: University of Calgary Press, 2002), pp. 1–17; Neil Clayton, "Weeds, People and Contested Places." *E and H* 9, (2003), 301–31; Taylor Spence, "The Canada Thistle: The Pestilence of American Colonialisms and the Emergence of an Exceptionalist Identity, 1783–1839," *AH* 90 (2016), 511–44. See also Peter Coates, *American Perceptions of Immigrant and Invasive Species: Strangers on the Land* (Berkeley: University of California Press, 2006).

steppes discussed in this book, however, tumbleweed encountered few barriers. It arrived by accident, established itself largely unnoticed, but then rolled inexorably and unstoppably across the plains in defiance of human attempts to control it.

Tumbleweed came to the attention of the United States Department of Agriculture (USDA) in 1891. The secretary of agriculture reported that the trespasser:

> is not known to be native in this country, but its home is in Russia, and it has more than likely been introduced from that country with forage seed or brought over in some way by emigrants. However it may have become introduced in the West, one thing is certain, . . . that it is rapidly spreading and threatens to be one of the worst weeds with which the farmer will have to contend.[12]

The USDA sent a botanist, Lyster H. Dewey, to the plains to investigate. He characterized the weed as indeed "one of the worst . . . ever introduced into the wheat fields of America." He continued:

> although rather pretty when reddened in the fall, and useful for forage when young, . . . it will take possession of a field to the exclusion of everything else, and it draws from the land a large amount of nourishment that might otherwise go to make useful plants. . . . it spreads and multiplies more rapidly [than other weeds] and hence takes up more space and more nourishment.[13]

It came to be known under various names, not just simply as tumbleweed, but also Russian thistle, Russian cactus, Russian tumbleweed, Tartar thistle, Hector weed, wind-witch, leap-the-field, prickly saltwort, among others. In the late nineteenth century, American botanists gave it the scientific name *Salsola kali* variety *tragus*,[14] and in the 1930s they sometimes designated it *Salsola pestifer*.[15] More recently, it has been reclassified as *Kali tragus*. Botanists have noted how it is distinguished

[12] United States Department of Agriculture, *Report of the Secretary of Agriculture, 1891* (Washington, DC: Government Printing Office, 1892), pp. 356, 357.

[13] Lyster H. Dewey, "The Russian Thistle and Other Troublesome Weeds in the Wheat Region of Minnesota and North and South Dakota," *USDA Farmers' Bulletin*, no. 10 (1893), 3–4, 7–8. Dewey's expertise was in fibre plants, including flax and hemp. Harry T. Edwards, "Lyster Hoxie Dewey," *Science* 101, no. 2610 (1945), 11–12.

[14] USDA, *Report of the Secretary of Agriculture, 1891*, p. 356; Dewey, "The Russian Thistle" (1893), 3–4; G. P. Clinton, "The Russian Thistle and Some Plants That are Mistaken for It," *University of Illinois Agricultural Experiment Station Bulletin*, no. 35 (1895), 87.

[15] Frank C. Gates, "Weeds in Kansas," *Report of the Kansas Board of Agriculture, June 1941* (Topeka: Kansas Board of Agriculture, 1941), p. 117.

by its "winged fruiting perianth,"[16] referring to its mechanism for spreading its seeds by tumbling.

Descriptions from the late nineteenth century by Dewey and Charles E. Bessey, a botanist at the University of Nebraska, give us a graphic impression of this curious immigrant from the steppes. It is an annual plant that grows from seed to a height of up to three feet. It branches profusely, forming dense, bush-like plants between two and six feet in diameter. Dewey noted: "When young it is a very innocent-looking plant, tender and juicy throughout, with small, narrow, downy green leaves." It blossoms in July and August. But, with the onset of dry weather in August, the "innocent disguise disappears, tender, downy leaves wither and fall, and the plant increases rapidly in size, sending out hard, stiff branches." The leaves are replaced by clusters of three sharp spines every half inch or so, "which harden but do not grow dull as the plant increases in age and ugliness." At the base of the clusters are small, papery flowers, and at bottom, the seeds. Fully-grown plants have thousands of seeds. When the frosts start, the plants change in color from dark green to crimson, and finally to white. In November, when the ground freezes, the root breaks and "the dense yet light growth," which is "circular or hemispherical in form," is carried by the wind. The detached part is dense, prickly, irritating, and impossible to pass a hand through without getting scratched. Tumbleweeds go "rolling across the country at racing speed," Dewey wrote, "scattering seeds at every bound, and stopping only when the wind goes down or when torn to pieces." Tumbleweed continues to roll around, spreading more seeds, for several months.[17]

Before the arrival of the intruder from the steppes, there were other types of tumbleweeds in the American plains, for example *Amarantus albus* and *Cycloloma atripicifolio*, but they were not as harmful, started to roll earlier, were less dense, and lacked the spines and other sinister features of the Russian immigrant. There was a similar tumbleweed, classified as *Salsola kali* L. (sic), found west of the Missouri River in South Dakota, in northeast Wyoming, as well as Wisconsin, and much further east along

[16] Hossein Akhani, Gerald Edwards, and Eric H. Roalson, "Diversification of the Old World Salsoleae S.l. (Chenopodiaceae), Molecular Phylogenetic Analysis of Nuclear and Chloroplast Data Sets and a Revised Classification," *International Journal of Plant Sciences* 168, no. 6 (2007), 931–56.

[17] Dewey, "The Russian Thistle" (1893), 3–4; Lyster Hoxie Dewey, "The Russian Thistle: Its History As a Weed in the United States, With an Account of the Means Available for Its Eradication," *USDA Division of Botany Bulletin*, no. 15 (1894), 10; Charles E. Bessey, "The Russian Thistle in Nebraska," *Bulletin of the Agricultural Experiment Station of Nebraska* (1893), 68–70. I got a nasty surprise, scratching my hands, the first time I picked up some tumbleweeds in a ditch by the side of highway 56, near Moscow, Stevens County, in southwestern Kansas in the spring of 2009.

the Atlantic coast. It looked like the Russian weed, but was confined to saline localities and had little tendency to spread. It was, Dewey noted, not the "real Russian thistle."[18] He looked into the origins of its more troublesome family member. He discovered that, since the mid eighteenth century, botanists had described it among the flora of eastern Russia and western Siberia. "The variety [*Salsola kali*] *tragus*," he wrote, "seems to have developed in the plains of southeastern Russia, where the conditions are very similar to ... the Great Plains region of the United States." In Russia, he continued, it had proved to be a destructive weed in fields of crops and had even caused cultivation to be abandoned in large areas of more arid steppe around the Caspian Sea.[19]

Tumbleweed (*Perekati-pole*) in the Steppes

In the late nineteenth century, Russian scientists used the same scientific name, *Salsola kali*, as their American counterparts for an identical plant in the steppes, where it was known in Russian as *perekati-pole* ("roll across the field").[20] Russian accounts of steppe vegetation over the nineteenth and early twentieth centuries mentioned many plants considered "weeds," including tumbleweed, but they did not single it out as especially pernicious. A survey of Kherson province in the steppes in 1863 specifically stated that *perekati-pole* was among the less harmful weeds.[21]

Studies of arable farming in the steppes over the same period regularly pointed to the trouble it caused in fields left fallow for several years, which was standard practice in parts of the steppe region. In the early years after fields were taken out of cultivation, the first plants to grow back were those farmers considered weeds. After a few more years, as more native grasses and other plants and shrubs returned, they kept the "weeds" under control. Thus, the natural vegetation prevented "weeds" from taking over

[18] Dewey, "The Russian Thistle" (1894), 6, 11. The plant classified as *Salsola kali tragus* was related to, but not the same as *Salsola kali* L. Thus, scientists realized that the plant that had arrived from the steppes was different from similar plants that were already in North America.

[19] Dewey, "The Russian Thistle" (1894), 11–12.

[20] See, for example, S. R [Semen Rostovtsev], "Perekati-pole," *ES* 23 (1898), 210–11.

[21] A. Shmidt, *Materialy dlia geografii i statistiki Rossii, sobrannye ofitserami general'nogo shtaba: Khersonskaia guberniia* (Spb: Departament General'nogo Shtaba, 1863), part 1, pp. 380–1. See also V. Mikhalevich, *Materialy dlia geografii i statistiki Rossii: Voronezhskaia guberniia* (Spb: Departament General'nogo Shtaba, 1862), pp. 81–2; A. Zashchuk, *Materialy dlia geografii i statistiki Rossii: Bessarabskaia oblast'* (Spb: Departament General'nogo Shtaba, 1862), part 1, p. 133; V. P. Semenov (ed.), *Rossiia: Polnoe geograficheskoe opisanie nashego otechestva*, vol. 14, *Novorossiia i Krym* (Spb: A. F. Devrien, 1910), pp. 87, 90–1.

completely.[22] The problems caused by weeds got worse as more of the steppe was plowed up and as long-fallow farming was replaced by crop rotations with shorter or no fallow periods. At the end of the nineteenth and start of the twentieth centuries, there were reports from different parts of the steppe region, including Samara province on the Volga and the Don Cossack region, of weeds, including *perekati-pole*, choking grain crops on land that was becoming exhausted and prone to erosion due to over-cultivation and lack of fallowing.[23]

Tumbleweed was especially troublesome during drought years, when it thrived while other plants could not. The German geographer and travel writer Johann Georg Kohl visited the Russian Empire in the late 1830s a few years after a catastrophic drought. While traveling across the southern steppes, including the region where the Mennonites and Black Sea Germans lived, he encountered a "weed that stands in very bad odor in the steppe" and "has been denominated wind-witch by the German colonists." His description is worth quoting at length:

> This is a worthless plant that expends all its vigor in the formation of innumerable thread-like fibers, that shoot out in every direction, till the whole forms a light globular mass. . . . It grows to the height of three feet, and in autumn the root decays, and the upper part . . . becomes completely dry. The huge shuttlecock is then torn from the ground by the first high wind that rises, and is sent dancing, rolling, and hopping over the plain, with a rapidity which the best mounted rider would vainly attempt to emulate. Hundreds of them are sometimes detached from the ground at once on a windy day, and when seen scouring over the plain, may easily be mistaken for a taboon [herd] of wild horses. The Germans could not have christened the plant more aptly, and . . . they no doubt thought of the national legends long associated with the far-famed witch-haunted recesses of the Blocksberg. The wild dances [of that mountain] are yearly imitated by the wind-witches on the steppe. Sometimes they may be seen skipping along like a herd of deer; . . . and sometimes rising by hundreds into the air, as though they were just starting to partake in the diabolical festivities of the Blocksberg . . . They adhere to each other sometimes like so many

[22] For examples from across the steppe region, see L. Cherniaev, "Ocherki o stepnoi rastitel'nosti," *SKhiL* 88 (1865), 2nd pagn., 33–48; Sergei Bulatsel', "Zemledelie v Slavianoserbskom uezde, Ekaterinoslavskoi gubernii," *Trudy Vol'nogo Ekonomicheskogo Obshchestva*, no. 1 (1866), 338; P. Kostychev, "Iz putevykh zametok: Perelozhnye khoziaistva s kratkosrochnymi zalezhami," *SKhiL* 161 (1889), 2nd pagn., 338.

[23] *Sbornik statisticheskikh svedenii po Samarskoi gubernii: Otdel khoziaistvennoi statistiki*, no. 6, *Nikolaevskii uezd* (Samara: Samarskoe gubernskoe zemstvo, 1889), p. 43; GARO, f.46, op.1, d.3282, l.18 (the Don Cossack region, 1898); GASO, f.5, op.11, d.102, l.87 (Samara province, 1909).

> enormous burrs, and it is not an uncommon sight to see some twelve or twenty rolled into one mass, and scouring over the plain like a huge giant in his seven league boots.[24]

At the time of another serious drought towards the end of the century, Russian scientist Nikolai Adamov (a colleague of Vasilii Dokuchaev) described the impact of the hot, dry, wind, known as the *sukhovei*, in eastern Khar'kov province in July 1892:

> From the morning a strong, violent, easterly wind began to blow, which at times raised a significant amount of dust from the road; the air became dry, far off there was a haze in the air, which foreshadowed an abrupt change. By midday . . . the entire horizon was covered with a very fine dust . . . The house trembled under the force of the strong wind, and from all sides broken branches of thistle, "*perekati-pole*" etc., were driven along.

As well as branches of tumbleweed, the wind left drifts of dust, sand and soil, fields stripped bare, and scorched crops.[25] Adamov's description unconsciously anticipated accounts of the Dust Bowl in the American plains four decades later. It also resembles a scene from a well-known American movie, set in Kansas. Tumbleweed could flourish, therefore, in drought conditions that killed off other plants, both cultivated and wild. Tumbleweed also did well in the steppe region at times of human devastation. Tumbleweed thus came to be associated with the worst depredations the steppe environment could pose its human inhabitants.

Russian agronomists and authorities sought ways to deal with tumbleweed and other unwanted plants. They came to associate the mounting problem of arable fields being taken over by weeds with little or "incorrect" cultivation of the land. This was the view, for example, of the authors of a survey of Simbirsk province on the Volga in 1868 and of a report on the causes of impoverishment among the Don Cossacks in 1898. In both cases, farmers were overburdened with other demands on their time and resources that prevented them giving due attention to controlling weeds by cultivating their fields.[26] Some farmers had found ways to combat the scourge of weeds.

[24] J. G. Kohl, *Russia: St. Petersburg, Moscow, Kharkoff, Riga, Odessa, the German Provinces on the Baltic, the Steppes, the Crimea, and the Interior of the Empire* (London: Chapman and Hall, 1842), pp. 473–4.

[25] N. Adamov, "Meteorologicheskie nabliudeniia 1892–1894 godov," *Trudy Ekspeditsii, snariazhennoi Lesnym Departmentom, Nauchnii otdel* 3, no. 1 (1894), 235–6. On recurring droughts in the steppes, see David Moon, *The Plough that Broke the Steppes: Agriculture and Environment on Russia's Grasslands, 1700–1914* (Oxford: Oxford University Press, 2013), pp. 65–8.

[26] Lipinskii, *Materialy dlia geografii i statistiki Rossii, sobrannye ofitserami General'nogo Shtaba: Simbirskaia guberniia* (Spb: Departament General'nogo shtaba, 1868), part 1, pp. 232–3; GARO, f.46, op.1, d.3282, l.18.

More prosperous and enterprising steppe farmers who had the means to do so, such as Mennonites, plowed their fields deeply and cultivated their fallow fields regularly in spring and summer to keep them free from all vegetation, including tumbleweed. The Mennonites introduced these practices in the 1830s following the drought earlier in the decade.[27] In the early twentieth century, Russian agronomists recommended frequent cultivation of the land to keep weeds under control.[28] They also recommended regular rotations of crops instead of the haphazard or absence of rotations that were replacing long-fallow farming due to population pressure. Leaving fields fallow for several years had allowed weeds to be controlled by the natural succession of native plants. The regular crop rotations recommended in the early twentieth century were designed in part to replicate this succession. Among the plants agronomists advised for use in rotations to control weeds were fodder crops known in English as crested wheat grass (*zhitniak, Agropyron cristatum*), brome grass (*koster, Bromus inermis* Leyss), and alfalfa or lucerne (*lutserna, Medicago sativa*).[29]

Tumbleweed has received far less resonance in the cultures of peoples of the Eurasian steppes than its wayward descendant in the American West. I struggled to find many examples. In 1887, Anton Chekhov entitled one of his lesser-known short stories "Perekati pole." The title was a metaphor for large crowds of pilgrims gathering at the Sviatigorsk monastery that provides the scene for the story. The monastery is located above the banks of the northern Donets River in southeastern Khar'kov province in today's eastern Ukraine (180 miles north of Chekhov's home town of Taganrog).[30] More recently, in 1991, a Russian musician from northern Kazakhstan, Aleksandr Lunev, composed a song called "Perekati-pole." A 2005 film, *Kolia – perekati pole*, was set in a small Russian provincial town.[31] Far to the east in Mongolia, the nineteenth-century poet Sandag the Fabler

[27] Filipp Vibe, "Khlebopashestvo Menonitov iuzhnoi Rossii," *Zapiski Imperatorskogo Obshchestva Sel'skogo Khoziaistva Iuzhnoi Rossii,* no. 4 (1853), 156–7; Alexander Petzholdt, *Reise im westlichen und südlichen europäischen Russland im Jahre 1855* (Leipzig: Hermann Fries, 1864), pp. 161, 169.

[28] N. K. Nedokuchaev, "Nastavleniia po zemledeliiu," in *Nastol'naia kniga russkogo zemledel'tsa*, 2nd edition, ed. P. N. Sokovnin (Spb: Sel'skii vestnik, 1914), p. 95.

[29] S. S. Bazhanov, "Novye sposoby vedeniia polevogo khoziaistva v iugo-vostochnykh stepnykh guberniiakh," in Sokovnin (ed.), *Nastol'naia kniga russkogo zemledel'tsa*, p. 35.

[30] Anton Pavlovich Chekhov, "Perekati pole," in *Polnoe Sobranie Sochinenii i Pisem*, vol. 6 (Moscow: Nauka, 1976), pp. 253–67.

[31] Discogs Music Database (2018) Aleksandr Lunev Discography, available online at https://bit.ly/2rbwWNo, accessed January 3, 2018; Nickolay Dostal (2005), *Kolia – perekati pole*. Russia: Marina Gundorina and Fyodor Popov (information available online at www.kino-teatr.ru/kino/movie/ros/3090/annot/, accessed November 6, 2019). Internet searches (in Russian) revealed few other uses of the term in contemporary Russian popular culture.

wrote an allegorical poem about the transitory nature of life entitled: "What the Tumbleweed Blown By the Wind Said."[32] The limited attention to tumbleweed in the culture the steppes suggests that it has not presented such serious problems as in the Great Plains and American West, and that it has been present for a very long time, rather than unexpectedly turning up in the late nineteenth century.

Tumbleweed attracted more interest among Russian scientists. In the 1890s, Dokuchaev looked back in time to the black-earth steppes of Russia and Siberia, before they had been plowed up, and imagined a sea of grasses, shrubs, and tumbleweed (*perekati-pole*). He projected his imagined virgin steppe on to grasslands around the world and thought, mistakenly, that the tumbleweed he was familiar with in his native steppes had also existed in the past in the unplowed plains of North America.[33]

The Introduction and Spread of Tumbleweed in the Great Plains

The German traveler Kohl, who provided such a dramatic description of the "wind-witch" in the steppes in 1830s, visited the prairies and plains of the United States two decades later. His American travelog contains no such evocation of tumbleweeds dancing and rolling over the plain for the very good reason that they had not yet hopped across the Atlantic.[34] Dewey's investigation in the early 1890s indicated that seed of the noxious tumbleweed had been brought, unwittingly, by immigrants from the Russian Empire with some impure flaxseed in 1873 or 1874. The immigrants sowed the flaxseed, and the tumbleweed seed, near the city of Scotland in Bonhomme County in the Dakota Territory (in the future South Dakota).[35]

As the extent of the problems created by tumbleweed became apparent, there was much speculation about who had brought it and what their motives may have been. In 1891, Mr. Norman S. French of Grand Rapids,

[32] R. Dor and G. Kara, "Literature in Turkic and Mongolian," in *History of Civilizations of Central Asia*, vol. 6, ed. Chahryar Adle, Madhavan K. Palat, and Anara Tabyshalieva (Paris: UNESCO, 2005), pp. 897–8.

[33] V. V. Dokuchaev, *K ucheniiu o zonakh prirody. Gorizontal'nye i vertikal'nye pochvennye zony* (Spb: Spb. Gradonachal'stvo, 1899), p. 8.

[34] See Johann Georg Kohl, *Reisen im Nordwesten der Vereinigten Staaten* (New York: D. Appleton & Co., 1857).

[35] Dewey, "The Russian Thistle" (1894), 12, 13. Most accounts agree with Dewey on its origins. There are other versions, but they are not supported by evidence. For example, without providing a source, a recent account stated that "Russian thistle" (sic) was "introduced to the Midwest around 1870 in a grain shipment from the steppes of Mongolia." Ceiridwen Terrill, *Unnatural Landscapes: Tracking Invasive Species* (Tuscon: University of Arizona Press, 2007), p. 94.

North Dakota (250 miles north of Scotland, South Dakota), sent a specimen to the USDA in Washington, DC. He reported: "Intelligent Russians have told me that the weed grows abundantly in southern Russia, in the vicinity of Odessa, where it is known locally as the Tartar thistle, and it is supposed to have been brought to America by Russian Jews." In the same year, S. W. Narregang, the president of the Dakota Irrigation Company, also sent a specimen to the USDA. He stated: "From the best authority I can obtain there is no question but that the Russians brought the seed to this country." He referred specifically to Scotland, South Dakota, where "the Russians formerly settled."[36] There was also speculation that "Russian Mennonite emigrants" had brought it deliberately "in revenge for social injustices they were receiving in their new homes."[37] In the early 1890s, Dewey tried to quell the ugly rumors: "There is evidently no foundation whatever for the theory, which is too often related as a fact, that [tumbleweed] was first sown in South Dakota by immigrants either for forage or to inflict an injury on an enemy."[38]

No groups of immigrants or their descendants have claimed responsibility for introducing tumbleweed to the Great Plains. This is in marked, if fully understandable, contrast to the many claims for bringing the more valuable (from a human perspective) Turkey Red wheat that arrived in the same way by the same route at the same time. It is difficult to identify which group of immigrants may inadvertently have brought the tumbleweed seed. Several of the groups who moved from the steppes to that part of the Dakota Territory in this period fitted the descriptions of the culprits reported to the USDA. There were agricultural communities of Russian–Jewish immigrants from the steppe region around Odessa who cultivated flax and grain crops in their new homes in the Dakota Territory. But, they arrived in 1882, a few years after the tumbleweed, and they settled in Aurora County, about seventy-five miles northwest of Scotland.[39] There were several groups of German Lutherans, Mennonites, and Hutterites who moved from the southern steppes to Bonhomme County and the surrounding area in the 1870s. They included a farming community of Hutterites

[36] USDA, *Report of the Secretary of Agriculture, 1891*, pp. 357–8. The story that it had been introduced by Russian Jews to Scotland, Bonhomme Country, in 1873 was repeated in Clinton, "Russian Thistle," 91.

[37] James A. Young, "The Public Response to the Catastrophic Spread of Russian Thistle (1880) and Halogeton (1945)," *AH* 62 (1988), 124. Mennonites from the Russian Empire were also blamed when the tumbleweed spread to the Canadian prairies. Evans, *The War on Weeds*, p. 97.

[38] Dewey, "The Russian Thistle" (1893), 5.

[39] Violet Goering and Orlando J. Goering, "Jewish Farmers in South Dakota: The Am Olam," *South Dakota History* 12, no. 4 (1983), 232–47.

from the Molotschna area, led by Preacher Michael Waldner, who settled near Tabor, Bonhomme County (sixteen miles south of Scotland), in 1874.[40] It may be that the unfortunate responsibility for accidently bringing tumbleweed to the Great Plains falls to this group from the steppes.

Bonhomme County is on the northern bank of the Missouri River, which forms the border between South Dakota and Nebraska. At first, tumbleweed spread slowly, hindered by the hilly, wooded ravines of Bonhomme County and by fields of corn. It appeared in Yankton County, the adjoining county to the east, in 1877. Five years later, it had spread to the counties north and west of its point of introduction.[41] Bessey noted that tumbleweed had crossed the Missouri River to Nebraska, but that it spread slowly, and was almost unnoticed there until around 1886.[42] From the late 1880s, however, tumbleweed started to spread more rapidly and in all directions. By 1888, it had infested a large part of the future state of South Dakota between the Missouri and James Rivers and south of the Huron, Pierre, and Deadwood Division of the Chicago and North Western Railway. By this time, it had spread beyond South Dakota to nearby northern Iowa as well as northeastern Nebraska.[43] It quickly tumbled right across Nebraska into northern Kansas. In his memoir, David Fairchild recalled that in the late 1880s, during his student days at the state agricultural college in Manhattan in northeastern Kansas, he "became immersed in the problem of tumbleweeds" under the "enthusiastic guidance of Professor Kellerman." "The barbed wire fences," he wrote, "were piled high in the autumn with a great variety of weeds which scattered their seeds as the continual winds rolled them across the prairies."[44]

Dewey suspected that in the 1880s, there may have been another accidental introduction of tumbleweed seeds into flax fields in Faulk or McPherson counties, north of the Huron, Pierre, and Deadwood railroad

[40] Douglas Chittick, *Dakota Panorama: Dakota Territory Centennial Commission* (Freeman, SD: Pine Hill Press, 1961), pp. 125–35; John A. Hostetler, *Hutterite Society* (Baltimore, MD: Johns Hopkins University Press, 1997), pp. 104–16; David Decker and Bert Friesen (September 2017) "Bon Homme Hutterite Colony (Tabor, South Dakota, USA)," Global Anabaptist Mennonite Encyclopedia Online, available online at https://bit.ly/32azJmR, accessed January 4, 2018. Hutterites and Mennonites both grew out of the Anabaptist movement of the Reformation.

[41] Dewey, "The Russian Thistle" (1894), 12. [42] Bessey, "The Russian Thistle in Nebraska," 73.

[43] Dewey, "The Russian Thistle" (1894), 12; Bessey, "The Russian Thistle in Nebraska," 67–8.

[44] David Fairchild, *The World was My Garden: Travels of Plant Explorer* (New York: Scribner's Sons, 1938), p. 12. Kellerman listed it in a textbook published in 1898. W. A. Kellerman, *A Text-book of Elementary Botany: Including a Spring Flora* (Philadelphia: Eldredge & Brother, 1898), p. 187. Fairchild graduated with a master's degree from Kansas State Agricultural College in 1889, before moving to Iowa. "David G. Fairchild," Notable Names Database (NNDB), available online at www.nndb.com/people/968/000160488/, accessed January 9, 2018.

in South Dakota.[45] If so, then this new introduction could explain the sudden appearance of tumbleweed in North Dakota. Norman French of Grand Rapids, North Dakota, wrote that tumbleweed had "first been seen in this vicinity" in 1886 or 1887. Warren Upham, who worked for the U.S. Geological Survey, noticed tumbleweed on a railway embankment near Clement, North Dakota and on a sandy, cultivated field near Towner in the same state in 1889.[46] Grand Rapids and Clement are in the southeast of the state, not far from the border with South Dakota. Towner, however, was far off in north-central North Dakota, nearer the border with Canada than South Dakota. The press recorded tumbleweed's progress. In the early 1890s, the *Daily Argus* in Fargo, North Dakota, reported:

> The vanguard of the invader is already north of Sheldon; with the fall winds it will make its appearance in our neighborhood – at Oakes the town is given over to it. LaMoure is gone with it. In the former town, the streets are lined with it. The sidewalks run between hedges of it. The chinks of the sidewalks are fringed with it, the yards and vacant lots are matted with it.[47]

USDA accounts from the early 1890s conveyed tumbleweed's inexorable spread across the plains and neighboring states. In 1891, the secretary of agriculture reported that tumbleweed "has already gained a strong foothold in North and South Dakota, extending eastward into Minnesota and Wisconsin, is common in northern Nebraska, and has recently been detected along railroad tracks near Denver, Colo."[48] It had arrived in northern Illinois by 1890.[49] Reports by Dewey in successive years from 1893 to 1895 reveal the speed of its advance. In 1893, he wrote that it was "more or less abundant" between the Missouri and James Rivers in South Dakota and extended into North Dakota. In total, around 30,000 square miles were infested. It was also present further east in St. Paul and Minneapolis, Minnesota, in northwestern Minnesota, and around Sioux City, Iowa, but not yet in sufficient quantities to cause harm.[50] In 1894, he reported that it had spread west across the Missouri River in at least four places in South and North Dakota, to more counties in western Minnesota, northwestern Iowa, and northeastern Nebraska, and had reached southern Nebraska and Madison, Wisconsin. He estimated that in the

[45] Dewey, "The Russian Thistle" (1894), 12.

[46] USDA, *Report of the Secretary of Agriculture, 1891*, p. 357. (The Secretary quoted from letters sent by French and Upham.)

[47] Quotes in Young, "The Public Response," 125.

[48] USDA, *Report of the Secretary of Agriculture, 1891*, pp. 356–7.

[49] Clinton, "The Russian Thistle," 93.

[50] Dewey, "The Russian Thistle" (1893), 6.

twenty years since its first introduction, tumbleweed had spread to an area of 35,000 square miles. "Very few cultivated plants," he asserted, "even, which are intentionally introduced and intentionally disseminated, have a record for rapidity of distribution equal to this weed."[51] By the following year, 1895, tumbleweed had spread further:

> During the twenty-one years since its introduction into this country it has spread with greater rapidity than any other weed. Ninety new localities have been reported during the present season ... It now occurs in places from eastern Ontario and western New York to the western border of Idaho, and from Manitoba to southern Colorado, being most abundant in South Dakota and adjacent states.

He warned that it was likely to cross the Rocky Mountains.[52] It had already reached Montana, where it was declared a "common nuisance" in that state's weed law of 1895. But, it was not initially as harmful as in other states. In 1901, Joseph Blankinship, a botanist at the Montana agricultural experiment station in Bozeman, wrote that "it did not appear to exhibit such dangerous characteristics," and had not yet made headway against native vegetation or invaded cultivated land. He noted that it was better adapted to conditions in the Milk and Yellowstone River valleys, and warned that it "may there prove a serious menace to the agricultural interests," and should be exterminated.[53]

The USDA warned the Canadian Department of Agriculture, which ordered the Royal Northwest Mounted Police to keep watch for tumbleweed. Even the famed Mounties could not stop it spreading across the 49th parallel into the Canadian prairies.[54] And, it had reached California before 1905.[55] By the 1930s, tumbleweed had long spread throughout the western United States. Describing his field trips in the 1930s, soil scientist Macy H. Lapham noted how near Pocatello, in southern Idaho: "In the cleared and farmed areas ... the sagebrush has been succeeded by the disliked Russian thistle and other introduced tumbleweeds that, borne along by the wind, gallop and bounce their way across the plains, scattering

[51] Dewey, "The Russian Thistle" (1894), 12.

[52] Lyster H. Dewey, "The Russian Thistle," *USDA Divison of Botany Circular*, no. 3 (1895), 4–5.

[53] J. W. Blankinship, "Weeds of Montana," *Montana Agricultural Experiment Station Bulletin*, no. 30 (1901), 20, 22–3, 55–6.

[54] James Fletcher, "The Russian Thistle or Russian Tumbleweed," *Dominion [Canada] Department of Agriculture Experiment Farm Notes*, no. 4 (1894); Evans, *The War on Weeds*, pp. 95–8.

[55] California State Commission of Horticulture, *The Russian Thistle: Its Introduction and Spread in California, With Laws and Measures for Its Control, With a Description of the Canada and Scotch Thistles* (Sacremento, CA: State Commission of Horticulture, 1905).

seeds as they go."[56] Back in 1893, Bessey had written that tumbleweeds "seem to have partaken of the conquering spirit of the west."[57]

The ways tumbleweed spread were examined by botanists working for the USDA and state governments. In 1895, G. P. Clinton of the Illinois Agricultural Experiment Station wrote:

> Nature has rather liberally provided the Russian thistle with means for the dissemination of its seeds. This fact, taken together with the great number of seeds that some of the plants produce, and the protection afforded the plant by its spiny leaves, accounts in great measure for the vigor with which the weed has usurped certain territory and spread to such an alarming extent over the country.[58]

Dewey identified three main modes of distribution: wind; with uncleaned crop seed; and railroads.[59] He elaborated on how tumbleweed spread with the wind:

> The Russian thistles are most wonderfully adapted for spreading and covering new territory. As tumbleweeds they are carried for miles by the wind, scattering seeds as they go. By this means alone they often advance 5 or 10 miles in a single season. Single stray weeds may doubtless be blown much farther.

Strong winds in the winter of 1887–8 and the dry summer of 1888 assisted the spread of the tumbleweed: it was in the late 1880s that, across the plains, people began to take note of the new arrival. The weed's general advance, Dewey noted, was in the direction of the prevailing winds or the most frequent high winds. The environment of the plains assisted the tumbleweed spread with the wind. In the winter, the ground was frozen and covered in hard, dry snow, creating ideal terrain for it to tumble across. The absence of trees and fences in much of the high plains meant there were few barriers to hold it up.[60] In Kansas, it became apparent over the following decades that the higher and more arid land in the west of the state was more conducive to the spread of tumbleweed.[61]

The second way tumbleweed spread was with seeds of crops that had not been properly cleaned to ensure their purity. By the early 1890s, tumbleweed seed had been found in flaxseed (which was how it had first arrived from the Russian Empire), and also among barley, oats, and wheat

[56] Macy H. Lapham, *Crisscross Trails: Narrative of a Soil Surveyor* (Berkeley, CA: W. E. Berg [1949]), p. 224.
[57] Bessey, "The Russian Thistle in Nebraska," 73.
[58] Clinton, "The Russian Thistle," 90.
[59] Dewey, "The Russian Thistle" (1894), 13.
[60] Dewey, "The Russian Thistle" (1894), 12–13, 15.
[61] Gates, "Weeds in Kansas," 117.

seed. Tumbleweed seeds were found in, and carried further by, threshing machines, when the weed got mixed in with crops.[62] In spite of efforts to prevent it, tumbleweed seeds continued to spread with uncleaned seeds. They were still being found in agricultural seeds in Kansas in the 1930s.[63] It is likely, moreover, that there were further unintentional introductions of tumbleweed with uncleaned seeds from the Eurasian steppes. Studies in the 1920s showed that seeds of tumbleweed (*Salsola pestifer*) were present in over ten percent of samples that were tested of alfalfa seed imported from Turkestan in Soviet Central Asia.[64]

The third way in which tumbleweed seeds were spread was by trains along the railroads. This facilitated the rapid dispersal of the noxious weed across the region, and also its sudden appearance in places far removed from the main areas of infestation. The seeds were carried in cars transporting stock and grain along the Northern Pacific Railroad from North Dakota, across northwestern Minnesota, to St. Paul and Minneapolis, where tumbleweed was found growing around grain elevators and stock yards. Railroads carried the seeds southeast to Sioux City, Iowa, and Madison, Wisconsin, and southwest to Denver, Colorado.[65] In the early 1890s, Bessey traced the tumbleweed's route from South Dakota to Nebraska following "the established lines of travel" along the Elkhorn railway and the Black Hills line of the Burlington railroad. He wrote that "[s]tock cars appear to be carrying the seed to many parts of the state," which probably explained its occurrence in the stock yards of south Omaha, from where it was "infecting" other parts of the state. It also appeared in five locations south of the Platte River, which were far from each other, but along railroads, strongly suggesting its mode of travel.[66] The railroad also brought the weed to Illinois, where it first appeared in southeastern Chicago in 1890. "It is believed," the state assistant botanist wrote, "that the plants were started here from seed that was carried in stock cars, such cars having been emptied of their contents along the side-track, where specimens were first found." Tumbleweeds were carried further along the railroads of Illinois and into neighboring Indiana.[67] And the railroads most likely explain the speed of tumbleweed's spread throughout much of the American Midwest and West in the late nineteenth and early twentieth centuries.

[62] Dewey, "The Russian Thistle" (1894), 13. [63] Gates, "Weeds in Kansas," 27.

[64] E. H. Hillman and H. H. Henry, "The Incidental Seeds Found in Commercial Seeds of Alfalfa and Red Clover," *Proceedings of the International Seed Testing Association* 6 (1928), 9.

[65] Dewey, "The Russian Thistle" (1893), 5–7; Dewey, "The Russian Thistle" (1894), 12.

[66] Bessey, "The Russian Thistle in Nebraska," 67–8. [67] Clinton, "The Russian Thistle," 93–6.

Scientists identified the conditions most conducive to tumbleweed's progress. It seemed to like high, dry places, where it was not crowded by other plants. It did less well in wet years, not because it did not like water, but because in such years other plants grew better than in dry years. It was found in different types of soils, alkaline and acidic, but was more vigorous in poorer soils. It did not do well, however, in light, sandy soils, such as the Sandhills of northwestern Nebraska. There seemed to be connection, moreover, between the dispersal of tumbleweed and the activities of the farming population. In 1891, the secretary of agriculture reported: "It is . . . found growing abundantly in cultivated fields, springing up among crops that are being cultivated, especially in wheat fields . . . It is never found in low wet places, and very seldom on the unbroken prairie."[68] Dewey also noted the link between human activity and tumbleweed's success:

> In many localities where a few plants were first seen four or five years ago every spot of land where the sod has been broken is now occupied. On every badger burrow and overfed spot in the prairie; on every roadside, railroad embankment, fire-break, and neglected garden; on every field or early-plowed land or stubble may be seen a patch of thistles.

The tumbleweed's nature, with its abundant seed supplies and methods of travel, allowed it to take advantage of the conditions. Dewey wrote that the seeds were not here and there "as with Eastern weeds," but everywhere.[69]

It became apparent to scientists that the arrival of the Euro-American settlers had created conditions conducive for the spread of weeds – i.e. plants in the wrong places – which tumbleweed was well adapted to take advantage of. Disturbing the land by clearing the native vegetation to create fields or gardens for crops or to build roads and railroads all provided opportunities for weeds, especially tumbleweed, to prosper. Tumbleweed found the crops farmers planted, especially wheat, to be less resistant than the native grasses the farmers had removed. The development of transportation and the mechanization of agriculture facilitated the more rapid spread of weeds in the early decades of the twentieth century. The weeds that caused the most trouble, however, were not those native to the plains, but those that had been brought by immigrants from Europe to the United States, where weeds such as tumbleweed found the conditions very hospitable.[70] Droughts and economic depressions that caused plains farms

[68] USDA, *Report of the Secretary of Agriculture, 1891*, pp. 357–8.

[69] Dewey, "The Russian Thistle" (1894), 15.

[70] See Gates, "Weeds in Kansas," 8.

to fail and their owners to abandon them left thousands of acres of disturbed, but bare, land to the tumbleweed. In Montana, where they had not initially been a serious hazard, farm failures during the 1917 drought created what M. L. Wilson of the state agricultural extension service described as "ideal conditions" for tumbleweed. In that year, he wrote, they became a "serious pest" in the state.[71]

As tumbleweed spread around the Great Plains it did great harm to the human population's efforts to modify the plains environment to create a viable and profitable farming economy.

Consequences

The secretary of agriculture noted in 1891: "This weed is a rapid grower and soon takes entire possession of neglected fields. . . . The Dakota farmers are much troubled by it, as it comes up after the wheat is cut and, overrunning the fields, it blossoms and matures seed for another year."[72] Tumbleweed had several harmful consequences for farmers and other inhabitants of the Great Plains: it damaged crops; provided a habitat for pests; its spines hindered people, horses, and dogs working on land infested by the weed; it damaged fences; contributed to soil erosion; and spread prairie fires.

In the early 1890s, Dewey noted that tumbleweed took possession of fields and used up nourishment that could otherwise have been used by plants more useful to the population. It was especially harmful to wheat and flax. If the growth of these crops was checked by droughts in the late spring or early summer, then tumbleweed – which grew well in dry weather – crowded out and starved the drought-weakened crops of much needed water. In 1893, many infested fields of wheat and flax were not worth harvesting. Tumbleweed damaged other grains, especially barley and rye. On the other hand, oats, millet, potatoes, and corn suffered little damage. Oats could choke out tumbleweed, and potatoes and corn matured before tumbleweed had the chance to get larger.[73] Dewey estimated that the average loss on wheat land was not less than five bushels an acre. In 1892, 640,000 acres of wheat land were infested. At fifty cents a bushel, this represented a financial loss of $1.6 million in wheat

[71] Quoted in Mark Fiege, "The Weedy West: Mobile Nature, Boundaries, and Common Space in the Montana Landscape," *WHQ* 36 (2005), 30.

[72] USDA, *Report of the Secretary of Agriculture, 1891*, p. 357.

[73] Dewey, "The Russian Thistle" (1893), 4–5; Dewey, "The Russian Thistle" (1895), 14.

production alone.[74] In 1893, the loss of crops in affected counties in North Dakota was estimated at fifteen to twenty percent of expected yields.[75] According to Bessey, farmers in parts of North Dakota suffering from drought considered tumbleweed to be "a more serious difficulty in the way of making profits out of their farms than the lack of rainfall." In some districts, real estate agents valued land with tumbleweed at a discount of $3 or more an acre. Farmers considered the "Russian thistle" to be "a thousand [times worse] in the frightfulness of its development and the trouble of its eradication" than other weeds, such as Canadian or Scotch thistle. "The Russian Thistle," Bessey concluded, "is the worst rolling tumble-weed on the prairies."[76] The harm caused to crops by tumbleweed persisted over the following decades. In 1941, the Kansas Board of Agriculture reported: "The weed uses large quantities of soil moisture, which accounts in part for its drought-resistant powers and which also makes it a competitor to crops sharing the same land."[77] The same report noted how tumbleweed harbored such pests as beet webworm, diamond-back moth, thistle grasshopper, and beet leafhopper, which damaged crops.[78]

Because of its sharp and irritating spines, tumbleweed made it awkward for people, and their horses and dogs, to work on land it had infested. Bessey reported that when "Russian thistles" had accumulated, it very difficult to cultivate, plow, and harvest the land as it was impossible to get gloves thick enough to protect peoples' fingers from the spines.[79] Dewey remarked:

> Some of its special characteristics render this thistle much more troublesome than other weeds. It is armed with spines quite as sharp and much stronger than those of common thistles. Because of these it is difficult to drive horses through a field where the plants are abundant.

Horses running in pastures covered with tumbleweed were injured as the spines broke off under their skin and caused festering sores. Farmers bound their horses' legs with leather to protect them. Hunters experienced problems getting their dogs to work well when hunting prairie chickens in stubble fields inundated with Russian thistles.[80] Tumbleweeds posed new problems with advances in transport technology: In 1941 the Kansas Board of Agriculture warned that the spines punctured rubber tires.[81]

[74] Dewey, "The Russian Thistle" (1893), 6–7. [75] Clinton, "The Russian Thistle," 97.
[76] Bessey, "The Russian Thistle in Nebraska," 72–3. [77] Gates, "Weeds in Kansas," 117.
[78] Gates, "Weeds in Kansas," 22–3. [79] Bessey, "The Russian Thistle in Nebraska," 72– 3.
[80] Dewey, "The Russian Thistle" (1893), 4–5. [81] Gates, "Weeds in Kansas," 12.

Tumbleweeds created no end of trouble. When the wind piled up large numbers against fences, they broke the fence posts.[82] Such piles of tumbleweeds also attracted deep drifts of dust blown by the same wind.[83] They infested railroad grades and prevented the growth of grass and other vegetation that would have protected the land against water erosion.[84] And tumbleweeds contributed to the spread of prairie fires. During wild fires they ignited and the balls of fire rolled in the wind, carrying the flames across fire breaks, thus allowing fire to take hold over larger areas and cause more damage to property. To make matters worse, the seeds were fire resistant and so continued to spread while all else around them was being consumed by flames.[85] Fire historian Stephen Pyne described how tumbleweeds became "wildblown firebrands."[86]

Besides the alarm caused by flaming tumbleweeds hurtling across the plains, the total cost of all the damage done by this unwanted immigrant from the Russian steppes has been enormous. Dewey estimated that for just the year of 1892 in the Dakotas alone, the value of losses of crops, injuries caused by the spines to humans and animals, and fires spread by them jumping across firebreaks amounted to $2 million. He stressed that this was a conservative estimate.[87] To put this figure in perspective: the combined population of North and South Dakota in 1890 was just under 540,000,[88] making a loss of around $4 a head. Or, in 2018 values, $2 million is the equivalent of around $55 billion,[89] or almost $100,000 per person in one year.

Tumbleweed did have some uses. Since it was drought resistant and grew in dry years when other plants did not, before its late-summer transfiguration into the spiny tumbling tumbleweed, the young, tender and juicy green leaves could be used as animal fodder. As such it proved invaluable to livestock farmers in the more arid western plains in the drought years of the 1890s and 1930s. During the Dust Bowl, farmers

[82] Dewey, "The Russian Thistle" (1894), 14–15; Gates, "Weeds in Kansas," 12.
[83] Frederic Clements, "Climatic Cycles and Human Populations in the Great Plains," *The Scientific Monthly* 47, no. 3 (1938), 202.
[84] Bessey, "The Russian Thistle in Nebraska," 72–3.
[85] Bessey, "The Russian Thistle in Nebraska," 72–3; Dewey, "The Russian Thistle" (1893), 4–5.
[86] Stephen J. Pyne, *Fire in America: A Cultural History of Wildland and Rural Fire* (Seattle: University of Washington Press, 1997), p. 98.
[87] Dewey, "The Russian Thistle" (1893), 6–7.
[88] See "Population of South Dakota State" (2016), available online at http://population.us/sd/, accessed January 16, 2018; Bernhardt Saini-Eidukat, "North Dakota Historical Population," available online at https://bit.ly/36ywetR, accessed January 16, 2018.
[89] CPI Inflation Calculator "$2,000,000 in 1892 → 2018, Inflation Calculator," US Official Inflation Calculator, Alioth Finance, https://bit.ly/36zI1b1, accessed March 16, 2018.

in western Kansas harvested tumbleweed to feed their animals. In the Oklahoma panhandle, out of desperation, people were reduced to eating the tumbleweeds they had gathered for their animals. Some families even claimed tumbleweeds were nourishing as they were high in iron. Cimarron County went so far as to proclaim a "Russian thistle week" and encouraged people on relief to harvest them. Others were less enthusiastic, complaining that the weeds tasted like cardboard or twigs, and were prickly like cactus. In January 1935, the *Hays Daily News* in western Kansas reported quips that for two years farmers had raised nothing and: "Only the rabbits and Russian thistles have thrived." It is a measure of the extent of the drought in the Oklahoma panhandle in the 1930s that even tumbleweeds were in short supply.[90]

In most years, however, tumbleweeds were all too abundant. The regional press contained stories, no doubt sometimes exaggerated, about the mischief brought about by these malefactors from an alien land. In the early twentieth century, journalists further east seemed to delight in alleged tumbleweed-related news from western Kansas. In 1909, the *Kansas City Star* reported that at Sharon Springs (near the Colorado border): "a great wall of thistles" struck the new courthouse and "the building was crushed beneath the weight." For hundreds of miles, reporters claimed, farmhouses, fences, and other buildings were covered by "the avalanche of thistles." In nearby Syracuse, Kansas, snowplows were needed to clear tumbleweed from the stalls in the roundhouse. Balls of thistles as high as the Washington monument were reported to be skipping over the prairies in the west of the state. They filled up the Republican River, it was reported, allowing heavy teams of horses to cross it as if on bridges made of concrete.[91]

Such stories became part of the folklore of the plains. The Works Progress Administration (WPA) guidebook to Kansas included accounts of tumbleweed in the notes for one of its driving tours. Oberlin, Decator County, in northwestern Kansas, was described as "a prosperous hillside village," whose "residents possess the friendliness and cordiality typical of the West." Turning right from Oberlin on U.S. highway 83, motorists were advised that "the most prolific form of vegetation [on the hilly pastureland] is the Russian thistle." The plant was "a blessing in time of

[90] Clinton, "The Russian Thistle," 97; Craig Miner, *Next Year Country: Dust to Dust in Western Kansas, 1890–1940* (Lawrence: University Press of Kansas, 2006), pp. 242, 274; Timothy Egan, *The Worst Hard Time: The Untold Story of Those Who Survived the Great American Dust Bowl* (New York: Houghton Mifflin Harcourt, 2006), p. 162; Evans, *The War on Weeds*, p. 4.

[91] Miner, *Next Year Country*, pp. 61–2.

drought when no other green succulent herb will flourish," and young plants could be eaten by livestock. But, the "mature thistles" were "too woody and thorny" for fodder, and caused a range of problems:

> Where the highway is sunken [tourists were to bear in mind] the thistles will drift into the cut, cling together like a barbwire entanglement, and make the road impassable till they are burned.
>
> In February 1909, residents report that a severe windstorm piled thistles high on the streets of Oberlin and buried one house to a depth of 20 feet. Only the chimneys were visible above the stack of tinder-dry thistles. Members of the family were afraid to light their breakfast fire, lest the sparks fly from the chimney and ignite the pile. After the storm had abated, neighbors with spades and corn knives chopped a tunnel through the thistles and rescued the marooned family.[92]

While some of these stories may have improved with retelling, there was no doubting the real and damaging consequences for farmers across the plains and western states caused by tumbleweed, and of the need to control, and if possible, eradicate it.

Efforts to Control and Eradicate Tumbleweed

In the late nineteenth century, as the problems caused by the newly arrived weed became evident, there were hopes that frosts or wet weather would kill it off. Some cases were noted of the weeds being choked by native grasses on good soil. But, these hopes soon evaporated as the "Russian thistle" established itself and became more troublesome.[93] In 1891, when the extent of the problems was becoming all too apparent, the secretary of agriculture advocated: "Some systematic and vigorous steps should be taken by the farmers of these Western States to eradicate this weed."[94] After it spread to Nebraska, Bessey noted "if means are not taken at once to check or destroy it, it will soon become a most serious farm pest."[95]

Devising methods to control tumbleweed proved difficult. It spread rapidly showing no respect for property boundaries and, in time, communities and states recognized the need to enforce collective action to tackle it.[96] Plains farmers proposed various ways to thwart the relentless spread of

[92] Harold C. Evans, *Kansas: A Guide to the Sunflower State* (New York: The Viking Press, 1939), pp. 322–3. (The guidebook mistakenly numbered the highway 183, which is further east.)

[93] Dewey, "The Russian Thistle" (1894), 17–18.

[94] USDA, *Report of the Secretary of Agriculture, 1891*, p. 356.

[95] Bessey, "The Russian Thistle in Nebraska," 67, 72–3.

[96] See Fiege, "The Weedy West."

the loathsome weed. In October 1893, just before tumbleweeds were due to break off and start to roll, E. T. Kearney from North Dakota suggested that the state legislature fund a wire fence across the state to hold them back.[97] But, H. L. Bolley of the North Dakota Agricultural Experiment Station pointed out that this would be waste of money, since his tests had shown that fences would be useless in stopping seeds of the "Russian cactus" from being carried for many miles by drifting snows and "our constant blizzard winds."[98] Dewey reported that in Russia "no effectual methods of exterminating the weed are known." Pasturing sheep on young plants helped keep it in check. But, it was continually growing more troublesome and spreading to new territories.[99]

In the early 1890s, Dewey made several recommendations. He advised plowing in August and September, before thistles had grown, stiffened, and gone to seed, and taking care that all thistles were well turned under. He urged cutting crop stubble together with tumbleweed before the latter had gone to seed, and then burning the stubble fields as soon as possible after the harvest. If thistles had got large and rigid, then he recommended raking them into heaps and burning them, but only as a last resort, and only when there was no wind, otherwise the burning thistles would spread, carrying fire and seeds with them. He noted that some crops, such as clover, millet, rye, or oats, were able to defy tumbleweed, and advocated sowing them on barren land to choke the weeds. He also noted that corn, potatoes, beets, or any cultivated crop, if well taken care of, would rid the land of the thistles. Sheep could be pastured on the young thistles to control them. He also urged that Russian thistles be exterminated along the sides of roads, railroad grades, fire breaks, and waste land that had once been plowed. These were all areas of land that had been disturbed, but were not cared for, and thus were ideal for tumbleweed to take hold. He recommended using mowing machines, scythes, hoes, rakes, and fire. To avoid creating conditions where tumbleweed could take over, he strongly urged inhabitants not to plow more land than could be cultivated properly, to take due care of all land, and to insist that neighbors did the same. To prevent the risk of repeating the mistake that had allowed tumbleweeds into the fields in the first place, he urged farmers to make sure their seed, especially flaxseed, was as pure and clean as possible. To bring tumbleweed

[97] Young, "The Public Response," 124–5.
[98] H. L. Bolley, "Distribution of Weed Seeds by the Winter Winds," *North Dakota Agricultural Experiment Station Bulletin*, no. 17 (1895), 102–5.
[99] Dewey, "The Russian Thistle" (1894), 12.

under control, Dewey acknowledged that the populations and governments in affected areas needed to take concerted action. From 1890, plains states enacted laws obliging people, companies, and officials to take measures to control and eradicate the weeds, and raised taxes to fund them. Railroad companies were starting to keep their rights of way clear of weeds. Road masters and constables were inspecting the roads for the weeds. Controlling tumbleweed was also a matter for schools. Dewey advised placing Russian thistles in school houses to familiarize pupils with the weed, and to "teach them to kill it wherever they find it as they would kill a rattlesnake."[100]

Across the border in Canada the authorities took similar measures. They found that keeping fallow fields clear of all vegetation during the summer proved effective. Weed inspectors were authorized, if necessary, to cut down crops if they were seriously infested with Russian thistle.[101] In the 1930s, the Kansas Board of Agriculture recommended:

> Good tillage methods will control this weed. Important to control on fallow fields and where land is prepared for crop. Its shallow root system makes the plant easy to eradicate. Small-grain stubble or fallow fields should be one-wayed or disked to tie thistles to the soil before producing seed or start tumbling.

Thistles that had accumulated in fences, along roads, and in ditches were to be burned. It was also noted that young plants could be freely eaten by cattle, sheep, and horses in pasture or when cut and cured as hay. During the Dust Bowl, tumbleweed proved to be valuable in "stabilizing land from blowing." But, the Board of Agriculture noted the problems caused if it was allowed to reach maturity and then break off and tumble, which was to be prevented at all costs.[102]

At this time, the USDA found another way of controlling tumbleweed. In 1936, surveys of range and pasture land in parts of the Great Plains, especially sub-marginal and abandoned land in the more arid western areas, showed that when land had been seriously over-grazed, Russian thistle replaced the grasses that had once grown there. Federal and state government agencies, including the Soil Conservation Service, recommended reseeding land that had been over-grazed, abandoned, eroded,

[100] Dewey, "The Russian Thistle" (1893), 8–10; Dewey, "The Russian Thistle" (1894), 18–26. Dewey's recommendations were developed by agricultural scientists in other states. See, for example, Bessey, "The Russian Thistle in Nebraska," 70–1, 74–7; Clinton, "The Russian Thistle," (Illinois), 98–9, 102–5. See also Blankinship, "Weeds of Montana," 22–3.

[101] Evans, *The War on Weeds*, pp. 101–3.

[102] Gates, "Weeds in Kansas," 117.

and thus liable to be inundated by tumbleweed, with "native or introduced grasses." The purpose was to prevent the soil blowing and to assist in restoring the land and keeping it free of tumbleweed. The grasses recommended included slender wheat grass and western wheat grass, which were native, and crested wheat grass and brome grass, which had been introduced.[103] South Dakota botanist Neils Hansen noted that one of the reasons crested wheat grass had proved so important was because when it was "sown in the fall right in the Russian thistles [it] entirely exterminate [s] the thistles in less than two years."[104] The USDA and Hansen had thus devised a way of tackling the infestations of fields with tumbleweed and other weeds that was similar to one proposed by Russian agronomists earlier in the century (see p. 361). Further, both crested wheat grass and brome grass had been introduced to the Great Plains from the steppes by Hansen. He had collected seeds of these grasses, and also alfalfa, on several visits to the Eurasian steppes for the USDA and the state of South Dakota in the 1890s and early twentieth century.[105] Hansen's biographer wrote: "Crested wheat grass yields twice as much forage than native grasses in extremely dry years, . . . chokes weed growth, and shows no bad effects of heavy grazing."[106] Thus, in the Great Plains, grasses introduced from the steppes were being used to control tumbleweed that had come from the same place in a similar way to which Russians were using these same grasses their native environment. The development of the same technique, however, may have been a coincidence, rather than a direct transfer, unless Hansen had picked it up from his Russian and Soviet contacts, for which I found no evidence.

[103] NARA KC, RG114, Entry 187 Dept of Agriculture. Forest Service. Region 2, General Records 1934–41, Box 10, G-Cooperation, AAA (Range Conservation Program), General, no. 1, USDA Forest Service, Shelterbelt Project, Lincoln, Nebraska, November 12, 1936, "Summary of Extensive Survey of Range and Pasture Conditions in the Great Plains Region in Relation to Land Use," 7, 16, 18, 24, 26, 27, 29, 32, 37–40, 45–6, 50–1; Henry Wallace, USDA Agricultural Adjustment Administration, 1936 Agricultural Conservation Program – Western Region, bulletin no. 2, Montana, issued Sept 26, 1936, 2; R. Douglas Hurt, "The National Grasslands: Origin and Development in the Dust Bowl," *AH* 59 (1985), 246–59. The lands taken out of use and reseeded were designated "National Grasslands" in 1960.

[104] SDSU ASC, UA 53.4, N. E. Hansen Papers, Series 2. Helen Hansen Loen Collection, Box 4, Folder 36, MS "Grains and Forage Plants," n.d.

[105] Niels E. Hansen, "Fifty Years Work as Agricultural Explorer and Plant Breeder," *Transactions of the Iowa State Horticultural Society* 79 (1944), 37–9; H. J. Taylor, *To Plant the Prairies and the Plains: The Life and Work of Niels Ebbesen Hansen* (Mount Vernon, IA: Bios 1941), pp. 23–34; H. L. Westover, "Crested Wheatgrass as Compared with Bromegrass, Slender Wheatgrass, and other Hay and Pasture Crops for the Northern Great Plains," *USDA Technical Bulletin* no. 307 (1932).

[106] Taylor, *To Plant the Prairies*, p. 23.

Conclusion

Tumbleweed, or Russian thistle, thrived in the Great Plains and American West, becoming a harmful, invasive species. It proved a more serious and intractable problem for the inhabitants of the Great Plains than the same plant, known as *perekati-pole*, did in its native environment in the steppes. When I looked for tumbleweed in the Korgalzhyn nature reserve in the steppes of northern Kazakhstan in the late summer of 2018, my local guides expressed surprise, and said that no previous visitors had ever enquired about it. The following spring, I found a few, forlorn tumbleweeds bunched up in a corner of the campus of Nazarbayev University in the nearby capital city. This contrasts with the attention attracted by and damage caused by tumbleweed in the Great Plains and American West.

Environmental historians have observed a connection between settler colonialism, in the Great Plains and other temperate parts of the globe, and the spread of introduced "weeds."[107] Weeds damaged the interests of the Euro-American settlers and the federal and state governments that were trying to transform the plains into European-style farmland. But, it was these very activities that contributed to tumbleweed's "success." Settlers from the steppes and the USDA's "agricultural explorers" brought over seeds of tumbleweed and other weeds, just as they introduced strains of wheat and other "useful" crops, some of the trees and shrubs planted in the shelterbelts, as well as the drought-resistant grasses sown to control erosion. The only difference, from a human perspective, is that the introduction of tumbleweed was an accident. By breaking the sod and plowing up the plains to grow crops, moreover, the settlers disturbed the environment and created conditions in which tumbleweed could thrive and spread. The settlers cleared native vegetation that could otherwise have controlled it. In their arable fields, tumbleweed competed successfully for nourishment and moisture with some of their crops, in particular wheat, and crowded them out. Further, federal and state governments promoted the construction of infrastructure for commercial farming, in particular railroads, roads, grain elevators, and stock yards, that also, and unintentionally, helped spread tumbleweed seeds. Thus, the human population unwittingly assisted the rapid expansion of territory infested by tumbleweed. The plant itself was

[107] Alfred W. Crosby, *Ecological Imperialism: The Biological Expansion of Europe, 900–1900* (Cambridge, England: Cambridge University Press, 1986), pp. 155–67, 228–9; Knoblach, *Culture of Wilderness*, 113–14; Timothy Neale (2016), "Settler Colonialism and Weed Ecology," The Engagement Blog, available online at https://bit.ly/2JNuLWI, accessed January 15, 2018.

well suited to thrive in the plains environment: it grew rapidly; was a prolific seed-producer; was well equipped, with prickly spines, to protect its seeds; and had evolved mechanisms to spread them effectively and quickly. In its new home, tumbleweed seems not to have encountered predators that harmed it or ate its seeds, and it was fire-resistant. It thrived, moreover, in dry years when the settlers and most of their crops struggled. Thus, the combination of the disturbed environment created by the settlers and the features of the plant itself facilitated its "success."[108]

In spite of efforts to control it, tumbleweed persisted and had become an established in part of the plains environment by the Dust Bowl of the 1930s. In late August and early September 1939, officials and scientists from the USDA embarked on an extensive tour through the western plains. When they visited Baco County in southeastern Colorado, they were advised that forty percent of the county, which received an average of only 14.77 inches of rain a year, was abandoned crop land, and that the farm population had fallen from 10,570 in 1930 to 3,920 in 1936. Nevertheless, there was good news for the visitors: "You are seeing Baco County at most favorable time of the year." And, the county had not seen a "black roller" dust storm for over a year. Many of the fields that had been blowing badly as recently as the spring of the previous year were "now" covered with weeds or cover crops. However: "The weed cover principally is Russian thistle, and if the thistles reach ordinary size, they will blow away next winter at the very time the ground needs protection."[109]

While this chapter has focused on the tumbleweed as the most celebrated of the "weeds" introduced to the Great Plains from the steppes, it was certainly not the only one.[110] In Kansas, the worst such weed was field bindweed (*Convolvulus arvensis* L.). It seems to have been introduced in infested wheat seed from the steppes of present-day Ukraine between 1870 and 1875. This was precisely the same time as the Mennonite migrants were arriving in Kansas with their Turkey Red wheat seed. Field bindweed was reported in 1877 near Topeka, around 100 miles northeast

[108] See Jacob Johnston (March 23, 2017), "Why Are Invasive Plants Invasive?" Habitat Network, available online at http://content.yardmap.org/learn/invasives/ accessed January 15, 2018; C. Perrings, H. Mooney, M. Williamson, "The Problem of Biological Invasions," in *Bioinvasions and Globalization*, ed. Perrings, Mooney, and Williamson (Oxford: Oxford University Press, 2010), pp. 1–16; William Beinart and Karen Middleton, "Plant Transfers in Historical Perspective," *E and H* 10 (2004), 15–16.

[109] NARA KC, RG114, Entry 187 Dept of Agriculture. Forest Service. Region 2, General Records 1934–1941, Box 10, USDA Conference Tour, Southern Great Plains, 1939, August 26 to September 3, 1939, 23–4.

[110] See Knobloch, *Culture of Wilderness*, 126–8.

of the Mennonite settlements. By 1888, it had spread to Nebraska, and continued to make its way around the plains. By 1937, it had infested 200,000 acres in Kansas, and half a century later, around 1.2 million acres. It is a creeping, deep-rooted, perennial weed. In contrast to the tumbleweed, which blows across the surface, the bindweed spreads its lateral root systems underground, and has taproots reaching as far as twenty feet below the surface. Once established, it is very hard to eradicate. It competes very effectively with crops for water, nutrients, and light, and entwines itself around crops, creating problems at harvest time. Tests in the late 1930s showed that it cut crop yields by 25–75 percent in eastern Kansas, but by 75–100 percent in the more arid west of the state. Real estate agents valued land infested with bindweed at around half that of uninfested land. By 1940, the value of the crop production lost to the bindweed in Kansas was put at $1.5 million. Losses were projected to rise to $100 million by 1975 if it was not controlled. The Kansas Noxious Weeds laws of 1937 enforced measures to try to do this, such as ensuring crop seed was clean of weed seed before it was sown, cultivating fields to try to eradicate the root systems, and the use of chemicals.[111] A Kansas botanist noted in 1941: "There are several other bad weeds in the state, such as Russian knapweed . . ., but they now exist on a smaller scale."[112] Russian knapweed (*Acroptilon repens*), however, which also spreads underground, was a very serious problem. It inhibited the growth of other plants and could be fatal for horses if they ate too much. In what was clearly part of a wider pattern, the Russian knapweed was introduced to the Great Plains from the steppes with alfalfa seed by Hansen in 1898. John Gerlach, an agronomist, pointedly entitled a paper on "plant invaders": "How the West Was Lost."[113]

This chapter opened with the iconic tumbleweeds becoming part of the scenery, real and imagined, of the Great Plains and the American West. The positive, even romantic, image in the cowboy ballad and some

[111] Gates, "Weeds in Kansas," 8–18; Dallas Peterson and Phillip W. Stahlman (March 1989), "Field Bindweed: Control in Field Crops and Fallow," Kansas State University Agricultural Experiment Station and Cooperative Extension Service, available online at www.bookstore.ksre.ksu.edu/pubs/MF913.pdf, accessed January 22, 2018.

[112] Gates, "Weeds in Kansas," 10.

[113] John D. Gerlach, (1997), "How the West Was Lost: Reconstructing the Invasion Dynamics of Yellow Starthistle and Other Plant Invaders of Western Rangelands and Natural Areas," California Exotic Pest Plant Council: 1997 Symposium Proceedings, available online at https://bit.ly/2NDDH1R, accessed January 17, 2018. See also "Russian knapweed, *Acroptilon repens* (*Centaurea repens*)," The Great Basin and Invasive Weeds website available online at https://bit.ly/2Cds1Od, accessed January 22, 2018. Field bindweed seeds were found in up to 10 percent of samples of alfalfa seed imported from Turkestan (in Soviet Central Asia) in the 1920s. Hillman and Henry, "The Incidental Seeds," 17.

other cultural representations belied the very real harm to the settler population and their agricultural economy caused by this invasive species from the steppes. Popular culture has also depicted tumbleweed in a more malevolent, and perhaps more accurate, light. The weed played a supporting role in a key scene of a Hollywood classic that has immortalized Kansas. Tumbleweeds blow all around Dorothy as she struggles in the storm, with a tornado ("cyclone") ominously on the horizon, in the opening scene of *The Wizard of Oz*.[114] (The scene, coincidentally, resembles Russian scientist Adamov's account of *perekati-pole* blowing in the steppes in 1892. See p. 360.) *The Wizard of Oz* was shot at the end of the drought years of the 1930s, albeit in a studio in California. The book on which it was based, *The Wonderful Wizard of Oz*, was published in 1900 and the author, Lyman Frank Baum, drew on personal recollections of South Dakota during the dry early 1890s.[115] In the movie, Dorothy, her dog, and her house are whisked away by the tornado from the stormy, drought-stricken, flat, treeless prairie, and away from the tumbleweeds dancing around her. They emerge in the peaceful, well-watered, mountainous, wooded, and weed-free land of Oz. The magical land is perhaps a mirror image of Kansas, but also a highly idealized version of thc lands from where many, but not all, of the Euro-American and European settlers in the Great Plains had come from. Oz is also a mirror image of the steppes, from where the tumbleweed, the migrants who accidently introduced it, varieties of wheat and other crops, agricultural sciences and techniques appropriate to such an environment had been transferred to the Great Plains since the 1870s.

[114] Victor Fleming (1939). *The Wizard of Oz*. USA: Metro-Goldwyn-Mayer. The storm scene starts after 14 minutes 30 seconds. She doubts she is still in Kansas at around 20 minutes.

[115] See Nancy Tystad Koupal, "On the Road to Oz: L. Frank Baum as Western Editor," *South Dakota History* 30, no. 1 (2000), 49–106. By chance, I found a Russian edition in the book store at Nur-Sultan airport in the similar environment of northern Kazakhstan in the spring of 2019. Laimen Frenk Baum, *Volshebnik iz stran Oz* (Spb: KARO, 2017).

Conclusion

Introduction

An episode in 1933–4 illustrates the extent to which American agricultural scientists had come to learn from experience in the steppes. On October 27, 1933, Knowles A. Ryerson of the United States Department of Agriculture (USDA) in Washington, DC, sought help from Soviet scientist Nikolai Vavilov in Leningrad. Ryerson wrote to him: "the question of erosion control has assumed a nation wide importance in this country," and that he would greatly appreciate information about Soviet work. He was interested in the types of plants used to check erosion, scientists involved in controlling erosion, and their publications. In return, he offered "native plant material" for use in experiments.[1] Over the following months, erosion control assumed even greater importance in the United States as the latest wave of droughts and dust storms, which had returned to the plains in 1932, developed into the ecological and human disaster of the Dust Bowl.[2]

The outcome was an expedition by USDA scientists Harvey L. Westover and Charles R. Enlow to Soviet Central Asia in 1934. It was organized with help from Vavilov, the Soviet Academy of Sciences, and the Soviet foreign tourist agency (Intourist). They also had support from the new U.S. Embassy in Moscow that opened after diplomatic relations between the two countries were established in November 1933. Westover and Enlow traveled through the steppes to the desert: environments the Americans considered analogous to their western plains. They

[1] TsGANTD Spb, f.318, op.1-1, d.530, l.115, Ryerson to Vavilov, October 27, 1933.

[2] R. Douglas Hurt, *The Big Empty: The Great Plains in the Twentieth Century* (Tuscon: University of Arizona Press, 2011), pp. 91–4. Droughts and dust storms were recurring phenomena. Kevin Z. Sweeney, *Prelude to the Dust Bowl: Drought in the Nineteenth-Century Southern Plains* (Norman: University of Oklahoma Press, 2016). On water erosion in the same period, see Paul S. Sutter, *Let Us Now Praise Famous Gullies: Providence Canyon and the Soils of the South* (Athens: University of Georgia Press, 2015).

visited experiment stations, including the Repetking experiment station (in the Turkmen Republic of the Soviet Union) where they learned about erosion control techniques, research institutes where they exchanged expertise with scientists, and farms where they saw the techniques in practice. They collected large quantities of seeds of plants used in restricting soil erosion. The Americans encountered some difficulties. On their way east from Moscow, they were robbed on the train in the night. Enlow lost his passport and large sums of American dollars. The unfortunate Enlow was later taken ill and hospitalized. And, some of their packages of seed went astray in transit. Vavilov intervened to help.[3]

After Westover returned to the United States, he wrote to Vavilov:

> I know you will rejoice with us to hear that the missing seed finally turned up at the [U.S.] Consulate in Moscow and has been forwarded to Washington. ... I can't begin to tell you how much we appreciate your kind and courteous treatment and wonderful assistance received through your efforts while we were in the USSR ... I don't know how we can ever repay you in full as so much was done for us.

Vavilov replied that he was glad everything had arrived, and signed off: "Best regards to Dr. Ryerson ... and to all our good friends in Washington. Kind regards to Dr. Enlow." Vavilov later informed them – rather alarmingly since this was the Soviet Union under Stalin – that the persons guilty of delaying their luggage would "undergo punishment."[4] Vavilov was glad also to return hospitality that had been extended to him on his

[3] TsGANTD Spb, f.318, op.1-1, d.698, ll.105–6, Ryerson to Vavilov, March 28, 1934; l.108, Vavilov to Ryerson, April 15, 1934; d.697, l.82, Vavilov to Westover, November 28, 1934; l.83, Westover to Vavilov, April 3, 1934; l.88-ob., Certificate (*Udostoverenie*) for Westover and Enlow signed by Vavilov, May 30, 1934; l.89, letter of introduction for Westover and Enlow to various institutions signed by Vavilov, May 30, 1934; ll.106–107 ob., Alekseev to Vavilov, July 11, 1934; NAL, Special Collections, Beltsville, Maryland, Call Number: 127 W52 R, H. L. Westover and C.R. Enlow, "Turkistan: Turkey Expedition," 3 vols., 1934 (Central Asia was also known as Turkistan or Turkestan); H. L. Westover, "Crested Wheatgrass," *USDA Leaflet*, no. 104 (1934). For the wider context, see Thomas D. Isern, "The Erosion Expeditions," *AH* 59 (1985), 181–91. Westover had traveled around the Soviet Union in 1929. "Explorers bring back new plants and seeds: Westover and Whitehouse find many alfalfas and fruits in Asia and Europe for trying out here," *The Official Record: United States Department of Agriculture* 9, no. 9 (27 February, 1930), 1, 3. Niels Hansen had visited Repetking in 1897. See p. 302.

[4] TsGANTD Spb, f.318, op.1-1, d.697, l.109, Westover, Washington, DC, to Vavilov, September 21,1934; l.110, Vavilov to Westover, n.d.; l.82, Vavilov to Westover, November 28, 1934. See also H. L. Westover, J. T. Sarvis, Leroy Moomaw, et al., "Crested Wheatgrass as Compared with Bromegrass, Slender Wheatgrass, and other Hay and Pasture Crops for the Northern Great Plains," *USDA Technical Bulletin*, no. 307 (1932); C. R. Enlow and G. W. Musgrave, "Grass and Other Thick-Growing Vegetation in Erosion Control," in *Soils and Men: A Yearbook of Agriculture 1938* (Washington, DC: USDA, [1938]), pp. 630–3.

visits to the United States by the USDA, universities, agricultural colleges, and experiment stations around the country.[5]

This episode shows how American and Soviet agricultural scientists cooperated because they had common interests in addressing similar problems in the similar environments of parts of their vast countries. It also shows them respecting each other as equals and enjoying good professional and personal relations. By the 1930s, as we have seen in this book, such relations between American and Russian/Soviet agricultural scientists had become well established since their origins in the late nineteenth century. For several decades, specialists in both countries had been exchanging seeds, research notes, scientific publications, experiences, and assistance. USDA plant explorers had been travelling to the steppes and other parts of Eurasia since Niels Hansen and Mark Carleton first visited in the late 1890s. As we have also seen, however, it took some time before many American scientists concerned with plains agriculture recognized that they could learn from the experience of their Russian and Soviet counterparts in settling, studying, and cultivating the steppes.

The Transfers and their Legacies

This book has demonstrated how, between the 1870s and the 1930s, aspects of Great Plains agriculture developed from roots in the similar environment of the steppes. The transfers from the steppes to the Great Plains discussed here – plants, agricultural sciences and techniques – were varieties of wheat, the science of understanding soils, the practice and techniques of planting shelterbelts of trees, and, unintentionally, an invasive species known as tumbleweed or Russian thistle. Following my visits to the Great Plains and conversations with farmers, I was unsure how lasting these transfers had been. I learned how agriculture in the plains and prairies has been transformed in recent decades. The USDA, which featured so prominently in this book, is now less important to farmers than agrochemical and biotechnology businesses. Many fields are sown with genetically modified crops, rather than new introductions from Eurasia. The fertile black earth is no longer designated "Chernozem" by the U.S. Soil Survey. Many shelterbelts have been removed as the rationale for planting them has overtaken by other developments.

A closer examination of plains agriculture, however, indicates that the Russian roots had been sunk more deeply. A role in some of the transfers

[5] On Vavilov's visits to the United States, see pp. 75–6, 86, 110, 116, 119.

was played by people who moved from the steppes to the North American grasslands. Many were of Germanic origin and included Mennonites. Some of the people I met in the American plains and Canadian prairies were their descendants. Commenting on the diversity of immigrants in the 1870s, Craig Miner wrote that "samovars were as much part of western Kansas as Levi overalls and sunbonnets."[6] The success of the migrants from the steppes attracted attention. In 1896, for example, William S. Harwood described the "Black Sea German farmers" bringing their grain to market in Eureka, South Dakota: "Out of the wagons of these German-Russians, who learned their trade of wheat-growing in the Old World, is unloaded more wheat year by year than at any other place in the world."[7] The immigrants and subsequent generations adapted to the customs and culture of their new country. At first, following the practice in the Russian Empire, they lived in houses grouped in villages and traveled to their fields. Quite quickly most adopted the American pattern of building their houses on individual family farms. The immigrants adopted other American customs, including aspects of dress and, in time, the English language. On the other hand, they retained elements from their cultures from the steppes, in particular their churches and cuisine.[8] Since the collapse of the Soviet Union in 1991, many of the migrants' descendants have visited the steppes on "homeland tours."[9] "Germans from Russia" who take a pride in their heritage are a living link between the Great Plains and the steppes.[10]

[6] Craig Miner, *West of Wichita: Settling the High Plains of Kansas, 1865-1890* (Lawrence: University Press of Kansas, 1986), p. 67.

[7] W. S. Harwood, "A Bit of Europe in Dakota," *Harper's Weekly* 40 (July 11, 1896), 690 quoted (in the original English) in Susanne Janssen, *Vom Zarenreich in den amerikanischen Westen: Deutsche in Russland und Russlanddeutsche in den USA (1871–1928: die politische, sozio-ökonomische und kulturelle Adaption einer ethnischen Gruppe im Kontext zweier Staaten)* (Berlin–Hamburg–Münster: LIT Verlag, 1997), p. 152. Harwood may have been exaggerating.

[8] There are large literatures, scholarly and popular, on the German–Russians, including Mennonites. See, for example, Bradley H. Baltensperger, "Agricultural Change Among Great Plains Russian Germans," *AAAG* 73 (1983), 75–88; Gordon L. Iseminger, "The McIntosh County German–Russians: The First Fifty Years," *North Dakota History* 51, no. 3 (1984), 4–23; T. J. Kloberdanz, "Plainsmen of Three Continents: Volga German Adaptation to Steppe, Prairie, and Pampa," in *Ethnicity on the Great Plains*, ed. F. Lubke (Lincoln: University of Nebraska Press, 1980), pp. 54–72; Royden K. Loewen, *Family, Church, and Market: A Mennonite Community in the Old and New Worlds, 1850–1930* (Urbana: University of Illinois Press, 1993), and on immigration by a wider range of ethnicities: D. Aidan McQuillan, *Prevailing Over Time: Ethnic Adjustment on the Kansas Prairies, 1875–1925* (Lincoln: University of Nebraska Press, 1990).

[9] See, for example, the "Journey to the Homeland Tour," available online through https://library.ndsu.edu/exhibits/homeland-tour, accessed July 23, 2018. I am grateful to Michael Miller of the North Dakota State University Libraries' Germans from Russia Heritage Collection for sharing his experience.

[10] The American Historical Society of Germans from Russia was founded in 1968. The AHSGR website available online at www.ahsgr.org/, accessed October 17, 2018.

Over the twentieth century, much of the wheat grown in the Great Plains was varieties from the steppes or strains derived from them. This was due in part to the immigrants, especially the Mennonites who moved to Kansas, in particular Bernhard Warkentin, who shipped over some of their wheat. Their hard red winter wheat (Turkey Red) did well in the Great Plains, especially in the dry 1890s, when other crops failed. The introduction of wheat from the steppes was due also to USDA agricultural explorers. Carleton traveled to the steppes in 1898–9 and 1900 to collect varieties suitable for the Great Plains. While he was there, Carleton visited relatives of the Kansas Mennonites. Carleton compared the environmental conditions in the steppes with the plains states:

> The region of Russia from which ... the hardiest winter wheats originate ... corresponds very fairly with that portion of our Great Plains, including Kansas, eastern Colorado, Nebraska, Iowa, South Dakota, and portions of Minnesota and North Dakota. It lies in the middle of the black soil (chernozem), and therefore includes the very richest lands, and has a climate marked by extremes of temperature and severe droughts.[11]

Some scholars have challenged the primacy of Carleton and the Mennonite migrants in introducing the hard red winter wheat. As I endeavored to demonstrate in Chapter 4, their roles were very significant, not least because they established their wheat as what was effectively a brand, and set the standards for such varieties. It has not been possible, moreover, to prove conclusively that the particular variety of hard red winter wheat that was grown in the steppes was cultivated in the United States prior to the Mennonites' arrival. The Turkey Red wheat was one of several wheat varieties introduced from the steppes, including hard spring durum wheats and a parent of the Marquis hybrid. Together, these introductions transformed wheat farming throughout the plains and Canadian prairies. Their legacy persists. At the start of the twenty-first century, an agronomist at Kansas State University wrote that hard red winter wheat and durum wheat were still major crops in the Great Plains. While the modern cultivars had changed greatly, about half the genes of the hard red winter wheat sown in the 1990s could be traced to the Turkey and Kharkof wheat the Mennonites, Carleton, and others had imported from the steppes. The Kubanka durum wheat Carleton brought from the steppes of today's

[11] Mark Alfred Carleton, "Successful Wheat Growing in Semiarid Districts," *Yearbook of the United States Department of Agriculture 1900* (Washington, DC: USDA, 1900), p. 535. See also Carleton, "Russian Cereals Adapted for Cultivation in the United States," *USDA Division of Botany Bulletin*, no. 23 (Washington, DC: Government Printing Office, 1900).

Kazakhstan was replaced long ago, but it was still the basis of modern durum wheats, and set the standards for quality.[12]

Mennonite communities in the Great Plains commemorate their ancestors' migration and role in introducing hard red winter wheat. The Kauffman Museum at Bethel College in North Newton, Harvey County, has a fine collection of artifacts on the "coming of the Mennonites from Europe to the central plains in the 1870s and their encounters with the prairie environment and its people."[13] In nearby Goessel, Marion County, the Mennonite Heritage and Agricultural Museum exhibits a chest, which they say is one of those used to transport Turkey Red wheat seed from the steppes.[14] Warkentin's house in Newton, now a museum, shows off the prosperous lifestyle he earned from importing and milling the wheat.[15] The story, which long ago gathered a life of its own, continues to attract attention. Warkentin's and the Turkey Red wheat's journey from the steppes to the plains of Kansas were showcased in a television documentary about the contribution of immigrants to the United States broadcast in May 2017.[16]

There is a postscript to this story. In 1921, the American Relief Administration (ARA) organized famine relief in Soviet Russia. Worst hit were the steppes in the Volga region. On the request of the Soviet People's Commissariat for Agriculture, the ARA supplied large quantities of spring wheat seed. The seed was selected in the northern Great Plains, where environmental conditions were closest to the Volga steppes, and, in the words of a Soviet scientist, "consisted of hard wheat (Kubanka, Arnautka), which had ... been received in America from Russia and is known there as 'durum wheat,' and a variety of soft wheat 'Marquis.'"[17] Purchasing seed for the famine-hit region was one of the motives for

[12] Gary M. Paulsen, "The Agronomic Legacy of Mark A. Carleton," *Journal of Natural Resources: Life Science Education* 30 (2001), 120–3; Gary M. Paulsen, obituary, available online at www.ymlfuneralhome.com/main/obituary/1562, accessed July 23, 2018. See also USDA Agricultural Marketing Service (January 20, 2012), "Hard Red Winter Wheat," available online at www.gipsa.usda.gov/fgis/commgallery/gr_hrw.aspx, accessed July 23, 2018.

[13] Kauffman Museum, North Newton, Kansas, details available online at https://kauffman.bethelks.edu/, accessed October 12, 2017. I visited in May 2017.

[14] Mennonite Heritage and Agricultural Museum, Kansas, details available online at www.goesselmuseum.com/, accessed October 12, 2017. I visited in April 2007.

[15] Warkentin House Museum, Kansas, details available online at https://bit.ly/2qsOAvr, accessed October 12, 2017. I visited in May 2017.

[16] Karen Schwatz, "The Seed Chest that Carried the Future, America: Promised Land," available online at www.history.com/the-promised-land/the-wheat-chest.html, accessed October 12, 2017.

[17] V. V. Talanov, "Predislovie," *Iarovaia pshenitsa* (Moscow: Novaia Derevnia, 1923), pp. 3–4. See also Bertrand M. Patenaude, *The Big Show in Bololand: The American Relief Expedition to Soviet Russia in the Famine of 1921* (Stanford, CA: Stanford University Press, 2002), pp. 146–7.

Vavilov's visit to the United States in 1921–2. He selected seed of varieties that had been introduced from the steppes.[18]

The next transfer from the steppes to the Great Plains considered in this book was the science of understanding the soils of such regions. By the 1930s, the U.S. Soil Survey classified the fertile soil of much of the plains as "Chernozem," black earth, using a scientific term, methodology, and understanding of soils that drew heavily on an innovation by Russian soil scientists led by Vasilii Dokuchaev during field work in the steppes in the 1870s–80s. The Russian scientists analyzed cross sections, or profiles, of soils that revealed a series of layers, or horizons. They showed how the underlying parent rock in the "C horizon" had been transformed into the soil in the "A horizon" at the top by the action of the climate and organic matter in the flat topography over time. The Russian science allowed the classification of soils into broad groups, based on the environmental conditions in which they had formed, for example, the Chernozem of the steppes and also the Great Plains.

The U.S. Soil Survey's adaptation of Russian soil science was overseen by Curtis Marbut, who led the Survey from 1913 until shortly before his death in 1935. Marbut learned about the Russian work from a book in German by Dokuchaev's former student, Konstantin Glinka. Marbut tested the Russian approach that had been devised in the steppes during field trips across the Great Plains. He analyzed soil horizons in soil profiles and found the fertile, black soils of the plains to be similar to those described by his Russian counterparts in the steppes. Marbut overcame considerable resistance to changing the scientific basis of the U.S. Soil Survey from its founder and chief of the USDA Bureau of Soils, Milton Whitney. Whitney stubbornly persisted in his older, and out-moded, physical and geological understanding of soils. Nevertheless, a patient and persistent Marbut had largely completed his "Russian Revolution" in U.S. government soil science by the time of Whitney's death in 1927. This was also the year of the 1st International Congress of Soil Science held in the United States. The "Russian delegation," led by Glinka, was celebrated for their achievements. The 2nd Congress in 1930 was held in the Soviet Union. At the opening ceremony Marbut graciously observed: "I am ... very interested in a pilgrimage to the land of Dokachayev [sic], ... and Glinka..." Marbut recognized the parallels between their countries: "America and Russia [are] [t]wo great continuous countries

[18] TsGANTD Spb, f.318, op.1-1, 1922, d.23, ll.7, 10-ob., 12–13; N. I. Vavilov, *Polevye kul'tury iugo-vostoka* (Petrograd: Red-izdat. Kom. Narodnogo Kommissariata Zemledeliia, 1922), pp. 5, 53.

which represent great bodies of land and ... bodies of soil and great numbers of kinds of soil."[19]

The U.S. Soil Survey revised its taxonomy again in 1975. Classes of soils were defined by soil properties that could be measured quantitatively. And, the nomenclature was changed from one that included Russian terms, such as Chernozem, to one based on Greek and Latin roots. Thus, Chernozems became "Mollisols."[20] The change in names did not, however, reflect a fundamental change in the Survey's scientific basis. Russian innovations, in particular the importance of horizons and profiles and the theory soil formation, were retained. In a statement broadly consistent with Dokuchaev's innovation, the current issue of the Soil Survey's *Soil Taxonomy* explains: "Mollisols characteristically form under grass in climates that have a moderate to pronounced seasonal moisture deficit. ... Mollisols are extensive soils on the steppes of Europe, Asia, North America, and South America."[21] More directly, the first page of the 2017 edition of the Soil Survey's manual states:

> The soil bodies contain a sequence of identifiable horizons and layers that occur in repeating patterns in the landscape as a result of the factors of soil formation as described by Dokuchaev (1883) and Jenny (1941). Soil scientists gain an understanding of the factors of soil formation in their area, along with the resulting expression of their interaction in the soil, and are then able to make maps of the natural soil bodies quite efficiently.[22]

Thus, the first reference in the current U.S. Soil Survey Manual is to Dokuchaev's monograph on the Russian black earth of 1883. Ronald Amundson, an American environmental scientist and specialist on soil science history, has noted: "Dokuchaev articulated that soil is a natural body on the landscape, one whose properties are determined by a set of

[19] ARAN, f.487, op.1, 1930, d.26, ll.143–4. Marbut's speech was translated into Russian for the benefit of his hosts, ibid., ll.137–40.

[20] *Soil Taxonomy, A Basic System of Soil Classification for Making and Interpreting Soil Surveys*, 1st edition, *USDA Agriculture Handbook, Soil Conservation Service* no. 436 (Washington, DC: USDA/U.S. Government Printing Office, 1975), p. 271, available online at https://bit.ly/2NgYAB3, accessed July 25, 2018.

[21] *Soil Taxonomy, A Basic System of Soil Classification for Making and Interpreting Soil Surveys*, 2nd edition, *USDA Agriculture Handbook, Natural Resources Conservation Service* no. 436 (Washington, DC: USDA/Government Printing Office, 1999), p. 121, available online at https://bit.ly/2CiavrW, accessed July 25, 2018.

[22] C. Ditzler, K. Scheffe, and H. C. Monger (eds.), *Soil Survey Manual: USDA Handbook* no. 18 (Washington, DC: USDA/Government Printing Office, 2017), p. 1, available online at https://bit.ly/2qnGwwj, accessed July 25, 2018.

factors ... This concept essentially is the modern definition of soil and forms the underlying theory of soil science ..."[23]

One of the measures adopted to address the Dust Bowl in the 1930s, planting shelterbelts of trees to protect the land from the wind, also drew on Russian and Soviet expertise in the steppes. Experience of shelterbelts among steppe farmers, including Mennonites, dated back to the early nineteenth century. Russian scientists began researching planting belts of trees and their influences on nearby fields shortly afterwards. Although Americans started to plant such windbreaks in the Great Plains in the late nineteenth century, a key factor in the decision by the U.S. Forest Service and federal government to launch the Shelterbelt Project in the summer of 1934 was the Russian experience. This was pushed by Edward Munns, head of silvics in the Forest Service in Washington, DC, and Raphael Zon, a Jewish émigré from tsarist Russia, who was director of the Lakes States Forest Experiment Station in St. Paul, Minnesota. Zon worked closely with American-born Carlos Bates, whose experience of shelterbelts in the Great Plains complemented Zon's knowledge of the Russian work. Munns and Zon provoked controversy with their claim, based on some Russian and Soviet studies, that shelterbelts could moderate the climate and increase rainfall.

By the end of the Shelterbelt Project in 1942, over 220,000,000 trees had been planted in belts on over 30,000 farms in six plains states.[24] John Guthrie of the Forest Service toured part of the Shelterbelt zone in Kansas in 1942. He downplayed the controversy and made no mention of the project's Russian roots. He concluded:

> I was refreshed, rejuvenated, and really uplifted by seeing on the Kansas ground the shelterbelt actually going ahead. No brainstorm, no crackpot idea, no impractical dream, but trees living, growing, and doing good to people through years of drought and almost constant wind, and largely because a few men had a vision. There isn't any theory here now; it's a very tangible reality today.[25]

[23] Ronald Amundson, "Soil Development: Conceptual and Theoretical Models," in *The International Encyclopedia of Geography*, ed. Douglas Richardson, Noel Castree, Michael F. Goodchild et al. (Chichester, UK; Hoboken, NJ: Wiley-Blackwell, 2017), DOI:10.1002/9781118786352.wbieg0798, accessed July 25, 2018. Evidence for the enduring international importance of Dokuchaev's work is the recent translation of his 1883 monograph into Japanese, available online at https://bit.ly/2NmRzi5, accessed December 10, 2018. I am grateful to Professor Kazutake Kyuma for drawing this to my attention.

[24] E. L. Perry, *History of the Prairie States Forestry Project* (Washington, DC: U.S. Forest Service, 1942), appendix, pp. 1, 2, 9, 15.

[25] John D. Guthrie, "Trees, People, and Foresters," *JoF* 40 (1942), 480.

In 1945, Bates reported the shelterbelts' success in cutting wind velocity and increasing crop yields in protected fields.[26] He took care to draw on his own research in the Great Plains, rather than re-ignite the controversy by citing Russian and Soviet studies claiming that shelterbelts could increase rainfall. Such claims were also controversial in the Soviet Union, where they were downplayed by leading scientist Georgii Vysotskii, who participated in the American debate over shelterbelts in 1935.[27]

Ever concerned for his reputation and proud of his own Russian roots, Zon promoted the Shelterbelt Project and its origins in Russian steppe forestry whenever the opportunity arose. In May 1945, Zon reported a rumor that at the Yalta Conference in February of that year, Franklin Roosevelt and Stalin had discussed shelterbelt planting. According to Zon, the president had expressed his pride in "our accomplishments," and Stalin had wanted to know more.[28] While Zon could not vouch for the rumor's authenticity, it may have been true. At the time of the Yalta Conference, Roosevelt asked Morris L. Cook for a report on the work of the drought relief committee he chaired. The committee's remit included the Shelterbelt.[29]

In 1948, Stalin approved a plan to plant extensive shelterbelts across the steppes. Contrary to an American scholar's assertion that the Soviet authorities sponsored "American-style shelterbelts to check wind erosion,"[30] the Stalin Plan for the Transformation of Nature of 1948 was based firmly on Russian and Soviet expertise. It was also, like the American project, accompanied by exaggerated claims for shelterbelts' influence on the climate and rainfall.[31] Soviet writers heaped praise on their achievements in comparison with the allegedly more limited results of the American Shelterbelt Project. One asserted that more trees had been planted in the Soviet Union in one year than in the United States over the previous century.[32] I. I. Kruglov criticized Zon for stating that cultivating trees

[26] Carlos G. Bates, "Shelterbelt Influences. I. General Description of Studies Made," *JoF* 43 (1945), 88–92.

[27] G. N. Vyssotsky [sic], "Shelterbelts in the Steppes of Russia," *JoF* 33 (1935), 781–8.

[28] Zon Papers, Box 13, Folder 1, Zon to Elmer McClain, Lima, OH, May 17, 1945.

[29] See Thomas R. Wessel, "Roosevelt and the Great Plains Shelterbelt," *Great Plains Journal* 8, no. 2 (1969), 71.

[30] J. R. McNeill, *Something New Under the Sun: An Environmental History of the Twentieth Century World* (New York: W. W. Norton, 2000), p. 43.

[31] See Stephen Brain, "The Great Stalin Plan for the Transformation of Nature," *EH* 15 (2010), 670–700; D. J. B. Shaw, "Mastering Nature through Science: Soviet Geographers and the Great Stalin Plan for the Transformation of Nature, 1948–53," *SEER* 93, no. 1 (2015), 120–46; V. G. Skorokhod, "Dokuchaevskie lesnye polosy v Donbasse," *Priroda*, no. 9 (1948), 64–6.

[32] "K novym uspekham v bor'be za preobrazovanie prirody," *Les i step'*, no. 10 (1950), 3–7.

required an annual average precipitation of 400 mm, since the Soviets had succeeded with half that amount.[33] (He omitted to mention that evaporation was higher in the warmer climate in the United States.) The Soviet shelterbelt project had mixed results due to faulty techniques for planting trees and ignoring advice from some of their own specialists. Stalin's plan was halted after his death in 1953. Denis Shaw has questioned whether the existing Soviet science was sufficiently developed to support such extensive shelterbelt planting.[34] Shaw's argument concurs with Bates' reaction in January 1935 when he realized that planting the belts in the Great Plains was going to be more complicated than they had thought after reading some of the Russian and Soviet studies (see p. 330).

Zon persisted in his enthusiasm for Russian shelterbelt expertise and its relevance for the American grasslands. In 1949, he wrote:

> The Russian steppe, like the American prairie, is the bread basket of the country. Yet both ... are largely flat regions with deficient rainfall ranging from 350 mm. to 525 mm. (14 to 21 in.) a year. They suffer from frequent droughts and are constantly exposed to strong winds. ... It is natural that the people in both regions should seek means to counteract the unfavorable climatic conditions and to improve living conditions. One method ... was ... planting ... trees, protective forest strips, windbreaks, or shelterbelts around farmsteads and cultivated fields.

He explained how the steppes had been settled "much earlier than the American prairies," and thus the Russians had prior experience in planting shelterbelts.[35]

In recent decades, many of the shelterbelts planted in the Great Plains in the 1930s and early 1940s have been removed. Farmers wanted the land to sow more crops. New techniques, in particular center-pivot irrigation, and new machinery required larger fields. And, the spread of no-till farming, in which weeds are controlled by herbicides rather than plowing, reduced the risk of newly cultivated soil being blown by the wind.[36] The farmer whose shelterbelt I visited in Nebraska in 2007

[33] I. I. Kruglov, "Opyt preobrazovaniia astrakhanskoi polupustuni," *Les i step'*, no. 2 (1949), 74.

[34] See Shaw, "Mastering Nature"; Brain, "The Great Stalin Plan."

[35] Raphael Zon, "The Volga Valley Authority," *Unasylva* 3, no. 2 (March–April 1949), n.p., available online at www.fao.org/docrep/x5349e/x5349e02.htm, accessed July 10, 2018. (There are no page references in the online version.)

[36] See James L. Van Deusen, "Shelterbelts on the Great Plains: What's Happening?" *JoF* 76 (1978), 160–1; Carson Vaughan (November 1, 2017), "Uprooting FDR's 'Great Wall of Trees,'" Food and Environment Reporting Network, available online at https://thefern.org/2017/11/uprooting-fdrs-great-wall-trees/, accessed July 25, 2018.

commented that while it was nice to have it on his land, it required a lot of maintenance and took up a lot of space. Nevertheless, the National Agroforestry Center in Lincoln, Nebraska, still advocates planting windbreaks of trees to help protect land and crops from the wind.[37] Mennonites still recall their experience in planting shelterbelts in the Great Plains.[38] In semi-arid grasslands around the world that suffer from wind erosion, trees are still planted in belts that trace their lineage to the Russian experience in the steppes. Since the late 1970s, the government of the People's Republic of China has been planting a "Great Wall" of trees along the southeastern end of the Eurasian steppes in Inner Mongolia and northwestern China.[39] A mega-shelterbelt of one million trees is being planted in the steppe around Kazakhstan's capital of Nur-Sultan (Astana), in the hope of the moderating the city's extreme and windy climate.[40] In North America, legacies of the Russian roots of the shelterbelts include species of trees and shrubs introduced from the Eurasian steppes. Caragana or Siberian pea tree (*Caragana arborescens*), Russian mulberry (*Morus alba tatarica*), and Siberian elm (*Ulmus pumila*), amongst others, can still be found in the Great Plains and Canadian prairies.[41]

Another immigrant from the steppes was an invasive plant species that spread uncontrollably across the Great Plains and the American West: the tumbling tumbleweed or Russian thistle that, rather bizarrely, has become a cultural icon of the American West. The tumbleweed was as unexpected, since it was brought accidently in sacks of flax seed, as it was unwelcome to

[37] USDA National Agroforestry Center, Practices, Windbreaks, available online at www.fs.usda.gov/nac/practices/windbreaks.shtml, accessed July 25, 2018. I had a guided tour of the Center and farms in the surrounding area in the spring of 2007.

[38] Larry Rutter, "Mennonites and Shelterbelts," *Kansas Canopy: Newsletter of the Kansas Forest Service*, no. 42 (Spring 2012), 8–9.

[39] See Susanne Stein, "Coping with the 'World's Biggest Dust Bowl'. Towards a History of China's Forest Shelterbelts, 1950s–Present," *GE* 8 (2015), 320–48.

[40] Almaz Kumenov (October 23, 2017), "Astana's Plan to Stay Warm in the Winter? Build a Ring of One Million Trees," Eurasianet, available online at https://bit.ly/36GaOLa, accessed October 15, 2018. I completed this book in the capital city of Kazakhstan and became accustomed to passing through the belt of young trees on trips out of the city.

[41] I counted several introduced species in Saskatchewan in the spring of 2018. David Moon "The Saskatchewan Steppe in a Comparative and Transnational Perspective," Network in Canadian History & Environment/Nouvelle initiative Canadienne en histoire de l'environnement, NiCHE, available online at https://bit.ly/2Nmd5n5, accessed July 25, 2018. I found some of the same species a few weeks later in the similar environment of northern Kazakhstan in the Astana Botanical Garden. The U.S. Forest Service donated seedlings from North Dakota to the botanical garden. USDA Forest Service: Apply Knowledge Globally (2017), "Forest Service Sends Bur Oak Seedlings to Kazakhstan for Astana's New Botanical Gardens," available online at https://bit.ly/33nJX4C, accessed October 17, 2018.

North American farmers. In the early twenty-first century, tumbleweed continues to harm farming communities and defy attempts to control it. When it infests large areas of low-value land, it is impractical and too expensive to use chemicals to control it. In the late 1990s, the USDA sent scientists to the grasslands of Russia and Hungary to look for fungi that attacked tumbleweed in its native environment. Tests on samples they brought back are still in progress, and other biocontrol organisms are being sought "that could help turn the tide on the loathsome weed."[42]

In the meantime, in dry years when it proliferates, tumbleweed continues to cause harm. In 2014, in Crowley County in southeastern Colorado, County Commissioner Tobe Allumbaugh was exasperated by the tumbleweed. They had tried plowing them, but it had been "like trying to round up balloons."[43] He described their appearance when they caught fire: "Have you ever seen a tumbleweed burn? It's like a dried-up Christmas tree on steroids – just PHOOO!"[44] The commissioner played a recording of the song "Tumbling Tumbleweeds" for a visiting reporter. He commented that some people dismiss tumbleweeds as a "harmless bit of nostalgia of a wide-open West." But, to emphasize the trouble they caused, continued: "What we have is not funny."[45] Nor has their Russian origin been forgotten. In 2014, the Smithsonian magazine published an article entitled: "America's Tumbleweeds Are Actually Russian Invaders."[46]

The other arrivals from the steppes – experienced farmers, appropriate crops, soil science, shelterbelts of trees – were mostly, if not always initially, welcomed in the Great Plains and the United States more widely. One of the most successful and welcome was Turkey Red wheat, which came at around the same time and by a similar route, in immigrants' baggage, as tumbleweed. The latter proved just as successful, from its point of view, as the Turkey wheat, if far less welcome from a human perspective.

[42] Jan Suszkiw (September 2014), "Fungi Readied for Weed Biocontrol," USDA Agricultural Research Magazine, available online at https://agresearchmag.ars.usda.gov/2014/sep/weed, accessed July 26, 2018.

[43] Quoted in Sarah Gilman (February 11, 2014), "Troubleweeds: Russian Thistle Buries Roads and Homes in Southeastern Colorado," *High Country News*, available online at https://bit.ly/34L9mpl, accessed January 16, 2018.

[44] Quoted in Gilman, "Troubleweeds."

[45] P. Solomon Banda (April 9, 2014), "Colorado Tumbleweeds Overrun Drought Areas," CBS Denver, available online at https://cbsloc.al/2JTkMiy, accessed January 16, 2018.

[46] Rachel Nuwer (August 15, 2014), "America's Tumbleweeds Are Actually Russian Invaders," SmartNews, Smithsonian Magazine, available online at https://bit.ly/2WQorQy, accessed July 26, 2018.

The Fate of Russian/Soviet–American Cooperation

One of the more surprising, indeed unexpected, aspects of this story has been the degree of cooperation between agricultural scientists in the United States and the Russian Empire that persisted after 1917 in the Soviet Union. They continued to cooperate between 1917 and 1933, when the U.S. government did not recognize the Soviet state. The episode at the start of this chapter, in which Vavilov assisted USDA scientists study erosion control in Soviet Central Asia in 1934, was just one of many examples of government scientists putting common interests ahead of politics. There were exceptions. In 1930, the U.S. government would not permit USDA scientists to attend the Soil Science Congress in the Soviet Union in their official capacities, but did allow them go as private citizens at their own expense.

Over the 1920s and 1930s, however, Soviet scientists were operating in an increasingly politicized environment. Most who collaborated with their American counterparts had been trained before the 1917 revolution. They retained their posts in the early Soviet period as the new authorities had not yet trained sufficient replacements from class backgrounds it considered more appropriate for their workers' and peasants' state. Until it could do so, the Soviet government had to rely on an older generation of "bourgeois specialists," such as Vavilov, whom they treated with suspicion.[47] For the Soviet government and ruling Communist Party, science was the key to building socialism in agriculture as well as industry. In the famine year of 1921, the Soviet leadership had allocated scarce foreign currency to Vavilov to visit the United States to re-establish contacts with American scientists, buy scientific literature and seed, and open a Bureau of Applied Botany in New York City. The Soviet authorities also allocated funds to Soviet soil scientists to attend the Congress in the United States in 1927 and host the following meeting in 1930. Far more resources were targeted at the collectivization and mechanization of agriculture and industrialization under the First Five-Year Plan (1928–32). The Soviet government hired American specialists, such as Thomas Campbell, to advise on farm organization, and American manufacturers, for example the Ford Corporation, to build factories to make tractors, trucks, and other

[47] See Michael Gordin, Walter Grunden, Mark Walker et al., "Ideologically Correct' Science," in *Science and Ideology: A Comparative History*, ed. Mark Walker (London: Routledge, 2003), pp. 39–43; David Holloway and Frederic J. Fleron, "The Politics of Soviet Science and Technology," *Social Studies of Science* 11 (1981), 259–74; Sheila Fitzpatrick, *Education and Social Mobility in the Soviet Union 1921–1934* (Cambridge: Cambridge University Press, 1979).

vehicles. The transfer of technology from the United States is the other side of the exchanges that have been the main focus of this book.

The drive to modernize Soviet agriculture and industry was accompanied by moves to replace the "bourgeois specialists" with scientists trained under Soviet power who would be more reliable politically. The tragic and unnecessary fate of Vavilov is all too well known. A leading scientist in his field, inside the Soviet Union and around the world, he was applying his knowledge to breed more productive and hardy crop varieties to improve food security in his country that had experienced repeated famines. But, he was attacked by younger, less talented, men. Trofim Lysenko, who came from a peasant family in Ukraine, claimed he could do the same more quickly (but with less scientific justification). In a dreadful paradox, Vavilov died of malnutrition in prison in Saratov in the steppe region in 1943.[48] Other Soviet scientists with American contacts also fell victim to Stalinist repression in the late 1930s. Nikolai Tulaikov, a leading Soviet specialist on American agriculture and dry-farming techniques, traveled to the United States several times from 1908 and attended the soil science congresses in 1927 and 1930. He perished in 1938.[49] Not all Soviet scientists with American connections lost their lives at this time. For example, forestry scientist Mikhail Tkachenko, who had visited the United States before 1917, was still alive in 1950, when he extolled the superiority of Soviet steppe forestry over American achievements in a propaganda publication for the Stalin Plan for the Transformation of Nature.[50] Nevertheless, by the end of the 1930s, as a result of Stalin's terror and official suspicion of contacts with the West, the "community of scholars" that had developed among American and Russian/Soviet agricultural scientists since the late nineteenth century had come largely to an end.

There were scientific contacts during the Second World War, when the two countries found themselves allies, and there was a brief possibility of renewed cooperation after the war. In 1943, Charles Kellogg (Marbut's successor as head of the U.S. Soil Survey), spoke positively about similarities between the soils of the two countries and opportunities for collaborative work at a meeting of the Congress of American–Soviet Friendship

[48] See Peter Pringle, *The Murder of Nikolai Vavilov: The Story of Stalin's Persecution of One of the Great Scientists of the Twentieth Century* (New York: Simon and Schuster, 2008).

[49] On Tulaikov's life, see K. P. Tulaikova, *Ot pakharia do akademika: Ob akademike Nikolae Maksimovich Tulaikove* (Moscow: Prosveshchenie, 1964). On his death, see S. P. Liapin and F. F. Perchenok, "Repressirovannye pochvovedy. Zapisko B. B. Polynova o 1937 g.," *Tragicheskie sud'by: repressirovannye uchenye Akademii nauk SSSR* (Moscow: Nauka, 1995), pp. 88–9.

[50] M. E. Tkachenko, "Sovetskii Soiuz: rodina stepnogo lesorazvedenii," *Les i step'*, no. 2 (1950), 21–8.

in New York City. He spoke in similar terms during and after a visit to the Soviet Union in 1945.[51] Prospects for a resumption of cooperation ended with the onset of the Cold War after 1945. In the Soviet Union, the authorities established firm political control over science and cut back contacts with the outside world.[52]

In the United States, Zon was caught up in the atmosphere of Cold War suspicion. In 1954, the Russian-born forestry scientist and persistent advocate of Soviet science, was sent a long, "interrogatory questionnaire" by the Civil Service Commission on account of "certain unevaluated information of a derogatory nature which, if true, might create a doubt concerning your loyalty to the Government of the United States." The questionnaire named associates of Zon and his wife, mostly Americans, believed to be Communist sympathizers. It listed organizations they were involved in that were considered Communist fronts, including the "American-Soviet Science Society," the "National Negro Congress," and even the Minneapolis "Saturday Lunch Club." The last, which Zon and his wife frequented, allegedly contained a pro-Communist and pro-Russian group (probably Mr. and Mrs. Zon). The questionnaire also listed publications, including some in Russian, that Zon subscribed to. He was requested to "furnish a statement with respect to your attitude, past and present, towards the Communist Party, the Communist cause, and the ideological conflict between the United States and the Union of Soviet Socialist Republics." The reason Zon fell under the purview of the Civil Service Commission was because after his retirement he had worked for the United Nations Food and Agriculture Organization (FAO). By 1954, Zon was approaching his eightieth birthday and was in poor health. He declined to reply and the demand was withdrawn since he no longer worked for the FAO. His son had these documents withheld for many years from his personal papers in the Minnesota State Historical Center.[53]

The Cold War atmosphere contributed to a sharp disagreement between American and Soviet soil scientists over priority in devising modern soil science. In 1961, Hans Jenny, of the University of California,

[51] NAL, Special Collections, Manuscript Collection 91, The Charles Edwin Kellogg Papers, Box 5, Folder 83, "Soviet Soil Technology and Agriculture"; Box 5 Folder 86, "Impressions of Soviet Science"; Box 5 Folder 87, "The Scientist Visits the Soviet Union, September 1945."

[52] See Ethan Pollock, *Stalin and the Soviet Science Wars* (Princeton: Princeton University Press, 2006).

[53] MHS, P1237, Raphael Zon Papers, 1887–1957 (hereafter Zon Papers), Box 14, Folder 3, 1953–4, International Organizations Employees Loyalty Board Interrogatory questionnaire; Pierce J. Gerety, Chairman, to Zon, March 18, 1954; T. Paul Fairbank, Exec. Sec., to Zon, April 2, 1954.

Berkeley, egged on by James Malin in Kansas, asserted primacy for Eugene W. Hilgard. Leading Soviet soil scientist Innokenkii Gerasimov responded, reasserting the case for Dokuchaev. At other times, however, there was respect among Russian and Soviet scientists for Hilgard's work (if not his primacy). By the 1920s and 1930s, as Milton Whitney's opposition was overcome, moreover, there was admiration among American soil scientists for the achievements of their Russian counterparts. Hilgard was ever interested in the work of Russian soil scientists and considered Whitney, and certainly not Dokuchaev, his number one enemy.[54]

An incident in the mid 1950s illustrates the change in relations between American and Soviet agricultural scientists since Stalin's terror and the start of the Cold War. In 1955, Dmitrii Borodin offered the archives of the Soviet Bureau of Applied Botany that Vavilov had established in New York in 1922 to the Kazakh branch of the Soviet Academy of Agricultural Sciences. The archives included correspondence between Russian, American, and Canadian organizations, information on American crop varieties introduced to the Soviet Union and crops of Russian origin in the United States. Borodin had been the head of the bureau and still lived in New York, where he made the suggestion to the Kazakh representative on the Soviet delegation at the United Nations. Borodin may have offered the archive to the Kazakh branch of the Academy because he had been born in 1890 in the north of what became the Soviet republic of Kazakhstan. His offer was declined. The Academy considered the materials interesting "from point of view of the history of ... American agriculture," but of "no practical significance at the present time." The Academy did express an interest any materials Borodin had on the "latest achievements of USA in agricultural technology," clearly indicating that the previously close contacts between Soviet and American agricultural scientists had ended.[55]

Another possible reason why Borodin offered the archive to the Kazakh academy was that in 1954, Soviet leader Nikita Khrushchev launched the "Virgin Lands Campaign" to grow grain in vast areas of unplowed steppe in

[54] Hans Jenny, *E. W. Hilgard and the Birth of Modern Soil Science* (Pisa: Collana Della Rivista "Agrochimica, 1961); KHS SA, Collection 183, Malin, James Claude, 1893–1979, Letters written by Malin, Microfilm MF 6306, Malin to H. Jenny, January 18, 1951 (and many subsequent letters); I. P. Gerasimov, "Pokushenie amerikanskogo uchenogo na prioritet V.V. Dokuchaeva v sozdanii nauchnogo pochvovedeniia," *Pochvovedenie*, no. 1 (1962), 124–8; V. A. Kovda and V. S. Muratova, "Professor E. V. Gil'gard (1833–1916)," *Pochvovedenie*, no. 3 (1958), 76–82. See also David Moon and Edward R. Landa, "The Centenary of the Journal Soil Science: Reflections on the Discipline in the United States and Russia Around a Hundred Years Ago," *SS* 182, no. 6 (2017), 209–10.

[55] TsGANTD RK, f. 103, op.1-6, d.582, ll.2, 3, 8, 10.

northern Kazakhstan and adjoining regions. Khrushchev and the campaign planners largely disregarded: the extensive experience and expertise in steppe agriculture that had been accumulated in the Soviet Union and Russian Empire since the eighteenth century; protests from the Kazakh Communist Party leadership, who were dismissed for stating that they needed the land for pasture and hayfields; and the knowledge of the Kazakh population who for many generations had used their grasslands to graze herds of livestock and plowed up only small areas with due caution. Starting in 1954, directed by the Soviet government, inexperienced and untrained people, mostly from outside Kazakhstan, equipped with plows and other machinery designed for different, more humid, conditions further west, plowed up over forty million hectares of "virgin and fallow land." They planted spring wheat year after year, leaving the upturned soil unprotected for several weeks from the heat and the wind until the wheat grew. The result, by the early 1960s, exacerbated by droughts and high winds, was soil erosion and dust storms similar in scale to those in the Great Plains three decades earlier, the lessons from which Khrushchev had also ignored. To mitigate the ecological disaster, and in a reversal in the direction of the flow of expertise that has featured in this book, Soviet agricultural scientists sought advice from their Canadian counterparts who had experience of farming in similar conditions in their prairies.[56]

In another reversal in the direction of the transfers discussed in this book, in the early 1970s, the Soviet government began to import grain from North America on a regular basis. It purchased hard red winter wheat that was a blend of modern varieties that could be traced back to those introduced from the steppes a century earlier.[57] On the other hand, today's Russian Federation has resumed a position as one of the world's leading grain exporters that had been held by the Russian Empire on the eve of the First World War. At that time, today's Ukraine and Kazakhstan, which share the steppe region, were part of the tsarist empire, and both are also now major grain exporters.[58]

[56] See Marc Elie, "The Soviet Dust Bowl and the Canadian Erosion Experience in the New Lands of Kazakhstan, 1950s–1960s," *GE* 8 (2015), 259–92; Zauresh G. Saktaganova, *Istoriia osushchestvlenniia sovetskogo opyta ekonomicheskoi modernizatsii v Kazakhstane (1946–1970)* (Karaganda: izd-vo KarGU, 2004), pp. 143–76. On Kazakh pastoralism, see N. E. Masanov, *Kochevaia tsivilizatsiia kazakov: osnovy zhiznedeiatel'nosti nomadnogo obshshestva* (Almaty: Sotsinvest, 1995).

[57] K. S. Quisenberry and L. P. Reitz, "Turkey Wheat: The Cornerstone of an Empire," *AH* 48 (1974), 98.

[58] Anatoly Medetsky (February 16, 2018), "Russia Is Exporting More Wheat Than Any Country in 25 Years," *Bloomberg*, available online at https://bloom.bg/330Xuc5, accessed June 25, 2019; Food and Agriculture Organization of the United Nations, GIEWS – Global Information and Early

How the Great Plains and the Steppes Resemble Each Other, but Differ from the Eastern United States and Western Europe

The success of the transfers from the steppes to the Great Plains demonstrated the similarities between the environments of the two flat, semi-arid grasslands in the interiors of their respective continents. The challenges farmers faced in cultivating crops in the Great Plains showed the extent to which the plains differed from the eastern United States and western Europe, where the topography was less uniform, the climate more moderate and more humid, the natural vegetation included far more trees, and the soils less fertile. Immigrants to the Great Plains from the eastern states and much of Europe struggled in the unfamiliar conditions. On the other hand, the migrants to the Great Plains who came from the steppes thrived in conditions that were familiar to them. The crop varieties American farmers grew back east, some of which were from western Europe, also struggled in the plains. On the other hand, plants from the steppes – crops and weeds – did well in conditions they found favorable. This is why American plant explorers seeking crops for their plains who had done their research traveled to the Eurasian steppes where the conditions were similar, if on average slightly harsher.

The older understandings of soils that persisted in the eastern United States and western Europe, where the land had been farmed for generations, proved little use in studying and classifying the soils of the Great Plains that had been plowed up only in recent decades, if at all. It was no accident that Russian scientists devised their new understanding soils in the steppes, where arable farming on a large scale was also a fairly recent development. The Russian scientists saw soils closer to their natural state that revealed how they had formed in the local environmental conditions (climate, organic matter, parent rock, relief) over time. Marbut chose to test the Russian science in the Great Plains where conditions were similar to those in the steppes where the Russians had devised it a few decades earlier. In both the Great Plains and the steppes, most settlers and the specialists who advised them were familiar with more humid, forested environments, and were part of traditions and scientific cultures in which such environments were the norm. It is not surprising that in both countries a "natural" solution to the challenge of farming and living in

Warning System, Country Briefs for Russia, Ukraine, and Kazakhstan, available online at www.fao.org/giews/countrybrief/country.jsp?code=RUS; www.fao.org/giews/countrybrief/country.jsp?code=UKR; www.fao.org/giews/countrybrief/country.jsp?code=kaz, all accessed July 17, 2018.

semi-arid, windy and dusty conditions was to plant trees to shelter the land. In both, planting trees made the land more familiar to the newcomers. In contrast, the indigenous populations – Plains Indians and steppe nomads – burned the grasslands to discourage the growth of arboreal vegetation.[59]

Thus, there were differences between the experiences of settlers in the Great Plains, where the particular environment required particular approaches to farming, and those of inhabitants of the eastern United States, most of whom were unfamiliar with the plains environment and the need to do things differently there. In her memoir of growing up on a Kansas farm in the late twentieth century, Sarah Smarsh looked back to the Homestead Act and maintained: "The federal government had given [settlers] land to work as though the arid plains were just like rich eastern soil."[60] Smarsh's main point was about differences of class and culture, rather than environments, between the "Heartland" and the coasts. Nevertheless, her powerful book encapsulates the persistence of an uncomfortable relationship, based on lack of understanding and imbalances of power, between two distinctive parts of the United States. In the 1930s, inhabitants of the humid, forested, eastern states, including members of Congress and the federal government, readily blamed plains farmers for the Dust Bowl, without full regard for the extreme climatic conditions in that decade, or the work farmers had invested in the land. Over time, the settlers had come to appreciate the contrast between their lives in the plains and those of people to the east, an appreciation that could develop into suspicion towards the eastern United States and the federal government. Suspicions of "liberal east-coast elites" at this time may explain, for example, the dismissive comments by a Kansas journalist about New Dealers as "crazy theorists" who wanted to test their "crackpot" ideas in the state (See pp. 344, 390). Regional pride and hostility to the federal government runs through some of the work of James Malin (the Kansas historian who combined history and ecology).

The differences between the eastern United States and the Great Plains contributed to the barriers to the innovations and transfers from the steppes. Many who resisted the Russian innovations had spent much of their lives in the east. For example, Whitney's obstruction to the new soil

[59] See David Moon, "Planting Trees in Unsuitable Places: Steppe Forestry in the Russian Empire, 1696–1850," in *Eurasian Environments: Nature and Ecology in Imperial Russian and Soviet History*, ed. Nicholas Breyfogle (Pittsburgh: University of Pittsburgh Press, 2018), pp. 23–42.

[60] Sarah Smarsh, *Heartland: A Memoir of Working Hard and Being Broke in the Richest Country on Earth* (New York: Scribner, 2018), pp. 86–7.

science from Russia can be explained in part because his experience was largely limited to the quite different soils of the eastern United States. Hostility to the Shelterbelt Project came from foresters based in eastern states with little, or little recent, experience of the plains. The debate over the Shelterbelt Project in the 1930s between Zon, who was born on the edge of the Russian steppes and was based in St. Paul, Minnesota, and Royal Kellogg, who did have experience of plains forestry, but had long lived in New York City, was expressed in regional terms in a song (sung to the tune of "Clementine") written by one of Zon's staff:

> Zon lives near the open prairie
> Where the winds are very bad.
> Farms are needing cattle feeding
> So he wants a shelterbelt.
> . . .
> Kellogg lives in New York City
> Far away from drought and wind.
> Broadway dandies never fancy
> Any need for shelterbelts.
> When he hears about the project
> He sits down and writes "The Times"
> In a letter: "I know better.
> They don't need a shelterbelt.
> "Planted trees will die on prairie,
> Eighteen inches [of precipitation] not enough.
> Suffocation, radiation
> Kill the trees in shelter belt.
> "Zon should move to New York City
> Live with me on old Broadway
> Here it's cozy, here it's rozy,
> And forget his shelter belt.[61]

Similar feelings that people from the eastern states did not understand the Great Plains lay behind hostility to federal government employees investigating the feasibility of planting shelterbelts. Some plains folk feared they would trample farmers' fields and let the stock escape by leaving gates open (See p. 345). Such hostility was despite the fact that the Democrat-controlled federal government was trying to help the plains population withstand the drought by drawing on Russian experience in similar conditions in the steppes. From the 1890s, inhabitants of the plains expressed

[61] Zon Papers, Box 1. Correspondence and Other Papers, undated, Folder 1, S. H., "The True Story about the Great Shelter Belt."

their disenchantment with the federal government and the mainstream parties, both of which seemed to represent the interests of other regions, by voting for populist and radical parties and, in the late 1930s, for the Republican Party that opposed Roosevelt's New Deal.[62] It was a big step, however, to move from awareness of differences between the Great Plains and the eastern United States to recognition of similarities with and possibilities of learning from Russia's grasslands. Yet, on further examination this is not as surprising or unexpected as it might seem at first sight.

Unexpected Russian Roots of Great Plains Agriculture?

Perceptions

Earlier in this book (see p. 5 and Chapter 2) I suggested several reasons why it would have seemed unexpected to many American agricultural specialists and scientists in the late nineteenth and early twentieth centuries that they could benefit from experience in the steppes of the Russian Empire and Soviet Union. When I have presented papers on this subject to various audiences in the United States and elsewhere in the western world, it was clear from some of the reactions that the idea of transfers, especially scientific, from Russia to the United States is still unexpected. I got different reactions from audiences in the Russian Federation, who were familiar with scientific innovations in their part of the world.

The Russian Empire/Soviet Union in the late nineteenth and twentieth centuries was very fertile in generating new ideas. In his imaginative book, *How Russia Shaped the Modern World*, Steven Marks argued that the collapse of the Russian Empire in the early twentieth century and of the Soviet Union towards the end of that century

> gave rise to the impression that it was an anachronistic monolith incapable of thriving in a world dominated by European and American technology and democratic ideals. This may well seem true if we consider Russia's

[62] See David M. Wrobel, *America's West: A History, 1890–1950* (Cambridge, England: Cambridge University Press, 2017), pp. 18–20, 125, 127, 131–2, 160–1; Charles Delgadillo, *Crusader for Democracy: The Political Life of William Allen White* (Lawrence: University Press of Kansas, 2018); Gerard W. Boychuk (May, 2017), "Populism and Democracy? The Northern Plains States and Prairie Provinces, 1915–1945," paper presented to the Canadian Political Science Association 2017, available online at https://cpsa-acsp.ca/documents/conference/2017/Boychuk.pdf, accessed July 17, 2018. (Cited with the author's permission.)

> material condition or the many ordeals suffered by its citizenry. Yet both the tsarist empire and the Soviet Union exerted significant impact on the world between the 1890s and the 1980s through the power of political and cultural ideas generated there.[63]

Marks was interested in Russian innovations in quite different fields to agriculture and agricultural sciences. He analyzed the global impact of terrorism, venomous anti-Semitism, anarchism, non-violent protest, and "remarkable" developments in art, ballet, literature, and theater that all had their roots in Russia.[64]

There were other innovations in fields closer to the topic of this book that arose in Russia in this period and made their way around the globe. In one of the most important discoveries in modern science, in 1869, Dmitrii Mendeleev, a chemist at St. Petersburg University and colleague of Dokuchaev, presented his periodic system for classifying elements by their atomic weight and chemical affinity. His "Periodic Table" is presented in textbooks and hangs on laboratory walls all over the world. Michael Gordin explained how Mendeleev came up with his discovery in the particular cultural and political environment of tsarist Russia.[65] Further, in the 1890s, scientists in the Russian Empire, including Dokuchaev and Nikolai Kuznetsov, pioneered the concept of protecting areas of what they believed to be pristine environments from all human activity besides scientific research.[66] The concept lay behind Dokuchaev's research stations of the 1890s to research ways (including planting shelterbelts) to make steppe agriculture more sustainable.[67] The principle of inviolable management underpinned a network of scientific nature reserves (*zapovedniki*) established in the main environmental regions of the Russian Empire/Soviet Union after 1916.[68] Jonathan Oldfield and Denis Shaw have pointed to original work by Russian and Soviet scientists in developing the concept of "sustainable development" and understandings of

[63] Steven G. Marks, *How Russia Shaped the Modern World: From Art to Anti-Semitism, Ballet to Bolshevism* (Princeton: Princeton University Press, 2004), pp. 1–2.

[64] Marks, *How Russia Shaped the Modern World*, p. 5.

[65] Michael Gordin, *A Well-Ordered Thing: Dmitrii Mendeleev and the Shadow of the Periodic Table* (New York: Basic Books, 2004).

[66] Douglas R. Weiner, *Models of Nature: Ecology, Conservation and Cultural Revolution in Soviet Russia*, 2nd edition (Pittsburgh: University of Pittsburgh Press, 2000), p. 12.

[67] David Moon, "The Environmental History of the Russian Steppes: Vasilii Dokuchaev and the Harvest Failure of 1891," *Transactions of the Royal Historical Society*, 6th series, 15 (2005), 149–74. See also pp. 196–7.

[68] Feliks Shtilmark, *The History of Russian Zapovedniks, 1895–1995*, trans. G. H. Harper (Edinburgh: Russian Nature Press, 2003).

anthropogenic climate change.[69] Another Russian scientific innovation that proved important in the United States and elsewhere was understanding the continuously frozen subsoil (mistranslated into English as "permafrost") in Arctic and sub-Arctic regions of the Russian and Soviet states. The Russian science was adopted in similar environments around the globe, including Alaska and northern Canada.[70]

Arguments concerning these genuine Russian and Soviet innovations are not to be confused with the plethora of claims for Russian scientific and technological inventions that have been advanced for propagandistic reasons at various times. In the Soviet Union after the Second World War in the last years of Stalin's rule, there were assertions of Russian primacy in several scientific and technical inventions. These included heavier-than-air powered flight, by Aleksandr Mozhaiskii in the 1880s, and wireless telegraphy, by Aleksandr Popov in 1895, amongst others. While there was some basis for these claims, they were exaggerated and promoted for nationalistic and political motives.[71] In recent years, in an atmosphere of renewed Russian nationalism, there have been claims for a wide range of inventions by Russians, some based in fact, others more speculative, from light bulbs to yacht clubs via caterpillar tracks. The authentic Russian and Soviet achievements in space technology and other fields of human endeavor are also trumpeted.[72]

The argument of this book about Russian innovations in agricultural sciences and techniques appropriate for semi-arid grasslands is not an exercise in neo-Stalinist Russian nationalism, but an attempt to suggest how innovations have been generated on the basis of practical experience in particular environmental and also social, political, economic, and cultural, circumstances beyond the western world.[73] In his book on how

[69] J. D. Oldfield and D. J. B. Shaw, "Revisiting Sustainable Development: Russian Cultural and Scientific Traditions and the Concept of Sustainable Development," *Area* 34 (2002), 391–400; Oldfield, "Imagining Climates Past, Present and Future: Soviet Contributions to the Science of Anthropogenic Climate Change, 1953–1991," *JHG* 60 (2018), 41–51. See also Loren R. Graham, *Science in Russia and the Soviet Union* (Cambridge: Cambridge University Press, 1993), pp. 229–33, 243.

[70] Pey-Yi Chu, "Mapping Permafrost Country: Creating an Environmental Object in the Soviet Union, 1920s–1940s." *EH* 20 (2015), 396–421; Chu, *The Life of Permafrost: A History of Frozen Earth in Russian and Soviet Science* (Toronto: University of Toronto Press, forthcoming), chapter 5.

[71] David Brandenberger, *National Bolshevism: Stalinist Mass Culture and the Formation of Modern Russian National Identity, 1931–1956* (Cambridge, MA: Harvard University Press, 2002), pp. 203, 222.

[72] See, for example, "Greatest Russian Inventors," available online at http://allrus.me/greatest-russian-inventors/, accessed July 12, 2018. An online search for "Russian inventions" will reveal many such sites.

[73] On the origins of modern science beyond western Europe, see Kapil Raj, *Relocating Modern Science: Circulation and the Construction of Knowledge in South Asia and Europe, 1650–1900* (New York: Palgrave Macmillan, 2007).

Russia shaped the modern world, Marks explained: "a nation so troubled that it imploded twice in less than a hundred years" could "produce ideas that swayed much of the globe" precisely because of its "seeming divergence from the path followed by England, France, or America."[74] The argument presented here builds on that of my previous book on Russia's steppe region. I argued there that when people, including farmers and scientists, who were familiar with a humid, forested environment encountered the semi-arid grassland of the steppes, it led them to develop innovative ideas concerning the ways people have conceptualized both the environment and the interaction between humans and the rest of the natural world. These new ideas enabled them to devise such innovations as sciences for understanding crops and soils and techniques for planting trees in the new environments. Since promoting steppe agriculture was a government priority in the Russian Empire, it funded and trained specialists in appropriate branches of science.[75] The Soviet government followed suit, although with better results before the terror of the late 1930s and the triumph of the pseudo-scientific ideas of Lysenko.[76]

The similarities between the environment of the steppes and the Great Plains and more particularly between the environmental histories of the two regions – in which farmers moved from different environments to the grasslands to engage in European-style arable farming – go a long way to explain why the Russian innovations were appropriate for the plains of North America. Thus, it was human choices, made first in the grasslands of the Russian Empire and later in North America, to engage in similar activities in similar environments, but not the environments themselves, that explain the "Russian roots" of elements of Great Plains agriculture. In such contexts, moreover, these roots were entirely to be expected.

Global and Longer-Term Contexts

The "Russian roots" of aspects of Great Plains agriculture also seem less "unexpected" if we broaden the focus from transfers between the Russian Empire/Soviet Union and the United States in the 1870s–1930s to exchanges throughout the globe over several centuries and millennia.

[74] Marks, *How Russia Shaped the Modern World*, p. 2.

[75] David Moon, *The Plough that Broke the Steppes: Agriculture and Environment on Russia's Grasslands, 1700–1914* (Oxford: Oxford University Press, 2013).

[76] See Zhores A. Medvedev, *The Rise and Fall of T. D. Lysenko*, trans. I. Michael Lerner (New York: Columbia University Press, 1969).

The transfers between the steppes and Great Plains occurred in the context of the plowing up of the grasslands of the globe to grow grain: one of the great transformations in global environmental history. The process began in the west of the Eurasian steppes, in present-day southern Russia and Ukraine, in the eighteenth century. In the nineteenth and twentieth centuries, the great plow up spread to vast areas of the prairies and Great Plains of North America, the Pampas of South America, the veldt of southern Africa, and lands in Australasia, northern India, north Africa, and the plains of Hungary and Romania, as well as southern Siberia and northern Kazakhstan in the east of the Eurasian steppes. The world's grasslands were conquered and colonized by states with settled, agricultural populations in more humid regions, sometimes immediately adjoining the grasslands, such as the Russian Empire or, in other cases, as part of Europe's overseas empires. These sedentary states encouraged arable farmers to move to the grasslands to create new agricultural lands. From the mid nineteenth century, transporting migrants and the grain they grew to and from the grasslands was aided by steam trains running on railroads, cargo ships powered by steam turbines, grain elevators, and port infrastructure. The grain grown in the grasslands was consumed by the globe's burgeoning urban populations, who were supplied by developing domestic and international markets. Farmers and agricultural scientists, meanwhile, wrestled with how to grow grain in regions with relatively low and unreliable rainfall. The soils that formed in such conditions were very fertile, but regular droughts caused crop failures. Farmers also came up against soil erosion that increased as more land was plowed up, and was worst in dry, windy, years.[77] In this wider context, it is not surprising or unexpected that agricultural scientists should learn from each other's experiences, nor, since the steppes were plowed up first, that Russian scientists and agricultural specialists were in the lead in devising appropriate techniques and the underlying sciences. In both Russia and North America, moreover, scientists looked to similar developments in the Argentinean Pampas: the other grassland that most resembled the steppes and Great Plains.[78]

[77] David Moon, "Plowing up the World's Grasslands," *GE* 11 (2013), 207–9; McNeill, *Something New Under the Sun*, pp. 38–43, 212–16, 276–81.

[78] See, for example, NARA CP, RG 54, Finding Aid A1, Entry 58 Division of Vegetable Pathology and Physiology: Correspondence of M. A. Carleton, 1891–1900, Folder M. A. Carleton – 1895, Carleton to Galloway, December 12, 1895; M. A. Carleton, *The Small Grains* (New York: Macmillan, 1916), pp. 242, 262, 283; Jacob S. Joffe, *Pedology*, 2nd edition (New Brunswick, NJ: Pedology Publications, 1949), p. 262; RGIA, f.91, op.2, 1910, d.355, "Perepiska s Departamentom zemledeliia Glavnogo upravleniia zemleustroistva i zemledeliia"; RGIA, f.403, op.1, 1912, d.204, "Perepiska po prosheniiam raznykh lits, zhelaiushchikh otpravitsia v Argeninu na zarabotki"; N. A.

Going back a few centuries, the transfers from the steppes and the Great Plains, and the other way, were part of the "Columbian exchange" and "ecological imperialism" identified by Alfred Crosby. In the aftermath of Christopher Columbus' voyages between Europe and the Americas and as a consequence of European overseas expansion, crops, weeds, people, ideas, technology, and diseases made their way from the Old World to the New, and some went the other way. People of European origin, with their plants, livestock, and other baggage, settled in temperate regions around the globe and established "neo-Europes." The Great Plains region of North America, with its Euro-American immigrant population, agricultural economy, and Eurasian crops, was just one of a number of such regions.[79] In this context, therefore, the movements of people, plants, agricultural sciences and techniques from the steppes to the Great Plains, from Eurasia to North America, from the Old World to the New, were part of longer-term global patterns. Since the steppes could be considered another "neo-Europe," the present book modifies the pattern identified by Crosby by focusing on transfers between neo-Europes in different continents.[80]

Going back a few millennia, the transfers examined in this book fit well into even longer-term patterns. Recent research by specialists in archeology, linguistics, and ancient DNA, has shown how, starting around 5,000 years ago, pastoral nomadic peoples, who consumed dairy produce, domesticated and rode horses, built wheeled vehicles pulled by oxen, and spoke proto-Indo-European languages migrated from the steppes of today's southern Ukraine, southern Russia, and northern Kazakhstan. Their innovations allowed them to dominate other people and to expand into northern India and throughout Europe, where, defined by genetics, their descendants comprise the largest single population groups. The arrival of arable farming and cereal crops from the fertile crescent of the Middle East added another string to the bows of inhabitants of the steppes and, in time, much of Europe, and later North America.[81]

Kriukov, *Argentina: Sel'skoe khoziaistvo v Argentine v sviazi s obshchim razvitiem strany* (Spb., 1911); D. G. Vilenskii, "Pochvy severnoi i iuzhnoi Ameriki," *Pochvovedenie*, no. 4 (1936), 562–87.

[79] Alfred W. Crosby, *The Columbian Exchange: Biological and Cultural Consequences of 1492*, 30th anniversary edition (Westport: Praeger, 2003); Crosby, *Ecological Imperialism: The Biological Expansion of Europe, 900–1900* (Cambridge, England: Cambridge University Press, 1986).

[80] For another example of transfers between "neo Europes," see Ian Tyrrell, *True Gardens of the Gods: Californian–Australian Environmental Reform, 1860–1930* (Berkeley: University of California Press, 1999), p. 13.

[81] See David W. Antony, *The Horse, The Wheel, and Language: How Bronze-Age Riders from the Eurasian Steppes Shaped the Modern World* (Princeton: Princeton University Press, 2007); David Reich, *Who We Are and How We Got Here: Ancient DNA and the New Science of the Human Past* (Oxford: Oxford University Press, 2018). Horses may have been first domesticated around 5,500

This present book is about all of these issues, but with a tighter focus. As a transnational history of transfers of plants, agricultural sciences, and techniques from the steppes of the Russian and Soviet states to the Great Plains of the United States, moreover, it challenges ideas of American and also Russian exceptionalism. It locates the grasslands of these parts of the world in wider, global stories of settler colonialism and agricultural expansion into semi-arid grasslands in continental interiors. It brings the exchanges and transfers between these two regions into histories of global exchanges and transfers of peoples and their ways of life between different continents over the last five centuries. And it traces them back to the development and movement of peoples, cultures, and practices from the Eurasian steppe around the world since prehistory. After several millennia, these movements came to include those by Mennonite farmers of Germanic origins and their crops and farming techniques from the steppes of present-day southern Ukraine to the plains of Kansas in the 1870s. The Russian roots of Great Plains agriculture this migration set in train may seem unexpected, therefore, only in the shorter-term context of notions of American "superiority" and Russian and Soviet "backwardness" in the late nineteenth and early twentieth centuries. In wider and longer-term contexts, the transfers from the steppes to the Great Plains are simply a part of larger patterns in global history that are to be expected.

Contrasting the Great Plains and the Steppes

In case broadening the focus has exaggerated the similarities between the two regions and blurred the differences, it is appropriate now to tighten the focus and contrast these two leading characters in our story. The two regions do have similar environments. Visitors familiar with one region who traveled to the other often noted the parallels. When Carleton explored the southern steppes at the very end of the nineteenth century, he had a feeling that he was in Kansas:

> A traveler on the plains of Kansas, if suddenly transported while asleep to southern Russia . . ., would discover very little difference in his surroundings except as to the people and the character of the farm improvements and livestock.[82]

years ago at Botai in the north of today's Kazakhstan. A. K. Outram Natalie A. Stear, Robin Bendrey et al., "The Earliest Horse Harnessing and Milking," *Science* 323 (2009), 1332–5.

[82] Mark Alfred Carleton, "Hard Wheats Winning Their Way," *Yearbook of the United States Department of Agriculture 1914* (Washington, DC: USDA, 1915), pp. 398–9.

He elaborated on his imaginary trip:

> Even [the people, farm improvements and livestock] ... would be of the same kind if he were transported from certain localities in Kansas where Russian immigrants [i.e. Mennonites from the Russian Empire] now live. It is therefore natural that the center of hard winter-wheat production in this country should be in Kansas, since in Russia it is in the Crimea.[83]

Seeing a recognizable environment elsewhere in the world could be disorientating. In 1927, the Soviet soil scientist Leonid Prasolov, who had two decades' experience of the steppes, took part in the excursion around North America that followed the soil science congress. Describing the journey across Kansas, he noted "southern chernozems," using the Russian term for American soils. Further west, beyond the 100th meridian: "before us lay the familiar landscape of our southeast: arid, short-grass steppe, salt flats... and mirages on the horizon."[84] Thus, on encountering a familiar environment in a foreign land, Prasolov had a feeling he was no longer in Kansas.

There were also contrasts between the two regions that were significant for the transfers between them. A key difference is that the start of the agricultural settlement and cultivation of the steppes preceded similar developments in the Great Plains by around a century. They began in the steppes in the mid eighteenth century, but did not start in the Great Plains until the mid nineteenth century. This was important as it meant there was prior experience in the steppes that American farmers, agricultural specialists and scientists could draw on, if and when they chose to do so. Another key difference is that, overall, the climate in the steppes is slightly harsher than in the Great Plains. Carleton noted: "The climate of the Russian district is a little more severe, which fact makes Crimean wheat all the more satisfactory for Kansas."[85] Plant explorers Carleton and Hansen deliberately collected plants from parts of the Eurasian steppe, including the north of today's Kazakhstan, with a more severe climate than the Great Plains, precisely because plants that could withstand the climatic extremes in the steppes would be able to survive all but the greatest fluctuations in the American plains (See pp. 95–7).

Such were the extremes in the steppes that Americans, including those with experience of the Great Plains, who visited the steppe region during

[83] Carleton, "Hard Wheats Winning Their Way," 398–9.

[84] L. Prasolov, "Kartografiia pochv na I mezhdunarodnom kongresse pochvovedov v Vashingtone (okonchanie)," *Pochvovedenie*, no. 3–4 (1928), 222–3. On Prasolov, see: "Osnovnye daty zhizni i deiatel'nosti Akademika L. I. Prasov," I. P. Gerasimov (ed.), *Voprosy genezisa i geografii pochv* (Moscow: izd. AN SSSR, 1957), pp. 6–7.

[85] Carleton, "Hard Wheats Winning Their Way," 398–9.

crises tended not to comment on the similarities between the two regions that were more apparent in "normal" years (see pp. 64–8). The extremes were noted by John L. Strohm, a journalist with agricultural experience from the prairie state of Illinois, who traveled around the Soviet Union in 1946 in the aftermath of the Second World War. He had the audacity to cable Stalin when he found it difficult to get a Soviet visa. In response, he was invited by the Soviet Minister of Agriculture, who advised him to "tell the truth" about what he saw. It served Soviet interests for an American to report on the wartime destruction of large parts of their country, including most of the steppe region.[86] Strohm's travels included the steppes of Ukraine and the Volga around Stalingrad. He was "staggered by the war devastation": people were eking out a living as best they could in their ruined farms and cities.

But, Strohm argued it was not just the war that was responsible for Russia's "weak food position":

> the Russian steppes extend from the Urals to Bessarabia [Moldova], a fertile black soil belt that would make any Iowa farmer start counting corn and soybean dollars. But the Mississippi Valley has the Rocky Mountains as a providential screen to shield it from the hot dry winds of our Great American Desert. The Soviet Union is not so lucky. Only the Caspian Sea stands between the desert and what should be a real Garden of Eden. But cold winds from the desert whip across the Ukraine in the spring, and hot, dry winds in the summer. "If we got one or two good crops out of five, we're lucky," a collective farm chairman in the Ukraine told me. The American farmer can expect four good crops out of five.[87]

Although the climate of Iowa is less extreme than the American Great Plains further west, Strohm's contrasts with the steppes are striking. Although it was not his point, his contrasts between the climate of the steppes and the American grasslands illustrate how crop varieties that could survive the extremes of the steppe environment were likely to thrive in the milder conditions of the Great Plains.

Another major difference between the steppes and the Great Plains is their orientation: the Great Plains run from north to south, while the steppes extend from east to west. Zon noted: "If a map of the United States or of the USSR is turned at right angles, the same succession of climate and vegetation is evident. In the United States this succession

[86] John L. Strohm, *"Just Tell the truth." The Uncensored Story of How the Common People Live Behind the Russian Iron Curtain* (New York: C. Scribner's Sons, 1947), pp. xi, 3–6, 9–10, 27, 61, 64; on Strohm, see, http://alumni.illinimedia.com/famers/view/44. accessed July 10, 2013.

[87] Strohm, *"Just Tell the Truth,"* pp. 55, 102–3. Strohm was referring to the deserts of the Great Basin.

proceeds from east to west, in the USSR from north to south." He was referring to the transition from forest, in the eastern United States and northern Soviet Union, to grasslands in the west and south of the two countries. His statement also holds for precipitation, which declines from northwest to southeast in Eurasia and from east to west in North America (as far as the Rocky Mountains).[88] Zon was oversimplifying, as temperature does not vary from east to west in North America, but as in Eurasia from north to south. The reason the belts of vegetation and moisture, but not temperature, proceed from east in the west in North America is the presence of the Rocky Mountains, running from north to south, that create a rain shadow to the east, where the consequent shortage of moisture favors grasses over trees. In contrast, in Eurasia, high mountain ranges are located south and southeast of the steppes and do not upset the neat coincidence of belts of vegetation, precipitation, and temperature (See p. 195).

The different geographical orientations create a further contrast: the central and southern plains of North America are a much further south, and therefore warmer, than the southern steppes. The Mennonites who moved from the steppes of today's southern Ukraine to Kansas left homes located around 47 degrees north, but relocated to lands at around 38 degrees north: a difference of nine degrees of latitude or about 600 miles. When Bernhard Warkentin was scouting out new lands in North America for his fellow Mennonites, he advised against the Canadian prairies in the north, as it was too cold. He also advised against Texas, in the south, on account of the heat (See p. 104). While the Great Plains extend a long way further south than the steppes, the northern peripheries of the two regions are at similar latitudes. Edmonton, Alberta, and Samara, Russia, are both at around 53 degrees north.

The contrasting orientations of the two grasslands, and resulting differences in the belts with similar precipitation and temperature, had implications of varying importance for three of the transfers between them: wheat; soil science; and shelterbelts. The larger area of warmer, more southerly, land in the Great Plains means that more winter wheat, which requires milder winters, is grown there than in the steppes, where spring wheat, which is planted and harvested later, is more important. In their previous homes in the southern steppes, Mennonites had only recently shifted to winter wheat, after they had devised ways to protect the young shoots from late frosts, on the eve of their migration to Kansas

[88] Zon, "The Volga Valley Authority." (There are no page numbers in the online version.)

in the 1870s (see p. 140). This was not such an issue in the milder climate of their new homes, where production of winter wheat, the famous Turkey Red, took off. Further, in comparison with the steppes, the far larger distance from north to south in the North American grasslands, and corresponding differences in temperatures, result in a longer vegetation period in the central and southern plains than in much of the steppes.

Another consequence of the north–south orientation of the American grasslands is a much longer harvesting season than in the steppe. At the present time, in northern Texas harvesting winter wheat begins in mid May and, 1,500 miles north in the Canadian prairies, the last of the spring wheat is gathered in by the end of August. The start dates of the wheat harvest in the southern plains are slightly earlier and the end dates in the northern prairies significantly earlier than in the recent past due to newer, faster-maturing, varieties.[89] In the mid 1990s, bringing in the winter wheat started in Texas in late May and harvesting spring wheat in the northern plains (North Dakota, Montana, and Minnesota) lasted until late September.[90] Thus, the march of the seasons from Texas to Alberta created a harvest season lasting over four months. Over the middle and later decades of the twentieth century, teams of custom harvesters with their combines made their way up the plains under contract to farmers in several American states and Canadian provinces.[91] In contrast, in the steppes with their east–west orientation and much shorter distances from north to south, the harvesting season is shorter, more intense, and requires a greater investment in harvesting machinery. In the second decade of the twenty-first century, the winter wheat harvest in southern Russia and Ukraine starts in mid July, while gathering the spring wheat in the more northerly steppes of Russia and northern Kazakhstan ends in late September: a harvesting season of two and a half months, less than two-thirds that of North America.[92] The short and pressured harvesting season, compounded by the near simultaneous ripening of winter and spring crops

[89] Thomas Isern, email message to the author, dated July 17, 2018.

[90] "Usual Planting and Harvesting Dates for U.S. Field Crops," *USDA National Agricultural Statistics Service Agricultural Handbook*, no. 628 (December 1997), [30–1], available online at https://bit.ly/2qwfO4k, accessed July 9, 2018.

[91] Thomas D. Isern, *Custom Harvesting on the Great Plains: A History* (Norman: University of Oklahoma Press, 1981).

[92] Food and Agriculture Organization of the United Nations, GIEWS – Global Information and Early Warning System, Country Briefs for Russia, Ukraine, and Kazakhstan, available online at www.fao.org/giews/countrybrief/country.jsp?code=RUS, www.fao.org/giews/countrybrief/country.jsp?code=UKR, www.fao.org/giews/countrybrief/country.jsp?code=kaz, accessed July 17, 2018.

in some areas, led Russian peasants to refer to harvesting time as the "suffering" (*strada*).[93]

A further consequence of the contrasting orientations of the steppes and the Great Plains is that the configurations of the environmental belts made it much easier for Russian soil scientists to see how soils formed. In the vast plain of Russia and Ukraine, with no high mountains to create rain shadows, belts with similar temperatures, precipitation, vegetation, and soils, all run from west to east. A couple of hundred miles south of Moscow, a belt of cool, humid, forest, with moderately fertile Podzol soils gradually gives way to a belt with warmer temperatures, less rainfall, covered with grasses over fertile Chernozem soils. When Russian soil scientists analyzed cross sections of soils in the 1870s, geography had helped them devise their theory to explain the formation of different types of soils in different environmental conditions. In North America, as a consequence of the Rocky Mountains, belts with similar precipitation, natural vegetation, and to some extent soils run from north to south, but belts with similar temperatures extend from east to west. The resulting checkerboard pattern made it harder for American scientists to see the connection between all the soil forming factors. It also made it more difficult to adapt the Russian theory to American soils.[94]

It was the similarities between the environments and environmental histories of the two regions that attracted the attention of both American and Russian/Soviet agricultural specialists and scientists to the possibilities of transfers between them. However, it was the differences between them that in context of human choices to pursue similar activities in similar environments facilitated the Russian innovations and the transfers. The cultivation of the steppes came first. This gave Russian farmers, specialists, and scientists more time to devise ways to cope with the challenges of arable farming in semi-arid grasslands, for example by planting shelterbelts, and meant there was prior experience Americans could draw on. Conditions in the steppes are slightly harsher, which meant crops, trees, and shrubs (but also weeds) introduced to the Great Plains were able to cope

[93] David Kerans, *Mind and Labor on the Farm in Black-Earth Russia, 1861–1914* (Budapest: Central European University Press, 2001), pp. 81–5.

[94] See Curtis F. Marbut , "Introduction," in Jacob Samuel Joffe, *Pedology*, 2nd edition (New Brunswick, NJ: Pedology Publications 1949), pp. ix–xii; J. S. Joffe, "Russian Studies on Soil Profiles," *BASSA* 12 (1931), 15; David Moon, "Scientific Innovation in the Russian Empire: The Case of Genetic Soil Science," in *Science and Empire in Eastern Europe: Imperial Russia and the Habsburg Monarchy in the 19th Century*, ed. Jan Arend (Munich: Vandenhoeck and Ruprecht, 2020), pp. 305–20).

with all but the most extreme fluctuations in the climate. And the accident of geography that created the coincidence of environmental belts in the steppes allowed Russian scientists to devise their new way of understanding soils that later made its way to the United States, where the fertile soil of the plains was classified by the U.S. Soil Survey as Chernozem.

Conclusion

This book has placed the environmental history of the Great Plains – and an analogous region part of the way around the globe – at its heart. I have avoided superficial and unsustainable arguments that the specific environmental conditions of the Great Plains and the steppes somehow determined their histories or the transfers between them. Instead, I have emphasized that agricultural settlers and their governments – first in the steppes, then in the Great Plains – chose to engage in European-style arable farming and that there were other possible ways of life that have been, and could be, pursued in such environments. It was the human choices that were made in the Great Plains, and in time the recognition that similar choices had been made in a similar environment in another country, that created the opportunities for sharing experience and expertise that led to the transfers.

Migrants from the Russian Empire played key roles in the transfers of plants (crops and weeds), agricultural techniques and sciences to the United States. The Germanic, including Mennonite, settlers brought with them their experience of farming in semi-arid grasslands that proved appropriate in the Great Plains. The varieties of wheat they introduced have attracted most attention, not least because Mennonites have retold the story many times. Immigrants have justifiably been less keen to claim credit for introducing the tumbleweed and other weeds that hitched rides with them from the steppes to the Great Plains. Just as important, if less numerous, were the Jewish migrants who made their ways to the United States to escape persecution in tsarist Russia. With their knowledge of the Russian language and scientific training, either in the Russian Empire or the United States, or in both, they served as conduits for Russian scientific knowledge in their new country at a time when few Americans knew Russian. The most prominent example in this book was the forestry scientist Raphael Zon, but several others translated publications by Russian scientists into English, making them accessible to American scientists. It was all these immigrants' experience, expertise, and language skills, rather than their ethnicity or religion, that were important in the context of this book, except insofar as the treatment of minorities in the Russian Empire explain their decisions to emigrate.

On the other hand, there have been few references in this book, since the brief discussions in the Introduction and Chapter 1, to the Native American peoples who inhabited the Great Plains prior to Euro-American settlement. This reflects the fact that the contemporary sources this book is based on made little reference to the indigenous peoples once they had been dispossessed and displaced. With the important exception of the enduring importance of one their main crops, corn, the Euro-American settlers and the authorities and scientists who supported and advised them largely disregarded Native American experience of living in the plains environment and the considerable local knowledge they had accumulated.

A rare exception came in the 1937 report on the Dust Bowl by the House of Representatives' Great Plains Committee. The Committee compared the Euro-American settlers unfavorably with the Plains Indians they had replaced. Echoing contemporary understandings that Native Americans had lived in balance with the natural world that have since been challenged (see pp. 44–5), the Committee asserted that the Plains Indians had hunted "buffalo" and set fire to the grass, but had not reduced the numbers of "buffalo" or destroyed the grass cover. Using the language of the time, the report explained how the first "White" settlers had found "the Red Man living in rude but productive harmony with Nature... the Indian could laugh at the burning sun, the strong but dustless winds. He had made his truce with them, and with the land." In contrast:

> The White Man knew no truce. He came as a conqueror first of the Indian, then of Nature. Today we see foothills shorn of timber, deeply gullied, useless or rapidly losing their fertile soil under unwise cultivation; the fertile earth itself drifts with the wind in sand hills and in dust clouds; where once the grass was rank, cattle nibble it to the scorched roots; the water of streams and the ground waters too often irrigate poor land, leaving the richer thirsty; men struggle vainly for a living on too few acres; the plough ignores Nature's "Keep Off" signs; communities, for all the courage of their people, fall into decay...[95]

Given prevailing attitudes among white Americans in the 1930s, however, the negative comparison between Euro-American farmers and Plains Indians is likely to have been a criticism of the former rather than praise for the latter. In any case, the Great Plains Committee saw the solution to

[95] U.S. Great Plains Committee, *The Future of the Great Plains* (Washington, DC: The House of Representatives, 1937), pp. 2, 14, 16.

the ecological crisis, not in drawing on Native American experience, but in relying on "science" to indicate the way for "man once more to make his peace with Nature."[96]

In the Great Plains, Euro-American farmers, agricultural specialists and scientists, and the local, state, and federal governments chose largely to ignore the knowledge of the indigenous populations. In part, this was because they wanted to recreate an agricultural way of life many were familiar with from the eastern United States and western Europe that would support far larger populations than the Native American lifeways. In part, it was due to stereotyped attitudes that considered the mobile ways of life and cultures of the indigenous peoples as "inferior." However, in the quite different environmental conditions of the Great Plains, some of the Euro-Americans' allegedly "superior" knowledge was redundant as many struggled in the unfamiliar conditions. The settlers' lot was hardest in dry years, such as the early 1890s, that periodically afflicted the plains. It was in these years, however, that the experience of the Mennonite settlers in the Great Plains, who were coping better due to their drought-resistant wheat and dry-farming techniques, came to wider attention, including to agricultural scientists, such as Carleton, working for the USDA. It was at this point, moreover, that Euro-American settlers, agricultural specialists and scientists, and their governments began to realize that there was prior experience they could usefully learn from in the Mennonites' previous homes in a similar semi-arid grassland in the steppes of the Russian Empire. In devising ways to develop arable farming in the steppes, the settlers, including Russians and Ukrainians as well as Germanic peoples, had also largely ignored the knowledge of the indigenous nomadic peoples, since like the Americans, they considered their mobile ways of life to be "inferior." Ironically, one of the barriers to American reception and adoption of transfers from the steppes was a sense of superiority over Russians, however, this turned out to be easier for them to overcome than their disdain for Native American experience: hence the "unexpected" Russian roots of aspects of Great Plains agriculture.

[96] U.S. Great Plains Committee, *Future of the Great Plains*, p. 18.

Archival Collections Cited

Kazakhstan

Tsentral'nyi gosudarstvennyi arkhiv Respublika Kazakhstan, Almaty (TsGA RK) [Central State Archive of the Republic of Kazakhstan]

f.30 Pravitel'stvennyi agronom Turgaiskoi oblasti pri Orenburgsko-Turgaiskom upravlenii zemledeliia i gosudarstvennykh imuchshestv, g. Orenburg

f.317 Zaveduiushchii pereselencheskim delom v Akmolinskom raione, g. Omsk

Tsentral'nyi gosudarstvennyi arkhiv nauchno-tekhnicheskoi dokumentatsii Respublika Kazakhstan, Almaty (TsGANTD RK) [Central State Archive of Scientific and Technical Documentation of the Republic of Kazakhstan]

f.103 Kazakhskii filial Vsesoiuzhnoi Akademii Sel'skokhoziaistvennikh nauk imemi V. I. Lenina

Russian Federation

Arkhiv Rossiiskoi Akademii nauk, Moscow (ARAN) [Archive of the Russian Academy of Sciences]

f.487 Dokuchaevskoe obshchestvo pochvovedov pri Rossiiskoi akademii nauk

f.582 Iarilov Arsenii Arsen'evich

f.687 Prasolov Leonid Ivanovich, akademik, pochvoved

Sankt-Peterburgskii filial Arkhiva Rossiiskoi Akademii Nauk (PFA RAN) [St. Petersburg branch of the Archive of the Russian Academy of Sciences]

f.184 Dokuchaev, V. V.

Gosudarstvennyi arkhiv Rostovskoi oblasti, Rostov-na-Donu (GARO) [State Archive of Rostov region]

f.46 Kantseliariia voiskogo nakaznogo atamana

Gosudarstvennyi arkhiv Samarskoi oblasti, Samara (GASO) [State Archive of Samara region]

f.5 Samarskaia gubernskaia zemskaia uprava

Rossiiskii Gosudarstvennyi Istoricheskii arkhiv, St Petersburg (RGIA) [Russian State Historical Archive]

f.91 Vol'noe Ekonomicheskoe Obshchestvo
f.382 Uchenyi komitet Ministerstva zemledeliia
f.383 Pervyi Departament Ministerstva Gosudarstvennykh imushchestv
f.398 Departament Zemledeliia Ministerstva Zemledeliia i Gosudarstvennykh imushchestv
f.399 Imperatorskii Sel'sko-khoziaistvennyi muzei
f.403 Obshchestvo "Russkoe zerno"
f.426 Otdel zemel'nykh uluchshenii Ministerstva Zemledeliia i Gosudarstvennykh imushchestv
f.515 Glavnoe upravlenie udelov
f.1281 Sovet ministra vnutrennikh del

Tsental'nyi gosudarstvennyi arkhiv naucho-tekhnichestkoi dokumentatsii, St. Petersburg (TsGANTD Spb) [Central State Archive of Scientific and Technical Documentation]

f.318 Vsesoiuznyi nauchno-issledovatel'skii institut rastenevodstva im. N. I. Vavilova

Ukraine

Derzhavnyi arkhiv Odes'koi oblasti, Odessa (DAOO) [State Archive of Odessa region]

f.22, Obshchestvo Sel'skogo Khoziaistvo Iuzhnoi Rossii

United States of America

Bancroft Library, Berkeley, California

The Hilgard family papers [c. 1848–1945], BANC MSS C-B 972

Georgetown University Library, Special Collections Division, Washington, DC

Robert S. Chilton Jr. Papers

Kansas Historical Society, State Archives, Topeka, Kansas (KSHS SA)

Ms. Coll. 13 Vivien Aten Long
Ms. Coll. 183 James C. Malin Collection
Ms. Coll. 789 Atchison, Topeka, and Santa Fe Railway Company records
Maps Unit no. 218477, acc. no. 2010-065.01 U.S. Bureau of Soils, Soil Maps of Kansas (7), 1902–15

Library of Congress, Manuscript Division, Washington, DC (LoC MD)

Gifford Pinchot Papers

Mennonite Library and Archives, Bethel College, North Newton, Kansas (MLA)

MS 68 Bernhard Warkentin (1847–1908) and Wilhelmina Eisenmayer Warkentin

Minnesota Historical Society, St Paul, Minnesota (MHS)

P1237, Raphael Zon Papers

Missouri State University, Duane G. Meyer Library, Special Collections and Archives, Springfield (MSU DGML SCA)

Ozarkiana Collection: Curtis F. Marbut Collection, Collection Number M014

National Agricultural Library United States Department of Agriculture, Special Collections, Beltsville, Maryland (NAL)

Manuscript Collection 91 The Charles Edwin Kellogg Papers
Manuscript Collection 127 W52 R, H. L. Westover and C. R. Enlow, Turkistan

National Archives and Records Administration, College Park, Maryland (NARA CP)

RG 54 Records of the Bureau of Plant Industry, Soils, and Agricultural Engineering [USDA]

RG 59 General Records of the Department of State
RG 95 Records of the Forest Service

National Archives and Records Administration, Kansas City, Missouri (NARA KC)

RG114 Records of the Natural Resources Conservation Service

Nebraska State Historical Society, State Archives and Manuscript Division, Lincoln (NSHS SAMD)

RG 3941.AM Paul Henley Roberts Papers, 1891–1971

North Dakota State University Archives, Fargo (NDSUA)

Mss 18 Henry Luke Bolley Papers

South Dakota State University Archives and Special Collections, Hilton M. Briggs Library, Brookings (SDSU ASC)

UA 53.4 N. E. Hansen Papers

State Historical Society of Missouri, Manuscript Collection, Columbia (SHSM MC)

Marbut, Curtis Fletcher (1863–1935) Papers, 1852–1963 (C3720)
Marbut, Curtis Fletcher (1863–1935), Papers, 1901–43, 1981–3 (C3721)

University of Minnesota Archives, Minneapolis (UMA)

Collection Number uarc 173 Department of Soils Records, 1911–90s

Wichita State University Libraries, Special Collections and University Archives, Wichita, Kansas (WSUL SCUA)

Federal Writers' Project

Index

Made in United States
North Haven, CT
08 January 2026